D1270117

DRUG DISCOVERY: THE EVOLUTION OF MODERN MEDICINES

DRUG DISCOVERY: THE EVOLUTION OF MODERN MEDICINES

WALTER SNEADER
University of Strathclyde
Department of Pharmacy
Glasgow
Scotland

A Wiley Medical Publication

JOHN WILEY & SONS
Chichester · New York · Brisbane · Toronto · Singapore

Library of Congress Cataloging in Publication Data

Sneader, Walter.
Drug discovery.

(A Wiley medical publication)
Bibliography: p.
Includes index.
1. Drugs. I. Title. II. Series. [DNLM: 1. Drugs. QV 55 S671d]
RM301.S6 1985 615'.1 85-702
ISBN 0 471 90471 6

British Library Cataloguing in Publication Data

Sneader, Walter
Drug discovery: the evolution of modern medicines.—(A Wiley medical publication)
1. Chemotherapy
I. Title
615.5'8 RM262

ISBN 0 471 90471 6

Printed and Bound in Great Britain

For Myrna

Contents

Preface

The past century has witnessed the realization of one of man's ancient dreams—the conquest of disease through the use of effective drugs. This book, which describes for the first time how this has transpired, attempts to convey some of the drama and excitement that must have been experienced by the doctors and scientists who were involved. Much of this drama arose from the opportunistic exploitation of unexpected observations, that combination of chance and sagacity known as serendipity. Recently, however, major new drugs have been developed through the application of rigorous scientific thought.

A major problem encountered in writing this book has been the acquisition of basic facts concerning the discovery of scores of drugs. This required many hours of painstaking searches amongst dusty volumes of old journals. Fortunately, this was to some extent offset by the availability of secondary literature dealing with the more celebrated drugs. Progress might have been faster, and the text more accurate in some details, had it been written by several hands, but it would have been difficult to communicate the overall integration and continuity of events in this manner. Hopefully, my readers will forgive the shortcomings that are inevitable in a single-author work attempting to cover so vast a field of knowledge; it might prove possible to rectify these in a future edition if readers will supply me with the relevant information.

The text is liberally interspersed with chemical formulae. Their presence should not deter either the general reader or the professional who harbours feelings of inadequacy so far as chemistry is concerned. These formulae supplement the information in the text so as to enable those with the appropriate knowledge to relate what appears in these pages to their existing understanding of the subject. Throughout this book the text should be comprehensible without reference to the chemical formulae.

It is my desire that this book should provide health-care workers in the medical, pharmaceutical, and nursing professions, as well as scientists whose work entails the handling of drugs, with an insight into how such a diverse range of chemical substances has been introduced into clinical practice. Many years of teaching experience have convinced me that this can create greater confidence in coping with an otherwise overwhelming plethora of pharmaceutical products. There is, however, a further dimension to this book insofar as it outlines the rise of one of the first so-called high technology industries. This may furnish a useful background to the public debate concerning the nature and funding of research in the pharmaceutical industry.

For the benefit of those readers who may wish to pursue the subject matter further, an extensive bibliography relating to drug discovery has been included. In many cases it will be evident as to which part of the text a given reference relates, but where this might not be the case an indication of the particular drug concerned is given after the reference. This may seem a little unconventional, but it has been done to avoid disrupting the readability of the text through the insertion of embedded reference numbers, bearing in mind that the majority of readers will consult the bibliography only infrequently.

Walter Sneader

Glasgow,
January 1985.

I

The Legacy of the Past

Before the advent of modern chemistry, physicians seeking drugs to treat their patients could only select natural products or inorganic materials. There was no shortage of the former. Herbals and pharmacopoeias filled with ancient plant lore were testimony to the effort that had been expended over thousands of years in the search for panaceas to cure all ills. The principles underlying the use of these herbal remedies were far from scientific, for primitive medicine had been firmly bound up with magic and religion. In a world where disease was held to be a result of either possession by demons or the wrath of the deity, it was the priesthood which was entrusted with the responsibility of discovering drugs. In Egypt, the priest-physicians concocted vile potions from dead flies, dried excreta, garlic, leek, onion, and bitter herbs, in heroic attempts to drive out the demons via any appropriate orifice in the body. They seem to have fought something of a losing battle, for the Ebers Papyrus, discovered by George Ebers in 1862, contained no less than 811 different formulae for fighting off disease. This was compiled around 1550 BC, but the legacy of the Egyptians remains with us, for many people still put their trust in herbal remedies or take purgatives regularly in the hope of maintaining good health! The influence of the Assyrians and Babylonians also persists amongst modern devotees of herbalism, but one wonders if they realize that the origins of many herbal remedies can be traced back to the astrologers of Chaldea who believed that each star could exert its influence upon a specific plant? With so many stars in the firmament, there was no shortage of plants that might cure the sick, provided a knowledgeable astrologer-priest was consulted.

Demonology was unacceptable to the Greeks, but they did not reject the herbs that were associated with it. Nor did they spurn the idea that the stars influenced the therapeutic qualities of plants. Indeed, so diverse was the range of herbs employed by Greek physicians that a corps of plant-gatherers emerged to satisfy the unremitting demand. Probably to protect their privileged position as much as out of devotion for the principles laid down by their predecessors, these forerunners of the apothecaries issued warnings that only by recitation of the correct incantation and due regard to the position of the heavenly bodies at the time of ingathering, could the therapeutic efficacy of herbs be guaranteed. Sometimes, even worse might happen if plants were gathered by those unauthorized to do so—as in the case of the mandrake, which could only be uprooted by a tethered dog as whomsoever wrested it from the ground died instantly!

The medical knowledge of the Greeks was enshrined in the writings of Galen (129–199), especially his voluminous *On the Art of Healing*. Born in Pergamum, he began his study of medicine at nearby Smyrna, and completed his training twelve years later in Alexandria. During his lifetime he achieved considerable fame as a clinician, and his success enabled him to devote much time to writing. He possessed a sound understanding of philosophy, his work being imbued with Aristotelian concepts. Thus, he asserted that imbalance of the four humours (blood, phlegm, black bile, and yellow bile) was the cause of disease. As these humours corresponded to Aristotle's four qualities (moist, dry, cold, and warm), of which all substances were constituted, Galen held that it should be possible to administer herbs with opposing qualities in order to cure disease. He expressed this succinctly with the maxim *contraria contrariis curantur*. Invariably, this necessitated the compounding of several plants in complex formulations that later became known as 'galenicals'. To discover plants with the appropriate properties, the so-called simples, has exercised the minds of his followers ever since. Despite this apparent recognition of the importance of the empirical testing of drugs, there was much speculation in Galen's books, particularly with regard to how medicines acted. He even managed to justify the age-old use of excrement and amulets. Galen's influence did not wane after his death, for his writings were translated into many languages. They exerted a stultifying influence on therapeutic practice for more than 1500 years, particularly through their incorporation into Arabic medicine, which came to the fore when the Arabs conquered the lands around the southern shores of the Mediterranean and also Spain.

The revival of interest in Greek culture during the Renaissance, hastened by the introduction of the printing press, focused attention on Galen's original texts, with the compilers of pharmacopoeias and herbals drawing freely from them. One of the earliest printed herbals, *Liber de Proprietatibus Rerum*, was published in Basle in 1470; it was written by Bartholomeus Anglicus, an English professor of theology in Paris. The first pharmacopoeia was the *Nuovo Receptario Composito*, a slim volume compiled in Venice in 1498 by the College of Physicians at the request of the Guild of Pharmacists. The next century saw the appearance of botanists in Germany, the most eminent of whom, Valerius Cordus, wrote a four volume history of plants.

The first major challenge to Galen's teachings came from the Swiss physician, Paracelsus (1493–1541). Rejecting the use of herbs, he urged alchemists to desist from the quest for gold and the compounding of worthless elixirs, and instead apply their skill and knowledge to the needs of the sick by developing chemical medicine from mineral sources. He sought the *arcanum*, the healing essence within all effective pharmaceutical preparations, be they animal, vegetable, or mineral. In this, he was ahead of his time, for it was not until the beginning of the nineteenth century that the first active principle was successfully extracted from a plant.

Despite Paracelsus, herbal medicine reached its zenith in the seventeenth century. Its subsequent decline was due to the emergence of physicians who rejected authoritarianism in favour of the experimental method. Opposing the magic and superstition that had dominated medical thinking, these pioneers demanded evidence for the effectiveness of medicinal preparations, be these the traditional galenicals or merely simples gathered from the hedgerow as domestic remedies for those who could not afford the expensive services of a physician. Gradually, other physicians came to share their views, with the result that towards the end of the eighteenth century, a mood of therapeutic nihilism developed amongst leading practitioners. Wise physicians came to the conclusion that, apart from cinchona bark for malaria and ipecacuanha for dysentery, both remedies having been introduced from the New World during the preceding century, opium and belladonna (the use of which had just been revived) were the only traditional drugs with any real value.

Desperate to find alternative therapeutic measures to satisfy their patients, physicians welcomed any new system that came into vogue, such as sea bathing, hydropathic spas, heliotherapy, electrotherapy, or diet therapy. It was against this background that English physicians began to take an interest in Joseph Priestley's experiments with 'fixed air', i.e. carbon dioxide, and 'dephlogisticated air', to which Lavoisier later gave the name oxygen. Patients were given carbonated drinks in the hope that any carbon dioxide that was absorbed would dissolve kidney stones. After Lavoisier had elucidated the role of oxygen in respiration, around 1785, this was administered by inhalation for emergency resuscitation.

In 1786, Lavoisier was visited by a young English doctor, Thomas Beddoes, who was keen to learn of the latest developments in pneumatic chemistry. After his appointment as reader in chemistry at Oxford two years later, Beddoes acquired the reputation of being the leading English exponent of pneumatic medicine. In an effort to establish a theoretical basis for the therapeutic inhalation of gases, Beddoes turned to the controversial Brunonian system of medicine which argued that patients were either asthenic, meaning their tissues required stimulation, or sthenic, being the opposite. As oxygen had undoubted stimulating properties, Beddoes administered air enriched with it to asthenic patients, whilst his sthenic patients were required to inhale air deficient in oxygen. Beddoes carried out his experiments at Oxford until 1792, when his outspoken views on the merits of the French Revolution finally forced him to resign his post. This turn of events persuaded Beddoes that it was time to open an institution where patients could be properly treated with specially manufactured 'factitious airs'.

Beddoes' father-in-law and several acquaintances were members of the elite Lunar Society, quaintly named because it met monthly in Birmingham when the moon was full, thus enabling its members to ride home by moonlight. When Priestley had lived in Birmingham during the 1780s, he greatly influenced the philosophical activities of this small group. Other distinguished members included Josiah Wedgewood, the pottery magnate, William Withering, the physician who introduced digitalis into medicine, his rival Erasmus Darwin, and the famous engineer James Watt, who had been persuaded by members of the Lunar Society to settle in Birmingham. It was to the Society, then, that Beddoes turned for patronage in 1793 after he published a pamphlet on the treatment of consumption (tuberculosis) by inhalation of factitious airs.

Whether Beddoes was aware of it or not at that time, several members of the Lunar Society had good cause to support his efforts, for members of their own families were dying of consumption. Not least amongst these was young Gregory Watt, the son of the engineer.

Intrigued by the imaginative nature of Beddoe's plans and their utility, the Society members backed him financially, Wedgewood contributing £1000 and others providing what they could afford. James Watt collaborated with Beddoes on the scientific side, the two men publishing the five volume *Considerations on the Medicinal Powers and the Production of Factitious Airs*, which appeared between 1794 and 1796. This marked the heyday of pneumatic medicine, and Beddoes forged ahead with his plans to open what was to become known as The Pneumatic Institution for Relieving Diseases by Medical Airs.

The Pneumatic Institution was established in Clifton, Bristol, in 1798. In the basement was a massive machine built by James Watt for the production of a variety of gases under the supervision of a young man who had been recruited from Cornwall, one Humphrey Davy. The latter was encouraged to experiment with new gases for the patients to inhale, and this led him to examine Priestley's nitrous oxide which an American, Samuel Mitchell, was claiming to be highly toxic. After establishing that small animals could be immersed in a jar of nitrous oxide without any apparent harm, Davy boldly inhaled the gas himself, only to experience what he later described as, '. . . the most vivid sensation of pleasure accompanied by a rapid succession of highly excited ideas.'

The reputation of nitrous oxide as a euphoriant spread quickly, earning it the popular name of 'laughing gas'. It was to remain as the most enduring product of the Pneumatic Institution, which itself soon developed into a mere nursing home as physicians and patients alike came to realize that factitious airs were not the panacea for all ills. Indeed, so rapid was the demise of pneumatic medicine that by the turn of the century only the inhalation of oxygen was considered to have any therapeutic merit. A few pioneers, however, did continue to experiment, including a general practitioner in Ludlow, England, who showed that animals could be rendered unconscious by carbon dioxide inhalation. In 1824, he published a pamphlet encouraging surgeons to experiment with this technique so that their patients might be spared the dreadful agonies of surgery. Despite pleading his case in London and Paris, Henry Hill Hickman died six years later at the age of thirty, without seeing the medical profession take up his proposals.

Richard Pearson, one of the Birmingham group of pneumatic physicians, discovered that when it was not practicable to use gases it was possible for patients to inhale ether as an alternative. He made no extravagant claims about the value of ether, but employed it for many years for respiratory disorders. In 1818, the *Journal of Science and the Arts*, published by the Royal Institution, carried a report that the effects of ether inhalation were similar to those of nitrous oxide. This report may well have been written by Humphrey Davy who was by then the director of the Royal Institution. The veracity of this report was confirmed by the wave of 'ether frolics' which swept through both Great Britain and America, ultimately leading to the discovery of its value as an inhalational anaesthetic.

While physicians were coming to the conclusion that pneumatic chemistry had little to offer them, an important development had been taking place in Paris. Antoine Fourcroy, the son of an apothecary, had been supported financially by influential members of the Société Royale de Médecine while he pursued medical studies. Such was his talent for chemistry that the Société permitted him to participate in its work even before he graduated in 1780. The Société was required to assess the medicinal value of mineral waters, and Fourcroy was given the responsibility of analysing these. He rejected the existing system of merely evaporating the waters to dryness, and instituted the use of specific chemical reagents to determine which minerals were actually present. From this modest beginning, Fourcroy proceeded to devote much of his distinguished career to the application of chemistry to medicine. Although he was primarily interested in the examination of the solids and fluids of the human body, in 1791 he published an analysis of St. Lucia and St. Domingo barks, which had been recommended as substitutes for cinchona bark in the treatment of malaria. This was for many years considered to be a model of vegetable analysis, and it stimulated others to examine cinchona and opium, the two most important vegetable drugs then in common use.

Following the decision of the Convention, on 8th August 1793, to suppress academic and professional bodies that had enjoyed privilege under the monarchy, new institutions of higher learning were established throughout France. The responsibility for medicinal analysis eventually passed to the Société de Pharmacie and the Ecole Supérieure de Pharmacie, which opened in Paris in 1803. Its first director was Nicolas Vauquelin, a close associate of Fourcroy, who had now become an important political figure. Vauquelin, an outstanding analytical chemist, encouraged the close association of chemistry with pharmacy in the curriculum. His response to growing concern in medical circles over the variable quality of plant products was to encourage his faculty members and their students to try to extract pharmacologically active principles from plants so that reliable chemical assays could then be established. He was inspired to suggest this through the work of the Swedish pharmacist Carl Gustav Scheele who, during the 1780s had isolated no leas than a dozen plant acids in pure form, including tartaric, malic, citric, oxalic, lactic, and uric acids. The fact that none of these were active principles did not seem to have discouraged Vaquelin.

Opium (Gr. *opos* = juice), obtained by drying the latex that exudes from the capsule of the poppy, *Papaver somniferum*, is probably the most ancient effective drug of all. Remains of the garden poppy have been found in the stone-age lake dwellings of Switzerland. The Ebers Papyrus refers to a mixture of poppy pods and flies as a sedative for children. In the Iliad, Homer mentions the poppy growing in gardens, which confirms its cultivation had been established by the eighth century BC. Whether the *nepenthe* that he wrote of in the Odyssey was opium is open to conjecture. The first accurate description of the poppy appeared in the *Historia Plantarum* written by Theophrastus (372–287 BC), the father of botany. Three centuries later, in his *De Universa Medicina*, Dioscorides, a Greek surgeon serving with Nero's army, explained how the poppy capsule should be incised in order to obtain its juice. Pliny the Elder (23–79) mentioned opium in his thirty-seven volume work on natural history. He shared some of his fellow citizens' contempt for the Greek physicians who monopolized medical practice in Rome, and did not lose the opportunity to warn of the dangers of opium. Nonetheless, its pain-relieving properties were by now clearly established, and its place in medicine was assured.

It was inevitable that opium should have been one of the first plant drugs to be investigated by means of the new system of plant analysis introduced by Fourcroy. Nevertheless, the French authorities had a long-standing interest in the constituents of the poppy. Wild rumours that poppy seeds and their oil, which were widely used for culinary purposes, had the same narcotic action as opium, forced the issuing of decrees, in 1718 and 1735, prohibiting their sale in France. Only after the matter was investigated by the Secretary of Agriculture in 1773 was the ban lifted.

In 1803, Jean-Francois Derosne, the owner of a fashionable Parisian pharmacy in the Rue St. Honoré, delivered a memoir to the Sociéte de Pharmacie, in which he reported that in the course of devising an assay for opium he had isolated a novel crystalline salt. A variety of tests revealed that this salt had alkaline properties, which Derosne attributed to contamination by the potash used to precipitate it from acid solution. He appreciated that he had been handling a peculiar substance that was certainly not a plant acid, and he described it as a salt because it crystallized readily. Nevertheless, he added the rider that this was a circumlocution to compensate for his inability to assign it to any known class of chemical compounds. In December 1804, Armand Séguin, formerly an assistant to Lavoisier and now the director of a highly successful tannery outside Paris, reported to the Institut de France that he had isolated a new plant acid and also a crystalline narcotic from opium. His findings were presented in a paper that he submitted to the Académie des Sciences, but it was laid aside. Séguin did not continue with his investigations, and by the time his paper finally appeared in the *Annales de Chimie* in 1814, similar observations had been reported elsewhere.

Friedrich Wilhelm Sertürner, the son of an Austrian engineer in the service of Prince Friedrich Wilhelm of Paderborn and Hildesheim, completed his apprenticeship with the court apothecary in 1803 when he was twenty years old. He remained at the Adlerapotheke in Paderborn for a further thirty months, during which period he was able to carry out a variety of chemical experiments. He turned to an examination of opium and was soon able to extract an organic acid that had not been reported in the literature. He named it meconic acid (Gr. *mekon* = poppy). When tested on dogs, it proved to be inactive. However, alkalinization of the mother liquors with ammonia caused precipitation of a substance that he collected and crystallized from alcohol. This time, when he administered it to a dog, it proved to be a narcotic. He published a preliminary report of his findings in 1805 in *Johann Trommsdorff's Journal der Pharmazie*. A detailed account of his isolation of the *principium somniferum* was printed in the same journal the following year, but scant attention was paid to it, possibly because it appeared in a journal read mainly by practising apothecaries. Sertürner included in his paper a footnote stating that he had not learned of Derosne's work until after his own had been completed. That he wrote of the 'almost alkali-like character' of this principle from a plant source should itself have caught the attention of chemists, for all plant principles isolated prior to this were acidic in nature. Sertürner explained that his *principium somniferum* could neutralize free acid, but he failed to recognize the great significance of this aspect of his work, and it was another three years before he briefly renewed his investigations after opening his own apothecary in Einbeck, Westphalia. In 1811, he published further papers in *Trommsdorff's Journal*, in which he confirmed that the narcotic principle of opium was an alkaline substance, or base, that formed salts with acids. However, it was not until 1815 that he carried out a more detailed investigation on it in the Ratsapotheke. He refined his earlier methods and identified the presence of carbon, hydrogen, oxygen, and possibly nitrogen in the narcotic material. In 1817, he published another paper, but this time it appeared in a prominent scientific journal, *Gilbert's Annalen der Physik*. The paper was entitled, 'On Morphium, a Salt-like Base, and Meconic Acid as Chief Constituents of Opium'. In this, Sertürner drew attention to the particular ease with which morphium reacted with acids to form readily crystallizable salts. He also described how he and three companions swallowed does of about 100 mg of morphium and experienced the symptoms of severe opium poisoning for several days despite recourse to strong vinegar to induce vomiting when these symptoms first appeared! This time, Sertürner was not ignored, Joseph Gay-Lussac, the doyen of French chemists, read the paper and immediately had it translated and re-published in the prestigious *Annales de Chimie*, the journal founded by Lavoisier and which Gay-Lussac now edited.

Gay-Lussac wrote an editorial to accompany the translation of Sertürner's paper. In this, he expressed surprise that Sertürner's work had been ignored for so many years, but not simply because the isolation of the active principle of opium was important. Of much greater significance, according to Gay-Lussac, was the discovery of a salt-forming organic plant alkali analogous to the familiar organic acids. He predicted that many other organic alkalis would be found in plants, for there was already some evidence to suggest that the few crude active principles isolated in the previous decade contained nitrogen and had alkaline properties. To ensure a degree of conformity in the naming of plant bases, Gay-Lussac proposed that their names should always end with the suffix '-ine'. This was the first time such standardization of nomenclature was introduced into organic chemistry. For this reason, Gay-Lussac altered Serturner's term 'morphium' to 'morphine'. In 1818 the German chemist Wilhelm

Meissner introduced the term 'alkaloid' to describe the plant alkalis, but several years passed before this was generally accepted.

Gay-Lussac asked Professor Robiquet of the Ecole Supérieure de Pharmacie to check Sertürner's experimental work. Robiquet noted the differences in the properties of the salts isolated by Derosne and Sertürner, and he concluded that they were different plant alkalis. He purified the base isolated by Derosne and gave it the name 'narcotine'; this is now generally known as noscapine. It had no narcotic properties, although it was later found to retain the cough suppressant properties of morphine, and is still prescribed for this purpose.

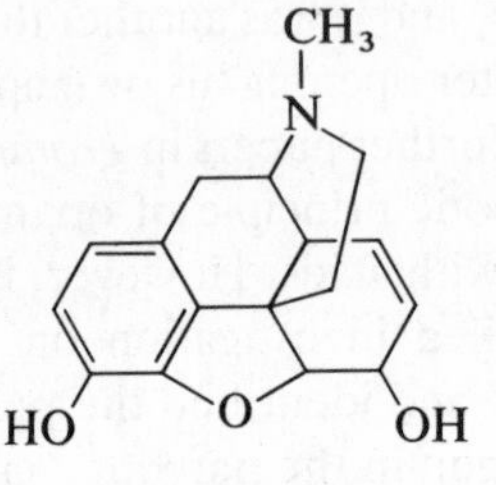

Morphine

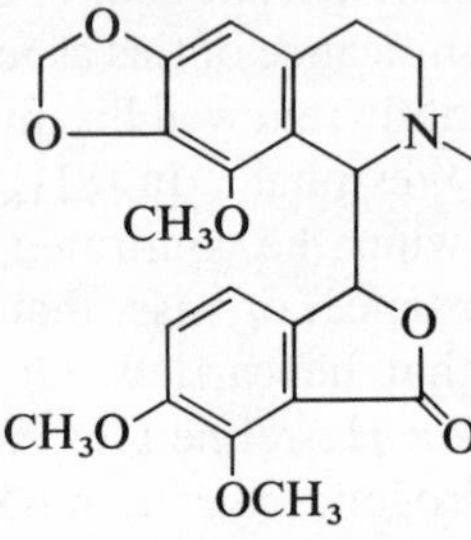

Noscapine

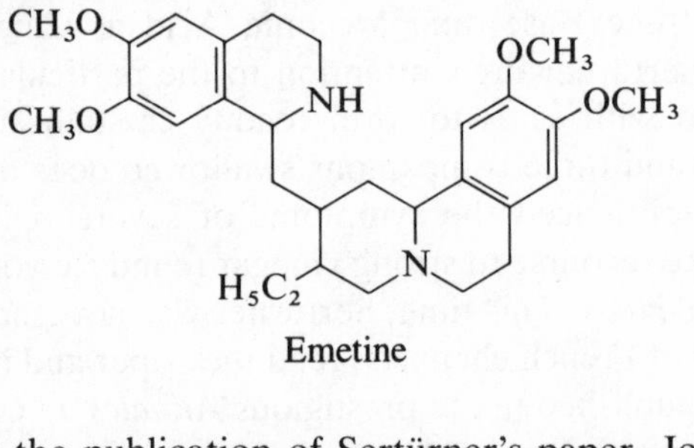

Emetine

Shortly before the publication of Sertürner's paper, Joseph Pelletier, the assistant professor in Robiquet's department at the Ecole Supérieure de Pharmacie, had collaborated with the brilliant physiologist François Magendie to isolate the emetic principle from ipecacuanha root. Early Portuguese settlers had found this root being used as an emetic by the natives of Brazil and Peru. An Amsterdam physician, Wilhelm Piso, who had spent several years in Brazil, described the root in his *Natural History of Brazil*, published in 1684. He stated that it was a specific remedy for dysentery. For some years the root was employed for this purpose in Spain and Portugal, but fell out of favour because it was thought to be too toxic. A few years later, a Parisian merchant who had imported the root, gave a sample of it to his physician, Afforty, as a sign of gratitude for treatment he had received. Afforty paid no attention to the merchant's claim that the root cured dysentery, but his assistant, Jean-Adrian Helvetius, tried the root and became convinced it was indeed a specific remedy for dysentery. He then placed placards around Paris, extolling the virtues of his secret formula that could cure dysentery. This came to the attention of the Court, and he was summoned to treat the dauphin and several courtiers. Louis XIV then ordered the remedy to be tested at the Hôtel-Dieu, and he was so impressed with the outcome that he paid 1000 louis-d'or to Helvetius for publication of his secret formula. After this episode, ipecacuanha was in constant demand, but considerable confusion existed over the nature of the root until Bernardino Gomes, a Portuguese naval surgeon, published a dissertation on it after returning from Brazil in 1800. He identified the root as *Cephaelis ipecacuanha*. Pelletier and Magendie isolated its active principle in 1817, and named it emetine once they realized it was a plant alkali.

Pelletier acted on Gay-Lussac's suggestion that further plant alkalies would be found. He was assisted by Joseph Caventou, a student who had shown considerable flair for chemical research. They attempted to provide evidence in support of Linnaeus's belief that plants of the same genus would exhibit the same pharmacological properties. In 1818, they examined different species of the *Strychnos* family, the most potent plant poisons then known. In 1540, Valerius Cordus had described *Strychnos nux vomica*, the poisonous seeds of an Indian tree. The following century it was in use to kill pestilent animals. The related *Strychnos ignatii* (Saint Ignatius bean) was first described in 1699 by Camelli, a Jesuit missionary who served in Manilla. Pelletier and Caventou managed to isolate the same plant alkali from Saint Ignatius beans and *nux vomica*, as well as from snake wood (*S. colubrina*). Tests were conducted to confirm that the pharmacological activity of the new principle was identical, irrespective of its source. In honour of the director of their faculty, Pelletier and Caventou named the new plant base 'vauqueline', but this was deemed inappropriate by the commissioners of the Académie des Sciences, on the grounds that such a distinguished name ought not to be associated with a harmful principle. The name strychnine was then substituted. Pelletier and Caventou expected to find strychnine in the bitter bark of the false angostura (*Brucea anti-dysenterica*), but instead isolated another new base which they called brucine.

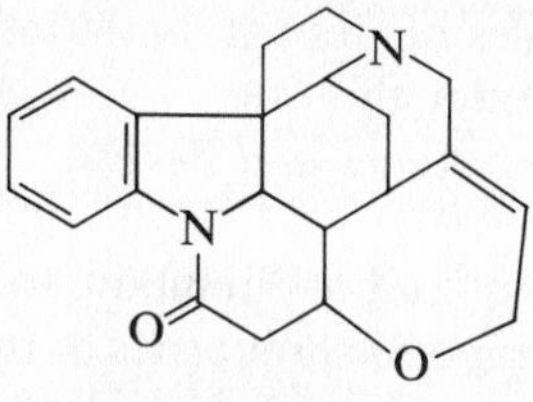

Strychnine

The work for which Pelletier and Caventou will always be best remembered was their isolation of quinine in 1820. The cinchona bark from which it was obtained had been introduced into Europe two centuries earlier. It was first mentioned in *The Chronicle of St Augustine*, written not later than 1633 by an

Augustinian monk who had lived in Peru, Father Antonio de la Calancha. He reported that the bark of a tree growing in the country now known as Ecuador could cure malaria. This was of importance not only to the conquistadors, for malaria was then the most widespread disease in the world. The city of Rome, in particular, had long been plagued by it, hence supplies of the bark were sent there from Seville by the family of Cardinal Joannes de Lugo. Seville at that time held a monopoly of the trade with Peru. Following its arrival in Rome, the bark became keenly sought after throughout Europe by pilgrims and princes alike. But it was not without its dangers, and disputes broke out over its merits. The fact that it was distributed by the Jesuits led to an element of religious bigotry adding to the confusion, and it was only the efforts of an English quack that finally ensured 'Jesuits' bark' a role in medicine.

Robert Talbor was an apprentice to an apothecary in Cambridge, where the Professor of Physic, Brady, began to prescribe 'Jesuits' bark' in 1658 when there was a serious outbreak of malaria. Talbor moved to Essex and then to London, where he treated cases of malaria with a secret remedy. In 1672, he published a small book called *A Rational Account of the Cause and Cure of Agues*. In this he warned of the dangers surrounding the administration of 'Jesuits powder' by the unskilled. Cunningly, he suggested that this remedy should not altogether be condemned, but nowhere did he even hint that he himself had used it! The book boosted his reputation, and he was knighted by King Charles II six years later. The king also issued letters patent appointing Talbor as his physician in ordinary. This was done despite fierce criticism from the College of Physicians, angered at the appointment of an unqualified practitioner. The physicians of Paris and Madrid were similarly outraged when, later that same year, Talbor cured the dauphin and the queen of Spain. The next year, King Charles' judgement was vindicated when he fell ill with tertian fever at Windsor and was then cured by Talbor's secret remedy.

Grateful for the curing of his son, Louis XIV persuaded Talbor to part with the secret of his remedy for a sum of 2000 louis d'or plus an annual pension of 2000 livres. Talbor agreed on condition that the formula would not be revealed during his lifetime. When he died in his fortieth year, in 1681, Louis arranged for this to be divulged in a small volume that appeared a year later, disclosing that it consisted of large doses of 'Jesuits' bark' infused in wine. An English translation was published under the title *The English Remedy: or, Talbor's Wonderful Secret for Cureing Agues and Fevers*.

The revelation of the nature of Talbor's remedy, followed by the publication, in 1712, of a book on the therapeutic properties of the bark by the professor of medicine at Modena, Francesco Torti, did much to popularize it. As demand increased, confusion arose over the botanical origins of the bark, especially as the colour varied in different samples. The situation was clarified after Charles-Marie de la Condamine, whilst engaged in an astronomical study near Quito in 1737, investigated the origins of the barks, and obtained specimens. These enabled the Swedish botanist Linnaeus to classify the family of trees from which the barks were obtained as *Cinchona*, a term derived from the Indian name for the fever bark tree.

Once the efficacy of cinchona bark had been established, controversy broke out over the question of how it should be administered, and which types of fever would respond. The problem here was that until the general acceptance of the germ theory of disease in the latter part of the nineteenth century, physicians considered fevers to be diseases in their own right, although William Cullen, who founded the Glasgow School of Medicine and later moved to Edinburgh to become the leading medical teacher in Europe during the second half of the eighteenth century, recognized that some fevers were caused by other illnesses. Much effort was expended by his contemporaries in categorizing the various types of fever, not on the basis of their cause, but rather on the basis of their temporal nature. The continuous fevers, usually associated with overcrowding in the slums of the cities or amongst the crews of ships, were distinguished from the periodic fevers. The latter were often to be found amongst those who lived in open spaces in the country or in warm climates. Such periodic fevers were held to be caused by marsh miasma, or malaria. There were two principal types, namely intermittent and remittent. The former, often called the ague, was milder and the fevers occurred in regular cycles that were described as quotidian, tertian, or quartan, depending on whether the attacks took place every second, third, or fourth day. Remittent fever was more severe and had a higher fatality rate, the patient experiencing variation in the intensity of the fever without any periods of relief.

John Brown, who at one time was William Cullen's secretary, subsequently became his bitter rival, and was responsible for the introduction of the Brunonian system of medicine that so strongly influenced Thomas Beddoes. Brown argued that intermittent fever was preceded by a state of debility, and those afflicted with it were asthenic. This meant that their nervous system required to be stimulated in order to effect a cure. The fact that cinchona bark frequently cured such patients was thereby interpreted to mean that it was a general stimulant of the nervous system, or tonic. By a logical extension of the Brunonian doctrine, the alleged tonic properties of cinchona were applied in the treatment of a wide range of diseases that today would be considered unrelated to malaria, but which Brown classified as asthenic. This misuse of medication was not confined to cinchona bark. Practitioners of the Brunonian system poisoned their patients with a wide variety of stimulants, ultimately inducing Samuel Hahneman of Leipzig to develop homeopathy in 1796. Hahneman's approach represented the opposite extreme, for he used infinitesimally small doses of drugs. The selection of these was based on the unproven principle *similia similibus curantur* (like cures like). This, of course, was the precise opposite of the principle enunciated by Galen sixteen centuries earlier, and symbolized Hahneman's rejection of allopathy, the use of drugs with opposing effects to those caused by the disease. He deduced this principle through his misunderstanding of what happened to him after he had taken a large dose of cinchona. This induced a paroxysm which he mistakenly believed was identical to a malarial paroxysm that he had experienced earlier. It is probable that the paroxysm was due to severe gastric irritation caused by the large dose that he had taken. Nevertheless, the experience led Hahneman to devise his system of medicine in which he justified the use of cinchona in intermittent fever on the grounds that large doses of it had caused malarial symptoms. He argued that therapeutic doses had to be minute (so much so that whether even a single molecule would be present is open to doubt!), apparently because the fever rendered the patient highly sensitive to the drug. Hahneman's ideas were promulgated in his book of 1810, entitled *Organon der rationellen Heilkunde*. One thing that can be said in favour of homeopathy is that it delivered many patients from the hazard of being exposed to the drugs favoured by adherents of the Brunonian system. Indeed, the contrast between the two opposing systems of medicine was highlighted by their respective use of cinchona, the one grossly overdosing and the other underdosing patients. Neither system could provide any justification for its proclaimed theoretical basis, or cite favourable results from objective clinical trials. Both systems were inherently sterile insofar as they attempted to reinterpret existing ideas, rather than develop an experimental approach. The proponents of each system adhered to the primitive belief that removal of disease required treatment of the patient rather than the disease. This was, of course, no longer a matter of expurging demons, placating the gods, or balancing the four humours. As in earlier years, this approach failed miserably, yet today it still finds support amongst practitioners of unorthodox medicine.

In 1802, Armand Séguin used nut-galls to precipitate one of the constituents of cinchona. Since nut-galls were known to precipitate gelatin, he wrongly concluded that gelatin was the active principle. He then proceeded to use this in the treatment of intermittent fevers. The following year, a Lyon pharmacist, Deschamps, obtained a 'febrifuge salt' from cinchona bark, but seven more years passed before there was further progress. Concern about the variable responses of his patients to treatment with cinchona bark led the Portuguese surgeon Bernadino Gomes to attempt to isolate the active principle. He managed to obtain silvery crystals by first extracting the bark with dilute acid, then neutralizing it with alkali. He named the crystals cinchonin (later changed to cinchonine), and assumed they were the sole active principle. They did not possess the bitter taste of cinchona. Gomes noted that the chemical properties of cinchonin were unlike those of any known plant product, but he did not have any reason to believe that it was a plant alkali. It was only after Gay-Lussac had drawn attention to the existence of such alkalies that cinchona bark was carefully examined for their presence.

In 1819, Friedlieb Runge, who had just completed his medical studies at the University of Jena and had decided to pursue a career in chemistry, isolated a base from cinchona. It was different from cinchonine, and Runge named it 'China base'. This was not the only alkaloid he isolated that year. During an interview with Goethe, then the principal of the University of Jena, Runge was invited to isolate the active principle of some coffee beans that he was then given. This led directly to the isolation of caffeine. Shortly after this, Pelletier and Caventou repeated Gomes' experiments on cinchona. By modifying the procedure somewhat, they isolated not only cinchonine from samples of grey bark, but also a more potent alkaloid, quinine, from yellow bark. Comparison of the properties of quinine and 'China base' showed them to have been identical, but such was the pre-eminence of the Ecole Supérieure de Pharmacie, that Runge's prior discovery was generally overlooked.

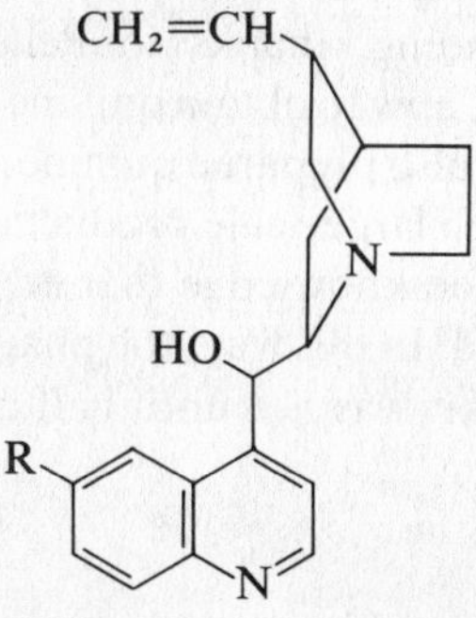

Cinchonine, R = —H
Quinine, R = $-OCH_3$

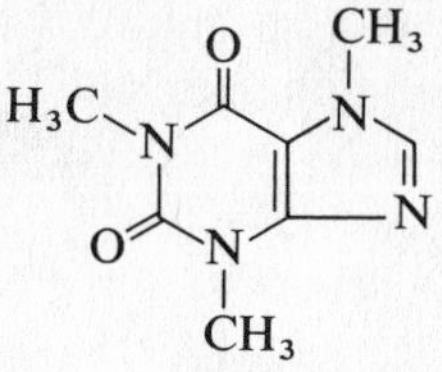

Caffeine

In their paper of 1820 announcing their isolation of quinine, Pelletier and Caventou urged medical practitioners to study the therapeutic properties of the pure cinchona alkaloids. This plea was immediately acted upon. It marked a new departure, for previously the active principles of plants had been isolated for scientific rather than therapeutic purposes. By urging their medical colleagues to study pure plant principles, Pelletier and Caventou set the course of drug discovery in a new direction. Amongst the first to experiment with quinine was François Magendie. He evaluated it in dogs, then went on to treat patients. By including it in his widely consulted *Formulaire* of 1821, he ensured that knowledge of quinine would quickly spread around the world. This small volume, of which there were many subsequent editions, listed single chemicals rather than the hitherto standard mixtures and plant products found in similar works. This was a significant advance, for it enabled Magendie's successors to consolidate the practice of experimental medicine on a firm basis, with the assurance that the drugs they studied were of unchanging constitution.

Several eminent Parisian physicians confirmed that quinine was able to arrest their patients' fevers and could be used instead of the unpalatable, nauseating powdered cinchona bark. To meet the immediate demand for supplies of quinine, Pelletier and Caventou manufactured it in their own pharmacies in the Rue Jacob and the Rue Gaillon. The following year, Pelletier rushed supplies of quinine to Barcelona, where an epidemic had broken out. Soon, factories had to be opened, and by 1826 Pelletier's factory was processing more than one hundred and fifty thousand kilograms of cinchona bark yearly, yielding around 3600 kg of quinine sulphate. Production elsewhere was also expanding, thanks to Pelletier and Caventou openly publishing the full details of how to obtain quinine from cinchona bark. Several manufacturers in Germany also prepared quinine. They quickly recognized the economies obtained through large-scale production of quinine and morphine, and also acquired specialized knowledge that enabled adulterated or inferior plant material to be rejected. In this way, the pharmaceutical industry gradually became established, but it was not until half a century later that synthetic drugs were developed.

II

Central Nervous System Depressants

VOLATILE ANAESTHETICS

In December 1841, Crawford Long, a young doctor in Jefferson, Georgia, was asked by friends to obtain nitrous oxide for inhalation at a party they were planning. Long supplied ether instead of nitrous oxide, and it proved a great success with the party-goers. So high-spirited and rumbustuous were the frolics that when, next morning, Long and his friends found they had collected several bruises, none of them could recollect having felt any pain during the party. Fortunately, Long had by now regained his wits sufficiently to realize the full significance of what had happened.

$CH_3CH_2OCH_2CH_3$ — Ether

N_2O — Nitrous Oxide

There had, at that time, been much interest in medical circles in Mesmer's claims that hypnosis could remove pain during surgery. Crawford Long had not been impressed by these claims, believing that mesmerism had only limited uses. Nevertheless, the idea of painless surgery came back to him after his experience at the ether frolic, and he made up his mind to pursue the matter further if a suitable opportunity arose. His chance came when, on the 30th March 1842, he placed a towel soaked in ether over the face of James Venables, a young man who had frequently inhaled ether. As soon as his patient was insensible, Long proceeded to excise a large growth on his neck. The youth felt no pain. History had been made, but the country practitioner did not bother to report his success until 1849, by which time he considered he had acquired enough experience to tell the medical profession what could be achieved with ether anaesthesia. Others were not to show such restraint.

Early in December 1844, Horace Wells, who was a dentist in Hertford, Connecticut, attended a public lecture at which Gardner Quincy Colton demonstrated the effects of laughing gas. Such lecture-demonstrations of popular science where then in vogue, and men like the ex-medical student Colton were able to earn a living touring towns around the United States. On this occasion, one of the audience who had volunteered to inhale the nitrous oxide accidently gashed his leg while staggering around under the influence of the gas. Noticing that the injury did not seem to hurt the man, Wells wondered if it might be possible to exploit nitrous oxide to ease the pain of dental extraction, a problem that he was particularly concerned with at that time. Two years earlier, he and his former partner in Boston, William Morton, had discovered the importance of the removal of entire decayed teeth rather than just the crowns. Although this was clearly in their patients' best interest, so great was the pain involved that few patients ever returned, and the partnership had to be dissolved.

After the lecture, Wells asked Colton if he would administer nitrous oxide to him whilst a troublesome molar tooth was removed. Colton duly appeared on the morning of 11th December 1844 and made history when he administered the first ever dental anaesthetic. Wells summed up the occasion succintly, 'A new era in tooth-pulling!'

Excited by his success, Wells wrote to William Morton, who still practised dentistry in Boston, and asked him to arrange a demonstration of anaesthesia in the presence of America's leading surgeons in the Massachusetts General Hospital. Sadly, this was to prove his downfall. The demonstration was a miserable failure because insufficient anaesthetic was administered. As the wretched patient struggled in agony under the surgeon's knife, Wells was jeered out of the operating theatre in disgrace.

Morton remained convinced that his former partner's approach to the problem of surgical pain was sound. He visited Wells and discussed the use of nitrous oxide before beginning his own experiments on his household pets in the summer of 1846. Convinced that nitrous oxide was not the ideal gas for the task in hand, he mentioned this to Charles Jackson, a chemist and geologist who was tutoring him for his medical qualification. As far as can be ascertained, Jackson suggested ether as an alternative, a suggestion which later provoked an unseemly row over the question of who deserved the credit and financial rewards for the discovery of surgical anaesthesia. In retrospect, it has to be said that many chemists and physicians were at that time aware of ether as an alternative to nitrous oxide. Had not Crawford Long offered to provide it when his friends had been seeking laughing gas? Most of the credit must go to Morton whose pioneering endeavours were to ensure the general acceptance of ether anaesthesia.

On the 30th September 1846, Morton successfully maintained anaesthesia with ether for three–quarters of a minute whilst a tooth was extracted from one of his patients. For the next three weeks he used it on all his patients, observing that their exhaled breath contaminated the ether, which was held in a rubber bag. This resulted in a lack of uniform response. To avoid this, Morton quickly designed an inhaler with a valve which vented exhaled breath into the atmosphere. The rubber bag was also replaced by a sponge-filled bottle, from which the patient inhaled ether through a tube held between the lips whilst the nostrils were pinched shut. This new inhaler was ready just in time to be used in the first successful public demonstration of surgical anaesthesia, held before a large gathering in the domed amphitheatre of the Massachusetts General

Hospital on the 16th October 1846. Morton administered the anaesthetic, whose nature he unsuccessfully tried to disguise by colouring it and calling it 'Letheon'. The patient, a twenty-year-old youth, was then presented to surgeon John Warren who swiftly proceeded to make a deep incision several inches long near the lower jaw. This time the Bostonians were deeply impressed, for the youth remained still most of the time.

It was now that the row over priority of discovery of surgical anaesthesia broke out. Not only did Jackson challenge Morton's claims, but so too did Horace Wells, without whose investigations on nitrous oxide Morton would never have become involved. This dispute forced Morton to protect his own interests by patenting his ether inhaler, a move that was to have far-reaching consequences for the practice of anaesthesia in different countries. The patent annoyed John Warren and his fellow surgeons so much that they immediately abandoned the use of the inhaler and instead anaesthetized their patients simply by pouring a few drops of ether at a time on to a sponge held over the mouth and nose. When London surgeon Robert Liston received reports of Morton's demonstration, he acquired an inhaler similar to Morton's for use in an operation the following December. Subsequently, English anaesthetists relied on inhalers which they were forever improving to overcome a variety of design shortcomings. In Edinburgh, however, the influence of the American surgeons was stronger, with the open mask technique prevailing. This continued to be used in Scotland, and on the Continent, well into the following century. Furthermore, it was a simpler technique to use safely, so that there was little demand for specialist anaesthetists in Scotland when the more dangerous chloroform was introduced, whereas the English preference for inhalers necessitated specialist training.

After James Young Simpson, professor of midwifery at Edinburgh University, witnessed ether anaesthesia in January 1847, he used it regularly in his obstetric practice, but two problems became apparent. Edinburgh was then the largest city in Scotland, with many of its inhabitants residing in high, crowded tenement blocks. Climbing up many flights of stairs on his domiciliary visits was bad enough for Simpson, but having to carry the large bottles of ether required by the inefficient open mask method, was too much! The inefficiency of the technique was a result of wasteful evaporation of the ether from the mask into the atmosphere, and it was this which also caused Simpson's second, more serious, problem. Gas lighting was by now installed in all the tenements; as Simpson proceeded to carry out his work, the concentration of ether in the air soon reached a hazardous level, posing the risk of a fire or even an explosion. Consequently, Simpson vigorously joined the quest for an alternative to ether. He was unaware that a solution of chloroform in alcohol had already been investigated at the Middlesex Hospital, only to be rejected, on account of its cost. However, at St. Bartholomew's Hospital, also in London, William Lawrence found this so-called 'chloric ether' to be a satisfactory alternative to ether. In Paris, Pierre Flourens, a former student of the great physiologist François Magendie, had already tested both chloroform and ethyl chloride on animals. The latter was subsequently used in Heyfelder's clinic in Erlangen, Germany, but it too was ultimately rejected because of its high cost.

Simpson began his search by testing a variety of volatile fluids on himself and his friends. These included Dutch Liquid (1,2-dichloroethane), acetone, ethyl nitrate, benzene, and iodoform vapour. In October he met a Liverpool pharmacist, David Waldie, with whom he discussed the problem. Waldie's company at that time supplied 'chloric ether' to local physicians, who told patients with respiratory disorders to inhale it. As it was made by dissolving chloroform in alcohol, Waldie knew that chloroform, besides being the sort of volatile liquid required by Simpson, was also safe to inhale. Unfortunately, as his laboratory had just been destroyed by fire, he was unable to supply any chloroform. Instead, Simpson acquired some from the Edinburgh druggists, Duncan, Flockhart, and Company, and at once proceeded to test it on himself and some friends. It met all his requirements, so on the 10th November 1847, Simpson reported to the Medical-Chirurgical Society of Edinburgh that chloroform had many advantages over ether: it was more potent, requiring only a small volume of liquid; the onset of anaesthesia was quicker; it was also pleasant to inhale; and it was simple to administer by impregnating a cloth held over the mouth. Moreover, it was also non-inflammable!

$$CHCl_3$$

Chloroform

When London surgeons used chloroform with inhalers they experienced difficulties not previously encountered with ether, and several deaths from cardiac irregularities were reported. The problem lay in the design of the inhalers which tended to exclude oxygen-containing air. The first specialist anaesthetist, John Snow, carried out animal and clinical experiments which established the importance of controlling the proportion of anaesthetic in the mixture of air and anaesthetic being inhaled. To this end, he designed improved apparatus, and sought out a safer volatile liquid from the increasing range of organic solvents then becoming available. Snow determined both the boiling point and the concentration of solvent when air was saturated with it, using these physical properties to compare one substance with another prior to animal experimentation. He sought a solvent with the physical properties of chloroform combined with the clinical safety of ether, but his early death denied him of his goal, although not before he had achieved worldwide fame by anaesthetizing Queen Victoria when she gave birth to Prince Leopold in 1853. Five years after his death, his pupil, Bernard Ward Richardson, continued the systematic investigation of organic compounds as anaesthetics and hypnotics, pioneering the idea that molecular modifications altered physical, and, consequently, physiological properties. Both Snow and Richardson took a different approach to the question of the influence of molecular structure upon biological activity from that of Thomas Nunnely of Leeds, who tried to show, in 1849, that the varying proportions of carbon, hydrogen, and other elements in anaesthetics were directly responsible for their differing actions. Snow and Richardson were correct in their assessment, and from these faltering beginnings the whole structure of contemporary drug research has been derived.

At the time Snow, Richardson, and Nunnely were exercising their minds regarding the influence of chemical structure on biological activity, it was generally recognized that certain clusters of atoms within various molecules could be associated with specific physical properties such as acidity or alkalinity, and also with certain types of reactivity. These clusters of atoms came to be known as 'functional groups', or as 'radicals' if they consisted solely of carbon and hydrogen atoms. The real breakthrough, however, came in 1858 with the publication of Friedrich August Kekulé's paper which explained that in organic molecules the carbon atoms were linked together in chains to form a backbone upon which the functional groups were mounted. Kekulé had begun to formulate his ideas in London four years earlier, supposedly whilst daydreaming on the Clapham omnibus, when he visualized carbon atoms dancing around forming pairs, then chains!

The development of chemical theory failed to facilitate the discovery of new inhalational anaesthetics until well into the twentieth century. In fact, the only anaesthetic other than ether or chloroform to be widely used during the nineteenth century was nitrous oxide. Gardner Quincy Colton tried it once again for dental anaesthesia in 1862, so impressing the dentist concerned, a Dr Dunham, that he was asked for information on how to manufacture the gas. On returning to Connecticut a year later, he heard that Dunham had successfully administered nitrous oxide to more than six hundred patients! Colton then persuaded him and a colleague to join him in partnership in New Haven. There, in a period of only twenty-three days, they extracted over three thousand teeth painlessly! Colton then moved to New York, where he built up a highly profitable dental practice. He administered nitrous oxide to thousands of patients without mishap. Word of this soon spread throughout the USA and Europe, and by 1868, nitrous oxide was regularly used in London hospitals. Not long after, it became available as a compressed gas in cylinders.

Although experimentation did not cease, no important new inhalational anaesthetic was introduced during the next half century until ethylene appeared in 1923. This was actually investigated by Thomas Nunnely in 1849, but he had rejected it as an anaesthetic, believing that its safety margin was too narrow and it would be too difficult to manufacture. Some years later, in 1876, Eulenberg mentioned in his *Handbuch der gewerbe Hygiene* that rabbits could be anaesthetized with ethylene, but its rediscovery as an anaesthetic the following century was a classic example of what Horace Walpole meant by the term 'serendipity'. He coined his term after reading a poem about the three princes of Serendip (Sri Lanka) who repeatedly made discoveries they were not seeking, thanks to a combination of chance and sagacity. The sequence of events in the case of ethylene anaesthesia began when William Crocker and Lee Knight at the botany laboratory in the University of Chicago were asked by carnation growers who shipped their products into the city to find out why their flowers closed up when stored in greenhouses and the buds also failed to open. Heavy financial losses had been incurred through this in the 1908 season, so Crocker and Knight began work at once. They traced the cause to ethylene in the gas used to illuminate greenhouses, confirming that as little as one part of ethylene in two million parts of air caused already open flowers to close! Further investigations revealed that ethylene damaged a large variety of plants.

$$CH_2{=}CH_2$$

Ethylene

Concern that ethylene might also prove toxic to humans led Arno Luckhardt and R. C. Thompson, colleagues of Crocker and Knight, to expose frogs, rats, and a dog to the concentration of ethylene (4%, v/v) present in illuminating gas. Unaware of the work done in the previous century, they were surprised when the animals were anaesthetized by the gas. A full investigation into the possibility of using ethylene clinically then began. It took five years, but before Luckhardt was ready to publish his findings, W. Easson Brown, an anaesthetist at the Toronto General Hospital, read a paper to the Academy of Medicine in Toronto in February 1923, reporting that animal experiments he had been carrying out for the last six months in the department of pharmacology at the university had indicated ethylene might be a promising anaesthetic. He, too, was unaware of the early studies on ethylene, having selected it for examination solely because it was 'one of the active constituents of a certain commercial ether'—a somewhat perplexing phrase! Easson's preliminary animal studies were published in the *Canadian Medical Association Journal* only a few days before the first patients began to be routinely anaesthetized with it in Chicago. This was after Luckhardt himself was safely anaesthetized in a demonstration before a distinguished gathering of medical men on the 14th March. Luckhardt established that ethylene induced anaesthesia very quickly and pleasantly, whilst recovery was also rapid and less eventful than with existing anaesthetics. It was used regularly until the early fifties, when there was mounting concern about the risk of explosion with anaesthetics, especially those that had no major advantage over newer agents.

Work continued at Toronto on the clinical evaluation of ethylene, both at the General Hospital and in the department of pharmacology where Professor V. E. Henderson extended the investigations to include the chemically related gas, propylene. This was also a promising anaesthetic when freshly prepared, but after storage in pressurized steel tanks a toxic impurity was generated, causing patients to experience nausea and even cardiac sequelae. The matter was not pursued and propylene was no longer used. Several years later, however, after a chemist, George Lucas, joined the research team, Professor Henderson asked if he could hazard a guess as to the likely nature of the contaminant. When Lucas suggested it might be cyclopropane Henderson asked that some be synthesized and tested on animals. So it was, that in November 1928 two kittens inside a bell jar were exposed to low concentrations of cyclopropane mixed with oxygen. To the amazement of all present the kittens quickly fell asleep, and on removal from the jar recovered with no hint of any toxic effects! Extensive investigations began at once with the result that a year later Easson Brown safely anaesthetized Professor Henderson, confirming the safety of cyclopropane in humans. Shortly after, a famous citizen of Toronto, Frederick Banting, who was co-discoverer of insulin, also received cyclopropane in a demonstration. The hospital authorities then announced that they would not permit patients to be anaesthetized with cyclopropane! Brown was forced to ask Ralph Waters of the University of Wisconsin to carry out the first clinical trials. These were completed by 1934, confirming the status of cyclopropane as a valuable and potent anaesthetic. It is still in use fifty years later. As for the toxic contaminant in propylene? Well, this was eventually traced to the presence of hexenes.

$CH_2{=}CH{-}CH_3$

Propylene

$CH_2{-}CH_2{-}CH_2$ (ring)

Cyclopropane

Cyclopropane was not the only anaesthetic to be discovered in the wake of investigations into ethylene. Chauncey Leake at the University of California Medical School felt that as ethylene lacked some of the desirable properties of ether as an anaesthetic, it would be reasonable to examine molecules containing both ether and ethylenic functions. The physical properties of five such compounds were examined. It was found that only vinyl ether, a novel substance synthesized for Leake by Randolph Major of Princeton University, had a boiling point and partition coefficient similar to that of ether. Predicting that vinyl ether would be an anaesthetic, Leake and Mei Yu Chen administered it by inhalation to mice. Their experiment was a success, perhaps too much so, for when vinyl ether was introduced clinically in 1930, anaesthetists found its main drawback was a tendency for it to produce too deep a level of anaesthesia very rapidly. On the credit side, it did not irritate the respiratory tract to the extent that ether did. Although eclipsed by the post-war development of the fluorocarbons, vinyl ether was an important landmark in drug design.

$CH_2{=}CHOCH{=}CH_2$

Vinyl Ether

$ClCH{=}CCl_2$

Trichloroethylene

The arrival of rationally designed anaesthetics did not end the opportunity for chance and sagacity to influence events. Trichloroethylene, which was first synthesized in 1864, was widely known as an industrial solvent. Its toxicological properties were once a cause for concern, for it was reported to cause a loss of facial sensation over an area corresponding to the distribution of the trigeminal nerve. A number of fatalities had occurred, with the cause eventually being traced to impurities, which included phosgene. By 1930, the problem had been overcome through an improved manufacturing process. Indeed, the liquid was then considered safe enough for it to be inhaled as a palliative remedy for the agonizing facial pain of trigeminal neuralgia, whilst one or two American anaesthetists even experimented with it clinically, only to abandon it soon after. In 1940, however, a London chemist wrote to the *Lancet* pointing out that he had observed youths in an aeroplane repair factory deliberately leaning over vats of trichloroethylene in order to become intoxicated. The chemist, a Mr Charmers, wrote that he had tried inhaling the fumes himself to see what effect this had, and, as a consequence, he was of the opinion that it should be tested as an anaesthetic. His suggestion was taken up at once by Christopher Hewer who safely anaesthetized a patient with a sample of trichloroethylene supplied by Charmers. Having used up the entire sample, Hewer approached ICI to obtain more, and he learned that they already produced it in a highly purified form for the cleansing of wounds and burns contaminated by grease. Subsequently, Hewer and others established that it had a clinical role in short, painful procedures requiring only a light level of anaesthesia. It became particularly popular for use during childbirth when the patient held an inhaler over her face until it fell away as the anaesthetic exerted its action.

FLUOROCARBON ANAESTHETICS

Early in 1930, Thomas Midgley was asked by the management of General Motors to develop a non-toxic, non-flammable, low boiling point refrigerant for the company's Frigidaire division. Ten years earlier Midgley, an engineer by profession, had successfully developed tetraethyl lead as an automobile engine anti-knock agent.

Together with his colleague Albert Henne, Midgley reasoned that the atoms on the right-hand side of the periodic table of chemical elements were the only ones that could permit sufficient volatility for a refrigerant, but most of those were likely to prove toxic. The inert gases, on the other hand, were already known to be too volatile. By a further process of elimination of elements that were likely to increase flammability and comparative toxicity, Midgley concluded that fluorine should be incorporated into hydrocarbons. He rejected the assumption that such compounds would necessarily be toxic, believing that the stability of the carbon-fluorine bond would be enough to prevent release of either fluorine or hydrogen fluoride. Once the first compound, dichloromonofluoromethane, was prepared, a guinea pig was immediately exposed to it in a bell jar and, to his delight, experienced no ill-effects. Thorough toxicological studies then confirmed that fluorocarbon refrigerants were non-toxic under normal conditions of use.

It was another eight years before H. C. Struck and E. B Plattner demonstrated that butane, propane, and hexane analogues in which all the hydrogen atoms had been replaced by fluorine, possessed weak anaesthetic properties. This could not be exploited clinically because the slightest disturbance was found to cause the anaesthetized animals to convulse.

After the war was over, a wider range of compounds became available as a consequence of the wartime development of stable fluorocarbons as solvents for uranium hexafluoride. Forty-six of these were synthesized at Purdue University by E. T. McBee, then sent for pharmacological evaluation to B. H. Robbins at Vanderbilt University, Tennessee. It was found that the more volatile compounds were convulsants, whereas those containing heavy atoms such as bromine had high boiling points and were apparently safer anaesthetics than ether or chloroform. The most promising compound then tested would today be recognized as the analogue of halothane in which the chlorine is replaced by a bromine atom. Scientists at the laboratories of Ohio Medical Products followed up Robbins' suggestion that more analogues should be examined. By this time there was considerable concern over the explosive nature of air–anaesthetic mixtures, especially in America where the nature of the problem had been exaggerated somewhat. This exaggeration was to reach a climax in 1962 when the Wall Street Journal alleged that before the war there had been fifty to sixty operating theatre explosions in America every week! The British Ministry of Health, however, made the more sober assessment that during the years 1947–1954 there had only been a total of around forty explosions in the seven million operations carried out in the United Kingdom. It is certainly indisputable that the risk of explosions caused considerable apprehension amongst operating theatre staff. Such was the extent of this, that it appeared to make good commercial sense for Ohio Medical to develop a non-explosive anaesthetic. This proved to be a difficult task for their chemists, led by J. G. Shukys, for the first non-inflammable fluorocarbon to be tested lacked anaesthetic activity, whereas the first promising anaesthetic prepared, which was the twenty-fourth fluorocarbon they synthesized, turned out to be flammable, albeit less so than ether. This was nevertheless considered by Professor John Krantz, of the University of Maryland, to compare so favourably with ether, that after extensive animal testing, he proceeded to administer it to a volunteer on the 10th April 1953. The volunteer was a leading American anaesthetist, Max Sadove, who subsequently carried out clinical trials of the new fluorocarbon, fluroxene ('Fluoromar'), at the University of Illinois. It was released for general use in 1956, but its inflammability rendered it inferior to halothane, which was then undergoing its first clinical trials.

$CF_3CH_2OCH{=}CH_2$	$CF_3CHClBr$	$CHCl_2CF_2OCH_3$
Fluroxene	**Halothane**	**Methoxyflurane**

Unaware of the developments in the United States, the director of the general chemicals division of ICI, James Ferguson, reviewed the company's range of fluorocarbons in 1950. As these had been developed to serve as solvents capable of resisting attack by uranium hexafluoride during the separation of uranium isotopes by diffusion, they were exceptionally stable compounds. It was this total lack of chemical reactivity which convinced Ferguson they should be examined as anaesthetics. Responsibility for the project was given to Charles Suckling, who selected fluorocarbons that satisfied a range of critical requirements drawn up by himself and James Raventos, a pharmacologist from the medicinals section of the company's dyestuffs division at Manchester. The prime objective was low toxicity, and the others were that the anaesthetic must be non-explosive, have the right degree of volatility, induce anaesthesia both rapidly and uneventfully, not cause respiratory irritation, and be highly potent. The exacting nature of these requirements was to pay handsome dividends because, unlike the American development of fluroxene, the product eventually introduced was unchallengeable even though it was only the sixth compound synthesized by Suckling. It was first prepared in January 1953, and came quickly to the fore by meeting all the physicochemical and animal criteria. After three years of exhaustive testing it was finally put on clinical trial at Crumpsall Hospital, Manchester, by Michael Johnstone. Patients were smoothly anaesthetized, their circulatory systems functioned normally, and recovery was free from nausea in most cases. The new fluorocarbon was given the approved name of halothane ('Fluothane'). It quickly became the most widely used inhalational anaesthetic throughout the world, a pre-eminence that has effectively remained unchallenged to this day despite the introduction of a few similar compounds by rival manufacturers. The most notable amongst these has been methoxyflurane ('Penthrane'), which was first synthesized at Cornell University in 1948 as part of the Manhattan Project to make an atom bomb. During a collaborative programme conducted in the 1960s between Abbott Laboratories, the Dow Chemical Company, and Cornell University, it emerged as a promising anaesthetic with good analgesic properties.

HYPNOTIC DRUGS

The early anaesthetics were discovered by physicians and dentists who wanted to relieve the agonies of the patients on whom they operated, whilst the first alkaloids were isolated by pharmacists striving to expand the frontiers of their science. It was, however, a new breed of scientists who introduced the hypnotic drugs, amongst others, during the second half of the nineteenth century, namely the pharmacologists. Their discipline was often described as experimental medicine, but this term blurred the distinction between physiology, pharmacology, and biochemistry. Whilst Magendie and Bernard carried out pioneering pharmacological experiments in France, they are best remembered as physiologists. This, however, cannot be said of Rudolf Buchheim, who was appointed to the first ever university chair of pharmacology in 1847, when he was only twenty-seven years old. This was at Dorpat in Estonia where, although it was an Imperial Russian university, the language of instruction and most of the staff were German, until a process of Russification began in 1891. Here, Buchheim's first task was to build a laboratory in the basement of his own home, the entire running costs of this, the first pharmacology laboratory anywhere, being met by Buchheim himself. In 1860, however, his reputation was such that his university finally provided him with a large purpose-built institute. It was from this institute that Oswald Schmiedeberg, a locally educated son of a forester, obtained his doctorate in 1866 for a thesis on the measurement of chloroform in the blood. He remained at Dorpat as a lecturer in pharmacology and, at the age of thirty-one, he succeeded Buchheim, who left for Giessen in 1869. Although Schmiedeberg's reputation ultimately outshone that of his mentor, it is to Buchheim that credit must be given for turning the purely descriptive and empirical Materia Medica into a soundly based deductive science.

$$CCl_3CH(OH)_2$$

Chloral Hydrate

$$CCl_3CH_2OH$$

Trichloroethanol

One of the first experiments carried out in Buchheim's larger laboratory was with chloral hydrate, then very much in the forefront of attention since it was used to generate chloroform. Buchheim wondered whether excessive alkalinity of the blood, which was then thought to be a complicating factor in some diseases, could be reduced by administration of chloral hydrate. The logic behind this was based on the liberation of chloroform and formic acid on treatment of chloral hydrate with caustic alkali. Buchheim believed that the chloroform released in the blood might be converted into hydrochloric acid, thus supplementing the neutralizing action of the formic acid. However, on taking a draught of chloral hydrate he quickly fell asleep, as did several of his colleagues. This was an even greater discouragement for Buchheim than was the foul taste of the chloral hydrate, since it appeared to indicate that the chloroform was not being broken down into hydrochloric acid. The investigation was thereupon abandoned.

It was eight years later that Oscar Liebreich, a research assistant in the chemical division of Berlin University's Pathological Institute, also tried to use chloral hydrate to liberate chloroform in the blood, unaware of the earlier investigation at Dorpat. Unlike Buchheim, he hoped the chloroform would induce unconsciousness, and so was delighted when animal experiments confirmed his expectations. Within a few months of clinical studies indicating the appropriate dose required for the safe induction of sleep in healthy individuals, chloral hydrate was in use all over the world as the first safe hypnotic, despite its unpleasant taste and the frequency with which it caused gastric irritation.

Soon after the acceptance of chloral hydrate into clinical practice in 1869, controversy arose over whether it was really releasing chloroform in the blood. It became evident that if any chloroform were present, it could only be in trace amounts. Today, of course, chemists realize that the alkalinity of the blood is so slight as to be unable to induce the decomposition of chloral, but in the 1860s there was no understanding of the subtleties of Sorensen's pH scale, which was not introduced until 1909. Although the controversy died down as a result of

Sorensen's new ideas, it was not until 1948 that an American researcher, Thomas Butler, provided unambiguous confirmation of Josef von Mering's contention, made seventy years earlier, that chloral was metabolized by the liver to form trichloroethanol, a hypnotic alcohol. Liebreich had indeed been lucky that his unsound hypothesis had nonetheless resulted in his achieving his declared objective. Such an outcome is by no means unique in drug research!

The significance of the discovery of the hypnotic properties of chloral hydrate lies not in the fact that it was the first safe hypnotic (indeed, it is still used in children), but rather in that it brought home to many people the potential of synthetic drugs as therapeutic agents. It also typified what was to be, for many years to come, the sole alternative to basing the structures of synthetic drugs on those of natural products, namely the idea of designing a molecule to give a slow release of an active agent.

Professor Schmiedeberg moved from Dorpat to the newly re-established University of Strassburg, which was opened when the city was returned to Germany after the Franco-Prussian war of 1870–1871. As no expense was spared by the Reichsland of Elsass-Lothringen (now Alsace-Lorraigne) in turning its university into one of the finest in Germany, Schmiedeberg was able to proceed with his plans to build a truly magnificent institute of pharmacology. Although it took until 1887 before the building was completed, the work of the department of pharmacology proceeded unhindered in the interim. As the medical faculty grew into one of the finest in Germany, Schmiedeberg had no difficulty in persuading the best students to work with him despite the low esteem in which his subject was held elsewhere. During the forty-six years he occupied the chair of pharmacology at Strassburg, Schmiedeberg trained most of the men who became professors in the subject at other German, and several foreign, universities. His efforts undoubtedly led to the overwhelming pre-eminence of the German pharmaceutical industry until the outbreak of the Second World War.

At Strassburg, Schmiedeberg maintained his interest in the subject on which he had worked for his doctoral thesis at Dorpat, namely the pharmacology of chloroform and related compounds. It was one of his protégés, Josef von Mering, who suggested that chloral hydrate was converted in the body into trichloroethanol. Later, in 1883, a former student who had returned to Italy, Vincenzo Cervello, followed the chloral hydrate approach by studying paraldehyde, which was known to decompose and release acetaldehyde. Some years earlier, a professor in Paris had advocated the use of acetaldehyde as a superior alternative to either chloroform or ether, but this suggestion was never taken up because of the marked bronchial irritation produced by its vapour. After experimenting on animals, Cervello used paraldehyde on patients in Milan hospitals and claimed that it caused less disturbance of the blood pressure than did chloral hydrate. It did, nevertheless, have a singularly unpleasant taste and imparted a foul odour to the breath for as much as twenty-four hours. This odour was due to paraldehyde itself, and although most of the drug was ultimately metabolized into acetaldehyde by the liver, the hypnotic effect was due to unmetabolized paraldehyde. Despite its unpleasant nature, paraldehyde has been in use for a century, but thankfully it is now reserved principally for treatment of status epilepticus, in which it remains a valuable drug.

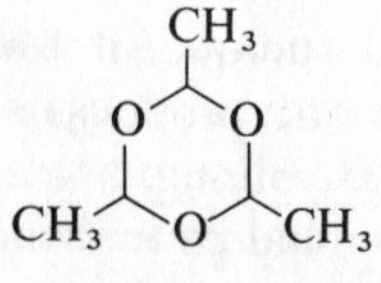

Paraldehyde

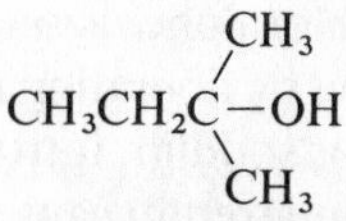

$CH_3CH_2C(CH_3)_2-OH$

t-Amyl Alcohol

In 1885 Josef von Mering, working in Schmiedeberg's department, followed up his earlier research on the nature of the active metabolite of chloral hydrate by taking advantage of his appointment as physician to a nearby prison. He established that tertiary alcohols, and in particular tertiary amyl alcohol (then known as amylene hydrate), were useful hypnotics in man. The tertiary amyl alcohol was not unpleasant to take, in contrast to both chloral hydrate and paraldehyde. It was used clinically for many years.

$NH_2C(=O)OCH_2CH_3$

Urethane

$NH_2C(=O)OCH_2CH_2CH(CH_3)CH_3$

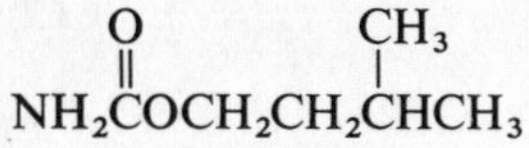

'Hedonal'

The popularity of chloral hydrate served to highlight its two major drawbacks, namely its ability to depress the brain centre which controls breathing, and its toxic action on the heart. Schmiedeberg sought to overcome the former problem with urethane, a chemical which could decompose into alcohol and ammonia. The latter was then classified as a respiratory stimulant—witness the continuing use of smelling salts to this day. An added bonus was that carbon dioxide, an established respiratory stimulant, was also released. So it was that, in 1885, he was able to demonstrate that urethane was a safe anaesthetic in small animals. Results on humans were disappointing, which is just as well since urethane is now known to be highly toxic to the cells of the bone marrow. As with chloral hydrate and paraldehyde, the original hypothesis of molecular decomposition resulting in the release of active drug was incorrect. Pharmacologists, however, could not overlook the efficacy of urethane on animals. To this day they use it as an anaesthetic. Numerous attempts were made to find an effective analogue, the first being introduced as a hypnotic in 1899 by Bayer's director of pharmacological research, Heinrich Dreser, under the proprietary brand name of 'Hedonal'. It resulted from an extensive research programme that involved the testing of a wide range of urethane analogues (carbamates) in which the alcohol moiety was varied. This was one of the earliest such programmes of research to be commercially successful.

Inevitably, the ploy of combining a hypnotic with ammonia in an effort to overcome the potential for respiratory depression had to be tried with chloral hydrate. In 1888, Nesbit found chloral ammonia was a useful hypnotic, but it was inferior to chloral formamide, which was introduced the following year by von Mering. This was the first chloral derivative that did not produce significant gastric irritation. It was followed by Heffter's chloralose, and Béhal and Choay's dichloralphenazone, which is still used today. All these chloral derivatives act by liberating trichloroethanol in the body, as does its phosphate ester, triclofos sodium, introduced by Glaxo in 1962. Attempts to use trichloroethanol foundered because of its unpleasant taste and its tendency to cause nausea.

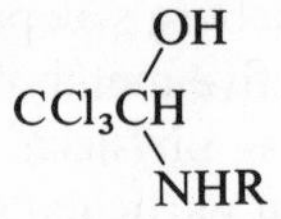

$$CCl_3CH(OH)NHR$$

Chloral Ammonia, R = —H
Chloral Formamide, R = —CHO

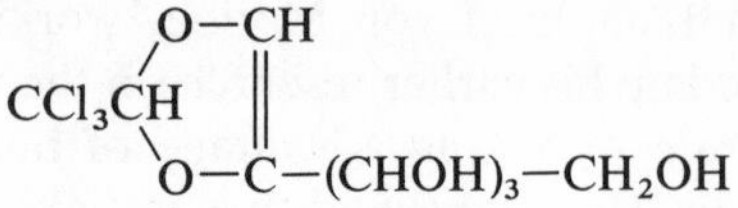

$$CCl_3CH\langle O{-}CH{=}\ ,\ O{-}C\rangle{-}(CHOH)_3{-}CH_2OH$$

Chloralose

$$CCl_3CH_2OP(OH)(=O)O\cdot Na$$

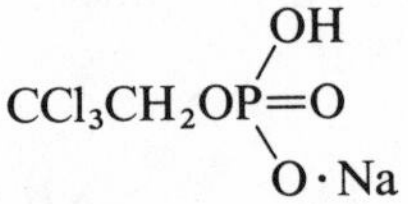

Triclofos Sodium

In the summer of 1887, Professor Alfred Kast was asked by his colleague, Professor Eugen Baumann, who held the chair of chemistry at the University of Freiburg, to see whether some novel sulphur compounds had any physiological activity. Kast injected two grams (!) of the first of these sulphones into a dog. Initially, there was no apparent reaction from the animal, but several hours later it began to stagger, then fall unconscious. The dog did not awaken until several hours later. The experiment was repeated on other animals, confirming that the new compound, sulphonal, was an effective hypnotic. After clinical evaluation, it was marketed the following year by the dyestuffs manufacturer F. Bayer and Company, of Elberfield. It was to be one of this famous German company's first profitable pharmaceutical products, as it combined palatability and absence of gastric irritancy with freedom from circulatory disturbance. Its chemical homologue, trional, was introduced soon after in an only partially successful attempt to avoid the delay of several hours before enough of the poorly soluble sulphonal had dissolved in the gut, from whence it could be absorbed to enter the circulation and induce sleep.

Consideration of the chemical nature of the hypnotics discovered during the last two decades of the nineteenth century convinced von Mering that a key feature was the presence in their molecular structure of a carbon atom containing two ethyl groups. Knowing of work already carried out by others on urea derivatives, he investigated diethyl acetylurea, finding it to be as potent a

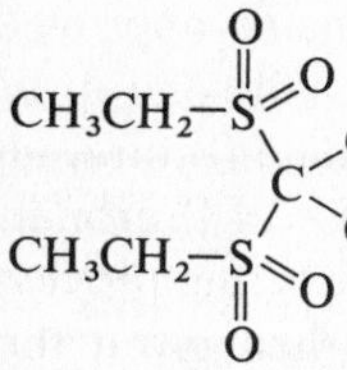

$$CH_3CH_2-S(=O)_2-C(CH_3)(CH_3)-S(=O)_2-CH_2CH_3$$

Sulphonal

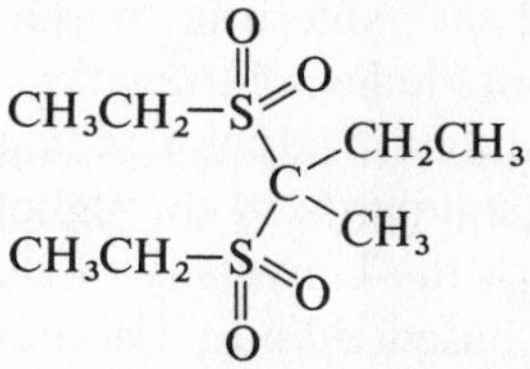

$$CH_3CH_2-S(=O)_2-C(CH_2CH_3)(CH_3)-S(=O)_2-CH_2CH_3$$

Trional

$$O=C(NH_2)-NH-C(=O)-CH(CH_2CH_3)_2$$

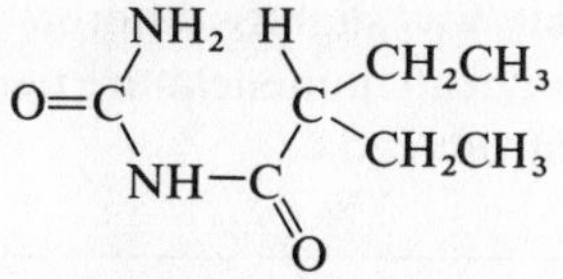

Diethylacetylurea

hypnotic as sulphonal. His conjectures then led von Mering to the conclusion that he should prepare 5,5-diethylbarbituric acid. The parent compound of this series, barbituric acid, had been synthesized by Adolph von Baeyer in 1864, and is variously said to have been so named after a young maiden with whom its discoverer was then in love, or, more prosaically, on account of its first preparation being on St Barbara's Day! Whichever version is true, barbituric acid had remained only a laboratory curiosity at the time von Mering first wanted to prepare any of its derivatives. Indeed, such was the lack of interest, that von Mering was unaware that the very barbiturate he required had already been synthesized by Conrad and Guthzeit twenty years earlier. He forged ahead, prepared the compound, and established that it was an hypnotic in animals.

Requiring confirmation of the structure of his new hypnotic, von Mering visited Berlin to call on his old friend from student days, Emil Fischer, the doyen of Germany's organic chemists. Fischer doubted the reliability of the synthesis and instructed his young nephew, Alfred Dilthey, to make the desired product. This was duly done. When tested on a dog it proved to be much more potent than von Mering's compound, provoking Fischer to remark that he now had the true compound—hence it was later given the proprietary name of 'Veronal' (L, *verus* = true). Fischer filed a patent on the new hypnotic at the end of January 1903, and a brief scientific communication appeared two months later. The new hypnotic was marketed by F. Bayer and Company. Overnight, all previous hypnotics, with the possible exception of chloral hydrate, were rendered obsolete.

When the United States entered the First World War in 1917, Congress passed the Trading with the Enemy Act to allow American firms to manufacture unobtainable German drugs covered by patents, such as 'Veronal'. Royalties were paid to the Alien Property Custodian for distribution to the American subsidiaries of the German companies when the war ended. The Act required the American products to be given a new name approved by the American Medical Association. This practice of giving a drug an approved, or generic, name in addition to that chosen by its original manufacturer ultimately became standardized throughout the world. In the case of 'Veronal', Roger Adams of the University of Illinois devised a manufacturing process for the Chicago-based Abbott Laboratories. The drug was then given the AMA approved name of barbital (to assist recognition, manufacturers' brand names will be cited within inverted commas throughout this book). In the United Kingdom, the approved name was slightly different, namely barbitone. This peculiar distinction from American nomenclature persisted in the names of all barbiturates subsequently introduced.

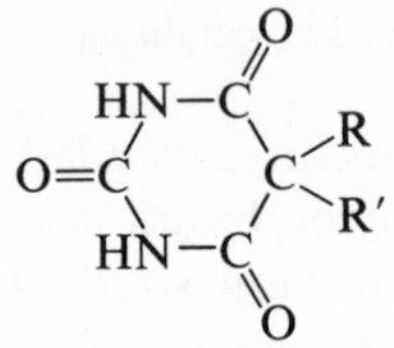

```
            O
           //
      HN—C
     /     \   R
O=C         C<
     \     /   R'
      HN—C
           \\
            O
```

	R	R′
Barbitone	$-CH_2CH_3$	$-CH_2CH_3$
Phenobarbitone	$-CH_2CH_3$	$-C_6H_5$
Butobarbitone	$-CH_2CH_3$	$-CH_2CH_2CH_2CH_3$
Amylobarbitone	$-CH_2CH_3$	$-CH_2CH_2CH(CH_3)_2$
Quinalbarbitone	$-CH_2CH{=}CH_2$	$-CH(CH_3)CH_2CH_2CH_3$
Pentobarbitone	$-CH_2CH_3$	$-CH(CH_3)CH_2CH_2CH_3$

Although von Mering, Fischer, and Dilthey reported on the activity of some eighteen analogues of barbitone, it was not until 1911 that Fischer synthesized a superior hypnotic. This, too, was marketed by F. Bayer and Company under the name of 'Luminal'. It later received the approved name of phenobarbitone. The discovery of its most important therapeutic role was a classical example of serendipity. A German physician by the name of Hauptmann had been assigned living quarters over a ward of epileptic patients for whom he was responsible. Unable to get a night of uninterrupted sleep because of their fits, he drugged his patients with the new hypnotic. To his delight, the remedy was successful, but to a greater extent than anyone could have anticipated because the incidence of fits among the patients decreased during the day as well as at night! Hauptmann correctly concluded that the phenobarbitone had a valuable anticonvulsant action at subhypnotic dose levels, but it was not until after the First World War that this was generally accepted. It is still widely used in treating grand mal epilepsy.

During the First World War, Chaim Weizmann, who was later to become the first president of Israel, discovered that a bacterium known as *Clostridium acetobutylicum* could convert cheap starchy materials to acetone and butyl alcohol. This was of immense military significance as the United Kingdom was desperately short of acetone for the production of naval explosives. Once peace was restored, the Weizmann process resulted in a sudden drop in the price of butyl alcohol, previously an expensive chemical. Roger Adams, the chemist who provided Abbott Laboratories with a highly profitable process for making procaine (a local anaesthetic sold by the German firm Hoechst) as well as barbitone, effectively exploited the situation. In 1920, he synthesized a procaine analogue in which butyl groups replaced ethyl groups, these groups being obtained from chemical intermediates prepared from the respective alcohols. The new compound, butacaine ('Butyn'), was an excellent local anaesthetic for application to the eye, since it did not dilate the pupil. It was one of the earliest, if not the first, commercially successful synthetic drugs to be developed in the United States. In the same year, Adams also synthesized 5-butyl-5-ethylmalonic ester. This proved to be an important intermediate as it permitted manufacture of the butyl analogue of barbitone by both Adams and also Arthur Dox of Parke, Davis and Company in Detroit. The Parke, Davis product was reported in 1922, and the following year Abbott marketed Adams' compound under the name 'Neonal'. It turned out to be almost three times as potent as barbitone, and had a somewhat shorter duration of action that reduced the likelihood of drowsiness on awakening. The new compound was introduced clinically under the name butobarbitone (butethal in the United States; 'Soneryl' is the best known brand in the United Kingdom). Today, we know that both its more rapid metabolic destruction and increased potency are due to its enhanced lipophilicity (fat solubility) caused by introduction of the longer butyl group that replaced the ethyl group, a feature which favours entry into the liver and brain.

Shonle and Moment of the Eli Lilly Company in Indianapolis announced their synthesis of amylobarbitone ('Amytal') in 1923. It contained an extra branched carbon atom on the butyl chain, but its physical and pharmacological properties were very similar to those of butobarbitone. Six years later, Shonle synthesized another similar compound, quinalbarbitone (Seconal'). This was at once followed by Volwiler and Tabern's pentobarbitone ('Nembutal') from Abbott Laboratories. Pentobarbitone had actually been synthesized previously by the Bayer Company in 1915, when part of the 'Hedonal' structure was incorporated into a barbiturate. The German company did not recognize its clinical potential.

No compounds to challenge the barbiturates emerged until the fifties, by which time there was growing concern about accidental overdosing by drowsy patients, as well as the use of these drugs in suicide attempts. In 1952, Tagmann and his colleagues at Ciba in Basle announced that they had found a potent hypnotic amongst a series of dioxotetrahydropyridines. These were related to sedatives investigated by their Swiss rival, Roche, just before the war. The new compound, called glutethimide ('Doriden'), was hailed as a safer hypnotic—a claim which has not withstood the test of time! As might have been expected, Roche quickly introduced a very similar compound, methyprylone ('Nodular'). Unfortunately, matters did not rest there because a small German company, Chemie Grunenthal, also became involved.

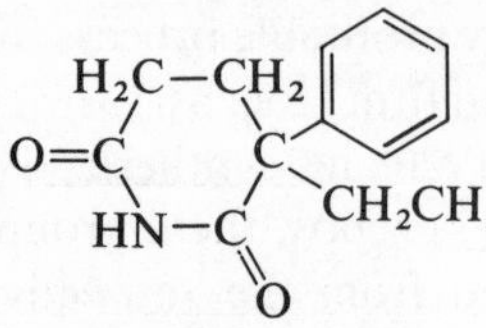

Glutethimide 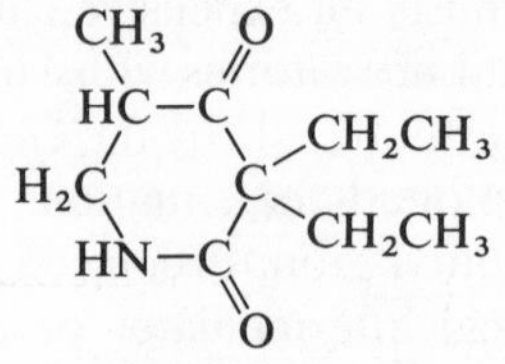 **Methyprylon** 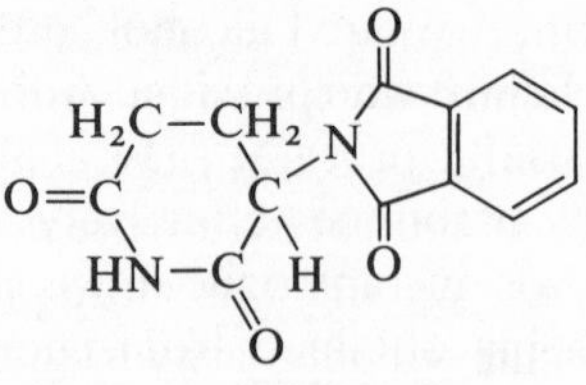 **Thalidomide**

Chemie Grunenthal was set up immediately after the war by a soap and toiletries manufacturer who wished to obtain a stake in the growing market for imported American antibiotics, then in desperately short supply. Heinrich Mueckter, who qualified in medicine before the war, was appointed as research director on the basis of his experience with the German army virus research group. In 1953, his assistant Wilhelm Kunz was given the task of preparing certain simple peptides required for antibiotic production. In the course of doing this he isolated a non-peptide by-product which was recognized by Grunenthal pharmacologist Herbert Keller to be a structural analogue of glutethimide. A series of related compounds were examined and one was tested further by Keller 'with special regard to its suitability as a hypnotic agent.' Strangely, however, it did not abolish the righting reflex of animals, a standard test for hypnotic activity. Keller conducted a series of studies on the motility of mice exposed to the drug, comparing it with several barbiturates and other central nervous system depressants. He also examined its toxicity in mice, rats, guinea pigs, and rabbits, concluding that it was a remarkably safe sedative. When Grunenthal approached manufacturers throughout the world it was eagerly taken up by firms anxious to secure part of a lucrative market for sedative-hypnotics. They each launched it before the public with their own brand name. This was the blackest day in the history of modern drug research, for the new sedative was the teratogen thalidomide. It led to thousands of children being born with appalling deformities. Whether the tragedy would have occurred had thalidomide been discovered in the laboratories of a large, reputable manufacturer is a moot point. One such company, when offered the drug by Chemie Grunenthal, is said to have rejected it on the grounds of its inadequate activity. Nevertheless, it is doubtful whether the standard tests used by most of the main manufacturers at that time would have detected the teratogenicity.

It was not until late in 1961 that there was general recognition of the teratogenicity of thalidomide. In the interim, several novel hypnotics were introduced. The first of these was discovered serendipitously when M. L. Gujral, R. P. Kohli, and P. N. Saxena of the department of pharmacology at the King George Medical College in Lucknow, India, were screening a series of quinazolones designed as analogues of the antimalarial febrifugine. This quinazolone alkaloid had been isolated for Lederle Laboratories by Benjamin Duggar of the University of Wisconsin (who was later to achieve fame for his isolation of chlortetracycline) during the desperate wartime search for an alternative to quinine after the Japanese had cut off supplies of cinchona bark from Java. Although febrifugine turned out to be a potent antimalarial, it could not be used clinically because it frequently caused vomiting. When synthetic analogues of it were prepared in Lucknow several exhibited hypnotic activity in rats, and one was subsequently marketed as methaqualone. This quinazolone achieved popularity in some quarters because it was a non-barbiturate hypnotic, but while there may have been a recognizable chemical differentiation, the same could not be said for its clinical profile. Its high abuse potential led to its withdrawal from the British market in 1980.

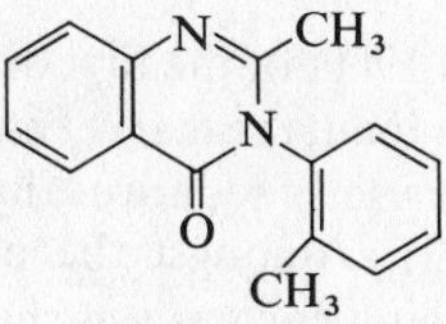

N CH_3 N O CH_3

Methaqualone

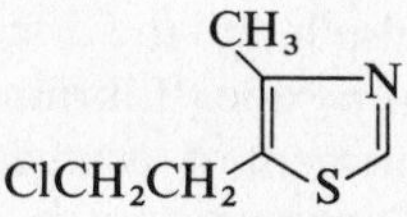

CH_3 N $ClCH_2CH_2$ S

Chlormethiazole

Methaqualone was not the only modern hypnotic to be discovered serendipitously. The knowledge that intravenous injection of vitamin B1 (thiamine) may induce convulsions led Charonnat, Lechat, and Chareton of the Pharmacie Centrale des Hôpitaux in Paris to investigate the pyrimidine and the thiazole moieties present in the molecule. The former was found to be responsible for the convulsant activity, whereas the latter afforded some protection against this by acting as a central nervous system depressant. By replacing a polar hydroxyl group in the thiazole fragment with a chlorine atom, the French investigators found the lipophilicity of the compound was increased to the extent that it exhibited potent hypnotic activity. This compound was originally prepared in 1935 as a potential thiamine analogue, but it was not until 1956 that its hypnotic activity was recognized and it was introduced as chlormethiazole ('Heminervrin').

Another serendipitous development involving a molecular fragment took place around the same time. Researchers at the US Schering Corporation laboratories were surprised when they found hypnotic activity in an acetylenic alcohol which represented part of the structure of the synthetic steroid hormone known as ethisterone. Seeking to optimize this activity, they prepared and examined more than 300 related compounds. Eventually, methylpentynol ('Dormison', 'Oblivon') emerged as the most suitable analogue for clinical use. It appeared to offer a wider margin of safety than the barbiturates, which meant it was unlikely to be used in suicide attempts. Careful examination of the patent literature then revealed that the Bayer Company had actually synthesized this alcohol in 1915! This prevented US Schering from taking out a patent. Nevertheless, they went ahead and introduced it, only to find that it was soon being rivalled when Pfizer marketed its analogue known as ethchlorvynol ('Arvynol'). The German Schering Corporation (separated from the US company since the First World War) then independently developed a similar drug called ethinamate.

$$CH_3CH_2-\overset{H_3C}{\underset{HC\equiv C}{C}}-OH$$

Methylpentynol

$$CH_3CH_2-\overset{ClCH=CH}{\underset{HC\equiv C}{C}}-OH$$

Ethchlorvynol

$$\text{(cyclohexane ring)}\underset{HC\equiv C}{\Big\rangle}-O\overset{O}{\overset{\|}{C}}NH_2$$

Ethinamate

Serendipity also led to the discovery, in 1960, of the first of the benzodiazepines, chlordiazepoxide ('Librium'). Since then, the market has been saturated with a vast assortment of its analogues for use as sedatives, hypnotics, muscle relaxants, and anxiolytics (see page 183). Amongst the earliest chemical modifications effected on the benzodiazepine nucleus was the introduction of the nitro group, almost certainly on account of the scope this offered the Roche chemists, led by Sternbach, for subsequent structural variation. Several such nitro compounds were tested. The one now known as nitrazepam proved to be much more potent than chlordiazepoxide in both mice and cats. Subsequent investigations showed that sleep could be induced by larger doses which were still well below the toxic level. Indeed, so wide was the margin of safety that self-poisoning with nitrazepam was most unlikely to occur. Naturally, this safety factor ensured worldwide acceptance of this new hypnotic which, to this day, retains its popularity. There is a tendency for it to persist in the body long enough to cause a hangover effect in some patients. Several analogues subsequently introduced are less likely to cause this problem because they are more susceptible to metabolic deactivation in the liver, e.g. lormetazepam ('Noctamine'), temazepam ('Normison', 'Euhypnos') and triazolam ('Halcion'). Such shorter-acting hypnotics may, of course, be less acceptable to patients who tend to awaken in the early hours. Unfortunately, the benzodiazepines can induce dependence and tolerance after prolonged use, as is the case with all other hypnotics.

There is still no such thing as an hypnotic which induces normal sleep. That this is so is merely a reflection of the manner in which hypnotics have been discovered. Without exception, their ability to disrupt brain function so severely as to render animals unconscious during waking hours has been considered sufficient reason to introduce them as agents to assist the onset and maintenance of sleep. The absurdity of this proposition seems to have been ignored for too long. Their continued widespread use is a matter for concern.

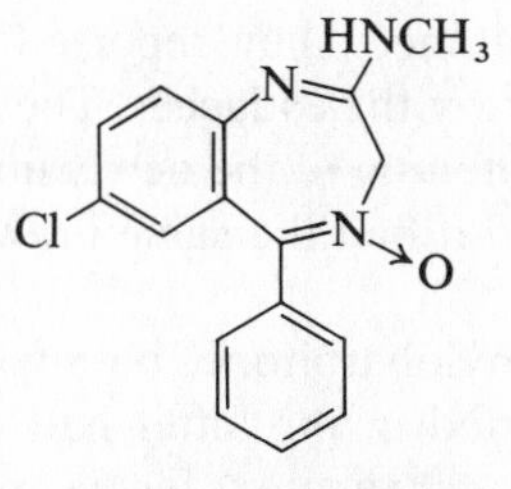

Chlordiazepoxide

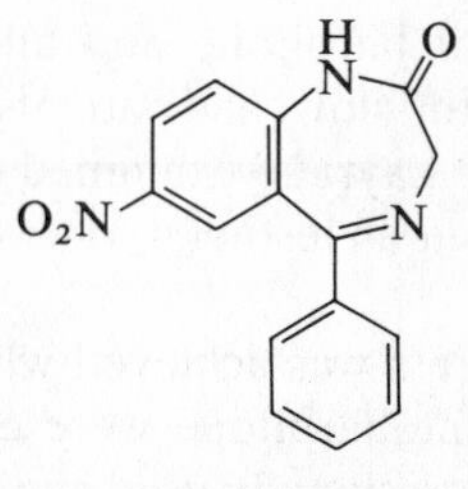

Nitrazepam

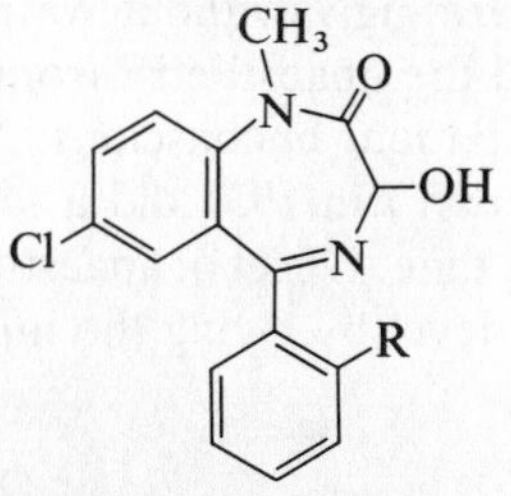

Lormetazepam, R = —Cl
Temazepam, R = —H

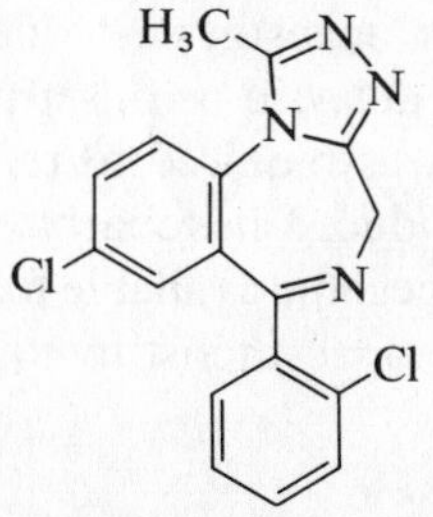

Triazolam

INTRAVENOUS ANAESTHETICS

In 1665, Sigismund Elholtz injected an opiate solution into a vein in an attempt to produce insensibility, but practical intravenous medication did not become feasible until after the development of the hypodermic syringe by Alexander Wood of Edinburgh in 1853. The earliest attempt at intravenous anaesthesia seems to have taken place in Bordeaux where Professor Oré, in 1872, injected a solution of chloral hydrate and achieved deep enough anaesthesia to remove a finger-nail. However, several post-operative fatalities dissuaded others from adopting the technique. It was not till 1905 that further development occurred when N. P. Krawkow of St. Petersburg successfully administered a saline solution containing the Bayer Company's recently introduced 'Hedonal' (see page 27). He subsequently used this method in more than five hundred operations. The technique was taken up in Russia and some parts of Europe, where it stimulated others to seek more suitable drugs. A few years later, Burkhart reported that the intravenous injection of chloroform or ether in normal saline was capable of inducing anaesthesia, although large volumes of solution were required. He achieved better results with a mixture of ether and the chloral analogue isopral (1,1,1-trichloroisopropanol). In 1913, Noel and Souttar tried using paraldehyde, well diluted with saline, for short operations.

The outbreak of war interrupted progress. When the war ended, French anaesthetists experimented with intravenous injections of alcohol after learning of encouraging reports from Mexico. Their results were most disappointing. However, this did not discourage Daniel and Gabriel Bardet from anaesthetizing surgical patients with injections of 'Somnifen', a water-soluble formulation of barbitone and allobarbitone. They reported, in 1921, that recovery was too slow, and patients awoke with headaches. The following year, Tiffeneau and Layraud examined the potential of the newly introduced butobarbitone as an anaesthetic, only to find it had the same drawback as 'Somnifen'.

Some progress was achieved when amylobarbitone, butallylonal ('Pernocton'), and pentobarbitone were each tried in the latter half of the decade. These were occasionally used clinically over the next few years, despite their many inadequacies as anaesthetics. Particularly disconcerting, however, was a tendency for anaesthesia to deepen alarmingly without warning. This was because the delay in its onset prevented the anaesthetist from knowing how much drug was required to render the patient unconscious. Not until I. G. Farben introduced hexobarbitone ('Evipan') in 1931 did a safe intravenous anaesthetic become available for the first time. Onset of anaesthetia was rapid, enabling the anaesthetist to control the level by giving the injection slowly.

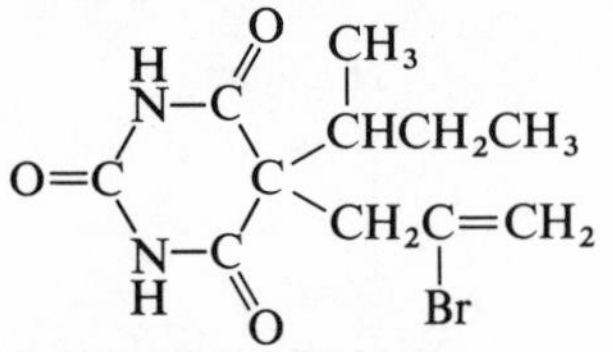

Butallylonal

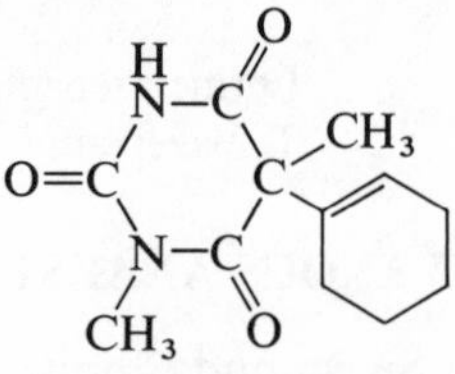

Hexobarbitone

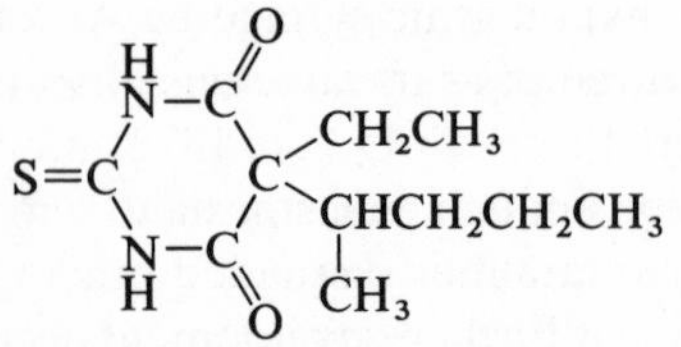

Thiopentone

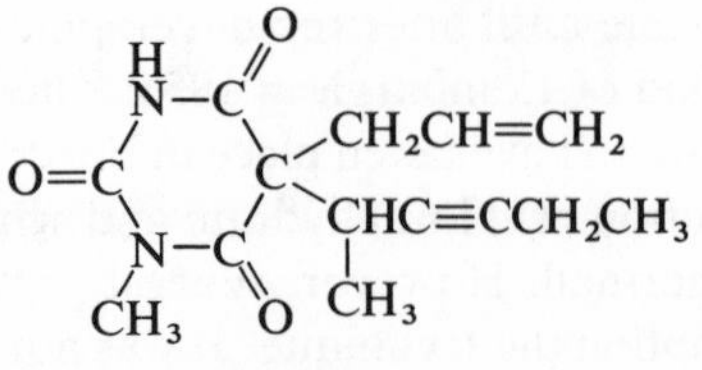

Methohexitone

Hexobarbitone was synthesized by the chemists Kropp and Taub at Elberfield, and it was first tried in man by Ernst Reinhoff. Its rapid onset of anaesthesia was due to the replacement of a hydrogen atom on one of the barbiturate ring nitrogen atoms by a methyl group. This rendered the molecule less water soluble and more lipophilic. Small though this molecular change may have been, it ensured almost instantaneous transposition of the drug from the blood into the brain cells. As a consequence, patients fell unconscious in the few seconds it took for the blood to carry the anaesthetic to the brain from the site of injection. Hexobarbitone was deservedly successful, and it is estimated that over the next twelve years some ten million injections of it were administered. Nevertheless, in 1937 the Council on Pharmacy and Chemistry of the American Medical Association warned that hexobarbitone had been introduced before adequate studies had been made of it under a great variety of conditions. Tragically, its appalling misuse in the aftermath of the Pearl Harbor bombing, four years later, proved the validity of this judgement.

Even before the first reports of the success of hexobarbitone had begun to circulate, Tabern and Volwiler of Abbott Laboratories were on the trail of the drug which was ultimately to render it obsolete. The work on pentobarbitone encouraged them to seek very short-acting barbiturates, probably with a view to introducing them as hypnotics free from any tendency to produce a hangover. They followed up old reports by Einhorn (who synthesized procaine) and others, stating that sulphur-containing thiobarbiturates were chemically less stable than the familiar oxobarbiturates. Thiobarbiturates were amongst the earliest barbiturates examined in 1903 by Fischer and von Mering. They were rejected after an oral dose of the sulphur analogue of barbitone had killed a dog. Dox and Hjort of Parke, Davis, and Company had also obtained unsatisfactory results with this same compound and other thiobarbiturates. Notwithstanding these unpromising reports, Tabern and Volwiler pursued their idea that a chemically unstable thiobarbiturate might decompose fast enough in the body to ensure that its effects quickly wore off. By 1934 they were convinced that thiopentone ('Pentothal'), the sulphur analogue of pentobarbitone, was a promising agent. Doubtlessly inspired by the recent success of hexobarbitone, they arranged for thiopentone to be investigated by Ralph Waters, who had just completed his pioneering investigations into cyclopropane anaesthesia, at the University of Wisconsin Medical School, Madison, and by J. S. Lundy of the Mayo Clinic in Rochester, Minnesota. Both confirmed the superiority of thiopentone over existing intravenous anaesthetics. Despite the major set-back following many deaths from the disastrous maladministration of it and also hexobarbitone at Pearl Harbor, thiopentone ultimately achieved recognition as the most useful agent for the induction of anaesthetia prior to the administration of an inhalational anaesthetic. It also became the most widely used intravenous anaesthetic for short operations.

Tabern and Volwiler's hope that thiopentone would rapidly break down in the body was not realized. Patients recovered only a few minutes after a single injection partly because of metabolic deactivation of thiopentone in the liver, but a high proportion of the lipophilic drug was absorbed into fat and muscle tissue where it then accumulated. If subsequent injections were given, these tissues became saturated with the drug, and so its concentration in the blood built up, reaching a dangerously high level. The anaesthetist had no advance warning of this. Attempts to design either thiobarbiturates or hexobarbitone-like *N*-methylbarbiturates such as methohexitone ('Brietal'), synthesized for Eli Lilly and Company by Doran in 1956, have not overcome this problem.

Failure to discover a barbiturate anaesthetic with a wide margin of safety has inspired the pharmaceutical industry to exploit alternative approaches. The first of these had its origin in work done in 1941 by Professor Hans Selye of McGill University, Montreal. He was investigating the effects of overdosage of steroid hormones, which were just then becoming available in large enough amounts for this type of study to be carried out. To his surprise, he found several hormones rendered rats unconscious when injected intraperitoneally (i.e. into the abdominal cavity, a convenient route in small animals). The rats later awoke unharmed. Knowing that many steroids were deactivated in the liver, Selye repeated his experiments in rats in which the liver had been removed. This time a fraction of the previous dose produced anaesthesia in more than half of the rats, indicating that the metabolites formed in the liver were inactive. The doses used were in the range of 5–50 mg for the steroids desoxycorticosterone acetate, progesterone, androsterone, testosterone, and methyltestosterone. The following year Selye extended his investigation to include many more steroids, in an attempt to relate anaesthetic activity to structural features in the molecules. He concluded that the anaesthetic effect had nothing to do with any specific type of hormonal activity. Indeed, the most potent steroid anaesthetic to emerge, pregnanedione, was wholly devoid of any such activity. For a steroid to possess anaesthetic activity, it appeared that an oxygen atom at each extremity of the molecule was essential. Selye did not pursue this work, presumably because the total lack of water solubility amongst the more active compounds ruled out their clinical application as intravenous anaesthetics.

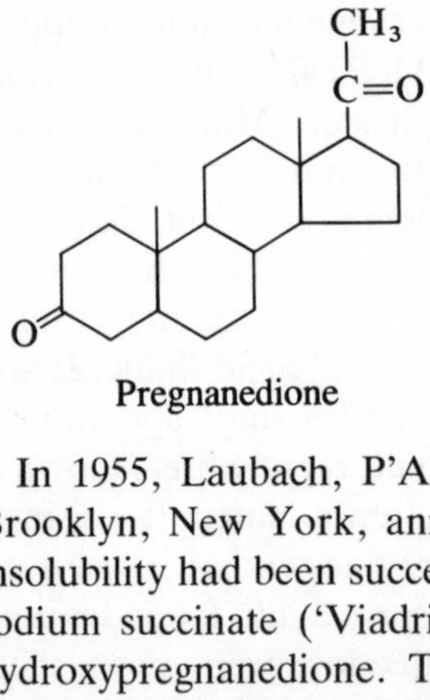

CH_3–C=O

Pregnanedione

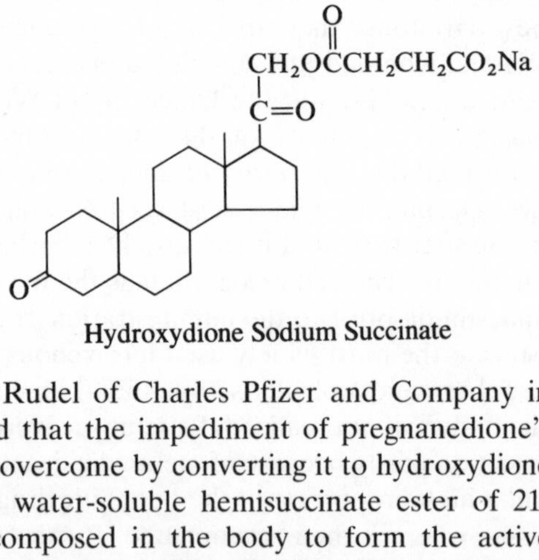

$CH_2OCOCH_2CH_2CO_2Na$–C=O

Hydroxydione Sodium Succinate

In 1955, Laubach, P'An, and Rudel of Charles Pfizer and Company in Brooklyn, New York, announced that the impediment of pregnanedione's insolubility had been successfully overcome by converting it to hydroxydione sodium succinate ('Viadril'), the water-soluble hemisuccinate ester of 21-hydroxypregnanedione. This decomposed in the body to form the active anaesthetic. Unfortunately, as it took several minutes for an effective concentration of the active metabolite to build up, anaesthetists were once again faced with the problem of a slow onset of action, leading to a dangerous tendency to overdose the patient. Under these circumstances, it may well be asked why the drug was introduced clinically at all. The answer presumably must be that it possessed a wider safety margin than did thiopentone. It also had less depressant activity on either the heart or the respiration, and it produced good muscle relaxation. Other advantages included a lower incidence of coughing and pleasanter recovery than with thiopentone. Although patients became ambulant much sooner than with other anaesthetists, the disadvantages outweighed the advantages, and the drug eventually ceased to be used.

At the Glaxo Research Laboratories in Fulmer, England, the potential advantages offered by drugs such as hydroxydione caught the attention of their chemists and pharmacologists, who had extensive knowledge of steroids, these representing one of the main fields of interest of the company. During the next nine years, Glaxo researchers examined the activity of around 170 steroids. One of the main leads they followed was a report that the delay in onset of the action of hydroxydione sodium succinate infusions might be due to the necessity for the drug to be metabolically converted to a steroid with a hydroxyl group at each end of the molecule. Dozens of such steroids were then duly synthesized by the Glaxo chemists and it was confirmed that they were very potent anaesthetics. None of them, however, has acceptable water solubility, so it became necessary to modify the structure by incorporating a phosphate ester function. Glaxo had recently done this with their trichloroethanol-derived hypnotic called triclofos sodium. The outcome was a promising water-soluble anaesthetic which was put on clinical trial in 1963 after completion of extensive animal testing. Unfortunately, patients reported a tingling sensation spreading downwards from the neck to the legs before they were rendered unconscious. This was a major set-back and it was another two years before Glaxo re-opened its investigations. Eventually, experiments on mice with a new steroid, alphaxolone, showed it to have most of the desired properties except adequate water solubility, a drawback which could be overcome by preparing a suspension of the drug. This, however, was not an acceptable way of using any intravenous anaesthetic in the clinic. Much time was spent in seeking a chemical derivative of alphaxolone possessing the required solubility, but invariably the greater the solubility, the less the potency.

Ultimately, it was found that when a small amount of a less potent steroid anaesthetic called alphadolone was added to alphaxolone in an aqueous solution of the surfactant cremophor EL, the solubility of both drugs was markedly increased, a rare phenomenon amongst steroids! The outcome was an acceptable formulation of alphaxolone-alphadolone acetate ('Althesin'). This proved to be safe when tested on animals, and was put on clinical trial in Glasgow Royal Infirmary in 1969, thirteen years after the project had been initiated by Glaxo.

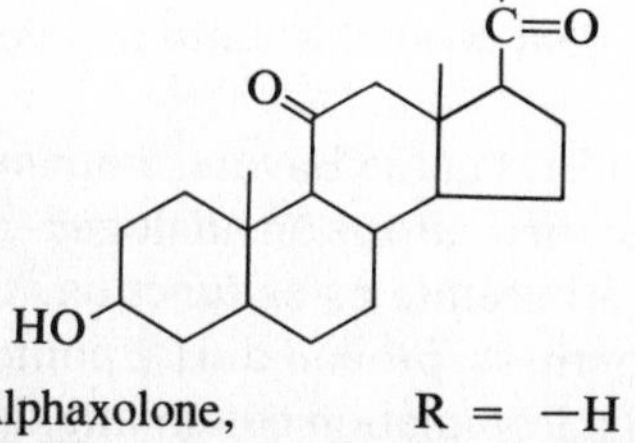

CH_2R

$C{=}O$

O

HO

Alphaxolone, R = $-H$

Alphadolone Acetate, R = $-OCCH_3$ (with $=O$ on the C)

After alphaxolone-alphadolone acetate came into general use a number of adverse reactions were noted. These were usually of an anaphylactoid (allergic) nature, resembling those reported with another intravenous anaesthetic, propanidid, which was also formulated with the surfactant cremophor EL. Wisely anticipating future difficulties ('Althesin' was withdrawn from the market in 1984), Glaxo renewed their investigations with the aim of developing an agent with greater water solubility, so as to avoid the need to incorporate cremophor EL in the injection. The outcome was the development of minaxolone, an analogue of alphaxolone in which the carbonyl substituent on the 11-position of the steroid ring was replaced with a dimethylamino group. Such a modification had been used at the beginning of the century by Einhorn to enhance the water solubility of local anaesthetics. Chemists knew that such aliphatic amines exist mainly as the water-soluble ionic species in mildly acidic injections.

Extensive tests on animals having indicated that minaxolone was likely to be suitable for both the induction and the maintenance of anaesthesia in man, volunteers were given single injections of it under strict supervision. Detailed monitoring of the response of all vital organs was carried out together with careful measurement of the disposition of the drug in the body. This is the standard procedure when any new drug is first given to humans; it is described as the Phase I trial. It is usually at this stage that a suitable dose range is determined. However, whilst Phase I trials of minaxalone were being completed, several centres around the world were beginning Phase II trials, where it was to be administered to patients who could, if necessary, receive multiple doses. Just at this point in time, disturbing signs observed by Glaxo toxicologists, during long-term studies in rats, made the company decide to withdraw the drug from all trials.

A totally different approach to find a new intravenous anaesthetic arose from work done during the early fifties by Ciba Laboratories in Basle. Here, research was being conducted into the pharmacological properties of several derivatives of eugenol, the local anaesthetic present in oil of cloves. Some of these derivatives were found to have general anaesthetic properties One in particular seemed to be more of a respiratory stimulant than a depressant such as were all anaesthetics then available. Another encouraging feature of this compound, estil, was its tendency to be rapidly metabolized, ensuring quick recovery from anaesthesia. Unfortunately, it was insoluble in water, and formulations designed to deal with this caused damage to the vein at the injection site.

Ciba abandoned the project, but the Bayer Company at Elberfield took it up in 1954. They concentrated their efforts on analogues of estil in which the allyl group was replaced with an acetate ester function, thirty-seven compounds being prepared. The best of these, propanidid ('Epontol'), was introduced into clinical practice in 1964 in a formulation containing the surfactant cremophor EL to solubilize the drug. As with alphadolone-alphaxolone, this was responsible for a high incidence of hypersensitivity reactions, all the more unfortunate because propanidid is still the only available ultra-short-acting intravenous anaesthetic ideal for brief operations. The short duration is due to cleavage of the ester function in the body.

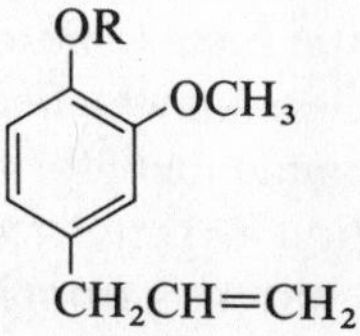

OR, OCH_3, $CH_2CH{=}CH_2$

Eugenol, R = $-H$
Estil, R = $-CH_2CON(C_2H_5)_2$

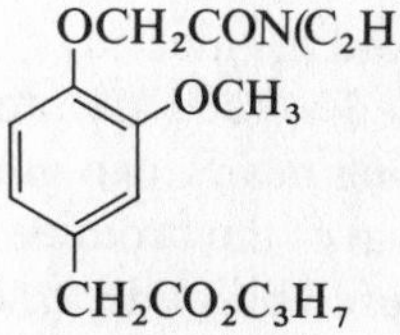

$OCH_2CON(C_2H_5)_2$, OCH_3, $CH_2CO_2C_3H_7$

Propanidid

Another serendipitous discovery occurred in the course of screening imidazoles for chemotherapeutic properties. The Janssen research laboratory in Beerse, Belgium, discovered that one of these induced a profound hypnotic state in rats, whether injected or administered orally. Nearly fifty analogues were then synthesized and screened. It transpired that the best intravenous anaesthetic was the ethyl ester analogue of the prototype methyl ester. Preliminary studies submitted for publication in 1964 reported this to be an extremely potent, short-acting anaesthetic. Full pharmacological studies were published in 1971, and two years later it was introduced into anaesthetic practice as etomidate ('Hypnomidate'). It permits rapid recovery and seems relatively free from cardiovascular effects, but pain on injection, possibly due to the acidity of its formulation, is an unwelcome feature.

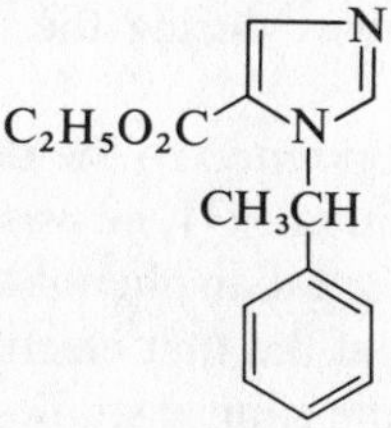

$C_2H_5O_2C$, N, N, CH_3CH

Etomidate

ANTICONVULSANTS

Bromine was discovered in sea-water by Balard, a pharmacist in Montpellier, in 1826. Once its chemical similarity to iodine had been recognized, French physicians attempted to use it as an alternative to iodine in the treatment of a wide variety of diseases, including syphilis and goitre. The highly corrosive nature of bromine water led them to experiment with the potassium salt, which they prescribed in massive quantities. No beneficial effects were reported, but the depressant effect of potassium bromide on the nervous system was inevitably recognized, being described as *ivresse bromurique*, viz. bromism.

On the 11th May 1857, Sir Charles Locock presided at a meeting of the Royal

Medical and Chirurgical Society of London at which the topic for discussion was epilepsy. He expressed the view, prevalent at the time, that a great number of cases of the disease were due to masturbation, particularly in young patients. He also considered one type of epilepsy to be associated with menstruation. Locock told the meeting that he had met with little success in treating this until he chanced to read an interesting report by a German who became impotent after taking nearly two grams of potassium bromide daily for two weeks. On withdrawal of the bromide, normal sexual function had returned. Locock then administered bromide to several young women suffering from hysteria. Large doses were given with good results, and he proceeded to treat a patient who had suffered from so-called hysterical epilepsy for more than nine years. The result was striking. The epilepsy ceased as long as medication was continued, and all but one of fifteen subsequent cases were said to be cured!

Despite Locock's high standing in the medical world, little attention was paid to his comments. He did not publish any further details, and it was not until 1868 that the use of bromide in epilepsy came to be accepted when Clouston carried out a carefully controlled clinical trial which established the correct dosage. This trial lasted for several months and clearly demonstrated that patients experienced a reduction in the number of fits whilst receiving bromide. It was the first drug to be of proven value, despite its obvious drawback of only providing benefit at almost incapacitating dose levels. Needless to say, the efficacy of bromide in epilepsy had nothing whatsoever to do with its effect on libido or sexual function!

The serendipitous discovery of the anticonvulsant activity of phenobarbitone in 1912 has already been described (see page 30). It, too, did not gain immediate acceptance. However, during the next decade its value became widely recognized.

When Tracy Putnam was appointed to the directorship of the neurological unit of the Boston City Hospital in 1934, he was able to initiate experiments to find a less sedating anticonvulsant than phenobarbitone. With the assistance of Frederick Gibbs, he established the first electroencephalographic laboratory in the world designed for routine clinical studies of brain waves. An important observation to emerge from the new laboratory was that epileptic seizures were accompanied by an electrical 'storm' in the brain. This led Putnam to conclude that it might be possible to induce convulsions in laboratory animals by applying an electrical current to the brain. Furthermore, it might also be possible to quantify the strength of current required, thereby affording a method of recognizing whether a drug was able to give some degree of protection to the animal. With the assistance of Paul Hoeffer, who acquired a commutator salvaged from a World War I German aircraft, a makeshift piece of apparatus was assembled. This enabled Putnam and Gibbs to demonstrate that phenobarbitone markedly raised the convulsive threshold in cats. Crude as the apparatus was, it was to be responsible for a major advance in the treatment of epilepsy.

Putnam ordered in a wide variety of phenyl compounds from chemical manufacturers, hoping that the phenyl group in phenobarbitone was somehow responsible for its efficacy. He also wrote to several pharmaceutical firms, but only Parke, Davis, and Company responded, providing nineteen chemical analogues of phenobarbitone, all of which were inactive as hypnotics. Putnam screened these, as well as over a hundred other chemicals. A few showed activity, but with one exception were too toxic for clinical use. The exception was one of the Parke, Davis compounds, phenytoin ('Dilantin'), which was more effective in protecting cats from electrically induced convulsions than was even phenobarbitone. As it was known to have no hypnotic or other disagreeable effects, this seemed to be just what Putnam had been waiting for.

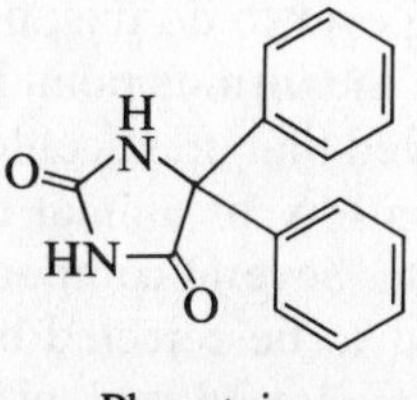

Phenytoin

Putnam gave the phenytoin to one of his young assistants, Houston Merritt, for clinical evaluation, in 1936. The first patient to receive the drug had suffered from seizures every day for many years, but as soon as his treatment began these ceased permanently! Subsequent studies confirmed that phenytoin was at least as effective as phenobarbitone, but had the important advantage of causing much less sedation. Paradoxically, this absence of marked sedation prejudiced many physicians against accepting the new drug. Although few would question its value today, phenytoin is far from being an ideal drug since it is difficult to ensure the optimum amount enters the circulation after oral administration. Slight changes in absorption from the gut can lead either to failure to control the epilepsy, or else to undesirable side-effects from overdosage. This is a pharmaceutical problem related to its insolubility.

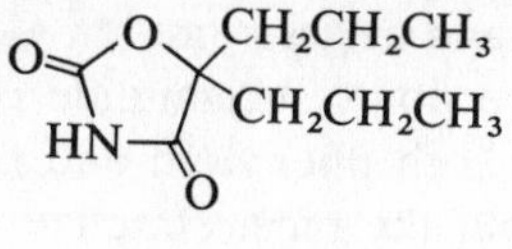

Propazone

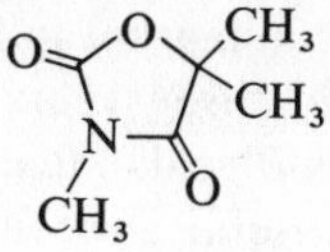

Troxidone

During the late nineteen-thirties there was some interest in a class of sedative-hypnotics known as oxazolidine-2,4-diones. One of these, propazone, proved to be an anticonvulsant, but its potent sedative activity ruled out any clinical application. At the Abbott laboratories in Chicago, tests confirmed that none of the known oxazolidine-2,4-diones had any analgesic activity. However, Marvin Spielman found that when the lipophilicity of these compounds was increased by substituting a methyl group on the nitrogen atom, analgesics comparable with aspirin were obtained. In the course of trials to establish its clinical value, the most promising of the new analgesics, troxidone ('Tridione'), was combined with a novel antispasmodic drug, amolanone. Toxicological studies in mice revealed that large doses of the antispasmodic which would normally induce convulsions, did not do so when troxidone was concurrently administered! The obvious conclusion was drawn, namely that troxidone was an anticonvulsant. This discovery was made in 1943, and after extensive animal investigations the new anticonvulsant was administered to children at the Cook County Hospital, Chicago, the following year. The paediatrician in charge of this trial, Meyer Perlstein, established that, unlike phenytoin or phenobarbitone, the new drug was capable of controlling petit mal absence seizures. It was the first drug ever to do this, but it caused many side-effects. Nevertheless, it proved to be the turning point in the development of anticonvulsant drugs because it showed that these could be selective in their spectrum of activity. Henceforth, a battery of animal tests was set up for screening potential new anticonvulsants. Several compounds subsequently found to be effective in such screens had to be rejected because they caused too much sedation. However, some were introduced instead as hypnotics or sedatives, e.g. glutethimide and chlormethiazole.

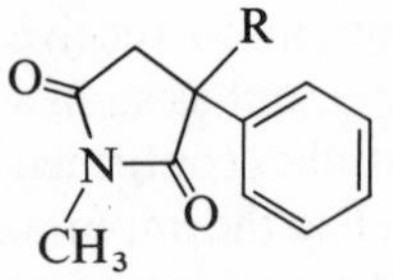

Phensuximide, R = $-H$
Methsuximide, R = $-CH_3$

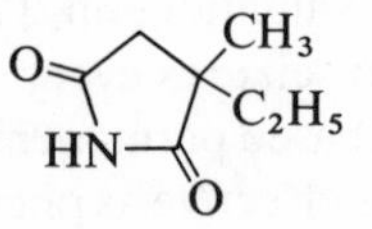

Ethosuximide

Parke, Davis initiated a major research project to find a less toxic drug to replace Abbott's troxidone in petit mal. This involved the synthesis and testing of over one thousand aliphatic and heterocyclic amides. The first success came after C. A. Miller and Loren Long synthesized phensuximide ('Milontin'), which G. M. Chen found to have promising anticonvulsant activity. It was put on the American market in 1954. This and methsuximide ('Celontin'), which was introduced three years later, were novel succinimide compounds that incorporated molecular features from both phenytoin and troxidone. They proved to be useful anticonvulsants, but the most effective member of this series was ethosuximide ('Zarontin'). It was introduced in 1958, and remains one of the best drugs for the treatment of petit mal absence seizures in children.

Another anticonvulsant that was discovered as a consequence of routine pharmacological screening of novel chemicals was found in 1951 by a group of researchers from the Pearl River laboratories of Lederle where several compounds containing a benzylamide residue had been found to exhibit anticonvulsant activity. By examining the change in potency with structural alteration, the Lederle researchers were able to obtain maximal activity with beclamide ('Nydrane'). It is occasionally used in the treatment of grand mal epilepsy.

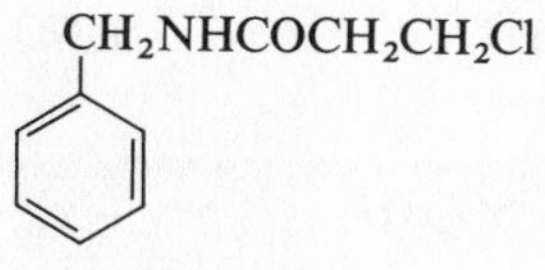

$CH_2NHCOCH_2CH_2Cl$

Beclamide

ICI scientists also tried to find improved anticonvulsants in the early fifties. H. C. Carrington of the Research and Biological Laboratories in Manchester considered that the hydantoins, although relatively free of sedating properties, produced too many side-effects. The barbiturates, in contrast, were usually free from these, but instead caused sedation. However, not all sedating barbiturates were anticonvulsants, so Carrington concluded that it should be possible to synthesize a barbiturate analogue in which this separation of activity was turned to advantage by affording a non-sedative anticonvulsant. Working with ICI chemists Boon and Vasey he developed primidone ('Mysoline'). This was given its first clinical trial in 1952 and results were satisfactory in grand mal epilepsy. How much of its efficacy is due to the phenobarbitone which is formed from it, and how much is due to the primidone itself is uncertain, but the drug is still widely used, especially where neither phenytoin nor carbamazepine is acceptable.

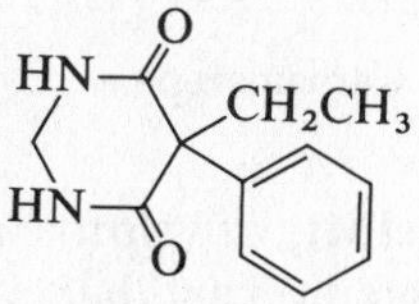

O
HN
CH_2CH_3
HN
O

Primidone

A novel approach to the management of epilepsy was provided by the discovery of the potent carbonic anhydrase inhibitor acetazolamide ('Diamox') in 1950 (see page 157). A few years earlier, it had been found that metabolic acidosis led to amelioration of epilepsy. Since acetazolamide could induce this metabolic change readily, it seemed to a group of American doctors, Bergstrom, Carzoli, Lombroso, Davidson, and Wallace, that a trial of it in epileptic children would be justified. The results of this were encouraging and were published in 1952. However, two years later doubt was cast on the hypothesis that acidosis was responsible for the beneficial effects. It was suggested that the inhibition of carbonic anhydrase resulted in a build-up of carbon dioxide within brain cells, resulting in interference with the propagation of nervous impulses. What made this interpretation particularly attractive was the knowledge that inhalation of carbon dioxide could prevent convulsions. It does, then, seem to be yet another case of a new therapeutic approach resulting from an erroneous hypothesis.

Not all carbonic anhydrase inhibitors are anticonvulsants. Most fail to penetrate into the brain. The only one used much nowadays is sulthiame

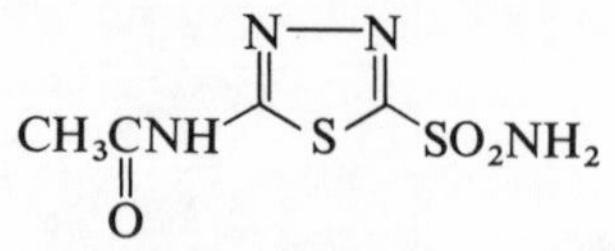

H_2NO_2S O_2 N S

N—N CH$_3$CNH S SO$_2$NH$_2$ O

Acetazolamide

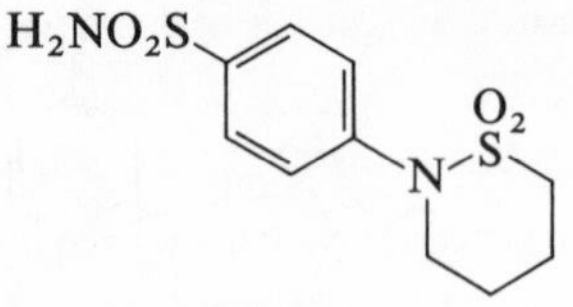

Sulthiame

('Ospolot'), which was introduced in the early sixties.

Carbamazepine ('Tegretol') was synthesized by Schindler of Geigy in 1953 when the company was investigating tricyclic analogues of the recently introduced chlorpromazine. It was only some years later that its anticonvulsant properties were recognized. The first clinical study was not carried out until 1963. Carbamazepine seems to have taken longer than most anticonvulsants to become established in clinical practice, but it is now considered by many doctors to be as effective as phenytoin in the control of grand mal epilepsy.

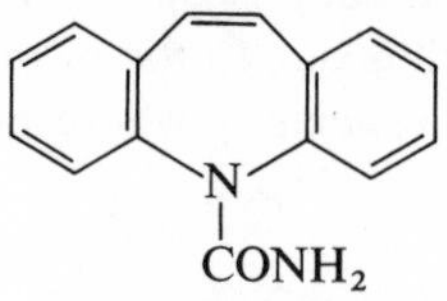

N CONH$_2$

Carbamazepine

When the anticonvulsant activity of valproic acid (dipropylacetic acid) was recognized in 1963, it represented yet another classic example of serendipity in drug discovery. Pierre Eymard, a research student at the University of Lyon, had synthesized a series of derivatives of khellin. This was a natural product said to have a variety of pharmacological properties. After completing his thesis, Eymard arranged to have his new compounds tested in Professor G. Carraz's pharmacology laboratory at the Ecole de Médecine et de Pharmacie at Grenoble. When Eymard tried to prepare a solution of the first compound to be tested, he could not get it to dissolve. He then sought advice from H. Meunier and Y. Meunier of the Laboratoire Berthier in Grenoble. They suggested that valproic acid might be a suitable solvent since they had used it in the past to dissolve bismuth compounds for clinical evaluation; it was certainly non-toxic. The valproic acid did dissolve Eymard's compound, and subsequent tests showed the khellin derivative to be physiologically active. Professor Carraz advised that it should be put through a general screen for a variety of possible actions, and this revealed it to have anticonvulsant activity. Shortly after this, H. Meunier used valproic acid to dissolve a coumarin compound. Although chemically unrelated to Eymard's compound, it too proved to have anticonvulsant properties. Meunier realized this could not be mere coincidence. He immediately tested the valproic acid and discovered it was an anticonvulsant. After detailed studies by Carraz and his colleagues, valproic acid was subjected to extensive clinical investigation before its sodium salt ('Epilim') was marketed in 1967 for the control of various kinds of epileptic seizures.

$$\begin{array}{l} CH_3CH_2CH_2 \\ \quad\quad\quad\quad\;\; \backslash \\ \quad\quad\quad\quad\;\; CHCO_2Na \\ \quad\quad\quad\quad\;\; / \\ CH_3CH_2CH_2 \end{array}$$

Sodium Valproate

III

Local Anaesthetics

During the retreat from Moscow in 1812, Napoleon's chief army surgeon, D.-J. Larrey, found that amputations performed at sub-zero temperatures caused the minimum of distress, with patients making rapid recoveries. This was duly reported and seems to be the earliest description of the achievement of localized anaesthesia through hypothermia. A quarter of a century later A.J. Thomson wrote in the *London Dispensatory* that ether acted as a refrigerant through its rapid evaporation when poured on to the skin, on which account it was applied to burns. In 1848, James Young Simpson suggested that if it were possible to induce anaesthesia without having to render the patient unconscious, many people would consider this a greater advance than his own discovery of chloroform. Within days of this statement, James Arnott of Brighton reported in the *Lancet* that he had overcome the pain caused by incision of the skin simply by the application of crushed ice in a pig's bladder. This was pursued further by John Snow who experimented with a variety of freezing mixtures and even solid carbon dioxide. He abandoned this line of enquiry to seek an anaesthetic that could be inhaled without causing unconsciousness. It was left to his protegé Bernard Ward Richardson to develop a practical method of achieving the required degree of refrigeration of the skin. His inspiration came from the newly introduced Eau-de-Cologne spray, which was popular with English ladies. Observing its cooling effect, he realized this would serve his purpose if the right volatile liquid were to replace the cologne. Initially, he tried ether, achieving sufficient cooling of the tissues for minor surgery to be carried out. Later, he and others improved the technique by utilizing first a petroleum distillate, then ethyl bromide, and finally ethyl chloride. The Richardson spray became popular on the Continent, where it ultimately gave a Viennese ophthalmologist the idea of finding a local anaesthetic which could be safely applied to the eye.

It was after he qualified in medicine in 1882 that Carl Koller decided to specialize in ophthalmology. He learned from his professor at the University of Vienna that vomiting caused by chloroform was often responsible for damage to the eye which had been operated upon. Furthermore, it was preferable for most patients to remain conscious during eye operations so as to assist the surgeon. Unfortunately, there was no method by which local anaesthesia could be obtained. It seemed to Koller that the problem was not insurmountable, so he began to experiment in the laboratory where, whilst a medical student, he had established an international reputation for valuable work in the field of embryology. He instilled solutions of chloral hydrate, sodium bromide, and morphine into the eyes of animals, only to find that all were harmful. The project was then abandoned, although Koller remained convinced of its importance by his own increasing experience of operating on unanaesthetized patients.

In the spring of 1884, Koller was asked by his close friend Sigmund Freud to assist in an experiment to establish how the Indians of Bolivia, Chile, Columbia and Peru were able to allay fatigue by chewing the leaves of the coca bush, *Erythroxylon coca*, which the Spanish colonists had called 'the divine plant of the Incas' as it featured in religious ceremonies. The earliest written account of the plant appeared in Nicholas Monardes' *Historia medicinal de Indias occidentales*, published in Seville in 1569. Despite the appearance of further accounts of the plant, it seems to have been ignored until its appearance in Europe in the middle of the nineteenth century. Several leading organic chemists tried unsuccessfully to isolate the active principle before Albert Niemann, an assistant to Professor Wöhler at Gottingen University, managed to obtain pure crystals of cocaine in 1860. For the next quarter of a century, the alkaloid was considered, as were coca leaves, to be no more than a mild stimulant comparable with the caffeine in tea or coffee, being added to proprietary beverages and wines. Puzzled by reports by British investigators who cast doubt on whether cocaine had any physiological activity at all, Freud wanted to find out whether or not it could really allay fatigue. When he asked Koller to participate in an experiment to determine its effect on muscular strength, the latter agreed, both men being unaware of the risk of addiction.

Freud has been credited with suggesting to Koller that cocaine would be an ideal local anaesthetic. This is not so. That Freud conducted important studies on the stimulant properties of the alkaloid, published as *The Coca Papers*, and drew Koller's attention to its existence is beyond dispute, but he did not alert Koller to its potential as a local anaesthetic. Perhaps the most significant connection between Freud and cocaine was that it undermined his faith in drug therapy as a result of his recommending it to wean a medical friend off morphine. The tragic consequence of thereby producing a cocaine addict prejudiced Freud against the use of drugs in psychotherapy.

Whilst continuing the experiment with Freud, a colleague who was also participating commented that the cocaine had numbed his tongue. Koller replied that this was experienced by everyone who had taken the drug. At that very moment he realized that he may have unwittingly stumbled upon a drug which could anaesthetize the eye. At once he returned to his laboratory in the Vienna General Hospital, where he and Professor Stricker's assistant, Dr Gaertner, dissolved some cocaine hydrochloride in a small quantity of distilled water, then instilled a few drops of the solution into the eye of a frog. At intervals of a few seconds the reflex of the cornea was tested by touching the eye with a needle. After about a minute the frog permitted his cornea to be touched and even injured without a trace of reflex action or any attempt to protect himself. The other eye responded with the usual reflex action to the slightest touch. With justifiable excitement, the experiment was continued with the same tests being performed on a rabbit and a dog, again with good results. Next, Koller and Gaertner trickled the solution under the upraised lids of each other's eyes, and looking in a mirror took a pin and tried to touch the cornea with its head. Neither felt anything, and they could make a dent in the cornea without the slightest awareness of the touch, let alone any unpleasant sensation.

Koller wrote up his findings for presentation at the next important ophthalmological meeting, which was to be held in Heidelberg on the 15th September 1884. He could not afford to purchase a rail ticket, so instead he asked Joseph Brettauer to read the paper and give a demonstration on his behalf. The paper was received with such enthusiasm that within one month cocaine was being used throughout Europe and in the United States. Koller chose to restrict himself to using it on the eye, but he recommended its use for anaesthetizing the ear, nose, and throat. Although his name became famous in medical circles throughout the world, within a few months he had to leave Vienna because of increasing antisemitism. Eventually, he settled in New York, where he worked for many years in private practice as an eye specialist.

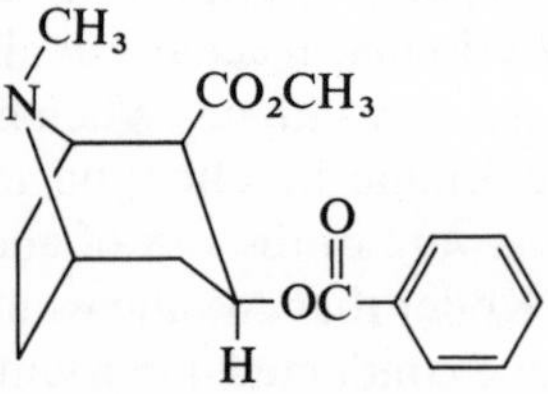

Cocaine

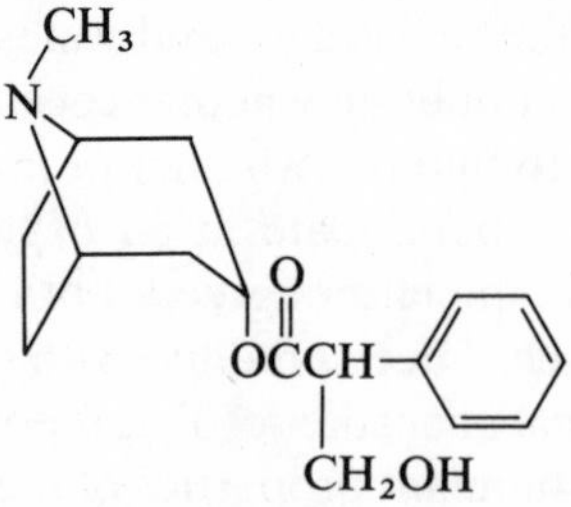

Atropine

In 1885, Calmels and Gossin suggested that a molecular fragment obtained on degrading atropine was also present in cocaine. Realizing that cocaine and atropine both had mydriatic activity, Professor Wilhelm Filehne at the University of Breslau (now Wroclaw, Poland) checked to ascertain whether atropine had local anaesthetic activity in the eye. He found that it did, although this was so feeble that there was no possibility of employing it clinically as the dose would undoubtedly cause toxicity after absorption into the circulation. Furthermore, it was irritant to the eye. From here it was but a logical step to test homatropine, a semisynthetic analogue which was less irritant to the conjunc-

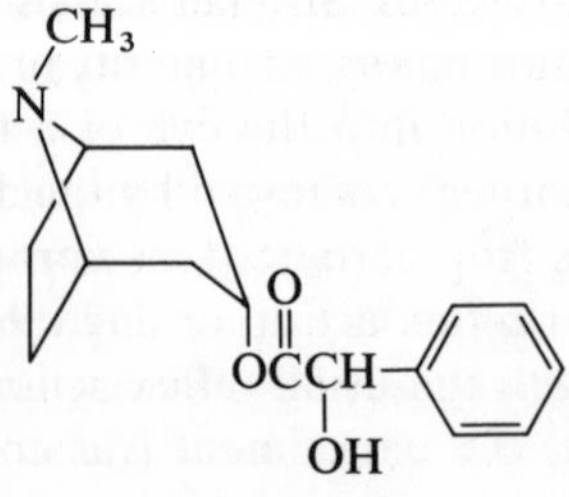

Homatropine

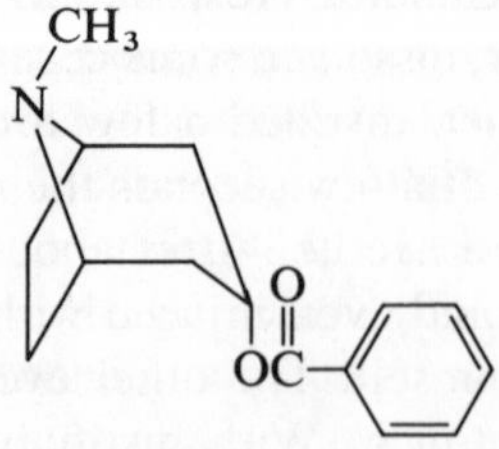

Benzoyltropine

tiva than was atropine (see page 122). It turned out to be more potent than atropine as a local anaesthetic, although in no way comparable with cocaine. Next, Filehne examined another of Alfred Ladenburg's semisynthetic analogues, benzoyltropine. He found it to be a powerful local anaesthetic, although too irritant for clinical use. Nevertheless, this established that just as the mydriatic properties of atropine could be varied by simple structural alteration, so could its local anaesthetic properties.

Filehne now attempted to correlate the change in potency as a local anaesthetic with structural alterations. It was already known that a benzoyl group was present in cocaine, and it was obvious that the most potent of the atropine analogues also contained this group. Filehne therefore tested the benzoyl derivatives of quinine, cinchonine, hydrocotarnine, and morphine, as well as that of methyltriacetone alkamine, a synthetic analogue of the tropine ring of atropine. These benzoate esters were prepared for Filehne by the Hoechst Dyeworks, which was already making considerable profits from its sales of its phenazone ('Antipyrin', see page 85). All were active, especially this latter compound. Unfortunately, it proved to be irritant.

The results of Filehne's pioneering investigation were published in the *Berliner Klinische Wochenschrift* on St Valentine's Day, 1887. By this time, the worldwide demand for cocaine had led to an escalation in its price. In addition, the toxic effects of the drug after it had entered the circulation were causing some concern, and so too was the increasing awareness of its addictive properties. Filehne's paper became the starting point for other investigators seeking a safe cocaine substitute. He himself had exhausted the supply of straightforward chemical analogues of atropine that were available. From here on it was up to organic chemists to design novel molecules in what was to become the first major investigation of its type. The task was to take a further fifteen years, in the course of which the pattern for most of the synthetic approaches followed by the drug industry in the twentieth century was established.

In 1892, Albert Einhorn, the professor of chemistry at the University of Munich, proposed a chemical structure for cocaine. This was based on that which Georg Merling had recently postulated for tropine. These proposed structures were wrong, but this was not realized at the time. Merling, who worked in Emil Fischer's laboratory at the University of Berlin, then speculated as to whether the two rings in the cocaine molecule were both essential for local anaesthetic activity. He therefore synthesized an analogue of cocaine containing only the piperidine ring. It was structurally similar to the methyltriacetone alkamine analogue of atropine, which Filehne had previously found to be a week local anaesthetic. However, Merling's compound proved to be a more potent local anaesthetic. In 1896, it was marketed under the proprietary name 'alpha-Eucaine' by a Berlin pharmaceutical manufacturer, Schering AG. When patients complained that it caused a burning sensation on application to the eye, it was quickly replaced by one of its analogues, 'beta-Eucaine'. This was marketed with the claim that it was not only less toxic than cocaine, but also non-addictive and more stable in injection solutions. However, it only achieved a modest degree of clinical acceptability since the problem of irritancy was not eliminated.

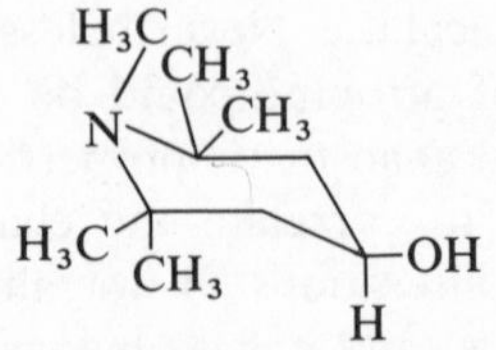

Methyltriacetone Alkamine

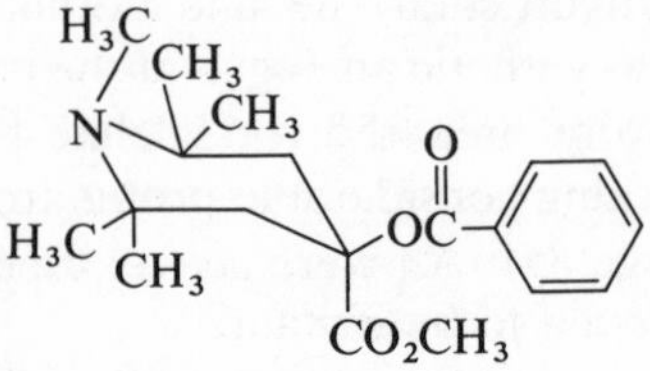

'Alpha-Eucaine'

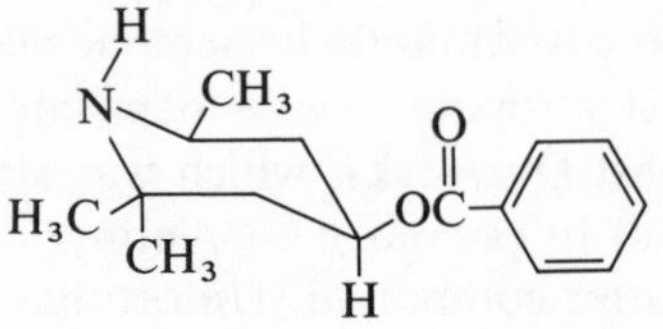

'Beta-Eucaine'

A young research worker in Einhorn's department at Munich, Richard Willstätter, queried the validity of the structure for cocaine proposed by his professor. He was curtly instructed not to pursue the matter, so he obligingly diverted his attention to atropine. The outcome was that he correctly determined the structures of both alkaloids in 1898, and went on to synthesize cocaine three years later! Meanwhile, Einhorn concentrated on the synthesis of analogues of cocaine. One of his earliest collaborators was Paul Ehrlich, who investigated the pathological liver changes induced by cocaine and its analogues. He found that the toxicity of the analogues did not always correspond with their potency as local anaesthetics. From this observation he subsequently derived the basis of modern chemotherapy by screening potential drugs to find those combining minimal toxicity with maximal curative properties in infected animals.

Although Merling had established that cocaine analogues featuring only the piperidine ring were active, Einhorn wondered whether the other ring in Merling's proposed structure also conferred local anaesthetic activity. To settle the matter he attempted to prepare appropriate compounds by a chemical reduction of benzene derivatives containing the carboxylate ester and benzoyl ester functional groups found in cocaine. To his dismay, the reduction process did not produce the desired hexahydro derivatives. Rather than abandon the compounds he was left with, he reasoned that as several anaesthetic or analgesic drugs contained a benzene ring, e.g. phenol, picric acid, methyl salicylate, phenacetin, methylene blue, etc, it would be worthwhile sending aromatic benzoyl esters to R. Heinz at the University of Erlangen to be screened for anaesthetic activity. Several turned out to be active. What was most surprising, however, was the discovery that activity was also present in the phenolic compounds obtained on removal of the relatively unstable benzoyl function by hydrolysis. This was the first clear indication of useful local anaesthetic activity in synthetic compounds which did not contain a benzoate ester function, and it was to turn research into the development of cocaine analogues in a new direction. Had Einhorn not chosen to use phenolic esters in his synthesis, this discovery may not have been made until years later, since only this type of ester is so unstable in water.

Einhorn exploited his lucky break by introducing one of his phenolic compounds into clinical practice in 1896. It was marketed by the Hoechst Dyeworks under the name 'Orthoform' (approved name = orthocaine) as a local anaesthetic with antiseptic activity (as it was a phenol derivative) for topical application to wounds. The lack of water solubility was turned to advantage by claiming that this ensured the drug would remain present on the wound to provide relief for many hours. Because 'Orthoform' was relatively expensive to manufacture, and also tended to form lumps, it was replaced by the isomer known as 'Orthoform New' the following year.

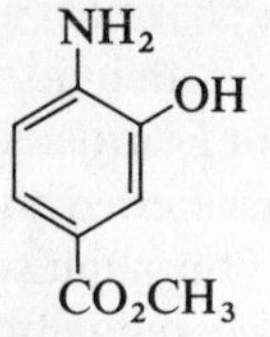

NH_2, OH, CO_2CH_3 — 'Orthoform'

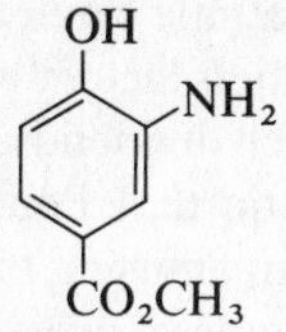

OH, NH_2, CO_2CH_3 — 'Orthoform New'

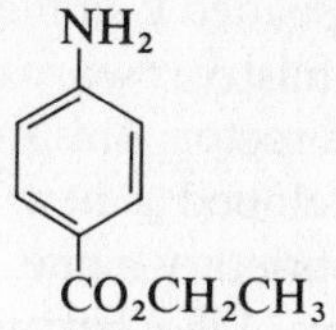

NH_2, $CO_2CH_2CH_3$ — Benzocaine

The success of the 'Orthoforms' led E. Ritsert to reconsider the clinical potential of the ethyl ester of 4-aminobenzoic acid which he had noticed, in 1890, was capable of numbing the tongue. Its lack of water solubility had seemed then to rule out its medicinal use, but after the introduction of 'Orthoform' the value of a topical anaesthetic in the treatment of wounds and sore throats had become firmly established. The ester was introduced in 1902 as 'Anaesthesine'. It later received the approved name of benzocaine.

Ritsert tried to formulate benzocaine as the water-soluble hydrochloride, but the resulting solution was too acid (since the amino group was aromatic). During the next few years he investigated other salts without success.

Encouraged by the success of 'Orthoform', Einhorn tried to prepare a water-soluble derivative of it, which would be suitable for injection. As he had already established that it was possible to substitute alkyl chains containing five carbon atoms on the amino group of his phenolic compounds, he now examined the effect of introducing other chains containing aliphatic amino groups. He hoped this would permit formulation of the resulting compounds as their water-soluble hydrochlorides. Such salts would not produce highly acidic solutions, as had been the case with the 'Orthoforms'.

In 1898, Heinz confirmed that one of the new compounds from Einhorn was effective as an anaesthetic, although its potency was low. This necessitated the use of fairly strong solutions which tended to be irritant. Notwithstanding this, it was introduced clinically as 'Nirvanin', principally because it was less toxic than either cocaine or the 'Eucaines'. It was soon rendered obsolete by superior agents.

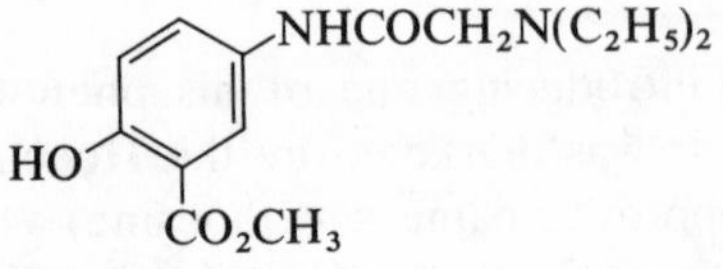

$NHCOCH_2N(C_2H_5)_2$ HO CO_2CH_3 'Nirvanin'

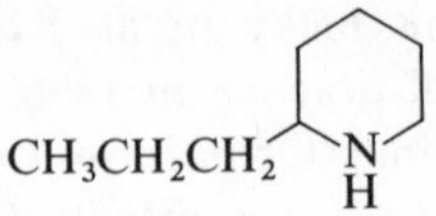

$CH_3CH_2CH_2$ N H Coniine

CH_3 $CO_2CCH_2N(CH_3)_2$ C_2H_5 Amylocaine

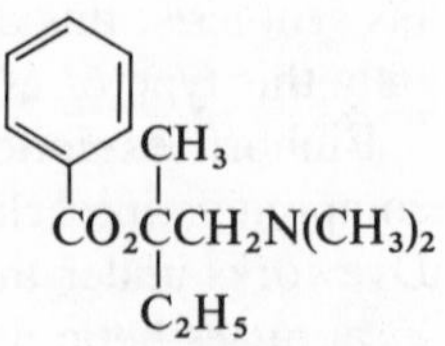

In 1902, Ernest Fourneau returned to France after having studied with several leading German chemists, including Emil Fischer and Richard Willstätter. He was particularly anxious that the French should not require to import drugs from Germany when they could easily make their own products. He persuaded Camille, Gaston, and Emile Poulenc to establish a pharmaceutical chemistry research laboratory in their factory at Ivry-sur-Seine, with himself as its director. This step was to pay rich dividends, and within a year Fourneau had developed a new local anaesthetic that Poulenc Frères marketed with the proprietary name of 'Stovaine', an amusing bilingual pun on Fourneau's own name! Like 'Nirvanin', it was capable of providing water-soluble salts through the incorporation of an aliphatic amino group. It had been deliberately designed to feature all the functional groups in the cocaine molecule, except for the piperidine ring. Fourneau considered this ring to be responsible for toxicity. After all, Socrates had killed himself by taking hemlock, the active principle of which was the piperidine alkaloid coniine. This was synthesized by Alfred Ladenburg at the University of Kiel in 1886, and was the first alkaloid ever to be synthesized.

'Stovaine' was an ester formed from benzoic acid and the simplest available amino alcohol resembling the amino alcohol moiety of the tropine ring. It was given the approved name of amylocaine. It was the first drug to achieve the long sought after goal of being a safe, non-irritant local anaesthetic that could be injected as a substitute for cocaine.

Albert Einhorn realized that a molecule combining the best features of amylocaine and benzocaine might be superior to either of these new agents. He selected esters made from aminobenzoic acids and amino alcohols and had them investigated at the Pharmacological Institute in Breslau by Biberfield. The most promising compound was sent for clinical evaluation to Professor Heinreich Braun at Leipzig, who was then completing what was to become the leading textbook on local anaesthesia. Braun began by testing the new drug on himself and volunteers. He immediately recognized that it was far less irritant than any existing agent, but its action was very brief unless strong solutions were injected. This, of course, negated much of the advantage of the new drug, since the larger amounts injected produced as much toxicity as cocaine. Fortunately, he had himself recently introduced a new technique which enabled this disadvantage to be dealt with.

In 1885, the leading exponent of local anaesthesia in America, Leonard Corning, demonstrated that the effect of cocaine could be intensified if the local circulation of blood was interrupted by application of a tourniquet. This prevented the anaesthetic from being washed away from the site of injection, and so enabled him to reduce the amount of cocaine to be injected. The idea spread to the Continent, but it was limited mainly to surgery of the fingers or toes. However, in the spring of 1900, Heinreich Braun read of the preparation of adrenal extracts which caused contraction of the blood vessels when locally applied to the tissues. He at once realized that adrenal extract might be applied in local anaesthesia to achieve much the same effect as using a tourniquet. Within a few days he obtained some extract, mixed it with cocaine, and injected it into his own arm. The resulting anaesthesia was of such intensity and duration as he had hitherto never experienced.

When, four years later, Braun began to investigate Einhorn's new local anaesthetic, he believed that by combining it with the recently introduced pure, active principle of adrenal extract, adrenaline, he might be able to use lower concentrations. The combination satisfied his expectations. Indeed, the safety, lack of irritancy, and the overall efficacy of the new drug when combined with adrenaline ensured that it dominated the field of local anaesthesia for nearly half a century. It was marketed by Hoechst as 'Novocaine', and later received the approved name of procaine.

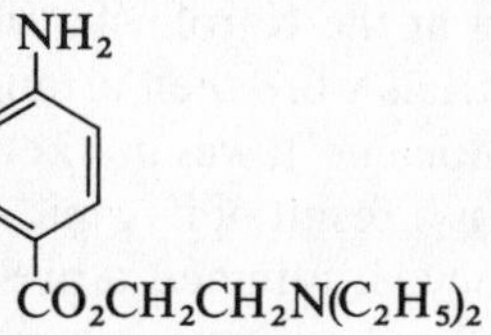

NH_2

$CO_2CH_2CH_2N(C_2H_5)_2$

Procaine

The only other important local anaesthetic to be introduced before the Second World War was discovered serendipitously, even though vast numbers of compounds were synthesized in attempts to find an agent superior to procaine. Karl Miescher of Ciba in Switzerland attempted to synthesize novel analogues of the antipyretic agent acetanilide by incorporating the anilide function into a second ring. The resulting dihydrocarbostyril proved to have sedative powers, so Miescher decided to synthesize amide derivatives of it in the hope that this would once again confer antipyretic activity. The new amides were routinely screened for a variety of effects. This revealed that some were potent local anaesthetics. Further synthetic variation provided an exceedingly potent anaesthetic which was longer acting than procaine, although more toxic. The new drug, cinchocaine ('Nupercaine', dibucaine), was very useful as a spinal anaesthetic because of its longer duration of action. It was also a valuable topical anaesthetic.

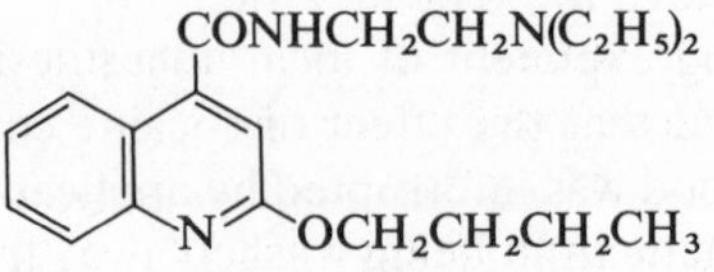

$CONHCH_2CH_2N(C_2H_5)_2$

N $OCH_2CH_2CH_2CH_3$

Cinchocaine

The discovery that the refusal of Central Asian camels to eat a certain type of reed was due to the presence in it of a toxic alkaloid called gramine, led scientists at Stockholm University to investigate the chemical nature of this alkaloid. To assist with the characterization of gramine, Holger Erdtman synthesized isogramine in 1935. As was his wont when making any new compound, Erdtman tasted a trace of it. It tasted bitter sweet, then his tongue went numb! The significance of this was not lost on Erdtman, who also noted that gramine did not have local anaesthetic activity. Next, he tasted the open-chain compound from which the bicyclic isogramine had been synthesized. It, too, was a local anaesthetic. Erdtman and his research student, Nils Löfgren, then set about synthesizing analogues to find the most effective. Their task proved daunting, with most of the active analogues being irritant. Lofgren's persistence paid off only after some fifty-seven compounds had been synthesized over a seven-year period! His colleague, Bengt Lundqvist, tested compound LL 30 on himself, then suggested it should be pharmacologically evaluated by L. Goldberg at the Karolinska Institute. The results were most encouraging and clinical trials were then arranged. Eventually the new drug, lignocaine (lidocaine, 'Xylocaine'), was marketed in Sweden by Astra in 1948. Over the next few years, as a result of its rapid onset of action, relative safety, and freedom from irritancy, it attained a pre-eminence over all other local anaesthetics which has never been seriously rivalled.

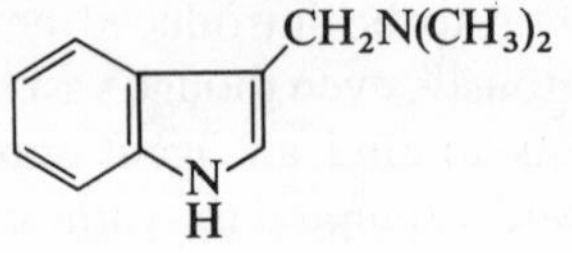

$CH_2N(CH_3)_2$

N
H

Gramine

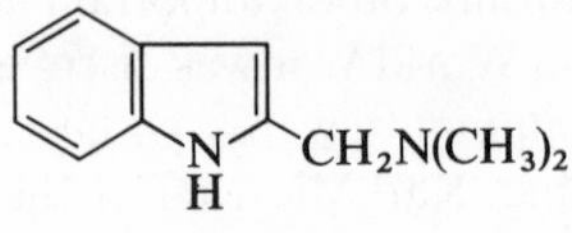

N $CH_2N(CH_3)_2$
H

Isogramine

CH_3

$-NHCOCH_2N(C_2H_5)_2$

CH_3

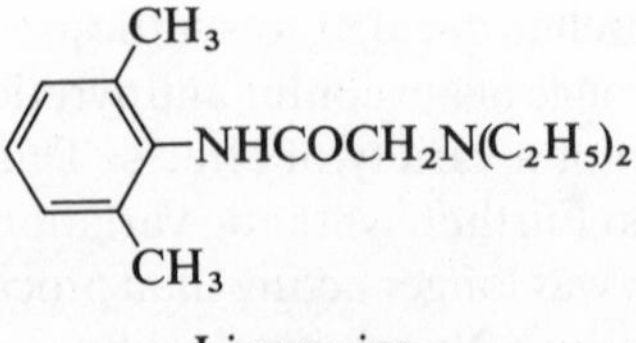

Lignocaine

In 1957, scientists from the Swedish pharmaceutical company AB Bofors reported the pharmacological activity of a series of lignocaine analogues in which the side-chain had been partially incorporated into a cyclic system. This molecular manipulation had, apparently, been instituted without any clear rational other than that it had chemical novelty. It did, nevertheless, provide two valuable local anaesthetics, mepivacaine ('Carbocaine') and bupivacaine ('Marcain'), the latter proving to be longer acting than any other agent, producing nerve blocks for up to eight hours. It is widely used for continuous epidural anaesthesia in childbirth.

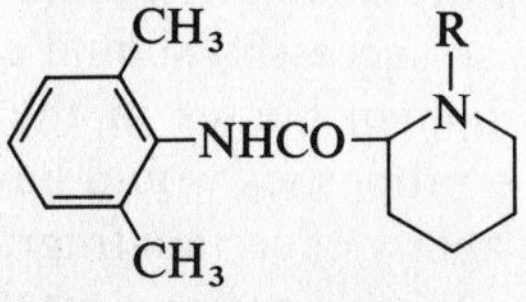

CH_3 R

—NHCO— N

CH_3

Mepivacaine, R = $-CH_3$
Bupivacaine, R = $-CH_2CH_2CH_2CH_3$

IV

Antiseptics

Ancient Egyptian mummies were embalmed with complex mixtures containing spices, aromatic plants, pitch, resins, cedar oil, and natron (sodium sesquicarbonate), which prevented putrefaction in a manner that withstood the test of time. That embalming was so successful should cause little surprise, for such was the rate of decay of human bodies in the Egyptian climate that any substance with preservative properties would have been quickly recognized. This is not to deny that the ability of certain fragrant oils and spices to prevent the foul odour of putrefying flesh or food was misconstrued. Had it been otherwise, medical antisepsis might have been introduced a long time ago. It would, however, be quite wrong to believe that antiseptics were initially developed solely as a consequence of Pasteur's brilliant discovery that infectious diseases were caused by bacteria.

The term 'antiseptic' was used in a series of short papers read to the Royal Society in London between 1750 and 1752 by the renowned military surgeon Sir John Pringle. He had held the chair of moral philosophy at Edinburgh prior to serving with the British Army on the Continent, where his suggestion that military hospitals should be regarded as sanctuaries was accepted by the opposing French forces. His researches on antiseptics were described in an appendix to his *Observations on the Diseases of the Army*. Pringle examined the ability of a variety of salts to prevent putrefaction of pieces of beef, taking a fixed concentration of common salt as a standard for comparison. He concluded that the most effective agent was alum, followed by salt of amber (sodium succinate) and borax. Next came salt of wormwood (potassium carbonate), nitre (potassium nitrate), then salt of hartshorn (ammonium carbonate), with which he was able to preserve the beef at room temperature for a year. Sal ammoniac (ammonium chloride) and several other salts were less efficient, although still superior to common salt. Camphor was found to be the best preservative amongst plant products, whilst a variety of infusions were also found to have preservative properties. Although Pringle's conclusions cannot be accepted unreservedly, his work represented the earliest attempt to compare the preservative properties of antiseptics.

In 1785, Berthollet observed the bleaching properties of aqueous solutions of the recently discovered chlorine gas. Three years later he recommended such a preparation as a bleaching and disinfectant solution, having noted that it prevented the foul odours of decaying organic matter. Eau-de-Javelle, a similar chlorine preparation, was marketed in 1792. However, it was not until 1825 that a chlorine-based formulation was employed for wound disinfection, when Labarraque, a Parisian apothecary, used his Eau-de-Labarraque for this purpose. Two years later, Alcock introduced this into the United Kingdom for the purification of drinking water, and it was subsequently included in the *London Pharmacopoeia* of 1836 under the name *Liquor Sodae Chlorinatae*. Eighteen years later, the British Royal Commission on Sewage recommended that sewage be deodorized with chloride of lime, a similar type of preparation.

In France, Eau-de-Labarraque was employed to disinfect hands and to treat gangrene. One of the first attempts to control an outbreak of the frequently fatal childbirth fever with it was that of Robert Collins in 1829. This was one year before Oliver Wendell Holmes, in America, asserted this was contagious, being transmitted by midwives. In 1835, Holmes reported total success in stemming puerperal fever simply by insisting on hand disinfection with chloride of lime solution. When he reported his conclusions to the Boston Society of Medical Improvement in 1843, he met considerable hostility from many of its members who felt his views to be insulting. Four years later, similar opposition was met by Ignaz Semmelweis, an assistant at the Lying-in Hospital in Vienna. He showed that the mortality from puerperal fever in the First Division maternity ward, where medical students were trained, was higher than that occurring in the Second Division ward, where nurses and midwives received instruction. He argued that this was due to the students infecting the patients with putrid particles adhering to their fingers after post mortem examinations. By insisting that students wash their hands in chloride of lime solution, Semmelweis ultimately reduced the mortality rate in his wards from twelve per cent to just over one per cent. Yet this did not satisfy the conservatism of some of his senior colleagues. They resented his interference and he became embroiled in bitter arguments, with the result that he was not reappointed when his post in the hospital was due to be renewed. He then had to move to Budapest. His classical study, *On the Cause and Prevention of Puerperal Fever* was published there in 1861. It included striking mortality statistics that should have convinced all but the most bigotted, yet even this failed to win over some of his critics, and he became insane through the aggravation this caused him. In 1865, the very year that Lister first used carbolic acid to prevent wound sepsis, Semmelweis died from a septic wound on his finger.

The effectiveness of hypochlorite solutions as disinfectants was never in doubt, but there were problems when they were applied to wounds and delicate tissues. Unless special precautions were taken both free alkali and chlorine were present, resulting in tissue irritation. This problem was overcome through work done by Henry Dakin in association with Professor J. B. Cohen of Leeds University. They knew that hypochlorite solution reacted with free amino groups, such as are found in proteins. On examining the products, known as chloramines, these were found to have antiseptic properties in their own right. Their instability precluded their clinical use, but when Dakin found that certain aromatic chloramines prepared in 1905 by Frederick Chattaway at St Bartholomew's Hospital, London, were able to form water-soluble sodium salts, he was able to exploit these with success. From the Berter Laboratory at the Rockefeller Institute in New York, he reported, in 1915, that chloramine-B (a

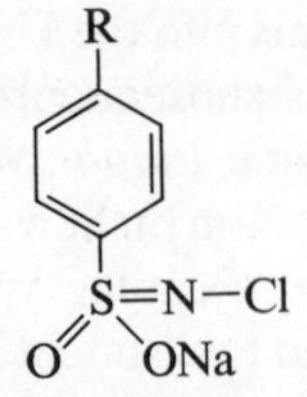

Chloramine B, R = $-H$
Chloramine T, R = $-CH_3$

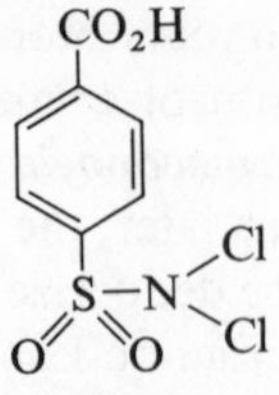

Halazone

benzene derivative) and chloramine-T (a toluene derivative) were powerful antiseptics that possessed all the advantages of hypochlorite whilst being stable compounds with low irritancy. They acted by slowly releasing hypochlorous acid over a prolonged period, this rendering them highly efficacious in wound disinfection. In 1917, Abbott Laboratories in Chicago marketed a related dichloramine known as halazone, which Dakin had developed to meet the wartime demand for tablets to sterilize drinking water.

CHI_3
Iodoform

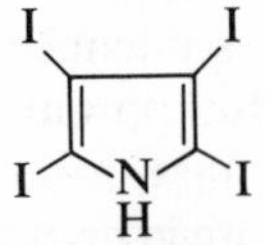

Iodol

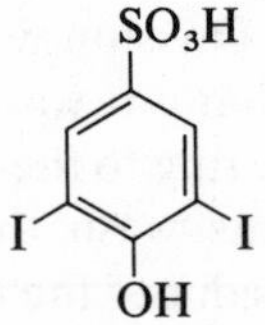

Soziodol

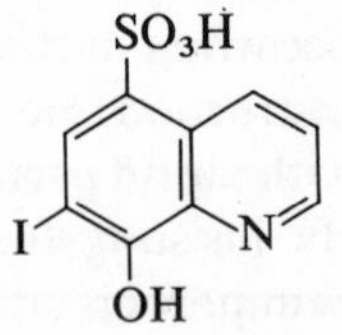

Chiniofon

Tincture of iodine was used as an antiseptic by Wallace in 1836. Three years later, a Dr Davies of Hertford applied it to disinfect wounds, a purpose for which it was later employed in the American Civil War. Its irritancy and the stinging sensation it produced led a London surgeon, Robert Glover, to replace it with iodoform, in 1847. This was a substitute for iodine that he had become acquainted with ten years earlier when he recommended it as a more acceptable form of iodine for internal administration in the treatment of goitre. Nevertheless, iodine disinfection failed to win popularity until after 1873, when C. Davaine, in France, reported that iodine solutions could kill anthrax bacilli. Others then demonstrated its efficacy against gonococci, staphylococci, and streptococci. This led Berkeley Hill, in 1878, to reinvestigate iodoform as an antiseptic for hospital use. From then until well after the Second World War, the characteristic smell of iodoform permeated through hospitals all over the world, despite general recognition of its potential toxicity after absorption from open wounds. The mild disinfectant action, as well as any toxicity, was caused by small amounts of iodine released on contact with the tissues. It is now considered doubtful whether the drug really had much value in the treatment of wounds. Indeed, its irritant properties and powerful odour led to the introduction of many substitutes. These were organic compounds that also released iodine, and included iodol (1886), soziodol (1887), chiniofon (1893), thioform (1894), sanoform (1896), clioquinol (1898), and many more introduced this century. As for tincture of iodine itself, detailed studies demonstrating its value as a skin disinfectant in surgery appeared from 1905 onwards. It was widely used for this purpose until the nineteen-thirties.

CARBOLIC ACID

The antiseptic properties of coal tar were noted by Chaumette in 1815, but its viscous nature discouraged him from using it for any practical purpose. In 1844, Bayard won an award from the Société de Encouragement for a clay-based powder that incorporated coal tar. This was recommended for disinfecting manure that was to be used as a fertilizer. In 1859, Demeaux suggested to the Academy of Sciences in Paris that a similar powder prepared from plaster and coal tar (patented by Corne the previous year) should be tested as a disinfectant for wounds. However, when it was tried in the hospitals of Paris, the tendency of the plaster to turn solid severely restricted the value of this product. It was replaced in August 1859, when Ferdinand LeBeuf and Jules Lemaire introduced emulsified coal tar. This was prepared by LeBeuf, the Bayonne pharmacist who had discovered, nearly ten years earlier, that water-insoluble substances could be emulsified by the addition of water and tincture of quillaia (obtained from Panama bark, rich in natural detergents known as saponins). He had sought Lemaire's assistance in evaluating his formulation as an antiseptic. Lemaire, a well-known Parisian pharmacist and physician, had considerable success in treating septic wounds with the emulsified coal tar, and his colleagues soon began to copy him. At their request, on the 25th April 1862, emulsified coal tar was officially authorized for wound disinfection in the civil hospitals of Paris. Meantime, Lemaire turned his attention to finding out which component of coal tar was responsible for its remarkable antiseptic properties.

In 1832, Friedlieb Runge, whose isolation of quinine in 1819 has already been discussed (see page 12), was appointed chief chemist in charge of a chemical works in Oranienburg, a small town near Berlin. The factory was sited in a former palace, and it was exceptionally well equipped. The most successful product at that time was sulphuric acid, but the management wished Runge to develop new processes. This he most certainly succeeded in doing, although the bank which owned the company failed to exploit most of Runge's discoveries. This was because of the furtive intervention of his immediate superior, who was jealous of his achievements. In 1833, Runge found a practical use for the coal tar waste that was being sent in large quantities to his factory from the Berlin gas works. He had tried to separate different fractions from it by means of acids, alkalies, and solvents, but met with no success. In desperation, he resorted to distillation, and obtained an oil with a pungent odour. Steam distillation of this oil afforded a lighter oil, part of which dissolved in milk of lime. This acidic fraction was named carbolic acid by Runge, and it was recovered from the lime solution by addition of acid. Runge was impressed by its ability to prevent the decay of animal tissue and wood, but considered it too expensive to market as a preservative.

The chemical constitution of carbolic acid was established by Laurent in 1841. His material, which he called 'acide phénique', was purer than that produced by Runge, although he also used a distillation process. The following year, Gerhardt called this purer material 'phenol', the name that eventually prevailed. Because of its historical significance, the original term 'carbolic acid' will be employed throughout this chapter.

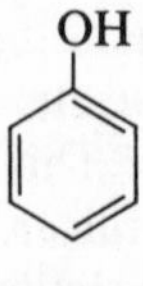

OH

Phenol
(Carbolic acid)

A Manchester chemist, Frederick Crace Calvert, manufactured a crude mixture of calcium salts and carbolic acid that had been patented in 1854 by Angus Smith and Alexander McDougall. This was used for purifying water and deodorizing sewage. In 1857, Calvert began large-scale production of reasonably pure carbolic acid to meet the demands of the newly established synthetic dyestuffs industry. Much of his output was taken by the French firm of Guinon, Marnas, and Bonnet. It may have been this business connection which brought Calvert into contact with the French Academy of Sciences after Demaux had recommended the use of coal tar powder as a wound disinfectant. Calvert suggested to the Academy that the principal disinfectant substance in coal tar was carbolic acid. This led Lemaire and other French investigators, including Parisel, Boboef, and Declat, to investigate its properties.

Jules Lemaire published his first paper on carbolic acid in 1861, and others followed. His book *De l'Acide Phénique* appeared in 1863 and was soon sold out. An enlarged second edition of 754 pages was printed two years later. This established Lemaire as the leading advocate of the use of carbolic acid in surgery at that time, but although his work was noted in Britain, it aroused little enthusiasm. This was despite the efforts being made by Calvert to interest the medical profession in carbolic acid by publishing papers in the *Lancet* and the *British Medical Journal*. A similar lack of enthusiasm was shown towards Gilbert Declat's lengthy volume entitled *Nouvelles Applications de L'Acide Phénique en Médecine et en Chirurgie* (New Applications of Carbolic Acid in Medicine and Surgery), which was published in 1865. Declat, who clearly understood the significance of Pasteur's work, referred to carbolic acid as a 'parasiticide'. He expressed the hope that carbolic acid would be used prophylactically to prevent infection, a different approach to that taken by his contemporaries. He even recommended washing of the walls and surroundings of the sick room with it!

Paradoxically, it was Calvert's success in convincing public authorities of the benefits of carbolic acid for treating sewage that ultimately led to its use in the system of antiseptic surgery introduced by Joseph Lister.

Lister was appointed as professor of surgery in the University of Glasgow at the age of thirty-three, in 1860. Over a year passed before he was able to take charge of three wards in the newly built surgical wing of the Royal Infirmary. Once he had begun to care for patients he was struck both by the large number of accident victims being brought to the hospital and by the high incidence of hospitalism. This was the term used to describe the often lethal infections following surgery in large hospitals. Soon, his main preoccupation became finding the cause of this with a view to remedying the situation. Knowing of Lister's interest in the cause of infection, his friend Professor Thomas Anderson, who held the chair of chemistry at the university, drew his attention to a paper published a year and a half earlier, in June 1863, by Louis Pasteur. Anderson had not seen the article sooner because the French journal in which it appeared, *Comptes Rendus Hebdomadaires*, had only just been returned from the binders. The article was the epochal account of how Pasteur had demonstrated that putrefaction was caused by microbes. Six years earlier he had established that fermentation was similarly initiated, but the implications for surgery were now much clearer. Lister did not fail to recognize them, and at once sought some means of preventing open wounds becoming infected with the microbes during surgery.

Lister spent the next few months seeking an agent powerful enough to ensure destruction of all the microbes that were likely to infect open wounds. He had been kept so busy with his surgical work at the infirmary that he was unaware of the studies being carried out in France with carbolic acid, but when his attention was finally caught by a newspaper report that carbolic acid was used in Carlisle to deodorize sewage, he asked Professor Anderson to obtain the chemical. This was duly done, but instead of Calvert's pure carbolic acid, Lister was provided with so-called German creosote, a preparation of inferior quality. The first case deemed appropriate for its application was a patient with a compound fracture of the leg, who was brought to Lister in March 1865. The prognosis for such patients was always poor, something in the order of fifty per cent dying since puncturing of the skin by the broken bone inevitably resulted in infection at, and also after, the moment of injury. Lister was unable to save the patient. He then modified his technique of covering the wound with dressings soaked in solutions of carbolic acid (now obtained from Calvert's firm), and he was ready for his next case five months later. On 12th August, an eleven-year-old boy who had been run over by a cart, was admitted to the infirmary with a compound fracture of the leg. This time the carbolic acid dressing successfully prevented infection, but another patient treated shortly after was not so fortunate. His leg had to be amputated. Several more cases were treated the following year, and by the time the first of five early papers by Lister on antiseptic surgery appeared in the *Lancet* in March 1867, eleven patients had been treated, with only one death. This was a truly remarkable achievement, and it inaugurated the era of modern surgery.

Although Lister was not the first person to apply carbolic acid as an antiseptic, he undoubtedly transformed surgical practice by employing it effectively to prevent wound sepsis. By basing his use of carbolic acid on a clear understanding of Pasteur's researches, Joseph Lister transformed surgery and rendered it safe. Even though German surgeons were soon to adopt aseptic techniques that did away with carbolic acid altogether, their aseptic surgery was, in reality, but an extension of Lister's principles.

That it was left to German surgeons to develop Lister's ideas is not altogether surprising, for his work provoked hostility from leading surgeons in Edinburgh and London. This was reminiscent of that previously encountered by Oliver Wendell Holmes in the United States, and Ignaz Semmelweis in Vienna, when they had advocated the use of hypochlorite solution in obstetric practice. This hostility towards these pioneers is best understood by the realization that they were each implying that infection was caused by intervention of the patient's surgical attendant rather than by the spontaneous generation of an infective agent within the body. Naturally, many surgeons resented the implication that their technique was directly responsible for the death of almost half of all the patients on whom they operated. They were unwilling to accept Pasteur's claim that there was no such thing as spontaneous generation and that the bacteria which caused fermentation and putrefaction could only be formed by the rapid multiplication of existing bacteria. The more fair-minded amongst Lister's critics did experiment with carbolic acid, but failing to grasp that its use was merely a means towards the end of preventing live bacteria from infecting open wounds, they adopted faulty techniques that resulted in sepsis. Some of Lister's opponents pointed out that neither he nor Pasteur had proved that microbes could cause wound infection or, for that matter, any known infectious disease. Indeed, this was true. Bacteriological and microscopic techniques had not yet advanced to the stage where pathogenic micro-organisms could be isolated, identified, and transmitted to an animal to induce the disease which they were supposed to cause. It was not until the arrival on the scene of Robert Koch and Carl Weigert, in the late eighteen-seventies, that this last remaining obstacle to the acceptance of Lister's approach was removed. But, of even more relevance to this present treatise, the work of Koch and Weigert established the theoretical basis upon which modern chemotherapy was founded by their protégé, Paul Ehrlich.

INTERNAL ANTISEPSIS

In 1857, Professor Giovanni Polli of Milan studied the effects of various sulphites on animals. He was surprised to observe that the carcasses of animals that had been killed at the end of his experiments did not decompose in the usual manner, putrefaction being entirely absent in those cases where the animals had received large doses of sulphites. Polli then carried out a series of experiments on no less than three hundred dogs(!), before concluding that sulphites protected them from infection after exposure to supposedly septic poisons. After satisfying himself that large amounts of sulphites could be administered safely to humans, Polli carried out trials on patients suffering from malaria, typhus, typhoid, septicaemia, puerperal fever, and the like. He then boldly claimed to have achieved an extraordinary amount of success with the sulphites. Nor was he the first to make such a claim, for the sulphites are converted to sulphurous acid in the stomach, and in 1698 Sir John Colbatch recommended the drinking of dilute sulphurous acid in cases of malignant fever.

When Professor Polli published his findings in 1861, they aroused much interest. During the next six or seven years, more than one hundred and fifty scientific papers commented on his results. Most of these supported his claim that sulphites could be taken internally to prevent various diseases, although Lister, who applied them to wounds before experimenting with carbolic acid, found them of little value in the prevention of sepsis. Polli actually believed that sulphurous acid, an antioxidant, denatured the tissues on which microbes fed, thus preventing their growth by starving them of usable nutrient. Pasteur's publications seemed to lend support to Polli's claims, with the result that the concept of internal disinfection became popular in many quarters, although subsequently modified to encompass the view that antiseptics acted directly on pathogenic organisms rather than on the tissues. Indeed, belief in internal antisepsis has not entirely died out one hundred and twenty years later. Many travellers firmly believe that antiseptics such as clioquinol ('Entero-Vioform'), recently withdrawn on account of its toxicity to the eye, can somehow prevent diarrhoea.

During the cholera epidemic of 1866, Arthur Sansom, physician to London's Royal Hospital for Diseases of the Chest, administered sodium sulphite to many victims, claiming that it showed 'unmistakably the advantage of the mode of treatment'. He attributed the failure of sulphite to effect an outright cure to its rapid absorption into the tissues from the alimentary canal, where it was required to act against the infecting organisms. However, four years earlier, Crace Calvert had written that carbolic acid was being used by physicians to treat diarrhoea, and since sewage throughout Britain was now being treated with it to prevent the spread of cholera, Sansom considered administering it to his patients. He refrained from so doing because only small quantities could be taken by mouth on account of its deleterious effect on the tissues. The following year, he learned from a mutual friend of his and Calvert, namely the eminent chemist William Crookes, of the existence of a new type of carbolic acid derivative known as sulphocarbolate of potash. Sansom wrongly understood this to be a 'compound salt' of carbolic and sulphuric acids with potassium that he presumably thought would decompose in the intestine and liberate sulphite and carbolic acid. He introduced it in 1868 as an internal antiseptic. Sansom originally used a mixture of the salts of ortho- and para-phenolsulphonic acids, but the former, known as solozic acid, was subsequently produced commercially as a one-in-three solution in water ('Aseptol'). Although it did not liberate carbolic acid, it was still widely prescribed until the turn of the century, especially is cases of diphtheria, scarlet fever, and puerperal fever.

The administration of internal antiseptics was not by any means limited to patients afflicted with infectious diseases. It was generally believed that organic compounds known to be associated with putrefaction could be produced in the intestines by bacteria. These ptomaines, it was held, could be the cause of ill health, especially as they were organic nitrogenous bases not unlike the physiologically active vegetable alkaloids. To maintain good health, many Victorians bravely underwent an exhausting ritual of purging, colonic irrigation, and enemas to rid their intestines of foul contaminants! They also consumed copious amounts of antiseptics and purgatives to eliminate residual bacteria that had withstood the former assaults. All that can be said for such determination is that it represented some advance on the ancient Egyptian belief that purging eliminated demons from the body.

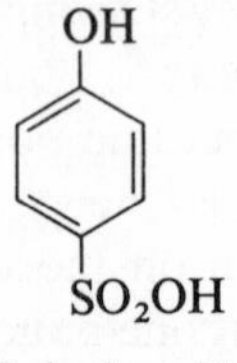

OH

SO_2OH

Solozic acid

The irritant nature of carbolic acid in the stomach persuaded investigators to seek alternative internal antiseptics in addition to sulphocarbolic acid. One of the earliest to be used was salicylic acid, introduced in 1874 by the Leipzig surgeon Carl Thiersch to replace carbolic acid in surgery. Thiersch had been the first Continental surgeon to adopt Lister's methods, but he was concerned about the damaging effects of carbolic acid on the tissues. This led him to approach his friend Hermann Kolbe, who was the professor of chemistry at Leipzig, in the hope that he might propose a suitable alternative. Kolbe suggested that the carbolic acid analogue known as salicylic acid might be appropriate, and he devised a synthesis that made this cheaply available for the first time. Although it never replaced carbolic acid in surgical practice, salicylic acid became a popular internal antiseptic. It was used in attempts to treat typhoid and rheumatic fever, whereupon its antipyretic activity was revealed (see page 81).

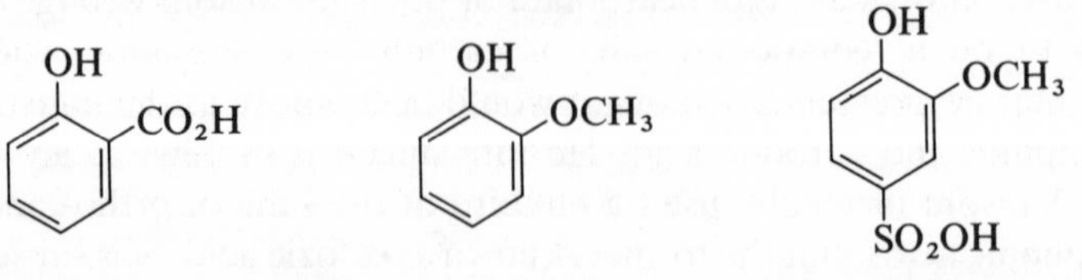

OH CO_2H — Salicylic Acid

OH OCH_3 — Guaiacol

OH OCH_3 SO_2OH — 'Thiocol'

K. Reichenbach distilled creosote from beechwood tar in 1830. Later, its physical similarity to the heavy oil distilled from coal tar led him to examine its antiseptic properties. Over the next few years numerous articles in medical journals extolled the virtues of creosote in the treatment of gangrene, cancerous lesions, eczema, impetigo, and foul-smelling wounds. Pieces of cotton soaked in creosote were even applied to tooth cavities to afford genuine relief. Notwithstanding its irritant nature, beechwood creosote was also taken internally from the 1880s onwards. It acquired an undeserved reputation for having extraordinary powers to arrest vomiting! Needless to say, responsible medical opinion questioned this, and also claims that it could relieve diarrhoea. Creosote was also inhaled, apparently to relieve excessive bronchial secretion. Pulmonary tuberculosis was treated by inhalation of creosote combined with administration by mouth of as much as one gram daily. All these applications resulted in a demand for a cheap substitute for the expensive genuine beechwood creosote. This demand was often met by the supply of coal tar creosote at around one-fifth of the price. As this contained a considerable proportion of phenol, it was much more irritant than genuine creosote, especially if inhaled or taken internally. However, when Béhal and Choay isolated crystalline guaiacol from beechwood tar in 1887, it quickly became popular since it was free of any unpleasant smell and could be cheaply isolated from coal tar creosote. In Berne, Professor Hermann Sahli recommended it for the treatment of tuberculosis. Being an analogue of carbolic acid, guaiacol irritated the stomach, hence the 'salol principle' was applied to it (see page 82). This resulted in the introduction of several of its esters, such as the products known as 'Duotal', 'Monotal', and 'Benzosol'. As these esters were insoluble and so could not be conveniently incorporated in liquid dosage forms, Emil Barell, a young chemist appointed in 1896 by the newly established firm of Hoffmann-La Roche, synthesized a soluble derivative modelled on sulphocarbolic acid, namely the potassium salt of guaiacol sulphonic acid. Under the proprietary name of 'Thiocol' it became the most widely used guaiacol derivative. The fortunes of the Hoffmann-La Roche Company were established by this product, and it remained on sale as an intestinal antiseptic until the outbreak of the Second World War, despite criticism that it lacked any appreciable antiseptic activity. Of greater theoretical significance, however, was Albert Einhorn's diethylaminoacetylguaiacol hydrochloride ('Guaiasanol'). This was the first drug to be rendered water-soluble by derivatization with an aliphatic amine function that permitted formation of a soluble salt. Einhorn subsequently applied this approach to his 'Orthoform' (see page 53), thus initiating a line of research leading ultimately to the development of procaine.

The first doubts concerning the effectiveness of internal antiseptics did not arise until bacteriological techniques had developed to the stage where it became possible to isolate and culture pathogenic organisms. It then became possible to assess the sensitivity of bacteria towards different drugs. The man who brought this about was Robert Koch.

In 1881, Koch solved the problem of how to grow bacteria in culture. His new techniques enabled the rational search for chemotherapeutic agents to begin, and he proceeded to test some seventy antiseptics against anthrax spores adhering to lengths of silk thread that had been dipped in liquid cultures. This showed that carbolic acid was by no means effective, the only agent that appeared to be sporicidal being mercuric chloride (also known as corrosive sublimate on account of its damaging effect on metals). He was, however, unsuccessful when he tried to use it as an internal antiseptic to protect guinea pigs subsequently inoculated with anthrax. Although the results of this experiment were disappointing, it served as the model on which the experimental basis of chemotherapy was later to be established. In the short term, however, it cast doubt on the value of internal antisepsis. This doubt was reinforced by experiments such as that carried out by Oscar Boer, who demonstrated that methyl violet had only one-hundredth the bactericidal activity against typhoid organisms that it had against anthrax. This suggested that even if non-toxic internal antiseptics were discovered, they would only be active against a limited range of bacteria. Today, this limitation is accepted with equanimity when certain antibiotics are prescribed, but the original concept of internal antisepsis rested on the belief that all pathogenic bacteria should be eliminated by a single agent. Thus in the early 1880s, many leading exponents of experimental medicine began to divert their efforts from apparently futile attempts to find new internal antiseptics. This coincided with the rise of immunotherapy, which promised to provide an effective therapy for all infectious diseases. This development came about serendipitously, when, in 1881, Louis Pasteur found that aging cultures of the causative organism of cholera had lost their ability to infect chickens. To his surprise, when these same chickens were subsequently injected with fresh, virulent cultures they were found to have developed complete immunity to cholera as a result of their prior exposure to the attenuated cultures. As Pasteur himself remarked later, chance favoured the prepared mind, with the outcome that the science of immunology was established by this unexpected turn of events.

MERCURIAL ANTISEPTICS

Koch's discovery of the apparent superiority of mercuric chloride as an antiseptic, a conception that was later to be challenged, seemed to justify the age-old belief in the value of mercury and its salts as 'alteratives' that could affect the course of diseases such as meningitis, bronchitis, pleurisy, pneumonia, dysentery, rheumatism, hydrocephalus, dropsy, digestive disorders, and much else. Indeed, it was not until the early 1950s that infants' teething powders containing mercurous chloride (e.g. 'Steedman's Powders') were withdrawn from the market because they caused pink disease, a wasting disease characterized by fever, insomnia, lethargy, loss of appetite, pain in the extremeties, and general misery for the infant. That this was not recognized earlier can be attributed to the belief that these symptoms were caused by teething, the very condition for which the powders were recommended!

In the sixteenth century, the turbulent Theophrastus Bombastus von Hohenheim, better known as Paracelsus, burnt the classic works of Galen during the inaugural lecture to mark his appointment to the chair of medicine at the Univeristy of Basle. Not all his acts were destructive, for he encouraged the use of inorganic compounds in medicine, and he recommended mercury as a specific remedy for syphilis. His views on this prevailed well into the twentieth century, interest being revived by Koch's demonstration of the apparent sporicidal activity of mercurous chloride. This resulted in mercury benzoate, carbolate, and salicylate being introduced in the late 1880s. The advantage of these organic mercurials lay in their freedom from the astringency associated with the inorganic salts. Their insolubility in water did not matter so long as they were formulated in pills or in ointments (an effective means of administering mercury as it was absorbed transdermally into the circulation). However, when mercury salicylate was to be given by subcutaneous injection in the treatment of syphilis, it had to be formulated in an oily suspension. This injection caused considerable pain, but around the turn of the century Schaerges introduced soluble sulphonic acid salts of phenylmercuric acid (e.g. 'Asterol'). The next two decades saw the introduction of phenylmercury hydroxide derivatives such as merbaphen ('Novasurol', see page 154). 'Afridol', mercurophen, and mercurochrome. These were followed in the late twenties by several mercurials that are still used as pharmaceutical preservatives, e.g. nitromersal, thiomersal, phenylmercuric acetate, and phenylmercuric nitrate.

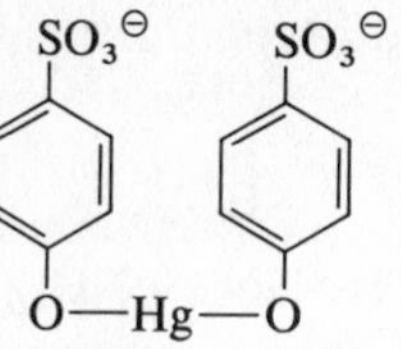

'Asterol' 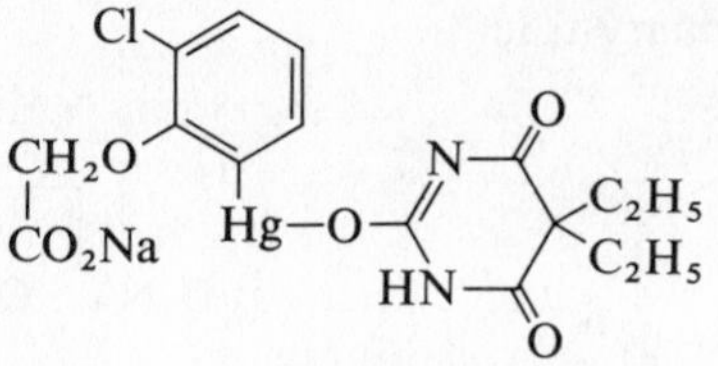 Merbaphen 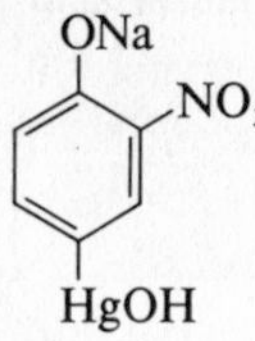 Mercurophen

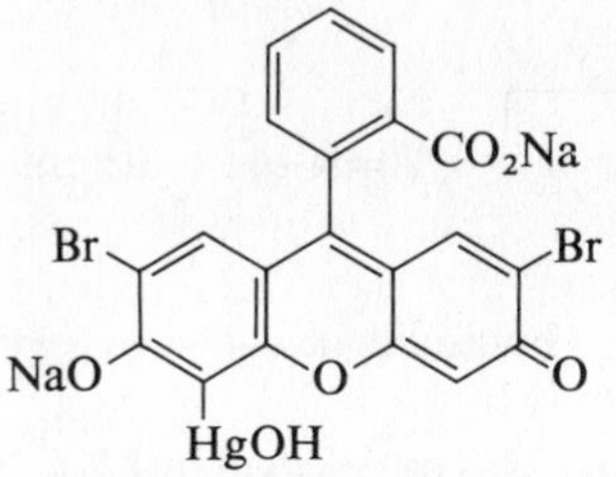

Mercurochrome 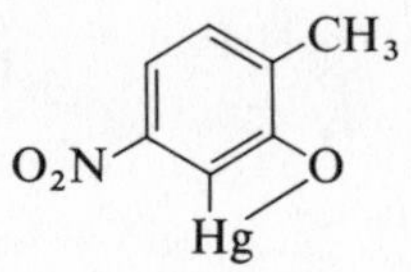 Nitromersal 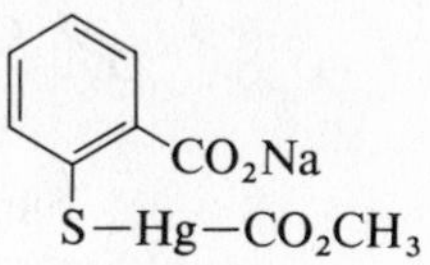 Thiomersal

URINARY ANTISEPTICS

Many investigators felt that internal antiseptics were ineffective for two principal reasons. Firstly, they were inactivated by proteins in the blood and tissues. Secondly, they could not be administered safely at the dose levels necessary to ensure attainment of therapeutic concentrations. This viewpoint, which is not inconsistent with current thinking, did not exclude the possibility of employing antiseptics to treat urinary infections, since the urine is normally free of protein. The only proviso here was that such drugs would have to be found amongst those antiseptics that were rapidly cleared from the circulation by the kidneys. This would ensure a bacteriostatic, or preferably a bactericidal, concentration in the urinary tract. Hexamine (methenamine, 'Urotropin') was the first drug to satisfy these criteria. The discovery of its value as a urinary antiseptic was serendipitous.

At the University of Breslau in 1894, Alfred Ladenburg discovered that piperazine formed a soluble salt with uric acid. He suggested this might dissolve the deposits of uric acid that caused much pain in patients with gout. Impressed by this suggestion, Arthur Nicolaier of the Medical Klinik at Gottingen tried the related hexamine in the hope that it might dissolve kidney stones. Soon, he noticed that urinary infections were being relieved by the hexamine treatment. Subsequently, it was shown that hexamine was only effective if the urine was sufficiently acidic to cause its decomposition, with the liberation of formaldehyde, the active antiseptic. In 1941, Kirwin and Bridges introduced the mandelate salt of hexamine following Rosenheim's suggestion that mandelic acid could be used as a urinary antiseptic. This is still prescribed, but it has the drawback that it merely acts as a bacteriostatic agent that stops the propagation of bacteria. This is undesirable since the immune system is unable to eliminate bacteria in the urinary tract.

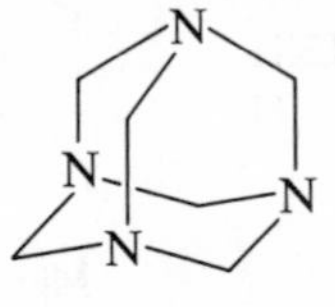

Hexamine

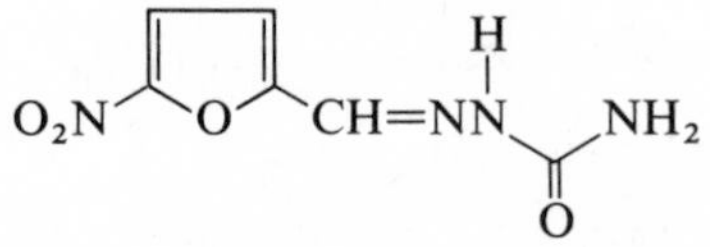

Nitrofurazone

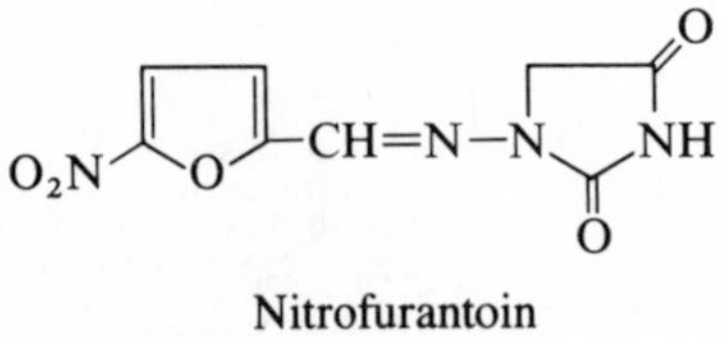

Nitrofurantoin

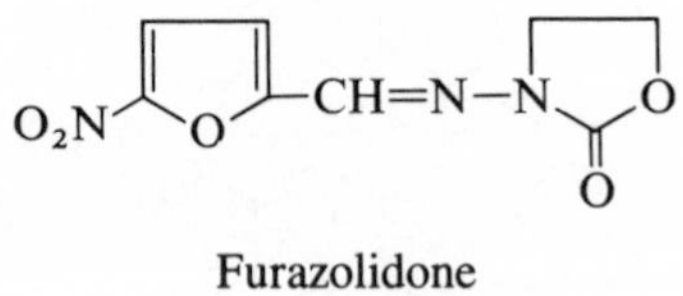

Furazolidone

Another class of urinary antiseptics, the nitrofurans, was developed by M. C. Dodds and W. B. Stillman of Eaton Laboratories (the ethical products division of the Norwich Pharmacal Company) of Norwich, New York. They followed up reports from Professor Chauncey Leake's laboratory in the University of California at San Francisco, which indicated that several furan derivatives were bactericides. These were synthesized from furfural, a product cheaply prepared by treating corn cobs with sulphuric acid. A systematic study of around forty compounds by the Eaton researchers revealed that the presence of a nitro group on the 5-position of the furan ring markedly enhanced antiseptic activity against a wide range of bacteria. A patent application was filed in 1943. Unfortunately, the efficacy of the nitrofurans was diminished in the presence of protein; this limited their clinical potential. Several years later, one of the twenty-five compounds that had proven active was introduced as a topical antiseptic with the approved name nitrofurazone ('Furacin'). Nitrofurantoin ('Furadantin') and furazolidone ('Furoxone') followed in 1952, the former being recommended as a urinary antiseptic, and the latter, which was poorly absorbed from the gut, as an antidiarrhoeal agent.

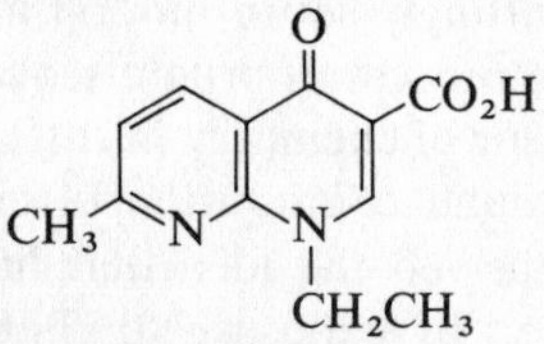

Nalidixic Acid

In 1946, Alexander Surrey and H. F. Hammer of the Sterling-Winthrop Research Institute devised a new synthesis of the antimalarial drug chloroquine. Some years later, Surrey and his colleagues discovered that a by-product (7-chloro-1,4-dihydro-1-ethyl-4-oxoquinoline-3-carboxylic acid) present in the mother liquors from this synthesis possessed antibacterial activity. On screening analogues of it, nalidixic acid ('Negram') emerged as a potent antibacterial agent. Since it was rapidly excreted unchanged in the urine, nalidixic acid was introduced as a urinary antiseptic in 1962.

V

Analgesic, Antipyretic, and Antirheumatic Drugs

In 1853, Henry How unwittingly set in motion a sequence of events which ultimately transformed the process of drug discovery. He was an assistant to Thomas Anderson, professor of chemistry in the University of Glasgow, who had determined the elemental composition of codeine, a mildly analgesic opium alkaloid. How conceived the idea that functional groups in natural products might be modified by exposure to chemical reagents. By heating morphine with methyl iodide in alcoholic solution, he hoped to convert the alkaloid into codeine. Instead, he obtained a novel substance, which he identified as the quaternary ammonium salt of morphine. He did not take matters further, but fifteen years later the professor of chemistry at the University of Edinburgh, Arthur Crum Brown, prepared this same compound, as well as quaternary ammonium salts of other alkaloids, for pharmacological examination by Thomas Fraser, professor of materia medica. This classic investigation was one of the first attempts to correlate chemical structure with biological activity. Fraser found that the quaternary ammonium salts of strychnine, brucine, thebaine, codeine, morphine, nicotine, atropine, and coniine had curare-like paralysing activity, despite the diversity of action of the parent alkaloids. This demonstrated that the quaternary ammonium function conferred curariform activity, thereby establishing the principle that certain clusters of atoms could confer specific types of biological activity upon molecules.

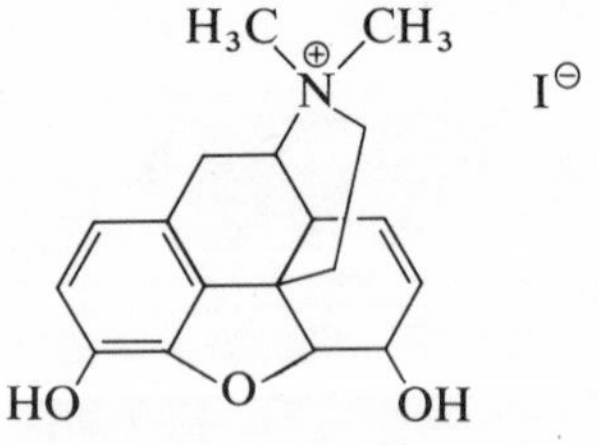

Quaternary Ammonium Salt
of Morphine

The work of Crum Brown and Fraser stimulated Alder Wright, the lecturer in chemistry at St Mary's Hospital Medical School in London, to treat morphine and codeine with organic acids, then submit the resulting esters for biological evaluation. Amongst these was diacetylmorphine, tested by F. M.

Pierce in 1874. Although it and several other compounds were active, Wright did not consider them to be worthy of clinical application. His work should not be overlooked, for it was one of the earliest attempts to prepare analogues of a natural product.

In 1887, Ralph Stockman, a colleague of Thomas Fraser, and David Dott, an Edinburgh pharmacist, established that the ethyl ether of morphine was pharmacologically almost identical to the methyl ether (codeine), whilst pure diacetylmorphine was more potent than morphine. He chose not to exploit these findings, doubtless having given the matter due consideration. His restraint was not followed by Josef von Mering a decade later. As a result of the successful introduction of the 'Eucaines' in 1896, as cocaine substitutes (see page 51), there was renewed interest in morphine analogues, especially in Germany. Having examined various ethers of morphine, von Mering persuaded the alkaloid manufacturer E. Merck of Darmstadt to market the ethyl ether. It was introduced as 'Dionin' in 1898, making it the first commercially available semisynthetic morphine derivative, promoted principally as a cough sedative for use in preference to codeine or other opiates. Nowadays, a similar ether, pholcodine, is preferred since it is slightly less toxic.

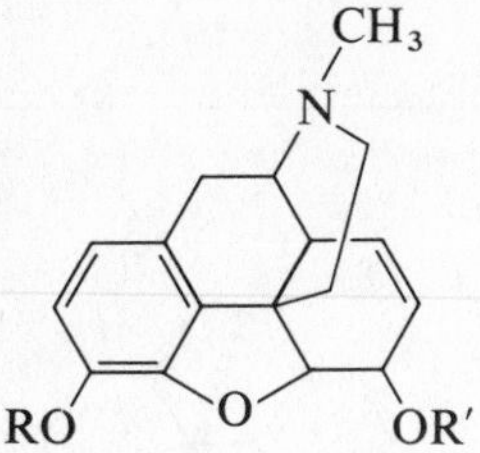

	R	R′
Morphine	$-H$	$-H$
Codeine	$-CH_3$	$-H$
Morphine Ethyl Ether	$-C_2H_5$	$-H$
Pholcodine	$-CH_2CH_2N \quad O$ 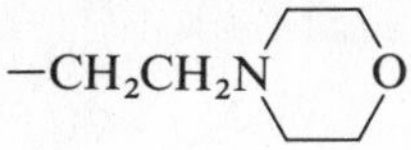	$-H$
Diacetylmorphine	$-COCH_3$	$-COCH_3$

Heinrich Dreser of Friedrich Bayer and Company was particularly concerned about the depressant effect of morphine on respiration. This was not infrequently a cause of death when morphine had been injected into those who had sustained severe injuries. Dreser devised new pharmacological methods to detect the safer morphine analogues he was seeking. In 1898, he enthusiastically introduced diacetylmorphine, which he described as a 'heroic drug' with the ability of morphine to relieve pain, yet safer. Bayer marketed this with the proprietary name of 'Heroin' (the approved name is diamorphine), and it quickly became popular throughout the world. Four years were to pass before the terrible truth began to emerge. Dreser had assumed that 'Heroin' was as non-addictive as the earlier morphine ether derivatives. It was, in reality, one of the most addictive drugs ever introduced into medicine. Amidst mounting anger and concern around the globe, governments were compelled to introduce laws controlling the supply of 'Heroin' and other dangerous drugs.

Until the correct structure of morphine was established in 1923 by John Gulland and Robert Robinson at the University of St Andrews in Scotland, its analogues had been prepared by modifying one or more of its functional groups. The failure of 'Heroin' to live up to its early promise discouraged most researchers from synthesizing morphine derivatives. Nevertheless, Martin Freund and Edmund Speyer at the University of Frankfurt, synthesized several hydrogenated derivatives of codeine in 1916, including one of the most potent codeine analogues, oxycodone ('Eucodal'). Four years later, they prepared hydrocodone ('Dicodid', dihydrocodeineone). These were originally made from the scarce opium alkaloid known as thebaine. An alternative synthetic route was devised by Knoll and Company, the biggest commercial producer of codeine in Germany at that time. Knoll also introduced hydromorphone ('Dilaudid') in 1923.

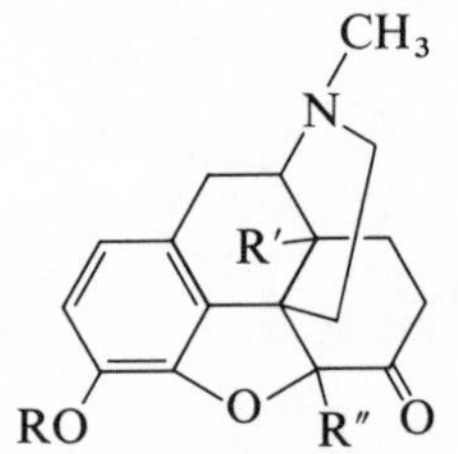

	R	R′	R″
Hydromorphone	—H	—H	—H
Hydrocodone	$—CH_3$	—H	—H
Oxycodone	$—CH_3$	—OH	—H
Metopon	—H	—H	$—CH_3$

In 1929, the Rockefeller Foundation, the Committee on Drug Addiction of the US National Research Council, the Public Health Department, and the Bureau of Narcotics, sponsored a research programme on non-addictive analgesics. For ten years they supported a team of chemists led by Lyndon Small at the University of Virginia, and one of pharmacologists led by Nathan Eddy at the University of Michigan. Promising agents were tried on volunteers at the Narcotic Prison Hospital in Lexington, Kentucky. Despite testing about

150 compounds, at a cost of millions of dollars, they failed to find a non-addictive agent. The most useful compound to come out of this research was metopon (methyldihydromorphinone), which was about three times as potent as morphine and less likely to cause drowsiness or nausea. Unfortunately, it proved difficult to produce cheaply for clinical use. These studies of Small and Eddy, however, established a firm basis for future work, with the influence of numerous molecular changes on activity and addiction liability being clearly established.

The correct chemical structure of morphine was first proposed by Robert Robinson and Gulland in 1923. Two years later Robinson suggested how morphine might be synthesized. In Switzerland, Rudolph Grewe of Hoffman-La Roche acted on this, and just before the war he succeeded in preparing a compound exhibiting the four rings in the morphine skeleton, known as *N*-methylmorphinan. This had only about one-fifth the analgesic potency of morphine. To increase its resemblance to morphine, the 3-hydroxyderivative was synthesized. This turned out to be more potent, more toxic, and less sedating than morphine, although still addictive. Nevertheless, it had one important advantage. When taken by mouth, it was longer acting than morphine and more reliable, principally because of a reduced tendency to be metabolically deactivated in the liver. Furthermore, the spatial relationships of the atoms surrounding one of its carbon atoms made it possible for chemists to separate two different stereoisomers of the new drug. The isomer which rotated plane polarized light leftwards, the laevorotatory isomer, bore a closer resemblance to the naturally occurring laevorotatory isomer of morphine, and was four times more potent as an analgesic. It was given the approved name of levorphanol ('Dromoran'). Its methyl ether was prepared, by analogy with codeine, and it too was resolved into its stereoisomers. The dextrorotatory isomer was devoid of analgesic activity, yet retained the ability of codeine to suppress coughing. It is still widely used in antitussive preparations under the approved name of dextromethorphan.

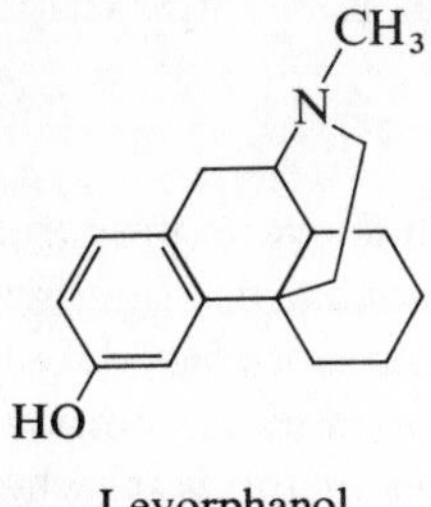

CH_3

N

HO

Levorphanol

A vast range of analogues of levorphanol have been synthesized. Changes such as substitutions on the nitrogen atom, or introduction of a hydroxyl group in the 14-position (cf. oxycodone), which were known to enhance the potency of morphine or its derivatives, were generally found to do likewise in the morphinans. Some of these analogues will be discussed later in this chapter.

During the nineteen-thirties, Swiss and German manufacturers examined large numbers of synthetic atropine analogues as potential antispasmodic drugs for use in the treatment of bladder or intestinal spasm. In 1937, at the Hoechst laboratories of I. G. Farbenindustrie (formed in 1926 by a merger of eight leading German chemical manufacturers, including Hoechst and Bayer) O. Schaumann studied the pharmacology of an antispasmodic synthesized in the company's laboratories seven years earlier by O. Eisleb. To his complete surprise, it caused a cat's tail to flip backwards in a rigid S-shape. Schaumann recognized this as the Straub reaction, characteristic of narcotic analgesics. Further tests confirmed it to have about one-tenth of the potency of morphine, with rapid onset of analgesia and a shorter duration of action. In the mistaken belief that since it bore no chemical resemblance to morphine it would be devoid of addictive properties, it was marketed in Germany as 'Dolantin' for the relief of moderate pain. Its British approved name is pethidine, but it is known as meperidine in the United States (where the leading proprietary brand is 'Demerol'). Eventually, it was found to induce tolerance and habituation just like morphine. However, it has been of great value in relieving pain during childbirth, as it does not depress the respiration of the baby in anything like the manner of more potent analgesics.

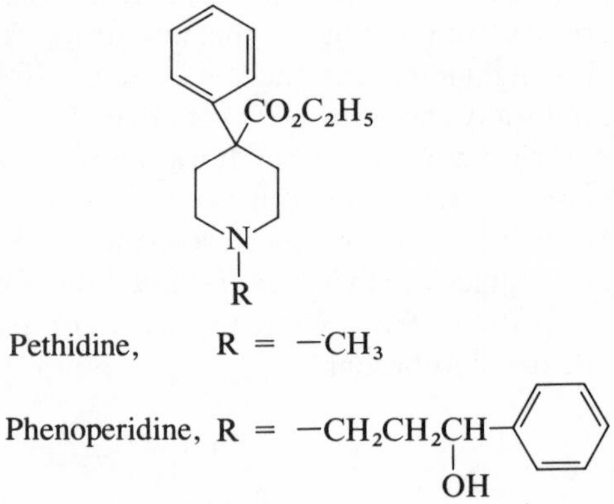

$CO_2C_2H_5$

N

R

Pethidine, R = $-CH_3$

Phenoperidine, R = $-CH_2CH_2CH(OH)-C_6H_5$

Many analogues of pethidine were prepared during the war by I. G. Farbenindustrie, but none showed any significant improvement. However, another series of antispasmodics synthesized by Eisleb was also found to have analgesic activity, the most outstanding compound being introduced in Germany as 'Amidon'. At the end of the war, when the US State Department investigated the wartime activities of the German chemical industry, reports about it were examined. Information was passed to the Committee on Drug Addiction and Narcotics of the National Research Council, resulting in details being published by Nathan Eddy in 1947. 'Amidon' (approved name methadone) was found to be an alternative to morphine in most types of pain, causing less sedation or respiratory depression. It was still addictive. Slow metabolic deactivation made it longer acting, a mixed blessing since it is difficult to avoid cumulative toxicity.

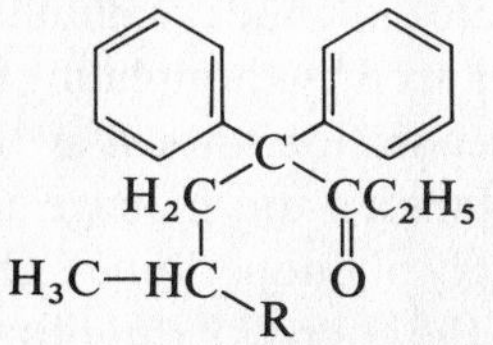

H_2C C CC_2H_5 H_3C-HC O R

Methadone, R = $-N(CH_3)_2$

Dipipanone, R = $-N$

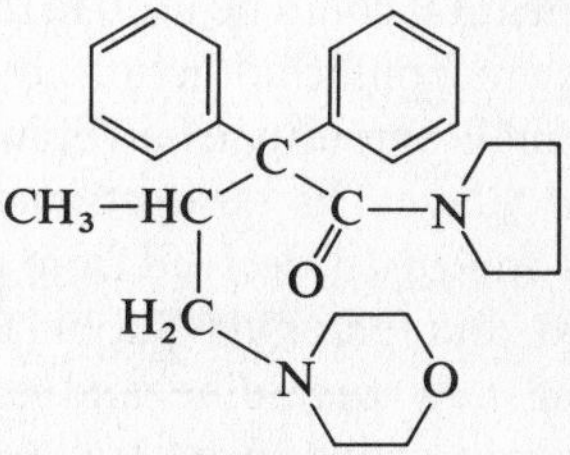

C CH_3-HC C—N O H_2C N O

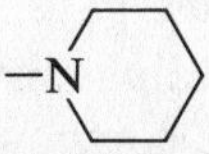

Dextromoramide

In 1950, the Wellcome Laboratories in the UK introduced an analogue of methadone known as dipipanone ('Diconal', 'Dimorlin'). This was both chemically and clinically similar to methadone, the dimethylamine group having been replaced by a piperidine ring. Further molecular variations on the methadone theme were introduced by the Belgian chemist Paul Janssen, with dextromoramide ('Palfium') in 1954, and piritramide ('Dipidolor') in 1961. Encouraged by his early success, Janssen proceeded to synthesize pethidine variants. One of the early ones combined the pethidine structure with that of his company's recently developed antispasmodic drug known as isopropamide. The resulting compound, diphenoxylate, lacked analgesic activity. It was

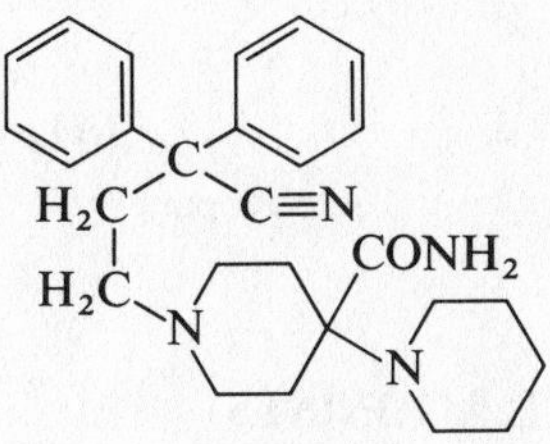

C H_2C C≡N $CONH_2$ H_2C N N

Piritramide

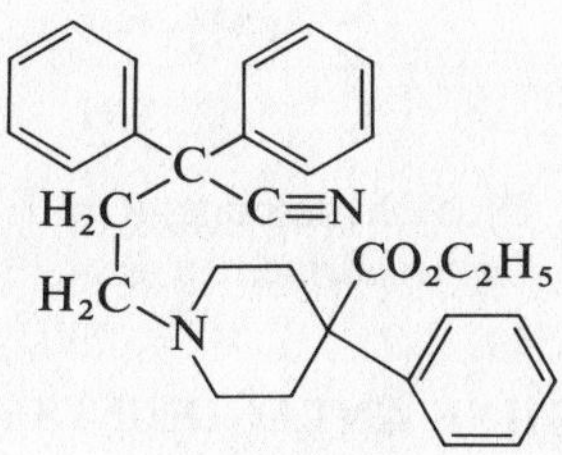

C H_2C C≡N $CO_2C_2H_5$ H_2C N

Diphenoxylate

CH_2 CH_2 COC_2H_5 N N

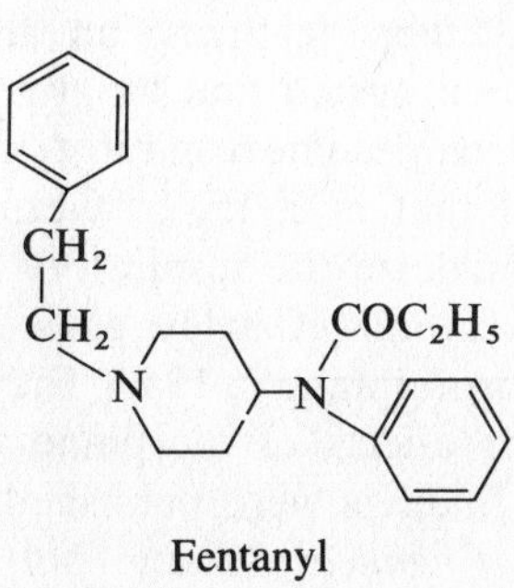

Fentanyl

found that this was due to rapid metabolism in the liver after the drug was absorbed from the gut. Making a virtue out of necessity, Janssen exploited the fact that diphenoxylate retained a morphine-like constipating effect on the gut. This meant it could be used in the control of acute diarrhoea, without the risk of causing morphine addiction. Its effect was enhanced by formulating it together with small amounts of atropine (as in the preparation known as 'Lomotil'), which delayed its absorption from the gut. Janssen did manage to prepare useful analgesics derived from pethidine, namely phenoperidine ('Operidine') in 1960, and fentanyl ('Sublimaze') in 1964. In the course of this programme of seeking new pethidine and methadone analogues, Janssen serendipitously discovered pronounced tranquillizing activity in one of his new compounds, leading to the development of haloperidol (see page 179) in 1958.

Orphenadrine ('Disipal') is a derivative of diphenhydramine ('Benadryl'), the widely used antihistamine. Both of these compounds have atropine-like anticholinergic effects, and this accounts for the value of orphenadrine in the treatment of Parkinson's disease. In an attempt to develop an analogue with fewer side-effects, the Riker Company synthesized nefopam ('Acupan') in 1966. At first, this was considered to be a centrally acting muscle relaxant, but eventually it was shown to be an analgesic. Its therapeutic value has not yet been fully established.

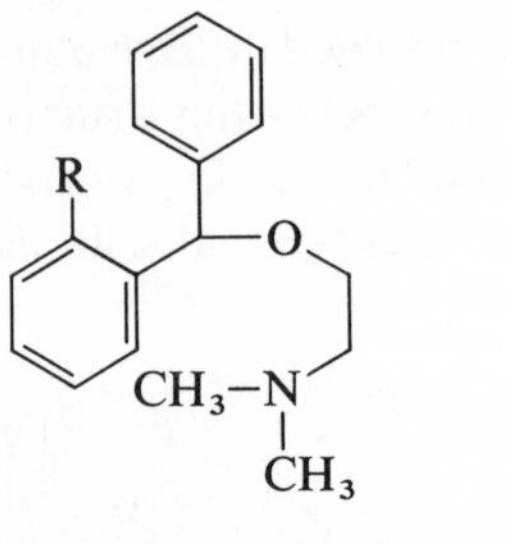

Diphenhydramine, R = —H
Orphenadrine, R = —CH_3

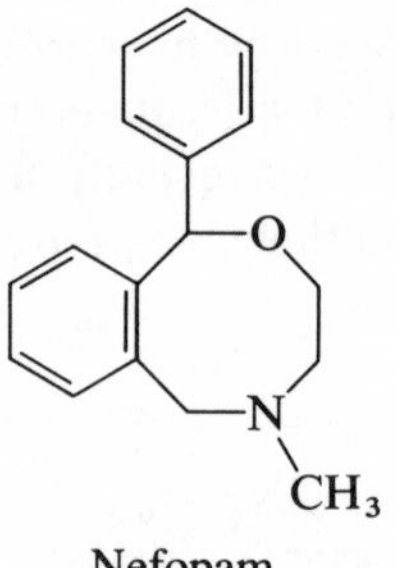

Nefopam

MORPHINE ANTAGONISTS AND PARTIAL AGONISTS

When Chauncey Leake was developing vinyl ether as a volatile anaesthetic (see page 21), he found that ethers featuring an allyl group had respiratory stimulant properties. This suggested that an allyl function in the morphine molecule might reduce its troublesome respiratory depressant effects. A search of the literature revealed that in 1915, in Germany, Pohl had found *N*-allylnorcodeine could antagonize the respiratory depressant action of morphine. In the light of this, Elton McCawley, assisted by Ross Hart and David Marsh, prepared *N*-allylnormorphine in 1940. They found that it antagonized the respiratory depressant effects of morphine more effectively than did *N*-allylnorcodeine. Their findings were published the following year in the *Journal of the American Chemical Society*, but a few months later John

Weijlard and A. E. Erickson of the Merck Laboratories reported to the journal that despite many attempts, they had been unable to repeat the published synthesis. Instead, they altered the method, obtained the desired compound, and proved it to have the correct structure. Its chemical properties were different to those of the earlier product, but it was certainly a potent antagonist of the respiratory depressant action of morphine. The parallel with the synthesis of barbitone by von Mering is obvious!

Detailed pharmacological and clinical studies resulted in the recognition of the new drug, now named nalorphine ('Nalline'), as a life-saving antidote in cases of morphine overdosage of accident victims. However, it was not until 1954 that Louis Lasagna and Henry Beecher of Harvard Medical School discovered nalorphine was a non-addictive analgesic in man, although it was ineffective in the standard rat tail-flick test. It turned out to be unsuitable for clinical use as it produced hallucinations. Its activity was eventually explained in terms of drug action at the brain receptors where morphine was considered to interact to produce analgesia. Firstly, the antagonism shown by nalorphine was seen to be due to its displacement of morphine from the brain receptors, with nalorphine preferentially binding to them. Secondly, the analgesia produced by the nalorphine was thought to be due to its ability to bind also to a second type of brain receptor, hitherto unrecognized, thereby triggering an analgesic response. This being done without producing respiratory depression was seen to be an important development. It meant, in principle, that it should be possible to find a related compound that could fit the nalorphine-accepting receptor to produce analgesia, without exhibiting the nalorphine tendency to cause hallucinations. These ideas were further refined in the seventies when a third group of receptors were considered to be responsible for nalorphine-like hallucinations. By means of sophisticated pharmacological experiments, it became possible to unravel which receptors different morphine analogues interacted with. Nalorphine acted at the so-called *kappa* analgesic-sedative receptors as well as at the *sigma* psychotomimetic receptors to produce a positive response, generally described as an agonistic response. At the *mu* receptor where morphine produces its analgesic, euphoriant, and addictive effects, nalorphine acted as an antagonist.

Ten years passed before Sydney Archer and his colleagues at the Sterling-Winthrop Research Institute in Rensselaer, New York, announced they had successfully exploited Lasagna and Beecher's discovery that nalorphine was a non-addictive analgesic. They reported the synthesis of pentazocine ('Fortral') in 1964. Despite its early promise, pentazocine proved to be only slightly more potent than codeine when used clinically at tolerable dose levels. When the higher doses required for potent analgesia were administered, it induced hallucinations. This indicated that it still had some activity at the *sigma* receptor. Nevertheless, pentazocine was less likely than codeine to cause addition. Understandably, it was seen to be an important step towards finding a potent, non-addictive analgesic.

In 1968, I. J. Pachter of Endo Laboratories in New York patented nal-

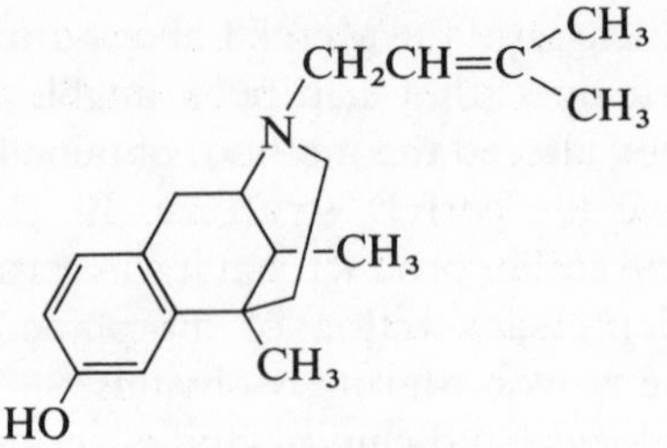

Pentazocine

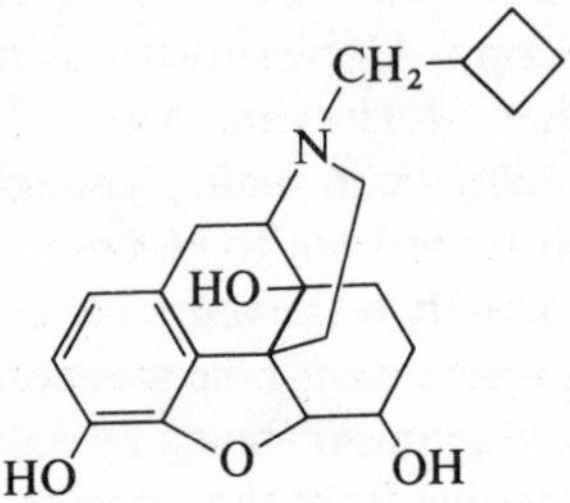

Nalbuphine

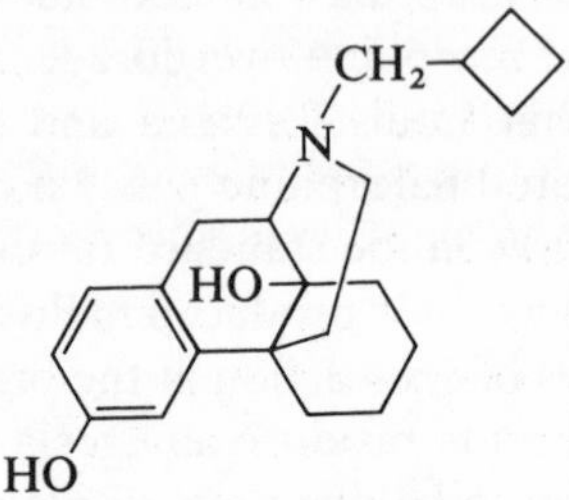

Butorphanol

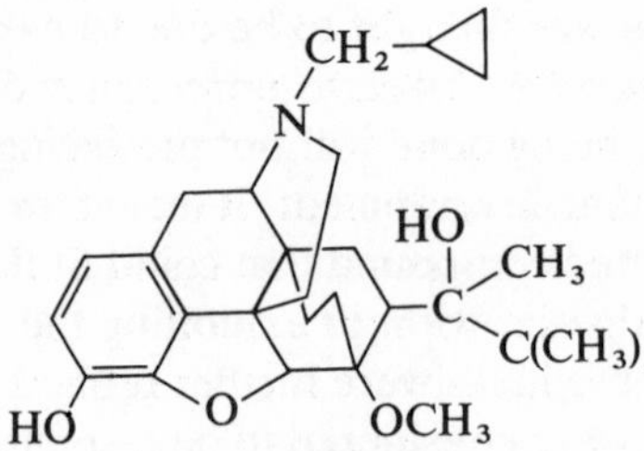

Buprenorphine

buphine ('Nubain'), an analgesic as potent as morphine, yet free of psychotomimetic activity. It had a low addiction liability, but was not orally active. Five years later, Pachter, Monkovic, and Thomas introduced butorphanol ('Stadol') for the Bristol-Meyers Company. This also had to be injected. The lack of oral activity was side-stepped with buprenorphine ('Temgesic') which could be taken sublingually to avoid deactivation in the liver. This was patented in 1968 for Reckitt and Coleman of Hull, in England, by K. W. Bentley. It had the additional advantages of a longer duration of action than morphine, coupled with a low dependence potential and diminished respiratory depression. Significantly, this moderate degree of respiratory depression reached a plateau and did not intensify when the dose was increased to afford stronger analgesia. Yet the drug still did not satisfy all the requirements for the ideal analgesic; like other new agents, it had side-effects such as drowsiness, nausea, or dizziness. In the absence of the undesired euphoriant-addictive properties of morphine, these side-effects were, paradoxically, more distressing to the patient despite being recognizably less severe than with morphine.

ANTIPYRETICS

The iatrophysicist Sanctorius Sanctorius of Padua (1561–1636) invented the clinical thermometer, but it was not adopted by the medical profession until a Scot, James Currie, introduced it at the end of the eighteenth century to monitor the progress of his typhoid patients whose fevers he reduced by immersion in cold baths. As an alternative to this unpleasant procedure, other physicians preferred certain vegetable products. These were described as 'antiperiodics', i.e. drugs that relieved malarial paroxysms, the commonest febrile condition at that time. Until the 1880s, quinine was the most frequently prescribed drug for lowering temperature in any febrile condition. It was the active constituent in 'Warburg's Tincture', a highly popular remedy which, during the mid-nineteenth century, made a fortune from the sales it amassed throughout the world for its inventor, Carl Warburg, a Viennese physician. Dr W. B. MacLean, Inspector General of the British Army, disclosed its hitherto secret formula in a paper in the *Lancet* praising its efficacy. He revealed that Warburg had told him his tincture contained aloes, rhubarb, angelica, elecamphane, fennel, saffron, prepared chalk, gentian, zedoary, cubeb, myrrh, camphor, agaric, and quinine. The result of MacLean's well-intentioned disclosure was financial ruin for Warburg.

The major obstacles to progress in clinical thermometry were the inconvenient length of the instrument and its cost; the thermometer used in the mid-nineteenth century was over a foot long, taking around twenty minutes to register the temperature. However, two developments in the 1860s made the thermometer cheap and convenient to use. First, German manufacturers discovered how to insert a printed scale between two glass tubes, drastically cutting production costs. Secondly, Sir Thomas Clifford Allbutt, a leading British exponent of clinical medicine, introduced the short clinical thermometer in 1867. It is, however, to Carl Wunderlich, professor of medicine at Leipzig, that credit must go for establishing the thermometer as an indispensible diagnostic instrument. Following the publication of his book on the subject in 1866, there was an initial over-reaction to clinical thermometry. Many physicians thought that at last they had found a new scientific method of treating disease—antipyresis! Traditional methods of reducing fever were reintroduced, either cold water treatments, or drugs such as quinine. Even though the germ theory of disease was generally accepted by physicians in the last two decades of the nineteenth century, and it became evident that fever was a symptom rather than a disease, the fascination with antipyresis continued well into the twentieth century.

At St Gallen in Switzerland, Carl Buss administered salicylic acid to typhoid patients as an internal disinfectant. The course of their disease was routinely checked by thermometry, fever charts being carefully compiled by the nursing staff. It became obvious that although the salicylic acid was an effective antipyretic, it did not lower body temperature by curing the typhoid infection. Widespread interest was aroused when Buss published his findings in 1875.

Doctors discovered that repeated doses could be administered to control fevers without causing the side-effects so frequently observed if quinine were used. For nearly ten years, salicylic acid remained a popular antipyretic agent despite its unpalatability.

An attempt to improve upon salicylic acid or phenol as internal disinfectants was made in 1883 by Professor Marcellus von Nencki in Switzerland. He reacted these two drugs together to form the ester which he called salol. This was so insoluble that after being swallowed it passed unchanged through the stomach. In the small intestine, however, it was more soluble, and the portion which dissolved subsequently decomposed to release small amounts of phenol and salicylic acid. For many years it was naively believed that some benefit was to be derived from these small amounts. The advantages over salicylic acid, namely that it was not unpleasant to take or irritating to the stomach, were offset by the delay in onset of activity and the variability of therapeutic response. Nevertheless, salol was a popular substitute for salicylic acid as an antipyretic and antirheumatic (see later) until the introduction of aspirin. It represented an early example of the gullibility of some members of the medical profession when presented with a drug exhibiting chemical novelty unsupported by any clinical validation of efficacy. Scientists, too, were taken in by the claims made for salol, even believing it to be the first example of a drug designed to release controlled amounts of an active agent. They spoke of 'the salol principle', ignoring the prior introduction of sulphocarbolic acid, chloral hydrate, salicylic acid, and iodoform. Perhaps the best testimony to the worthlessness of salol was the amount of effort put in by chemists to synthesize more effective analogues.

The large doses of salicylic acid which were being recommended in the 1890s for the control of rheumatism (see page 88) tasted unpleasant, and frequently caused the patient to vomit. One such patient asked his son, Felix Hoffman, to seek out a more acceptable drug at the Bayer laboratories in Elberfield, where he worked as a chemist. Hoffman searched through the literature on salicylic acid derivatives and came across acetylsalicylic acid, which had been synthesized in 1853 by Charles Frederick von Gerhardt at Strassburg. It is not clear whether Hoffman then gave some to his father or whether he reported straight to Bayer's chief chemist, Arthur Eichengrun, but the outcome was that Heinrich Dreser put the compound through a series of searching tests on animals. These gave most encouraging results. Samples of acetylsalicylic acid were sent, in 1898, to Kurt Witthauer at the Deaconess Hospital in Halle, and to Julius Wohlgemut in Berlin, for clinical trial. Their results were published the following year, revealing that acetylsalicylic acid was as effective as salicylic acid in rheumatism, whilst free from its unpleasantness. Witthauer enthused over the new drug and persuaded the Bayer Company to market it despite the fact that it was difficult to manufacture and could not be protected by patents as it was not a new compound. The company overcame this problem by patenting the processes used in its large–scale manufacture. It also sought protection in as many countries as possible for the proprietary name bestowed on the product.

This was 'Aspirin', a term derived from 'a' for acetyl, and 'spirin' from the official name of the plant from which salicylic acid was first obtained, viz. *Spirea ulmaria*. To ensure the success of the new drug, the company circularized more than 30 000 doctors in what was probably the first mass mailing of product information.

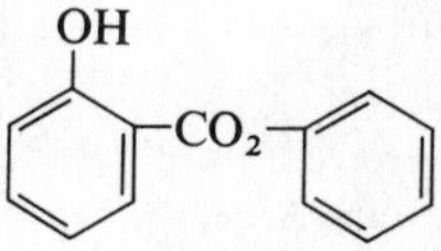

Salol

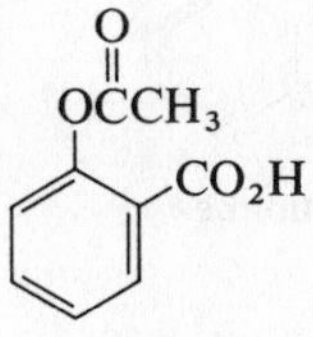

Aspirin

Since the beginning of this century, acetylsalicylic acid has been the most widely used drug in the world. When the First World War broke out the British were no longer able to obtain supplies from Germany. The government then offered a £20 000 reward to anyone who could develop a workable manufacturing process, albeit infringing the still valid Bayer patents. American firms were unable to do this as their country was not at war with Germany, whilst British manufacturers were involved too deeply in the war effort. It was left to a Melbourne pharmacist, George Nicholas, to come up with the prize-winning solution in his 'Aspro' tablets which, after the war, rivalled Bayer's 'Aspirin' throughout the world. To add insult to injury, the Bayer Company also lost the exclusive right to the use of the name 'Aspirin' in Britain and the Commonwealth after the British Custodian of Enemy Property sequestered it. Many years later, this actually became the approved name in the *British Pharmacopoeia*.

The astonishing popularity of aspirin brought to the fore its disadvantages. Dr Witthauer had originally warned that the tablets should be allowed to disintegrate in sugar-water, flavoured with lemon juice, before swallowing. Yet, many people swallowed the tablets whole, with the result that lumps of aspirin lodging against the stomach wall sometimes caused ulceration. To avoid this problem the use of the soluble calcium aspirin was recommended in Germany in 1913, but this presented manufacturing problems. It was not until British gastroenterologists used the newly developed gastroscope in 1938 to demonstrate the marked difference in damage to the stomach wall between ordinary aspirin and calcium aspirin that any serious attention was given to this problem. Today, however, the superiority of dispersible and effervescent formulations is widely recognized.

The salicylates were not the only antipyretics introduced around this time. Organic chemists had begun to unravel the complex structure of quinine, and it was out of this effort that the first substitutes came. The British investigators James Dewar and John McKendrick reported to the Royal Society in 1875 that certain bases obtained from decomposition of quinine retained some of the biological effects of quinine. Later, McKendrick suggested that these bases, which included quinoline, might find a place in the materia medica. A Hungarian, Julius Donath, subsequently introduced quinoline as an antipyretic, but others found its effects to be too feeble for practical application.

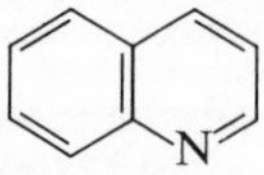

Quinoline 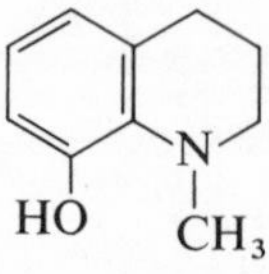 'Kairin' 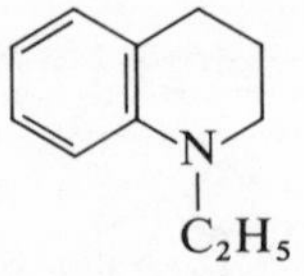 'Kairoline A' 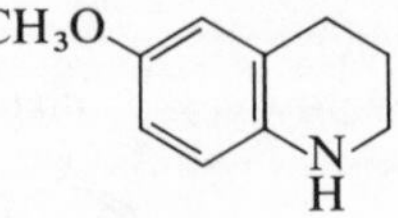 Thalline

It was not until Wilhelm Koenigs, at the University of Munich, had published his mistaken proposal that quinine was a derivative of tetrahydroquinoline, that any advance was made towards the development of an effective quinine substitute. Koenigs and Otto Fischer proceeded to synthesize a variety of tetrahydroquinoline compounds, which they sent in 1881 to Wilhelm Filehne, professor of pharmacology at the University of Erlangen. One of them proved to be an antipyretic. It was put on the market the following year by Hoechst Dyeworks under the name of 'Kairin'. When reports that it was toxic appeared, it was quickly replaced by the structurally related 'Kairoline A'. The most successful of the tetrahydroquinoline antipyretics was Professor Skraup's 'Thalline', introduced around 1885, but soon eclipsed by other developments.

In the spring of 1875, whilst working in Adolph von Baeyer's laboratory in Strassburg, twenty-three-year-old Emil Fischer unintentionally synthesized a novel chemical known as phenylhydrazine. After moving to the University of Munich with von Baeyer later that year, Fischer initiated his celebrated researches into the reactions of phenylhydrazine, which permitted the facile synthesis of a variety of cylic compounds incorporating nitrogen atoms in their ring structure. Fischer continued his pioneering studies in heterocyclic chemistry after transferring to the University of Erlangen. One of his assistants, Ludwig Knorr, studied the reaction between phenylhydrazine and acetoacetic ester, and he mistakenly concluded that the product of this reaction was a tetrahydroquinoline. Knorr told Professor Filehne who was currently testing new tetrahydroquinolines as antipyretics following the reports of the toxicity of 'Kairin'. Filehne found Knorr's compound to be devoid of antipyretic activity. Nevertheless, he advised Knorr to modify its structure by attaching a methyl or ethyl group to the ring nitrogen. This had proved beneficial in the series of compounds related to 'Kairin'. When Knorr did so, he found that the N-methylated derivative no longer possessed the acidic properties of the parent compound. This change in properties would not have occurred had the compound possessed the structure ascribed to it. With hindsight, one can now see that the unplanned repression of the acidity was responsible for the derivative being a potent antipyretic since the lipophilicity was thereby increased. The new compound turned out to be far more palatable than either quinine or salicylic acid. Knorr granted the patent rights to Hoechst Dyeworks, who marketed the compound under the proprietary name chosen by Knorr whilst on his honeymoon in Venice, viz. 'Antipyrin' (the approved name is phenazone). By the time it reached the market, Knorr had discovered that it was actually a pyrazolone derivative.

Knorr synthesized phenazone in 1884. Five years later it was put to the test when a major influenza epidemic swept through Europe. For the next fifteen to twenty years 'Antipyrin' was the most widely used drug in the world, until aspirin began to outsell it. At the turn of the century the Hoechst Dyeworks were producing upwards of 17 000 kilos annually. If any single product can be said to have established the commercial viability of synthetic drugs, it was surely phenazone. Today, however, it is rarely used because of a risk of a serious blood condition known as agranulocytosis.

The popularity of phenazone inevitably led to someone discovering its value in relieving headache. One early report of this appeared in the *British Medical Journal* in 1887. All antipyretics developed since then have routinely been considered for use as mild analgesics.

Professor Filehne was not content to rest on his laurels. He believed that a superior antipyretic to phenazone could be developed and he made suggestions to this end to Friedrich Stolz, the chief chemist at the Hoechst Dyeworks. It happened that one of the suggested compounds had already been synthesized by Knorr, so it was tested in the Institute of Physiological Chemisty at Strassburg University by Filehne and Karl Spiro in 1896. They found it to be three times as potent an antipyretic as phenazone. It was marketed by Hoechst Dyeworks under the name 'Pyramidon' (the approved name is amidopyrine). Whether it possessed any significant advantage over the less potent phenazone is a moot point, but it became one of the best selling drugs in Europe until, in 1934, it was incriminated in several cases of agranulocytosis.

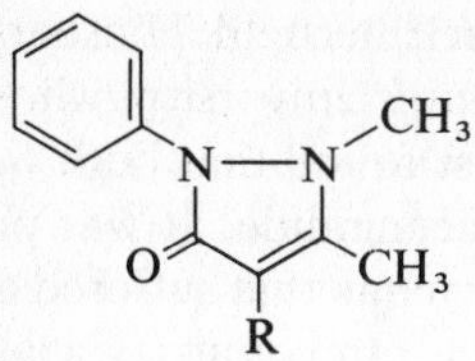

Phenazone, R = —H
Amidopyrine, R = $-N(CH_3)_2$

Strassburg University was also the scene of the discovery of yet another antipyretic in the 1880s. The department of internal medicine was noted for its investigations into intestinal worms. Professor Kussmaul, the director, asked two young assistants, A. Cahn and P. Hepp, to treat patients with naphthalene, which had been used by Professor Rossbach of Jena as an internal antiseptic. The young doctors were disappointed with the initial results, but Hepp persevered with the naphthalene treatment in a patient suffering from a variety of complaints besides worms. Surprisingly, the fever chart revealed a pronounced antipyretic effect from the treatment. This had not been observed before, but further investigation revealed that Hepp had wrongly been supplied by Kopp's pharmacy in Strassburg with acetanilide instead of naphthalene! Cahn and Hepp lost no time in publishing a report on their discovery of a new antipyretic. This appeared in August 1886, and a small factory, Kalle and Company, situated outside Frankfurt and not far from the Hoechst Dyeworks, set up in competition with 'Antipyrin', mischievously calling their product 'Antifebrin'. In 1908, however, the Hoechst Dyeworks obtained control of Kalle and Company, which had grown considerably in size largely due to the success of their antipyretic. Acetanilide was cheaper to manufacture than other antipyretics, so it remained in use for many years, despite the fact that it inactivated some of the haemoglobin in red blood cells, a medical condition known as methaemoglobinaemia. Sometimes acetanilide was used illicitly as a cheap adulterant of other antipyretics.

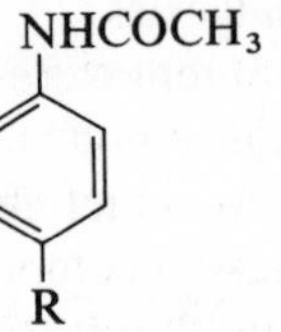

$NHCOCH_3$

R

Acetanilide, $R = -H$
Phenacetin, $R = -OC_2H_5$
Paracetamol, $R = -OH$

Immediately after the publication of the report of the antipyretic activity of acetanilide, the 4-methoxy and 4-ethoxy derivatives of it were prepared by 0. Hinsberg, an assistant to Carl Duisberg, chief research chemist at the dyeworks of F. Bayer and Company in Elberfield. Hinsberg took the two new compounds to Professor Kast at Freiburg University, whose 'Sulphonal' was then being investigated by Bayer. Kast found that both were antipyretics, but the ethyl ether was less toxic than acetanilide. It was promptly put on the market as 'Phenacetin', a proprietary name that suffered a similar fate to 'Aspirin' at the end of the First World War. In countries where this name continued to be recognized as the property of the Bayer Company, the approved name was acetophenitidin. It was a highly successful product, establishing F. Bayer and Company as a leading pharmaceutical manufacturer. It remained popular for about ninety years until mounting concern about kidney damage in chronic users led to restrictions on its supply in some countries, including the United Kingdom.

Many attempts were made to find an antipyretic superior to phenacetin. Joseph von Mering collaborated with the Bayer Company in a clinical trial of paracetamol in 1893. He found it to be an effective antipyretic and analgesic, but wrongly believed that it had a slight tendency to produce methaemoglobinaemia. This could conceivably have been due to contamination of his paracetamol with 4-aminophenol, from which it was synthesized. Such was the reputation of von Mering that nobody challenged his observations on paracetamol until half a century later, when Lester and Greenberg at Yale University, and subsequently Flinn and Brodie at Columbia University, New York, confirmed that paracetamol was formed in humans as a metabolite of phenacetin, a claim originally made by Treufel and Hinsberg in 1894, albeit without adequate substantiation.

In 1953, paracetamol ('Panadol') was marketed by the Sterling-Winthrop Company. It was promoted as preferable to aspirin since it was safer in children and anyone with an ulcer. Time has shown that it is not without its disadvantages, for it is far more difficult to treat paracetamol poisoning than that caused by aspirin. Indeed, so worrying is the risk of insidious liver damage from overdosage, that consideration is now being given to the incorporation of an antidote in paracetamol tablets! It is probably fair to say that were either paracetamol or aspirin to be introduced into medicine today, they might be denied a license by responsible government agencies. Nevertheless, when used in moderation they remain a boon to mankind.

ANTIRHEUMATIC DRUGS

In November of 1874, Thomas MacLagan, a Dundee physician, first used the active principle from willow bark, salicin, to treat a patient afflicted with rheumatic fever. A year and a half later, on reporting to the *Lancet* that he had attained successful results in around one hundred patients, he expressed his conviction that there was a similarity between intermittent fever (malaria) and acute rheumatism. Having noted that cinchona bark was to be found in tropical areas where the ravages of malaria were most severe, he was led to believe that a remedy for acute rheumatism would be found in a low-lying damp locality, with a cold rather than a warm climate. Such were the conditions under which rhematic fever was most readily produced. MacLagan concluded that the plants whose haunts best corresponded to his requirements were those belonging to the natural order Salicaceae, the various forms of willow.

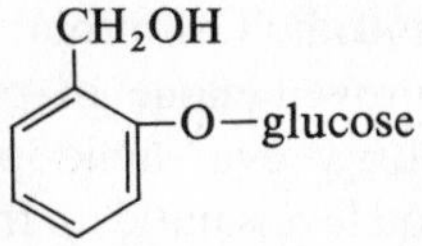

CH_2OH

O—glucose

Salicin

MacLagan had been influenced by the herbalists' belief that antidotes were to be found in the vicinity of poisons, a belief exemplified to this day by those who seek to relieve nettle stings by turning to a nearby dock leaf. Unscientific as the principle behind MacLagan's approach may have been, his patients undoubtedly benefitted. He was to learn shortly after that the eminent biochemist Professor Marcellus von Nencki of Basle had shown, in 1870, that salicin is converted in the body into salicylic acid. When the first reports of the antipyretic action of salicylic acid appeared, he tried this too, and again achieved success in easing the affliction of his patients, without actually effecting a cure. By the time his report was published in the *Lancet*, another one had already appeared in a German medical journal, extolling the virtues of salicyclic acid in rheumatic conditions. This was written by Franz Stricker of Berlin, who having read Carl Buss's report of the antipyretic activity of salicylic acid, immediately tried out the new drug on his patients with rheumatic fever. To his delight he found that not only did it act as an antipyretic, but it also had an unquestionable value as an antirheumatic agent. His results were published in January 1876, two months before MacLagan's first paper on salicin appeared in the *Lancet*. Others quickly confirmed the value of salicylic acid in rheumatic conditions, and Kolbe no longer had to promote salicylic acid as an internal antiseptic. Its value as an antipyretic and an antirheumatic established its place in therapeutics. Soon, a factory had to be built in Dresden to meet the demand from all over the world.

As new antipyretics were introduced in the 1880s they were examined for their antirheumatic properties. Phenazone was recommended for use in acute rheumatic fever as early as 1885, when its ability to ease joint pain was specifically commented upon. Within two or three years its analgesic properties were generally recognized and physicians then began to associate antipyretic with antirheumatic and analgesic activity. Acetanilide, phenazone, and amidopyrine were all used for many years in rheumatic conditions. However, after the Second World War there was mounting concern about the risk of the blood disorder known as agranulocytosis resulting from treatment with amidopyrine. One way to minimize the risk was to administer smaller amounts of the drug by injecting it directly into the circulation, but the insolubility of amidopyrine had first to be overcome. The Geigy company in Switzerland formulated the basic amidopyrine as a soluble complex with an acidic analogue of it, this formulation being known as 'Irgapyrine'. This was introduced clinically in 1949 and found to be very effective as an antiarthritic preparation. Subsequently, it was discovered that blood levels of the acidic analogue were much higher than those of the amidopyrine, an observation which led directly to clinical trials of this analogue in both Scotland and California. It was marketed in its own right in 1952, receiving the approved name of phenylbutazone ('Butazolidin'). Possibly its clinical superiority over amidopyrine was initially overlooked because only in man do blood levels build up through an unusually slow rate of clearance of the drug from the body. This would not have been apparent from animal experiments.

Being an analogue of amidopyrine, it was not altogether surprising that phenylbutazone was still found to carry the risk of causing agranulocytosis without any prior warning signs. As this can occur even within the first few days of treatment, it was clearly a drug that should be used only in patients who did not respond to other remedies. It could also induce another serious blood condition known as aplastic anaemia, besides sometimes producing fluid retention to such an extent that it precipitated cardiac failure in predisposed patients. The active metabolite formed in the body has been available since the late fifties for clinical use as oxyphenbutazone ('Tanderil'), but it offered no obvious advantage. In 1984, the UK Committee on Safety of Medicines advised the government that product licenses for oxyphenbutazone should be revoked, whilst phenylbutazone should only be available for the hospital treatment of ankylosing spondylitis. The CSM released figures which revealed that oxyphenbutazone had caused 131 deaths in Britain during the previous twenty years, and the more widely prescribed phenylbutazone had been responsible for 445 deaths in the same period. At the height of its popularity in the late sixties, general practitioners were issuing more than three and a half million prescriptions annually for various brands of phenylbutazone. Two related compounds, azapropazone (apazone, 'Rheumox'), and feprazone ('Methrazone') so far seem to be much safer, although the latter has been incriminated in a few cases of certain less serious blood disorders.

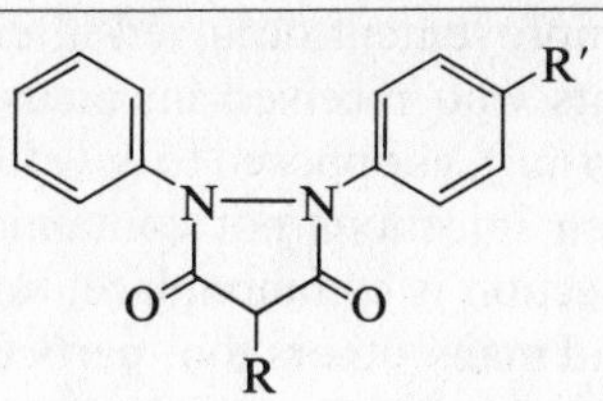

	R	R′
Phenylbutazone	$-CH_2CH_2CH_2CH_3$	$-H$
Oxyphenbutazone	$-CH_2CH_2CH_2CH_3$	$-OH$
Feprazone	$-CH_2CH{=}C(CH_3)_2$	$-H$

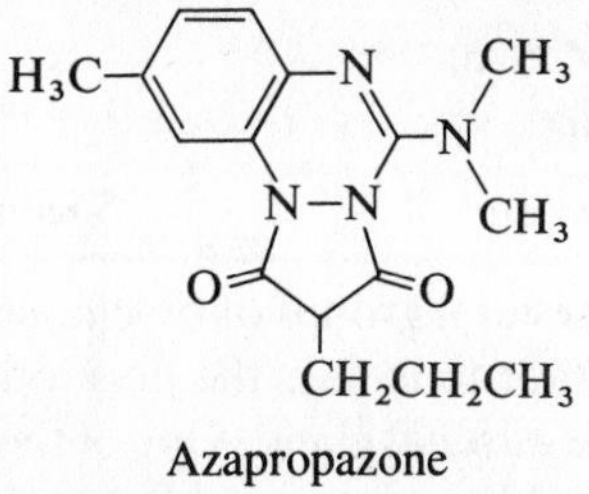

Azapropazone

Not all antirheumatic drugs were developed from antipyretics. In 1890 the great bacteriologist Robert Koch told an international congress in Berlin that gold–cyanide complexes were the most effective of all known antiseptics against the tuberculosis bacillus, at least when tested in the test tube at high dilution. Unfortunately, they were unable to cure infected animals. Apart from sporadic work reported from Germany, little further attention was paid to gold treatment until 1924, when a Danish veterinarian demonstrated that gold sodium thiosulphate ('Sanochrysine') has a beneficial effect in bovine tuberculosis. Soon, physicians were trying this and similar compounds in humans, albeit with little success. Believing (wrongly!) that gold had a general antiseptic effect in addition to any possible antitubercular activity, Lande in Germany administered aurothioglucose ('Solganal', a gold compound prepared by the Schering Corporation), to thirty-nine patients suffering from a variety of complaints attributed to bacterial infection, but in most cases actually due to rheumatic fever. The most notable outcome of this ill-defined clinical trial was the relief of joint pain. Lande concluded that a trial of the drug in arthritis would be worthwhile. Four years later J. Forestier in Paris began to use another gold compound, gold-thiopropanol sodium sulphonate ('Allochrysine') in rheumatoid arthritis. Gradually, the use of gold preparations in arthritic conditions began to spread, but it was not until 1944 that the results of a properly controlled clinical trial were first published by Fraser in Glasgow. He reported that clinical improvement occurred in eighty-two per cent of fifty-seven rheumatic patients who received intramuscular injections of sodium aurothiomalate ('Myocrisin'), as opposed to only forty-five per cent of forty-six patients who were given injections not containing the drug. This placebo response to a blank injection is commonplace, and makes the evaluation of antirheumatic drugs, and many others too, particularly difficult. In the Glasgow trial neither patients nor their attendants knew whether the injection contained the drug or not; this constitutes the so-called 'double-blind trial'. Sodium aurothiomalate had been investigated in 1939, but only after the Glasgow trial was its value in therapeutics generally accepted. Nowadays, the blood of the patient is monitored regularly, with treatment continuing indefinitely unless side-effects intervene, which they often do.

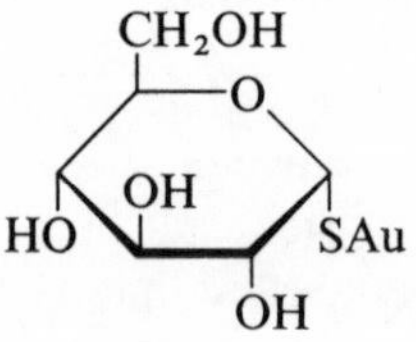

Aurothioglucose

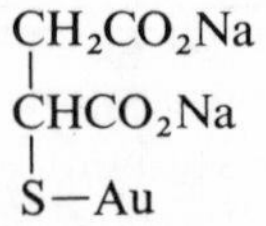

CH_2CO_2Na | $CHCO_2Na$ | $S-Au$

Sodium Aurothiomalate

During the late thirties several groups of biochemists in different parts of the world were involved in a race to isolate the hormones known to be present in extracts of the cortex of the adrenal gland. One of these groups was based at the Mayo Clinic in Rochester, Minnesota, and led by Edward Kendall, who had previously isolated the thyroid hormone, thyroxine. He managed to isolate five pure compounds, of which two were hitherto unknown, namely Compound A (11-dehydrocorticosterone) and Compound E. Kendall was enthusiastic about the potential of the new hormones in therapy. He believed that apart from being of value in the treatment of the adrenal cortex deficiency condition known as Addison's disease, the hormones might be of use in shock, burns, and traumatic injuries. When, early in 1941, he discussed the matter with a colleague from the Mayo Clinic, Philip Hensch, another potential application was drawn to his attention. Eleven or twelve years earlier Hensch had noticed that if patients with rheumatoid arthritis contracted jaundice then their painful symptoms were markedly diminished for a time. A similar state of affairs often occurred when women with arthritis became pregnant, the pains returning after childbirth. Believing these observations to be instances of stress inducing release of a protective hormone, Hensch sought the views of Kendall. Both men agreed that once adequate supplies of a hormone such as Compound E became available, it should be tried in rheumatoid arthritis.

In the summer of 1948 Hensch began treating a young woman who was desperately ill with arthritis. Nothing was able to help her. By September, her condition was giving rise to such concern that Hensch begged Kendall to supply some Compound E, which had become available in April that year thanks to its synthesis by Lewis Sarett of Merck and Company, Rahway, New Jersey. Kendall advised Hensch to approach Merck, who had already sent out three hundred samples of Compound E on clinical trial. When Sarett was contacted by Hensch he decided that as the clinical results had been disappointing, he would release the one gram which remained of the original five grams that had been synthesized. Hensch received the sample on the 21st September 1948 and at once injected some into his desperately ill patient. She showed no signs of improvement until after her third daily injection when the response was truly dramatic. She awoke the following morning to find herself totally free of pain on moving, something she had not experienced for over five years! After a further week of treatment the young woman was able to leave the hospital unaided and walk around Rochester. She had by then received one quarter of Hensch's supply of Compound E, so an urgent request for more was sent off to Merck. The company responded magnificently. The thirty-six-stage chemical synthesis was repeated on a hugely increased scale to produce one thousand grams for Hensch within only a few weeks. As a result he was able to proceed with a limited trial of the hormone in a further thirteen arthritic patients who received it for six months. Once again, remarkable results were achieved, but only as long as the injections continued. It became clear that the new treatment could not cure rheumatoid arthritis. It could only suppress the distressing symptoms. For patients who did not respond to any other therapy this in itself was, nevertheless, truly miraculous. Today, this type of treatment is reserved solely for such patients.

Hensch reported his results with Kendall's Compound E to a meeting of physicians held in April 1949. He received wide acclaim from his peers, while the press seized on the dramatic turn of events with much zeal. Some newspapers reported that the new drug was 'Vitamin E', thereby creating an undeserved reputation for this vitamin as a miracle drug! To avoid further confusion Compound E was renamed 'cortisone'. Within a short period there was a world-wide cortisone famine which the pharmaceutical industry was hard pressed to overcome. Merck scaled up Sarett's process still further so that by the end of the year they were able to offer limited amounts of cortisone acetate to physicians at a price of $200 per gram. A year later, however, they produced one thousand kilograms and reduced the price to $35 per gram.

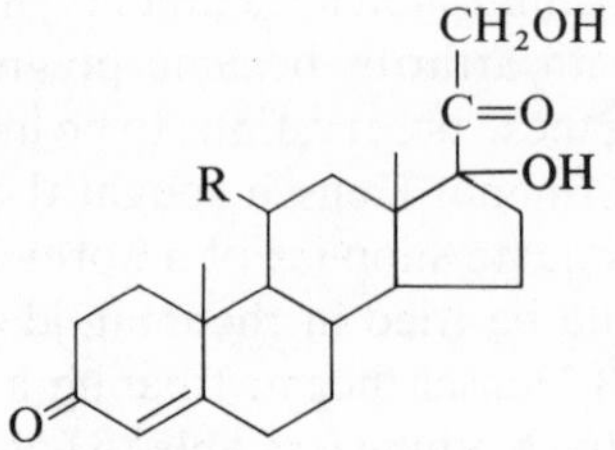

Cortisone, R = =O
Hydrocortisone, R = −OH

Before Merck were able to set up their production line to meet the sudden demand for cortisone, a rival company was able to step in. G. D. Searle and Company set up a new process which had been developed by Drs Gregory Pincus and Oscar Hechter at the Worcester Foundation for Experimental Biology. This consisted of a huge array of containers, each holding cattle adrenal glands immersed in a serum-like solution into which was pumped cheap biochemical precursors. From these the glands synthesized a potent analogue of cortisone known as hydrocortisone. Searle ignored the costs involved in this unique production line consisting of literally hundreds of these perfusion cells. The company distributed the first clinical supplies of hydrocortisone without charge. Meanwhile, yet another novel approach was adopted by a third American company, Upjohn of Kalamazoo. Here, more than 150 chemists were put to work to come up with a quick answer to the problem of cortisone production. They succeeded by using a mould known as *Rhizopus nigricans* to convert cheaply available progesterone to 11-hydroxyprogesterone which, in turn, was easily made into hydrocortisone.

One of the major disadvantages of using either cortisone or hydrocortisone in rheumatoid arthritis was their effect on the sodium ion levels in the body. Today prednisolone (see page 224) is used instead since it does not have this effect.

NON-STEROIDAL ANTI-INFLAMMATORY AGENTS (NSAIs)

Concern mounted over the incidence of unacceptable gastric irritancy from aspirin preparations used in high dosage to control rheumatic conditions. This inspired the pharmaceutical industry to seek out better alternatives after the limitations of cortisone therapy became apparent during the 1950s. New animal screening procedures were developed, including one in which oedematous swelling was produced in the foot of a rabbit by injecting an irritant material. The ability of test compounds to relieve this oedema was considered to be an indication of anti-inflammatory efficacy. That a quarter of a century later no drug markedly superior to aspirin has emerged must surely be taken as an indication of the inadequacy of the screening procedures.

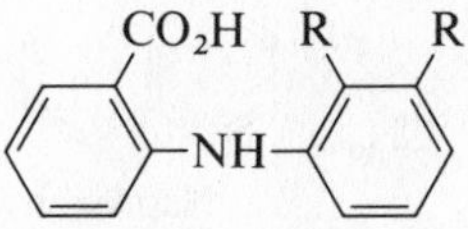

Mefenamic Acid, R = $-CH_3$, R′ = $-CH_3$
Flufenamic Acid, R = $-H$, R′ = $-CF_3$

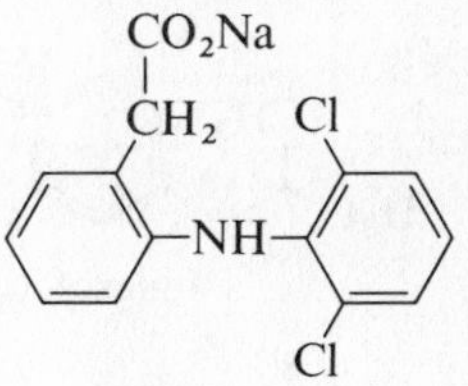

Diclofenac

The first of the alternatives to aspirin was mefenamic acid ('Ponstan') which bears a tenuous chemical relationship to salicylic acid insofar as the acidic phenolic group can be considered to have been replaced by a substituted nitrogen moiety. Since the latter is basic rather than acidic, any resemblance to salicylic acid is really fanciful. It was developed in the early sixties by Parke, Davis, and Company and is still in use, as are the related flufenamic acid ('Meralen') and diclofenac ('Voltarol').

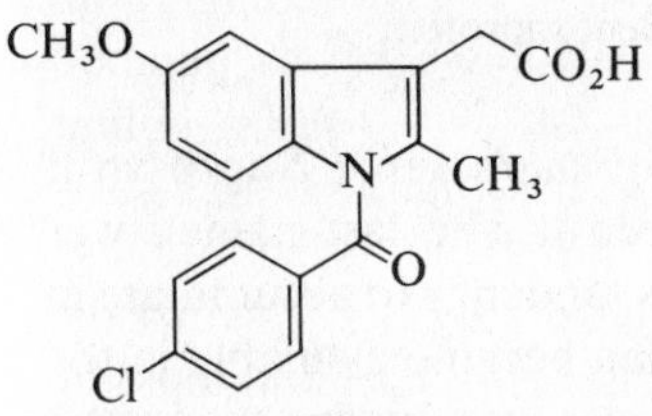

Indomethacin

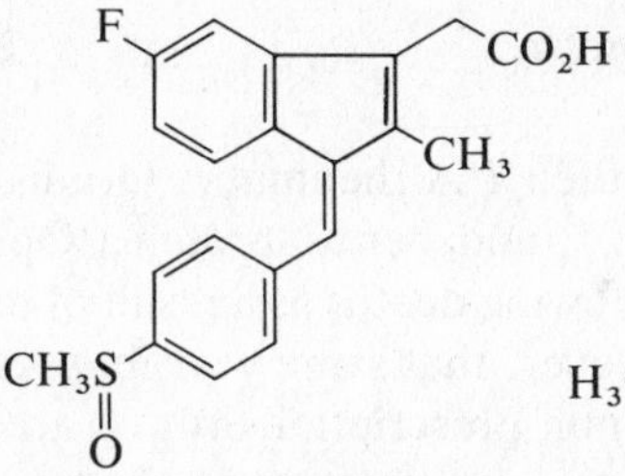

Sulindac

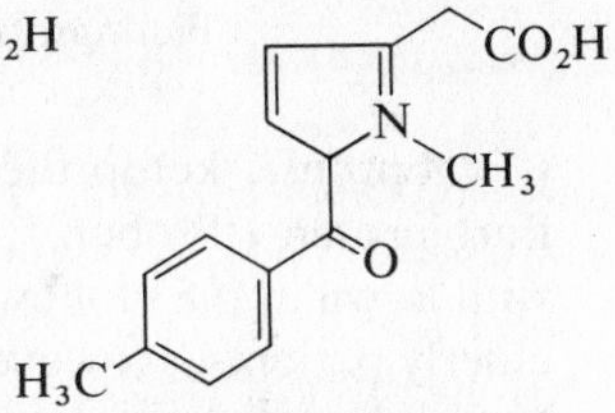

Tolmetin

In 1963, indomethacin ('Indocid') was introduced by the Merck Institute for Therapeutic Research, West Point, Pennsylvania. It had been the most efficacious amongst three hundred and fifty indole compounds screened in animals because of the possibility that the indolic hormone known as serotonin (or 5-hydroxytryptamine) might be involved in the inflammatory process. Attempts to design analogues with a lower incidence of side-effects have led to the introduction of drugs such as sulindac ('Clinoril'), and tolmetin ('Tolectin').

During the early sixties, researchers at the Boots Company in Nottingham were convinced that the presence of a carboxylic acid group was responsible for the anti-inflammatory activity of aspirin and some of its newer analogues. The Boots pharmacologists screened carboxylic acids which had already been made by the company, including a range of herbicides synthesized during the previous decade. One of these proved to be twice as potent as aspirin, initiating a search for superior analogues amongst this series of phenoxyalkanoic acids. More than 600 compounds were synthesized and tested before ibuprofen ('Brufen') emerged as a useful drug. Since its introduction in 1964 numerous phenylpropionic acid derivatives have been introduced, including fenoprofen

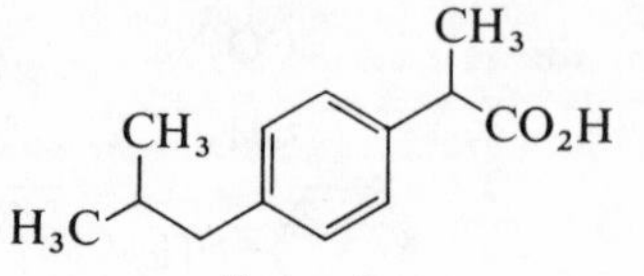

Ibuprofen

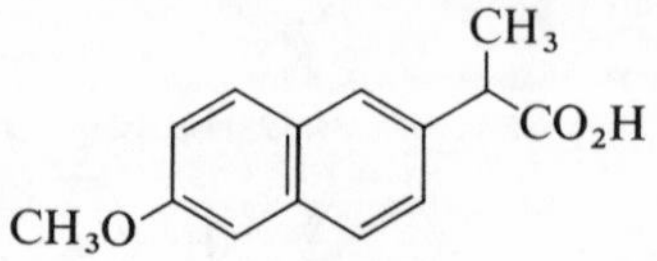

Naproxen

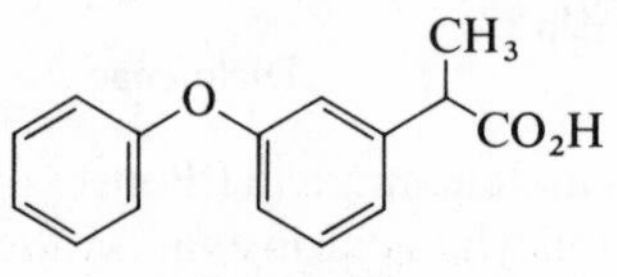

Fenoprofen

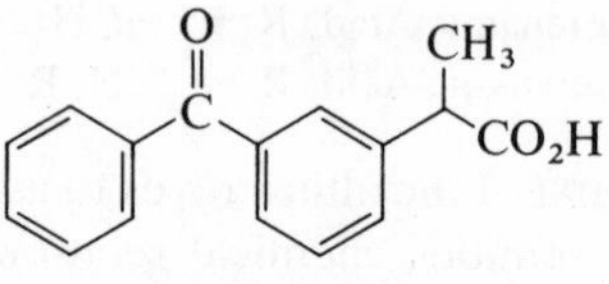

Ketoprofen

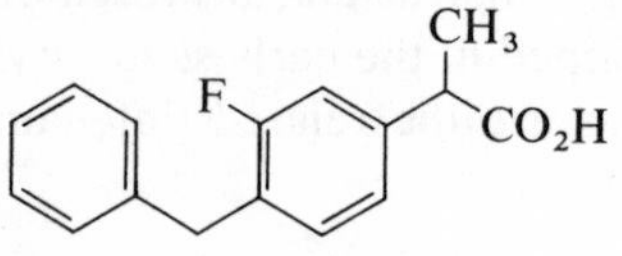

Flurbiprofen

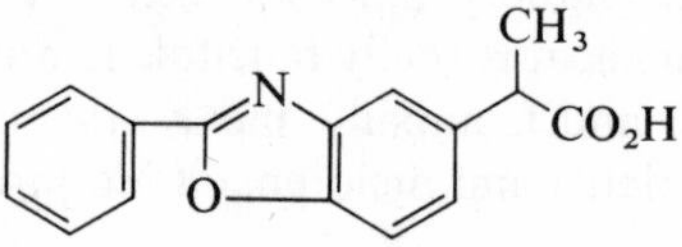

Benoxaprofen

('Fenopron'), ketoprofen ('Alrheumat', 'Orudis'), naproxen ('Naprosyn'), flurbiprofen ('Froben'), and benoxaprofen ('Opren'). The last named was withdrawn in 1983 following deaths as a result of its tendency to accumulate in elderly patients. However, that same year ibuprofen became available in the United Kingdom as a non-prescription drug on account of its having the lowest overall rate of reporting of suspected adverse reactions among the non-steroidal anti-inflammatory agents, some twenty million prescriptions having being issued over a fifteen-year period.

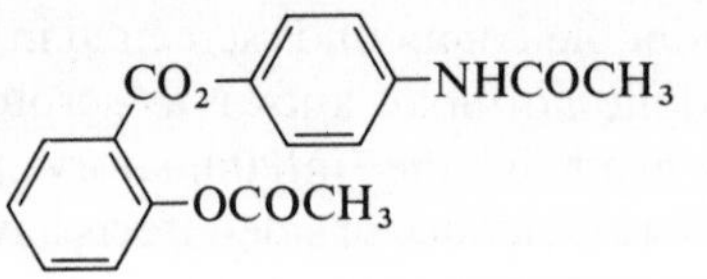

Benorylate

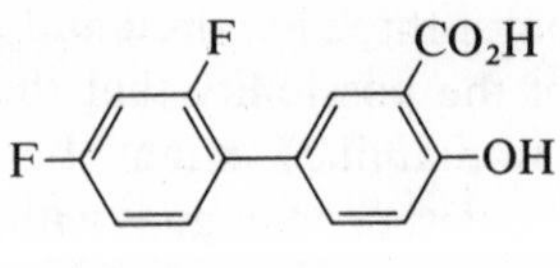

Diflunisal

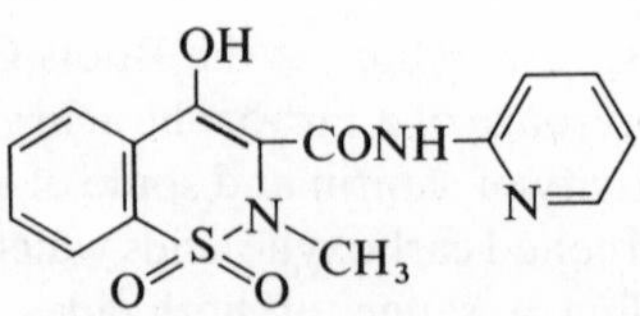

Piroxicam

Salicylic acid has continued to be a basis for the design of new antirheumatic drugs. Modern derivatives include the ester formed by reacting aspirin with paracetamol, known as benorylate ('Benoral'). It is broken down into its original components after absorption from the gut and is reputed to be less irritant to the stomach than is aspirin. Another salicylic acid derivative is diflunisal ('Dolobid'), introduced in 1979 by Merck, Sharp and Dohme after a fifteen–year search for a longer-acting, safer analogue of aspirin, in which 500 compounds were synthesized and evaluated. Pfizer introduced Piroxicam ('Feldene') around the same time. It is administered once daily.

DRUGS USED IN GOUT

Dioscorides, a surgeon in Nero's army, has been described as the founder of Materia Medica on account of the legacy he left in his herbal entitled *De Universa Medicina*. *Colchicum autumnale*, which was originally found growing in the Colchis region of Asia Minor and is known today as the meadow saffron, was classified by him as a highly poisonous plant, and this served to deny one of the few sources of relief to those unfortunate enough to be afflicted with gout. However, the most influential of the Byzantine physicians, Alexander of Tralles, recommended it in the sixth century for 'podagra'. The German botanist Hieronymous Tragus mentioned in the 1552 edition of his *Kreuterbuch* that the root was employed by Arabian physicians in cases of gout and rheumatism. Colchicum was one of the simples favoured by country folk in England, but its misuse earned it the nickname *Colchicum perniciosum* within the medical profession. Only after Baron Antonius Storck of Vienna wrote in his *Libellus de Radice Colchicum Autumnale* in 1763 of its value in dropsy (oedema), did it gain acceptance in medical circles. Around this time there appeared a popular medicine known as Eau Medicinale, which acquired a reputation for alleviating the pain and inflammation of gout. This was carefully investigated in the early years of the nineteenth century by Sir Edward Home, surgeon to St. George's Hospital in London. His favourable conclusions established the root as a valuable remedy. In 1820, Pelletier and Caventou extracted the active principal and named it colchicine. This proved to be considerably safer than the root which, in addition, contained veratrine, a cardiac poison. Until the introduction of allopurinol in the 1960's (see page 349), colchicine remained the mainstay of gout therapy.

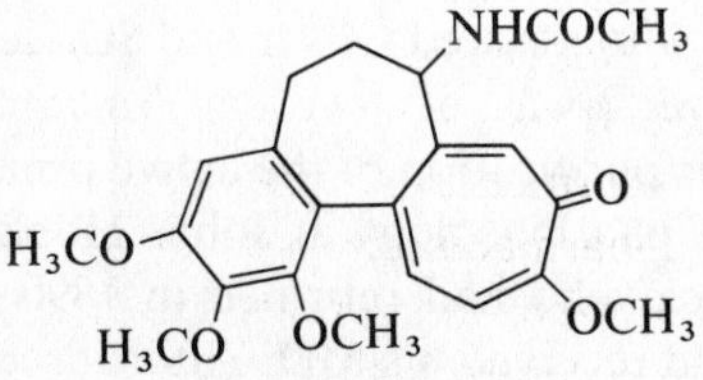

$NHCOCH_3$ O H_3CO H_3CO OCH_3 OCH_3

Colchicine

VI

Drugs Affecting Nervous Transmission

ADRENALINE

In the autumn of 1893, George Oliver, a physician in the popular spa town of Harrogate in the north of England, spent his spare time investigating the action of glycerine extracts of various glands upon the diameter of the radial artery, using a new instrument which he had recently designed. Organotherapy was then very much in vogue because of the remarkable claims being made by Edouard Brown-Séquard about the supposed rejuvenating properties of testicular extracts (see page 192). Oliver found that when a volunteer swallowed an extract of sheep adrenal glands, constriction of the artery could be detected. It says much for the sensitivity of his apparatus that he was able to detect this, since around ninety per cent of the active principle present in the extract must have been deactivated as it passed through the gut wall. Oliver decided to pursue matters further. He called on Professor Edward Schäfer, at the physiological laboratory of University College in London, with the happy outcome that during the winter months the two physiologists investigated the glycerine extracts prepared by Oliver. Only the adrenal and, to some extent, the pituitary extracts were found to be active. Detailed studies were then carried out on the adrenal extracts, with the findings being reported to the Physiological Society the following March. Oliver and Schäfer told the Society that water, alcohol, or glycerine could extract from the adrenal gland a substance with a powerful action upon the blood vessels, the heart, and the skeletal muscles of a variety of animals. Within one year of this report appearing in print, an adrenal liquid extract was on sale in Germany, although considered safe only for topical use. In 1897, two colleagues of Schäfer, Bayliss and Starling, coined the word 'hormone' to describe a substance like adrenaline which is carried in the blood from one locus in the body to stimulate another organ.

The first correct assessment of the chemical nature of the active principle in adrenal extract was made by S. Frankel of Vienna who, in 1896, stated that its chemical reactions pointed to it being a nitrogenous derivative of catechol. Mr B. Moore, the chemist who assisted Oliver and Schäfer, wrongly concluded it was a pyridine derivative. It was, however, an American who was to make the fastest progress towards purification of the active principle. He was the newly appointed professor of pharmacology at Johns Hopkins University in Baltimore, John Jacob Abel, who had returned in 1890 from the University of Strassburg where he had received his MD. After three years at the University of Michigan he moved to Baltimore, where he read the reports of the work done in England on the adrenal extracts and decided to attempt his own isolation process. In 1897 he isolated the active principle in impure form, but the following year he told the American Physiological Society that he had obtained it almost pure as its benzoate derivative. He confirmed that the hormone, which he named epinephrine, was a base which was very prone to oxidation, a feature which made its isolation difficult. In the same year, Otto von Furth of the University of Strassburg introduced a similar crude preparation of the hormone, this time as a complex with iron, which was marketed in Germany under the name 'Suprarenin'.

In the autumn of 1900 Professor Abel was continuing with his efforts to purify epinephrine, when he received a visit from Jokichi Takamine, an industrial chemist based in New Jersey. Takamine had studied the chemistry of brewing at Anderson's College in Glasgow (later to become the University of Strathclyde) before establishing the superphosphate fertilizer industry in Japan. After settling in the United States he tried to introduce a starch-digesting enzyme to the distilling industry, but failed to interest any of the distillers. In 1895, however, he persuaded Parke, Davis and Company of Detroit to sell it, under the name of Taka-diastase, as a remedy for digestive disorders. He thereby established a close link with the company. It is doubtful that Professor Abel suspected the motives of the Japanese visitor to his laboratory, but he soon had cause to regret the visit. Within weeks of his trip to Baltimore, Takamine had realized how he could overcome the stumbling block to the final purification of epinephrine. This involved using the same chemical procedure adopted by Serturner a century earlier when the Paderborn pharmacist isolated morphine from an aqueous extract of opium by addition of ammonia, thus avoiding exposure of the alkaloid to strongly alkaline solutions with which it would form a salt that could not be isolated. The new process worked admirably. Four grams of pure crystalline base were obtained, making it the first hormone ever isolated as a pure substance.

Takamine immediately patented his process for isolating the adrenal hormone, arranging for it to be marketed by Parke, Davis, and Company, one of whose chemists, T.B. Aldrich, had already been attempting to prepare it. The proprietary name used by Parke, Davis was 'Adrenaline', a term used outside of the United States as an approved name ('epinephrine' has been retained as the American approved name).

In 1903, Friedrich Stolz, director of chemical research at the Hoechst Dyeworks, began his attempts to synthesize adrenaline. At that time, two possible structures had been proposed. Stolz readily prepared a precursor of one of these, and sent it and several related compounds to Professor Hans Meyer and Otto Loewi at the University of Marburg for evaluation. They found these ketonic compounds to have many of the physiological properties of adrenaline, but only to a much lesser degree. The compound most closely resembling adrenaline was given the name adrenalone. Meanwhile, at the University of Leeds, Henry Dakin also synthesized adrenalone as one of a series of adrenaline analogues required for a study of the influence of chemical structure on hormonal activity. He obtained similar pharmacological results.

In principle, it should have been a straightforward matter for Stolz and Dakin to convert their ketonic compounds to alcohols in order to obtain adrenaline and its analogues with different nitrogen substituents. On carrying out the required chemical reduction process, however, they had difficulty in isolating pure compounds. Both men probably separated pure material on a laboratory scale in 1905, but it was another year before Stolz and Franz Flaecher managed to solve all the problems involved in large-scale production of synthetic adrenaline. The Hoechst Dyeworks began production in 1906. This immediately cut the cost of adrenaline as a vasoconstrictor to prolong the activity of 'Novocaine', the company's newly introduced local anaesthetic (see page 55). Adrenaline was also marketed in its own right as a haemostatic to control bleeding, and as a pressor agent to reverse the severe drop in blood pressure associated with surgical shock. One problem remained. Synthetic adrenaline had only about half the potency of adrenaline extracted from ox adrenals. This was due to the fact that natural adrenaline consisted solely of the laevorotatory stereoisomer, whilst the synthetic material contained equal parts of this and the feebly active dextrorotatory isomer. Flaecher eventually solved this problem by his discovery that the acid tartrate salt of the unwanted isomer readily dissolved in methyl alcohol, leaving behind the laevorotatory adrenaline acid tartrate to be filtered off. Its activity was identical to that of natural adrenaline. Hoechst replaced the earlier synthetic material with the pure isomer.

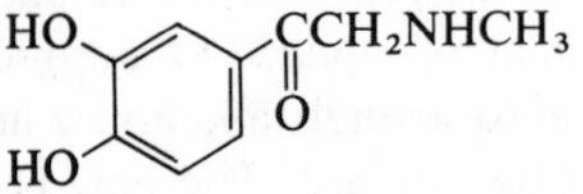

HO, HO (benzene ring) $C(=O)CH_2NHCH_3$

Adrenalone

HO, HO (benzene ring) $CH(OH)CH_2NHCH_3$

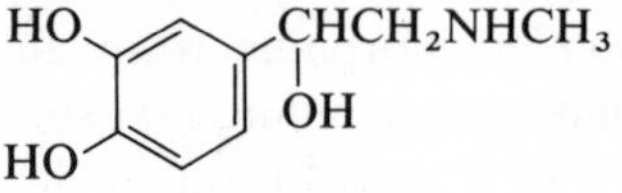

Adrenaline

ADRENALINE ANALOGUES

Friedrich Stolz synthesized several amines related to adrenaline, including noradrenaline (arterenol; the prefix 'nor' was an acronym derived from *nitrogen ohne radikal,* indicating the absence of a methyl group). Noradrenaline was not used clinically until after the discovery that it was the principal chemical transmitter mediating electrical impulses in the sympathetic nervous system, rather than adrenaline—as had been suggested in 1904 by Thomas Elliott of Cambridge University. Elliott had noted the parallels between the actions of adrenaline and electrical stimulation on the bladder, although he subsequently had reservations over his proposal that adrenaline was released as a result of nerve stimulation. The studies that showed noradrenaline, rather than adrenaline, to be the principal neurotransmitter were carried out at the Karolinska Institute in Stolkholm by Ulv von Euler, his first paper on the subject being published in 1946. Two years later, Maurice Tainter, director of the Sterling Winthrop Research Institute in Rensselaer, New York, introduced the laevorotatory isomer of noradrenaline ('Levophed') for treating clinical shock. Nowadays, the use of such vasoconstrictors for this purpose is viewed with some concern since the rise in blood pressure is often gained at the expense of perfusion of vital organs.

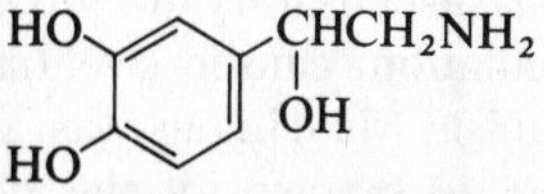

HO, HO, $CHCH_2NH_2$, OH

Noradrenaline

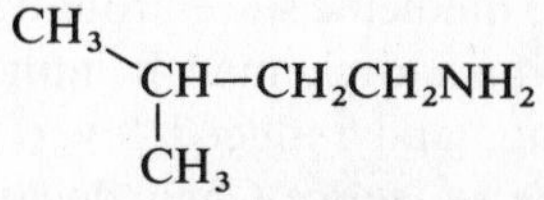

CH_3, $CH{-}CH_2CH_2NH_2$, CH_3

Isoamylamine

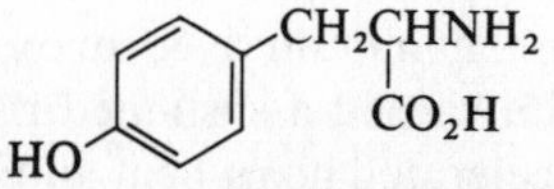

CH_2CHNH_2, CO_2H, HO

Tyrosine

$CH_2CH_2NH_2$, HO

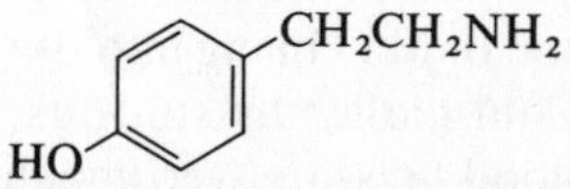

Tyramine

In 1905, Abelous and his colleagues in France reported that an intravenous injection of an alcoholic extract of putrefied meat could cause a rise in the blood pressure of an animal. An impure hydrochloride salt was shown to be responsible for this pressor effect. On repeating the experiment at the Wellcome Physiological Laboratories in London, George Barger and G.S. Walpole identified it as the hydrochloride of isoamylamine. Matters did not rest there. Professor Walter Dixon of King's College. London, and Frank Taylor of Cambridge University reported to the Royal Society of Medicine, in 1907, that an alcoholic extract of human placenta had adrenaline-like effects when injected intravenously. This extract was further examined at King's College by Otto Rosenheim, who established that adrenaline was not present. He then discovered that the pressor agent was only present in material that had undergone bacterial decomposition. When the extract was investigated by Barger and Walpole, they identified the active substance as tyramine, an analogue of adrenaline. It was formed from the amino acid tyrosine, which was present when protein decomposed. The Wellcome researchers also isolated tyramine from the putrefied meat extract. Their colleague Henry Dale confirmed that it was a stronger pressor agent than was isoamylamine, although it had only a fraction of the potency of adrenaline. However, as the duration of its action was much longer than that of adrenaline, it was put to clinical use in the treatment of shock, for which purpose it was synthesized and marketed for some years by Burroughs Wellcome.

The chemical and physiological similarities between tyramine and adrenaline persuaded Barger and Dale to investigate a wider variety of amines than had previously been studied. Many were found, like adrenaline, to produce physiological effects similar to those observed on stimulation of the sympathetic nervous system. For this reason, Dale described them as 'sympathomimetic amines'. Despite this early progress, it was not until the late twenties that the clinical potential of these adrenaline analogues was recognized.

In 1923, Ku Kuei Chen, who had just completed his PhD in physiology and biochemistry at the University of Wisconsin, returned to China to take up a lectureship in pharmacology at Peking Union Medical College, an institution then supported by the Rockefeller Foundation. He took an interest in Chinese herbal medicine, and his uncle told him of Ma Huang. This was a herb used in Chinese medicine since around 3100 BC. In his dispensatory of 1596, Pen Tsao Kang Mu had claimed it improved the circulation, caused sweating, eased coughing, and reduced fever. To the botanist, Ma Huang was known as *Ephedra vulgaris*. Chen decided to investigate extracts of the plant. For assistance, he turned to Carl Schmidt, who had taken up a post at Peking Union after obtaining his MD from the University of Pennsylvania. Schmidt injected a decoction of Ma Huang into the vein of a dog that had survived an experiment carried out earlier by students. The injection caused a sustained rise in the dog's blood pressure, accompanied by an accelerated heart beat and constriction of the blood vessels in the kidney. These effects were similar to those of adrenaline and tyramine. Chen then used solvent extraction techniques to isolate crystals of an alkaloid. On consulting the literature, he was surprised to find that this had previously been isolated by Professor Nagajosi Nagai at Tokyo University in 1897. It was known as ephedrine. After further tests demonstrated that, unlike adrenaline, ephedrine was effective by mouth, a supply of it was given to T. G. Miller at the University of Pennsylvania, and L. G. Rowntree at the Mayo Clinic. Their clinical investigations culminated in the approval of the drug by the Council of Pharmacy and Chemistry of the American Medical Association, in 1926. It then went on to become one of the most widely used drugs in the pharmacopoeia, particularly because of its value in asthma, where the prolonged action on the bronchi helped to ward off attacks of bronchospasm.

The remarkable thing about ephedrine was that it had previously been in use in Western medicine as a mydriatic to dilate the pupil, but had been considered too toxic for other applications. This had followed the original isolation of the impure alkaloid in Japan by G. Yamanashi in 1885. He died shortly after, but the work was continued by Nagai. After obtaining pure crystals of ephedrine, he elucidated its structure and synthesized it. Nagai then asked a friend, K. Muira, to examine its physiological properties. These were reported to be comparable to those of atropine, with the potential advantage that the mydriatic effect lasted only two hours. Several years later, in 1917, two other Japanese researchers actually reported that the actions of ephedrine were essentially similar to those of adrenaline, yet Western workers overlooked the potential of the alkaloid as an orally active sympathomimetic until it was rediscovered by Chen.

An attempt by one manufacturer to monopolize supplies of *Ephedra vulgaris* from China led to a scarcity of ephedrine from August 1926 until well into the following year, resulting in high prices being charged on the drug market. Fearing that many of his patients would be unable to afford regular prescriptions for ephedrine, a Los Angeles allergy specialist, George Piness, asked

Gordon Alles, a young chemist who worked in his laboratory preparing purified proteins for allergy tests, to synthesize a cheap substitute. Alles synthesized several compounds which he tested in the pharmacology department of the University of California at San Francisco. By the summer of 1928, phenylethanolamine sulphate had emerged as a promising ephedrine analogue, but it was not active by mouth. Since its chemical structure differed from ephedrine by the absence of two methyl groups, Alles, Piness, and another clinician, Hyman Miller, examined the influence of these methyl groups by comparing three compounds, one of which (phenylethylamine) did not have either of the methyl groups, whilst the remaining compounds had one or other of them. One of these produced a prolonged rise in blood pressure in both animals and man when given by mouth. It was passed to Smith Kline and French Laboratories in Philadelphia for further investigation, where it was discovered that the company had been examining the same substance as a possible nasal decongestant. The drug was not a new compound, having been synthesized in 1897. It had been examined in 1910 by Barger and Dale during their study of sympathomimetic amines, but the pharmacological tests carried out at that time did not indicate any superiority over adrenaline. However, Smith Kline and French astutely exploited the volatility of its free base, an oily liquid, by formulating it in a plastic inhaler that was inserted into the nostril to afford relief from nasal congestion. In 1932, it was marketed as the 'Benzedrine Inhaler'. The amine received the approved name of amphetamine.

Although the 'Benzedrine Inhaler' became the most profitable application of amphetamine, it was not the only way in which the drug was used. Early clinical experience with it as an oral decongestant in patients with hay fever revealed its stimulant properties. Later, Alles showed that these were particularly associated with the dextrorotatory stereoisomer, dexamphetamine ('Dexedrine'). This was introduced in 1935 for the treatment of the abnormal sleepiness of narcolepsy.

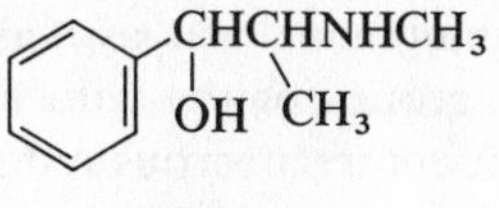

$C_6H_5CH(OH)CH(CH_3)NHCH_3$

Ephedrine

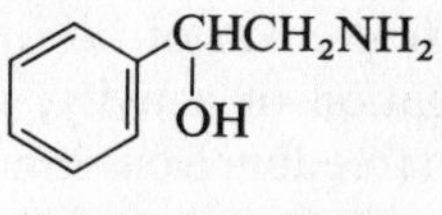

$C_6H_5CH(OH)CH_2NH_2$

Phenylethanolamine

$C_6H_5CH_2CH(CH_3)NHR$

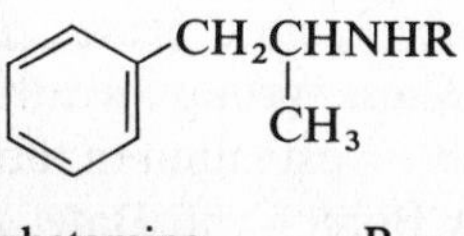

Amphetamine, R = —H
Methylamphetamine, R = $-CH_3$

During the Second World War, either dexamphetamine, or the even more potent methylamphetamine ('Methedrine'), was given to troops to ward off fatigue. Once the war ended, these amphetamines were sold without prescription in Japan for use by the general public. Young people consumed them to keep alert during long hours at work, a habit that eventually spread across the Pacific to the very city where amphetamine had been discovered, San Francisco. Amphetamine abuse soon reached alarming proportions, with addicts injecting themselves intravenously. The problem was further aggravated by the combination of amphetamines with barbiturates, a practice developed as a consequence of the wartime discovery that amphetamines curbed the appetite. When the war was over, clinicians began to employ amphetamines to treat obesity, but the stimulant effects were troublesome at night. In an attempt to alleviate this problem, Smith Kline and French combined dexamphetamine with amylobarbitone in a preparation known as 'Drinamyl'. This approach was followed by other leading manufacturers during the fifties and early sixties. These combination preparations were abused by teenagers, the problem quickly getting out of hand. Ultimately, physicians refused to prescribe amphetamines and manufacturers withdrew them from the market. They are now used solely for treating a limited number of narcoleptic patients.

Several analogues of amphetamine were introduced as anorexic agents to treat obesity. They were selected on the basis of their relative freedom from stimulant effects, but even those that satisfied this criterion were quite useless in the long-term treatment of obesity. Fortunately, most of the developments arising from the discovery of ephedrine have been beneficial.

By the end of 1928, medicinal chemists in several laboratories had recognized the opportunity afforded by the existence of four structural variants amongst sympathomimetic amines of clinical interest. Adrenaline was a derivative of catechol, i.e. dihydroxybenzene, whereas tyramine was derived from phenol, i.e. monohydroxybenzene. Both ephedrine and amphetamine were derivatives of benzene without any hydroxyl substituent. Furthermore, the side-chains differed in each of these drugs. It became possible to graft one or other of these side-chains on to benzene, phenol, or catechol to obtain novel sympathomimetic amines which featured desirable properties associated with one or other of the original drugs. For example, the extra methyl group present in the side-chain of both ephedrine and amphetamine was associated with prolongation of activity, whilst the hydroxyl groups of the catecholamines ensured freedom from stimulatory effects on the central nervous system. Many permutations were explored during the 1930s by pharmaceutical manufacturers, including Eli Lilly and Company (whose effort was directed by Ku Chen) and Sharp and Dohme in the United States, and I. G. Farben and C. H. Boehringer in Germany. These groups synthesized new compounds and re-examined those previously investigated in the early years of this century by Stolz, Meyer and Loewi, by Barger and Dale, and by Carl Mannich and his colleagues at the University of Berlin. Several compounds introduced in the thirties are still marketed as non-prescription decongestants, including phenylpropanolamine and phenylephrine. This latter drug, unlike adrenaline, can be safely used to maintain blood pressure during anaesthesia.

Just as the war began, H. Konzett of C. H. Boehringer in Ingelheim

$C_6H_5CH(OH)CH(CH_3)NH_2$

Phenylpropanolamine

$HO-C_6H_4-CH(OH)CH_2NHCH_3$

Phenylephrine

recognized that a new analogue of adrenaline, its *N*-isopropyl derivative, was a promising drug for the relief of bronchospasm in asthma. Knowledge of this development did not become generally available until the US State Department was investigating the work carried out by German chemical manufacturers during the war. It was then learned that the drug retained the desirable bronchodilating action of adrenaline, but was more or less free of troublesome pressor activity. Furthermore, being a catecholamine it did not cause insomnia as did the more lipid-soluble ephedrine, which penetrated into the brain. The compound was introduced clinically in 1951, with the approved name of isoprenaline (isoproterenol in the United States). For the next twenty years, isoprenaline was considered to be the drug of choice for the relief of acute asthmatic attacks. Tragically, its introduction in convenient aerosol form during the sixties led to an estimated three thousand deaths amongst asthmatic teenagers in the United Kingdom alone. This was attributable to the effects of overdosage on the heart.

$(HO)_2C_6H_3-CH(OH)CH_2NHCH(CH_3)_2$

Isoprenaline

$(HO)_2C_6H_3-CH(OH)CH_2NHCR(CH_3)_2$

Orciprenaline, R = —H
Terbutaline, R = $—CH_3$

The success of isoprenaline served to highlight an obvious disadvantage, long before its cardiac toxicity became a matter of concern. Because it was a catecholamine like adrenaline, it was susceptible to enzymes in the body which rapidly deactivated catecholamines. As a result, isoprenaline was too short acting to be employed in the prophylaxis of asthma as ephedrine had been. The obvious way round this difficulty was to replace the catechol system with one which, while similar enough to act on the catecholamine receptors in the bronchi, would be unable to fit on to the active site on the destructive enzyme, viz. catechol *O*-methyltransferase. In 1961, C. H. Boehringer researchers achieved this objective with a drug called orciprenaline ('Alupent'). In this, one of the two hydroxyl groups was moved to an adjacent (meta) position on the benzene ring. Although this slight change in molecular topography markedly reduced the potency of the drug, it also conferred resistance to enzymic destruction. Furthermore, as the hydroxyl substituents remained on the ring, the drug was too polar to enter the brain, hence patients did not experience the problem of insomnia, as was the case with ephedrine.

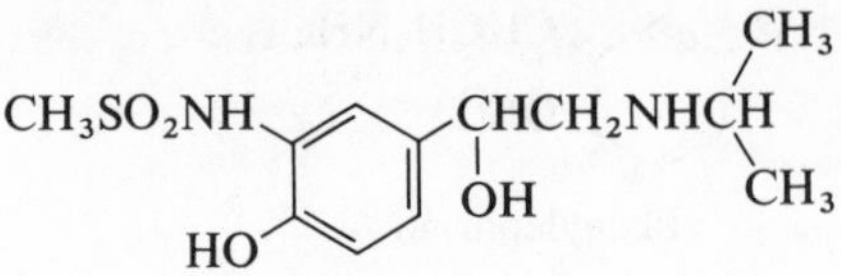

Soterenol

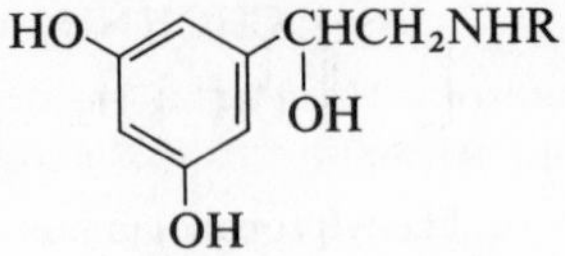

Fenoterol, R = $-CH(CH_3)CH_2-C_6H_4-OH$

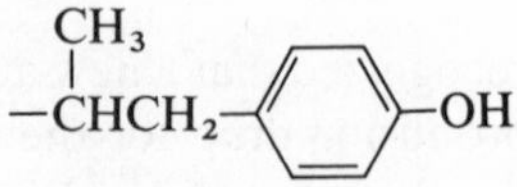

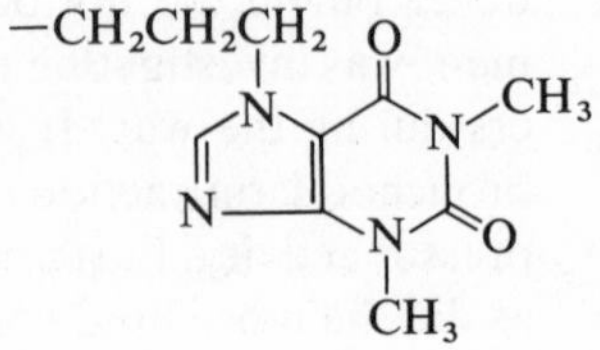

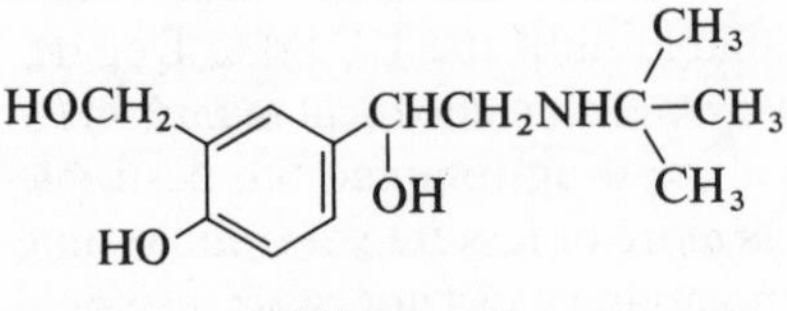

Salbutamol

Reproterol, R = $-CH_2CH_2CH_2-$

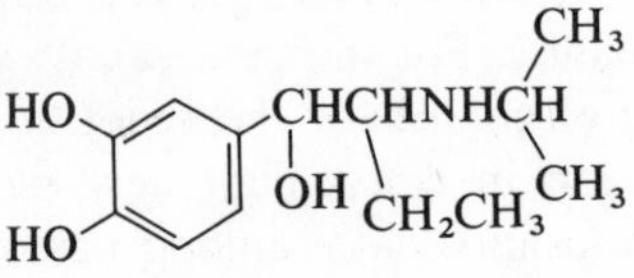

Isoetharine

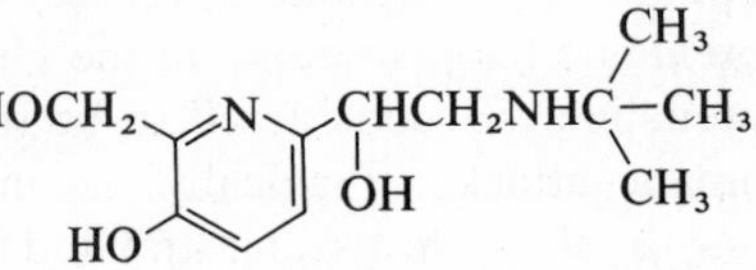

Pirbuterol

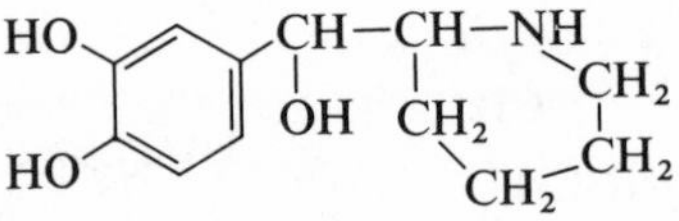

Rimiterol

An alternative approach to changing the position of one of the catechol hydroxy groups was used in 1964 by A. A. Larsen and P. M. Lish of the Mead Johnson Company in the United States. They replaced the hydroxyl group in position 3 of the benzene ring by the chemically similar methanesulphonamide group to form soterenol, an effective bronchodilator. It was, however, a British company which had greatest success with this approach when Hartley, Jack, Lunts, and Ritchie of Allen and Hanbury's research division in Ware (now part of Glaxo) introduced salbutamol ('Ventolin') three years later. This time, the hydroxyl group was replaced with a hydroxymethyl group to ensure resistance to destruction by catechol *O*-methyltransferase. As anticipated, the duration of action was lengthened by this manoeuvre, whilst potency remained high. However, another modification consisted of the introduction of a tertiary butyl group into the molecule in place of the isopropyl group attached to the side-chain nitrogen atom. This increased affinity for the receptors for the drug in the lung and reduced that for heart receptors. So marked was this selectivity in salbutamol, that within only a few years it had displaced isoprenaline as the drug of choice for the control of asthma attacks. It was, nevertheless, soon rivalled by terbutaline ('Bricanyl'), developed by the Swedish company Astra, whose chemists replaced the isopropyl group in orciprenaline with a tertiary butyl group. This enhanced potency and selectivity. Similar drugs include fenoterol ('Berotec'), isoetharine ('Numotac'), pirbuterol ('Exirel'), reproterol ('Bronchodil'), and rimiterol ('Pulmadil').

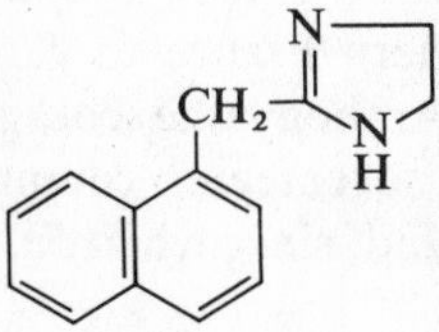

Naphazoline

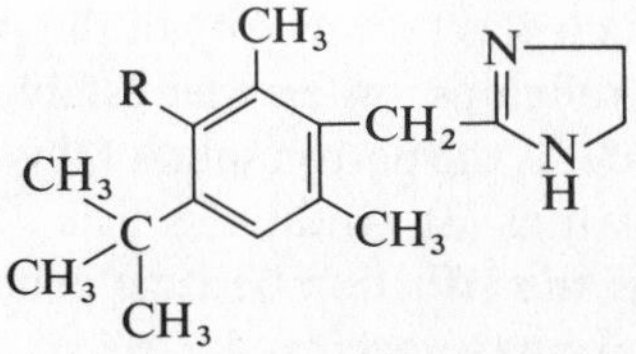

Oxymetazoline, R = —OH
Xylometazoline,R = —H

Not all adrenergic drugs are structurally related to adrenaline. There are several imidazolines. This type of compound was first examined pharmacologically in 1894, when Alfred Ladenburg reported that injection of 2-methylimidazoline had relieved chronic gout. No details were given. Matters rested there until 1935, when Henry Chitwood and Emmet Reid of the chemistry department at Johns Hopkins University decided to reinvestigate 2-methylimidazoline. Together with ten of its homologues, it was given to David Macht for pharmacological assessment. He found that only the methyl homologue had any effect on the acidity of the urine, a pointer to increased excretion of uric acid—which was the usual mode mode of action of drugs that relieved gout. He also reported that toxicity decreased as the methyl group was replaced with longer alkyl groups. This unusual observation prompted Fritz Uhlmann of the Ciba laboratories in Basle to examine the series of compounds for himself. Uhlmann obtained the opposite effect so far as the influence of chain length on toxicity was concerned. In the course of this investigation, however, it was noted that the imidazolines dilated the peripheral blood vessels. This was shown by a drop in blood pressure. When Ciba chemists subsequently introduced cyclic substituents at the 2-position of the imidazoline ring, the toxicity decreased. Benzene and naphthalene rings were then incorporated into the molecular structure. Unexpectedly, several of the resulting derivatives increased blood pressure by acting like sympathomimetic amines. This was especially marked with the naphthyl derivative, which was introduced into clinical practice in the early forties as naphazoline ('Privine'). The structure of this and related vasoconstrictors, such as xylometazoline and oxymetazoline, exhibited some of the molecular features previously associated with other sympathomimetic amines. Nowadays, these drugs are employed mainly as nasal decongestants.

ADRENALINE ANTAGONISTS

During the Middle Ages, tens of thousands of people died after eating bread made from rye contaminated with a fungus known as ergot (*Claviceps pur-*

purea). The outbreaks of ergotism reached epidemic proportions in the corn-producing areas of France and Germany, death ensuing after gangrene of the limbs had set in. As this manifested itself as an overall blackening of the diseased extremity, superstitious minds attributed it to charring by the Holy Fire! Such was the severity of the problem at the end of the eleventh century, that a religious order was established in southern France to care for the afflicted. As the patron saint of this order was St Anthony, the scourge came to be known as 'St Anthony's Fire'. Not until the seventeenth century did the cause of the affliction become generally recognized, since when only isolated outbreaks have occurred.

Early attempts to use ergot medicinally preceded the realization that it was such a dangerous substance. In his '*Kreuterbuch*', published in Frankfurt in 1582, Adam Lonicer mentioned that it was used by midwives to hasten labour. The dose he recommended was consistent with contemporary usage. Although some European midwives continued to administer ergot, it was not until after a letter was written in 1808 by John Stearns of Saratoga County, New York State, to the *Medical Repository*, that orthodox practitioners paid any attention to the therapeutic potential of ergot. Stearns stated that he had been successfully using ergot, either as a powder or as a decoction, for several years in order to hasten prolonged labour. His attention had originally been drawn to ergot by an old woman who had immigrated to America from Germany, where she first learned of its value. (Barger, in his authoritative monograph '*Ergot and Ergotism*', has pointed out that the toxicity of ergot was so feared in Europe, that only in the New World would it have secured recognition amongst orthodox practitioners.) On the 2nd June 1813, one hundred members of the Massachusetts Medical Society heard Oliver Prescott deliver his '*Dissertation on the Natural History and Medicinal Effects of the Secale cornutum*'. This was published later that year in pamphlet form, being reprinted in London and Philadelphia. This finally ensured the general acceptance of the drug. At first, enthusiastic practitioners ignored Stearn's warning to avoid using ergot if prolongation of labour was due to obstruction. So many still-births and maternal deaths were reported that Professor David Hosack of Columbia University in New York suggested, in 1822, that its Latin name should be changed from *pulvis ad partum* to *pulvis ad mortem*! Fortunately, his advice that ergot be restricted to treating post-partum haemorrhage was, thereafter, generally accepted.

Ergot presented a formidable challenge for organic chemists, and it was not until 1875 that Charles Tanret, a French pharmacist, isolated the first of its alkaloids, ergotinine. Disappointingly, this proved to be inert, another thirty years passing before a potent substance was finally isolated.

Shortly after joining the staff at the Wellcome Research Laboratory in London, George Barger began to investigate problems associated with handling and standardizing the ergot extracts manufactured by the company. Two years later, in 1905, he and Francis Carr isolated a complex of pure alkaloids, which they called 'ergotoxine' (in 1943 it was shown by Stoll and Hofmann to be a mixture consisting mainly of ergocornine together with some ergocristine and ergocryptine). Their success was due in no small measure to Henry Dale's assays of the changes produced by ergot extracts on the blood pressure of decerebrate cats. This classical study revealed that ergotoxine could not only stimulate smooth muscle such as that of the uterus, the arteries, and the pupil, but it could also reverse (i.e. block) many of the actions of adrenaline, notably that on the blood pressure.

In 1917, the Sandoz Company of Basle entered the field of pharmaceutical research by appointing a distinguished Swiss chemist, Arthur Stoll, as director of research. He had previously worked with Richard Willstätter at the University of Munich on the isolation and identification of chlorophyll, for which Willstätter received a Nobel Prize in 1915. The two men remained close friends, and in 1938 Stoll managed to rescue Willstätter, a Jew, from Nazi Germany. When Stoll moved to Basle he chose to work on the isolation of the active principles of medicinal plants, including ergot. Within a year, he had isolated ergotamine, the first active homogeneous alkaloid. For a while, it was thought that ergotamine was chemically identical to ergotoxine since its pharmacological properties were similar, but this was soon disproved. It took Stoll another thirty-three years before he was able to establish the full chemical structure of ergotamine, a highly complex alkaloid. His colleague and successor, Albert Hofmann, assisted by a team of chemists at the Sandoz Laboratories, finally achieved the total synthesis of ergotamine in 1961.

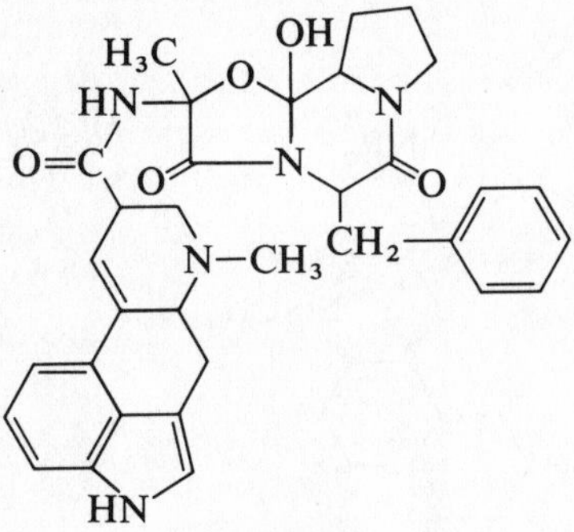

Ergotamine

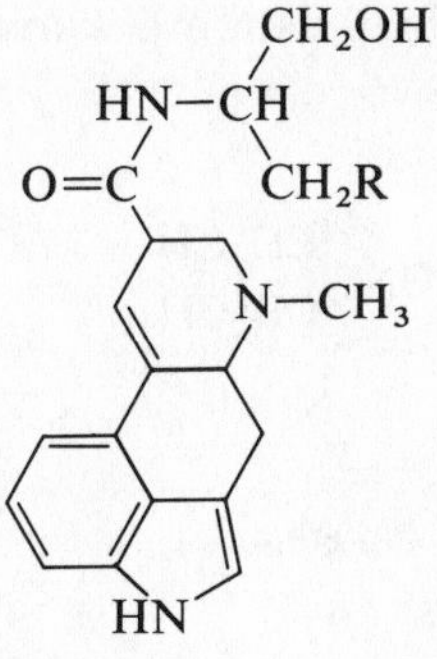

Ergometrine, R = $-H$
Methylergometrine, R = $-CH_3$

The similarity between ergotoxine and ergotamine caused clinicians to query whether either had any particular advantage over the other. Eventually, the Therapeutic Trials Committee of the Medical Research Council, which was chaired by Sir Henry Dale, invited F. J. Browne of University College Hospital, in London, to organize a clinical trial to compare the drugs. One of Browne's assistants, John Chassar Moir, devised a sensitive method of detecting their effects. He inserted, under sterile conditions, a small balloon into the uterus of patients undergoing routine pelvic examination one week after giving birth. The balloon was connected to a manometer which indicated the slightest pressure change arising from drug-induced uterine contractions. Once the initial study had confirmed that both ergotoxine and ergotamine were reliable, though slow-acting, uterine stimulants when injected, it was decided to investigate the controversial liquid extract of ergot which many authorities had decried. Their objection to it was based on the belief that its method of preparation meant it could not possibly contain the alkaloids known to be active. To Moir's amazement, the response produced when the liquid extract was administered by mouth was wholly unprecedented, both in the intensity of the contractions and the rapidity of their onset. Moir promptly drafted a report of his observations and sent it to the *British Medical Journal,* where it was published in June 1932.

Henry Dale, as director of the National Institute for Medical Research, asked Harold Dudley, the Institute's chief chemist, to collaborate with Moir in an attempt to isolate the new substance now believed to be present in the liquid extract of ergot. Three years passed before pure crystals of active material were isolated. It was then agreed that the full method of isolation would be openly published, with no attempt being made to secure patent cover. This was duly done, the new alkaloid being called 'ergometrine' when it was described in the *British Medical Journal* in March 1935. By now it was clear that the traditional actions of ergot had been due to ergometrine rather than the alkaloids previously isolated. A particularly desirable feature of the new alkaloid was the absence of either vasoconstrictor activity or adrenaline antagonism, such as was obtained with ergotamine.

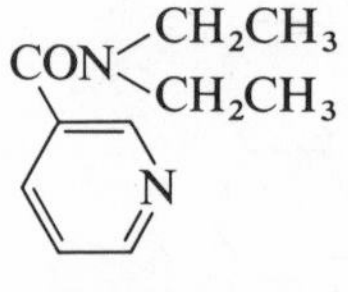

Nikethamide

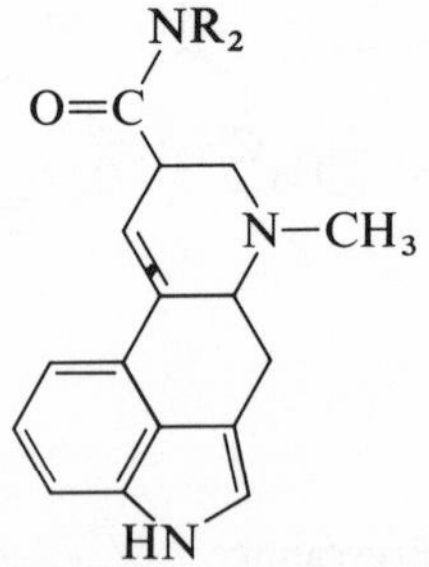

Lysergic Acid Diethylamide, R = $-CH_2CH_3$
Lysergamide, R = $-H$

A few weeks before the publication of the British success, Morris Kharasch and his colleagues at the University of Chicago reported their isolation of an active substance from ergot, naming it 'ergotocin'. Subsequently, this was found to be identical to 'ergometrine'. Shortly after, Marvin Thompson, of the University of Maryland, gave details of his extraction method, the patent rights to which were acquired by Eli Lilly and Company in 1934. The Sandoz group were also successful, Stoll and Burckhardt publishing their results around the same time as the other three groups. Because each of the four research teams gave a different name to the new alkaloid the American Medical Association felt it necessary to adopt the name 'ergonovine'. This so angered Sir Henry Dale that he resigned as British correspondent of the AMA. To this day, the *British Pharmacopeoia* retains the name 'ergometrine'.

After Walter Jacobs and Lyman Craig of the Rockefeller Institute, New York, identified lysergic acid as the common component of the ergot alkaloids, Stoll and Hofmann set about synthesizing ergometrine from this acid. They achieved their goal in 1937, then proceeded to prepare a series of analogues. Amongst these were methylergometrine ('Methergine'), which was at first thought to be superior to ergometrine for obstetric use, and several hydrogenated derivatives of the naturally occurring alkaloids, including dihydroergotamine. Since this was a potent adrenaline antagonist which, unlike the natural alkaloids, did not stimulate smooth muscle, it was used in certain circulatory disturbances. In an attempt to enhance its pharmacological activity, Hofmann decided to incorporate molecular features found in a respiratory stimulant known as nikethamide. The resulting molecule, lysergic acid diethylamide (LSD 25), not only was a powerful uterine stimulant, but it also caused excitement in some animals, and a cataleptic condition in others. Hofmann decided to re-examine LSD 25 in the spring of 1943, some five years after he had first synthesized it. One afternoon, however, he was forced to stop work because of dizziness and a peculiar restlessness. On reaching his home, he experienced extraordinary hallucinations for almost two hours. After he had recovered, Hofmann realized that the LSD 25 might somehow be responsible for his experience, even though he could only have ingested the most minute traces of it by accident. Subsequent experiments on himself confirmed his suspicions, revealing the potent psychotomimetic properties of LSD 25. Other workers at the Sandoz laboratories tried the drug, then a study was carried out at the psychiatric clinic of Zurich University. In February 1947, Professor Stoll revealed publicly that a few micrograms of lysergic acid diethylamide, taken by mouth, could cause profound alterations in human perception, reminiscent of those experienced by schizophrenics.

In 1957, Albert Hofmann received samples of hallucinogenic mushrooms collected in Mexico by Gordon Wasson, a retired New York financial journalist who, with his wife, had made a lifelong study of the role of mushrooms in human society. Wasson had now uncovered the story of the cult centered around the *Psilocybe mexicana*. From these mushrooms, Hofmann isolated crystals of psilocybin, which, like LSD 25, was an indole alkaloid. He swallowed the drug himself, and thus confirmed that it was hallucinogenic. With this success behind him, it was inevitable that Hofmann would proceed to investigate *ololiuhqui,* another Mexican psychotropic plant also brought to his attention by Wasson. This was identified as the seeds of two types of morning glory, viz. *Turbina corymbosa* and *Ipomoea violacea*. On analysis, these were unexpectedly found to contain alkaloids closely related to LSD 25, as well as ergometrine! The lysergic acid amide and hydroxyethylamide were already known to be hallucinogenic, but it was not until 1976 that any consideration was given to the possibility of ergometrine itself being psychotomimetic. This was revealed through self-experimentation by Hofmann after Wasson had queried whether the Ancient Greeks might have used ergot as the hallucinogen which featured in the celebration of the Mysteries at Eleusis. Hofmann knew that Professor Ernst Chain and his colleagues in Italy had, in 1960, demonstrated the main constituents of the Mexican morning glory seeds were to be found in a variety of ergot (*Claviceps paspali*), which often infected a wild grass that grew all around the Mediterranean basin. He then confirmed that the same alkaloids were present in ergot grown on wheat or barley as were to be found in that grown on rye. This was an important point so far as Wasson's speculation about the Eleusian Mysteries was concerned, for the Ancient Greeks did not cultivate rye. Hofmann now had to establish whether ergometrine, the only alkaloid readily extracted from ergot of rye by water, was hallucinogenic. His discovery that it was, begged the question of why this had not been revealed earlier. In answer, he pointed out that its oxytocic effects were exerted by an oral dose of only one-tenth to one-quarter of a milligram, whereas the hallucinogenic dose was in the order of one to two milligrams. This revelation not only supported Wasson's contention that ergot was used in the celebration of the Mysteries, but also explained the occurrence of apparent madness in many victims of ergotism in the Middle Ages.

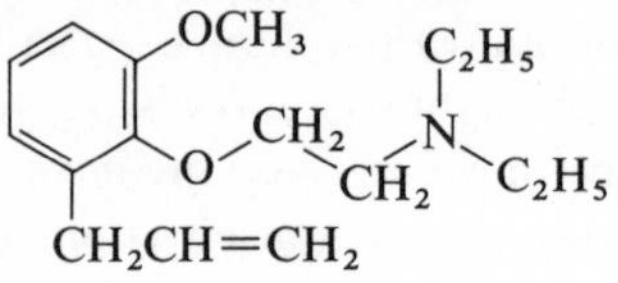

'Gravitol'

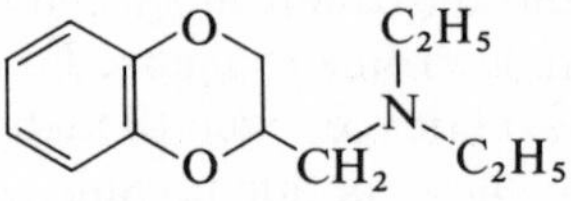

'Prosympal'

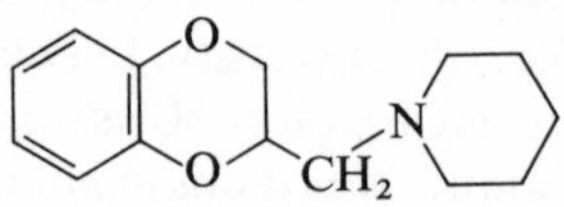

Piperoxan

The first simple synthetic compound to exhibit oxytocic activity was discovered in the I. G. Farben pharmacology laboratories by Eichholtz. This compound was a phenolic ether named 'Gravitol' that had been synthesized in 1923 by Hans Hahl in the Elberfield laboratories and subsequently submitted for screening. It was of little clinical value, but several analogues were synthesized. The most important of these were the benzodioxans, the first of which were prepared by M. Maderni for his doctoral thesis under the supervision of Ernest Fourneau at the Pasteur Institute. After reading reports from Japan, by S. Anan and T. Okazaki, indicating that an analogue of 'Gravitol' possibly had some adrenergic blocking activity, Fourneau became convinced that this type of cyclic phenolic ether would be worth investigating. When, in 1933, Fourneau and Daniele Bovet examined one of Maderni's benzodioxans, code-named 883 F, they found it could indeed antagonize some of the actions of adrenaline. It was named 'Prosympal'. Soon after this, its piperidine analogue, piperoxan (933 F), was introduced clinically as an adrenaline antagonist, but its duration of action was too brief for any useful clinical application. The principal importance of this compound was that it was ultimately selected as an adrenaline antagonist for screening by Bovet and his colleagues in their search for an effective antihistamine (see page 166).

Fritz Uhlmann's investigations into the physiological activity of the imidazolines led to the discovery that some of these compounds caused dilation of peripheral blood vessels. The most potent of these, tolazoline ('Priscol'), was introduced clinically before it was found to have weak adrenergic blocking activity. In the early fifties, Ciba introduced an analogue with more marked adrenergic blocking activity, namely phentolamine ('Rogitine').

During the Second World War, Louis Goodman had been involved in the development of nitrogen mustards as cancer chemotherapeutic agents (see page 335). The active mustards contained two chloroethyl groups, but after the war ended Goodman and Mark Nickerson, working in the pharmacology department at the University of Utah, reported that an analogue containing only one chloroethyl group, dibenamine, had been found to be a strong adrenergic blocking agent with a very long duration of action. This was because it decomposed into a highly reactive ethyleneimonium ion similar to that formed by the nitrogen mustards. This ion attacked the adrenergic receptor, becoming permanently attached to it by a covalent chemical bond. Receptors affected in this manner were no longer able to react with the adrenergic hormone. Such was the interest aroused by this development that more than 1500 analogues of dibenamine were screened in various laboratories, resulting in the clinical introduction of a more potent analogue known as phenoxybenzamine ('Dibenyline'), which was developed by Smith Kline and French Laboratories in Philadelphia. It was at one time used to treat hypertension, but its tendency to slow the heart and cause orthostatic hypotension (failure of compensatory mechanisms to maintain adequate blood pressure when rising from a resting position) has rendered it obsolete.

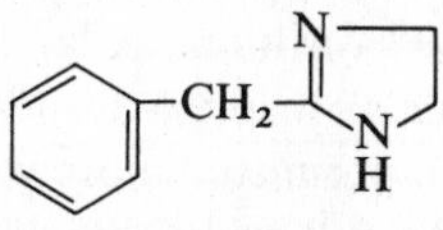

Tolazoline

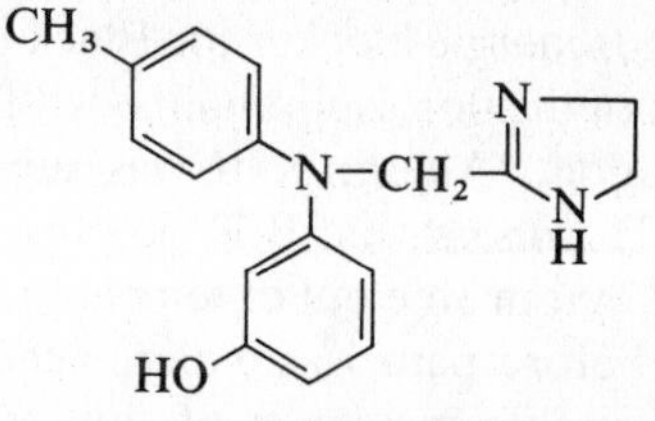

Phentolamine

The adrenergic blocking agents so far described could not antagonize the effects of adrenaline on the heart. The first drug to do this was serendipitously discovered by Irwin Slater, of Eli Lilly in Indianapolis, whilst testing analogues

$C_6H_5CH_2$–N(CH_2CH_2Cl)–$CH_2C_6H_5$

Dibenamine

$C_6H_5CH_2$–N(CH_2CH_2Cl)–$CH(CH_3)CH_2OC_6H_5$

Phenoxybenzamine

of isoprenaline as potential long-acting bronchodilators. The compounds were screened for their ability to relax tracheal strips contracted by exposure to pilocarpine to simulate asthmatic bronchoconstriction. The strips were checked for responsiveness between test runs by relaxing them with adrenaline. However, after exposure to dichloroisoprenaline, the adrenaline had no effect on them. When this antagonism of adrenaline by dichloroisoprenaline was reported by Dr Slater at a scientific meeting in 1957, Neil Moran of Emory University in Atlanta, Georgia, requested samples of the new drug to investigate its effects on the heart. He confirmed that dichloroisoprenaline antagonized the changes in heart rate and muscle tension produced by adrenaline. Unfortunately, it still mimicked the activity of these hormones to some extent.

3,4-$Cl_2C_6H_3$–$CH(OH)CH_2NHCH(CH_3)_2$

Dichloroisoprenaline

Moran's report interested James Black, newly appointed to ICI Pharmaceuticals Division at Alderley Park in Cheshire. He saw the clinical potential in a drug that protected the hearts of patients with coronary disease against adrenaline release provoked by physical or emotional stress, and realized it should be possible to synthesize an analogue of dichloroisoprenaline devoid of intrinsic action of its own when bound to the receptors in the heart (Moran had claimed that these receptors belonged to the type described by Ahlquist in 1948 as beta-adrenergic receptors). John Stephenson synthesized the first effective beta-adrenergic blocker for Black in February 1960, by replacing the chlorine atoms of dichloroisoprenaline with a second benzene ring to form pronethalol (nethalide, 'Alderin'). It was subsequently tested on humans by Professor A.C. Dornhorst and B.F. Robinson at St. George's Hospital, London. They found that in nine out of ten patients with angina it permitted more work to be done before pain was experienced. When put on full-scale clinical trial, an unacceptable incidence of side effects was noted. Shortly after, long-term toxicity testing in mice revealed that pronethalol could cause cancer of the thymus gland. It was immediately superseded by propranolol ('Inderal'), an analogue synthesized by Leslie Smith and shown to be non-carcinogenic as well as ten times as potent. This satisfied all the clinical criteria and became the first beta-adrenergic antagonist on the market when launched in 1964. Later, it was learned that it had previously been synthesized by H. Koppe of C.H. Boehringer and Company, Ingelheim, shortly before pronethalol had been discovered. Its clinical potential was not recognized, hence no patent was claimed at that time.

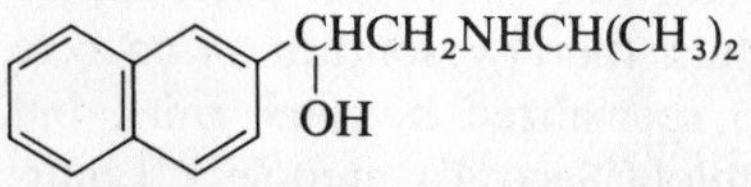

$CHCH_2NHCH(CH_3)_2$
OH

Pronethalol

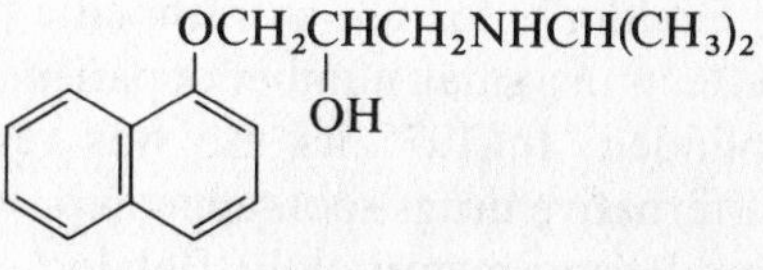

$OCH_2CHCH_2NHCH(CH_3)_2$
OH

Propranolol

Rival companies introduced structural analogues of propranolol, most of which had no important advantage over it. A possible exception was sotalol ('Sotacor', 'Beta-Cardone'), synthesized by A.A. Larsen of Mead Johnson in October 1960. He had previously worked with sulphonamides and synthesized sotalol in the belief that it would be worth replacing the acidic phenolic group of isoprenaline with an acidic sulphonamide group. Due to its polar nature, sotalol did not enter the brain to cause the vivid dreams sometimes experienced after taking propranolol.

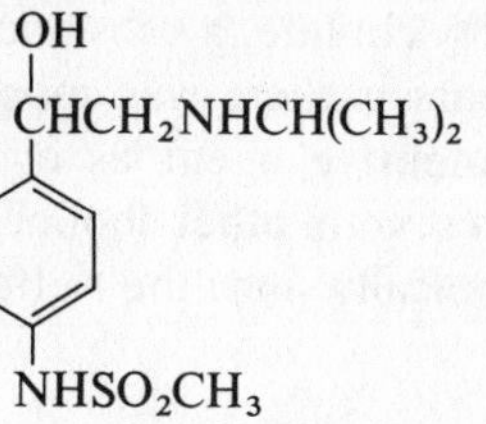

OH
$CHCH_2NHCH(CH_3)_2$

$NHSO_2CH_3$

Sotalol

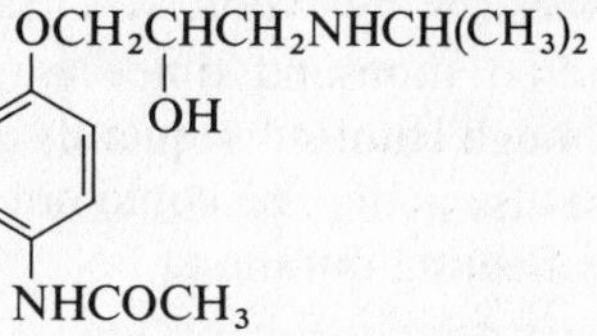

$OCH_2CHCH_2NHCH(CH_3)_2$
OH

$NHCOCH_3$

Practolol

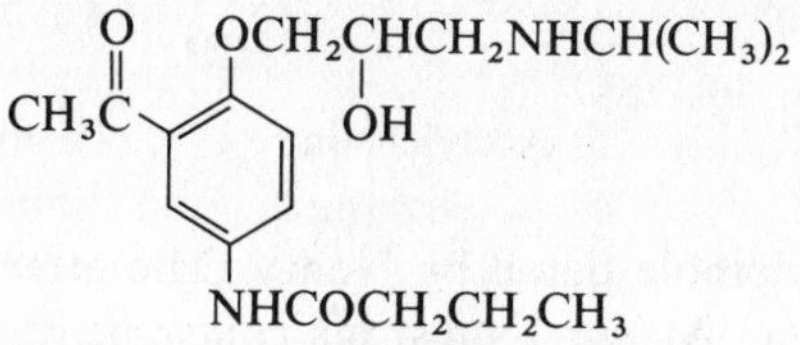

O
$OCH_2CHCH_2NHCH(CH_3)_2$
CH_3C
OH

$NHCOCH_2CH_2CH_3$

Acebutolol

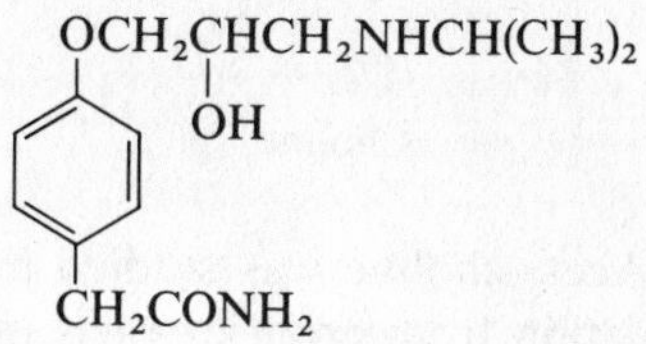

$OCH_2CHCH_2NHCH(CH_3)_2$
OH

CH_2CONH_2

Atenolol

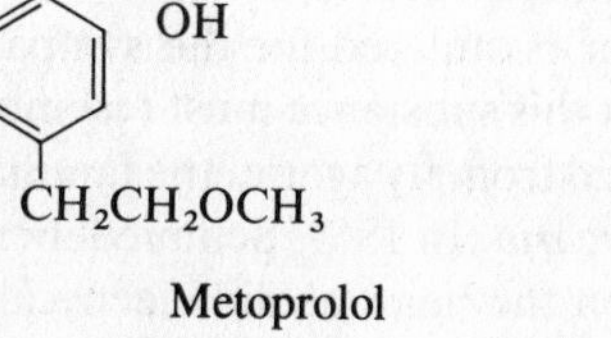

$OCH_2CHCH_2NHCH(CH_3)_2$
OH

$CH_2CH_2OCH_3$

Metoprolol

The early beta-blockers acted at all beta-adrenergic receptors. This was of little consequence for most patients, but it did trouble asthmatics because of the increased risk of bronchospasm. However, in 1966, D. Dunlop and R.G. Shanks of ICI unexpectedly discovered that an analogue of sotalol acted selectively on the heart receptors. In 1970, this was marketed as practolol ('Eraldin'), for use in asthmatic patients. Tragically, it caused serious side effects in a small number of patients on long-term therapy, some of whom were blinded. In 1975 its use was restricted to specialized hospital units, but alternative drugs soon appeared, viz. acebutolol ('Sectral'), atenolol ('Tenormin'), and metoprolol ('Betaloc', 'Lopressor').

CHOLINERGIC DRUGS

In 1898, Reid Hunt joined John Abel's department at Johns Hopkins Medical School, where he became particularly interested in the nature of the blood pressure lowering principle that remained in adrenal extracts after the adrenaline had been removed. He found that choline was responsible for some of this activity, although it was certainly not the principal hypotensive substance present. Reporting his findings to the American Physiological Society in 1900, Hunt suggested that either a precursor or derivative of choline was likely to be the main active principle. He never succeeded in isolating the choline derivative, but in 1906 he did demonstrate that acetylcholine, a substance originally synthesizes by Adolf von Baeyer forty years earlier, was more than one hundred thousand times as potent a hypotensive agent as choline itself. Although Hunt subsequently established that several other choline derivatives were also active, he could not exploit this clinically since their effects were of only fleeting duration.

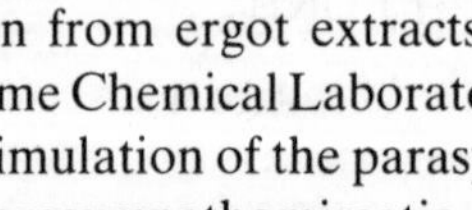

$$HOCH_2CH_2\overset{\oplus}{N}(CH_3)_3$$

Choline

$$CH_3CO_2CH_2CH_2\overset{\oplus}{N}(CH_3)_3$$

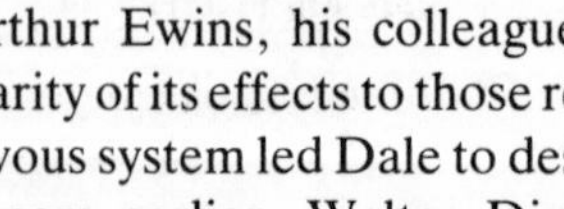

Acetylcholine

Acetylcholine was studied in considerable detail by Henry Dale after its isolation from ergot extracts in 1914 by Arthur Ewins, his colleague at the Wellcome Chemical Laboratories. The similarity of its effects to those resulting from stimulation of the parasympathetic nervous system led Dale to describe it as a parasympathomimetic agent. Eight years earlier, Walter Dixon had suggested that parasympathetic nerves must release a chemical substance in a manner analogous to that postulated for the sympathetic nerves by Thomas Elliott. He presumed that this substance must resemble muscarine, an alkaloid that Schmiedeberg isolated from fly agaric, the familiar red toadstool known to botanists as *Amanita muscaria*. In 1869, Schmiedeberg showed that muscarine evoked the same effect on the heart as did electrical stimulation of the vagus nerve. This was the first time a drug had been shown to mimic the action of nervous stimulation.

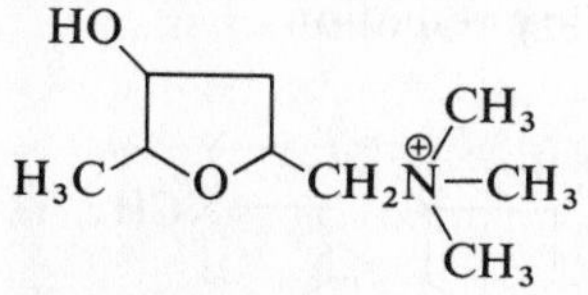

$$\text{Muscarine: } H_3C\text{-(tetrahydrofuran ring with HO)-}CH_2\overset{\oplus}{N}(CH_3)_3$$

Muscarine

When the First World War was over, Professor Otto Loewi, at the University of Graz in Austria, took up the challenge offered by the possibility that acetylcholine was present in the body as a chemical mediator of nervous transmission. In 1921, he demonstrated that repetitive stimulation of the vagus nerve supplying an isolated heart immersed in saline solution caused the release of a chemical which slowed down the rate of a second heart with no nerve supply. Loewi finally isolated this '*Vagusstoff*' in 1926, after discovering that its metabolic destruction, which Dale and Ewins had proposed was due to the action of an esterase enzyme, could be prevented by the alkaloid physostigmine, a drug known to produce cholinergic effects. Loewi thus showed the *Vagusstoff* to be acetylcholine. The Nobel Prize for physiology or medicine was awarded to Dale and Loewi in 1936 for their work on chemotransmission.

The identification of acetylcholine as the parasympathomimetic transmitter occurred around the same time as adrenaline analogues were being introduced. The possibility of finding useful acetylcholine analogues was considered by Randolph Major and Joseph Cline of Merck and Company, Rahway, New Jersey. They noted considerable confusion in the chemical literature over the properties of derivatives of choline which had been synthesized over the preceding fifty years. In 1931, they synthesized a variety of choline derivatives, including several examined twenty years earlier by Reid Hunt. These were tested at the University of Pennsylvania, first by Andre Simonart, a research fellow sponsored by Merck, and thereafter by Professor Isaac Starr and his colleagues, who were able to extend the study to the clinic. It was found that the acetylcholine analogue with an extra methyl group adjacent to the metabolically sensitive ester linkage of the molecule was somewhat resistant to enzymic degradation. This was in agreement with a prediction by Simonart, being due to steric hindrance to adequate fit to the active site of the enzyme responsible for the destruction of acetylcholine, viz. cholinesterase. This enabled the compound to be used clinically to stimulate the parasympathomimetic system so as to dilate peripheral blood vessels in certain types of circulatory system spasm, as well as in a variety of other conditions where its role was often of dubious merit. It was given the name of methacholine. Carbachol, for which Kreitmair secured a patent in the same year that the value of methacholine was recognized, proved to be even more resistant to metabolism by cholinesterase, as was its close analogue, bethanechol. In both of these drugs, the sensitive acetyl group of acetylcholine was replaced by the more robust carbamate function. Their action on the gastrointestinal and urinary tracts was more pronounced than that of methacholine, enabling them to relieve post-operative intestinal atony and urinary retention.

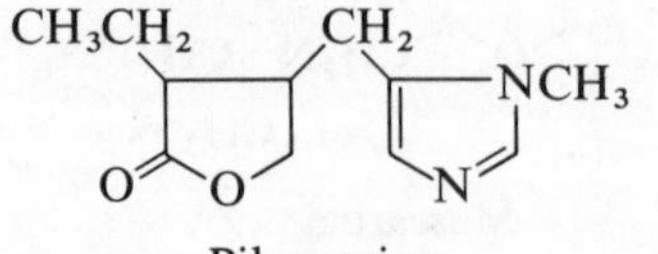

CH_3CH_2 CH_2 NCH_3 O O N

Pilocarpine

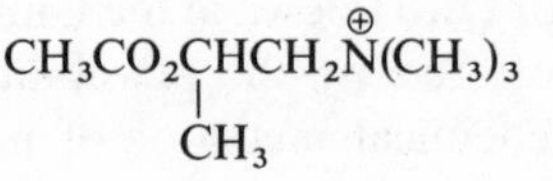

$CH_3CO_2CHCH_2\overset{\oplus}{N}(CH_3)_3$ (with CH_3 on the CH)

Methacholine

$H_2NCO_2CHCH_2\overset{\oplus}{N}(CH_3)_3$ (with R on the CH)

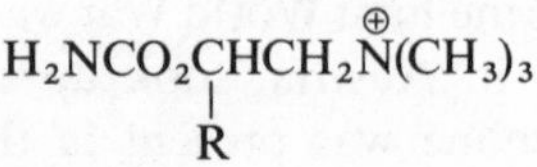

Carbachol, $R = -H$
Bethanechol, $R = -CH_3$

A drug frequently employed to mimic the action of acetylcholine on the eye is pilocarpine, an alkaloid obtained from the South American shrub, *Pilocarpus jaborandi*. Coutinhou, a Brazilian physician, observed that the leaves caused excessive salivation when chewed by the natives. In 1874, he prepared a simple extract of the leaves and found that it induced profuse sweating (diaphoresis), an action then generally considered to be of value in dropsy and other oedematous conditions as it removed excess fluid from the body. Subsequent experiments with jaborandi leaves showed that 250–400 ml of sweat could be voided after taking a modest dose. Reports of Coutinhou's experiments led to the isolation of pilocarpine the following year by A. W. Gerrard and, independently, by M. Hardy. The alkaloid was so potent a diaphoretic that it extended the scope of this type of therapy until the introduction of potent diuretics in the 1920s. Nowadays, pilocarpine is used to decrease the intraocular pressure in glaucoma.

ANTICHOLINESTERASES

In 1846, William Daniell, a British Army surgeon, reported to the Ethnological Society of Edinburgh that he had observed the use of Calabar beans as an ordeal poison in Old Calabar, Nigeria. Criminals convicted of capital offences were made to swallow a watery extract of the beans. If they vomited and survived, they were set free (it has been suggested that those with a clear conscience boldly swallowed the drug and this caused them to vomit). Daniell's report naturally interested Robert Christison, professor of materia medica at Edinburgh and author of *A Treatise on Poisons*. He arranged for a missionary to send him a Calabar bean, which he then had cultivated in Edinburgh. Christison carried out animal experiments which showed extracts of the bean killed by stopping the heart beat. Uncertain about the authenticity of the bean which he had received, he chose to experiment on himself. He experienced extreme weakness and was fortunate to survive, for his heart had probably gone into auricular fibrillation! Christison published a report of his experience with the Calabar bean in 1855. His distinguished student, and future successor, Thomas Fraser, won a gold medal in 1862 for an MD thesis on the pharmacological actions of the bean. The following year, he isolated its active principle as an amorphous powder, calling it 'eserina' after the native name for the bean, *esere*. Jobst and Hesse obtained it pure in crystalline form in 1864, and named their product physostigmine.

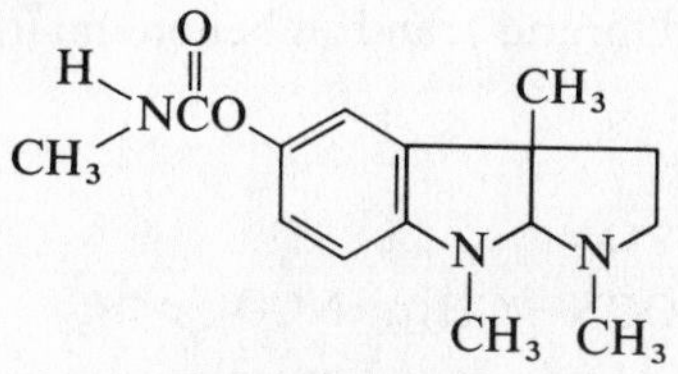

Physostigmine

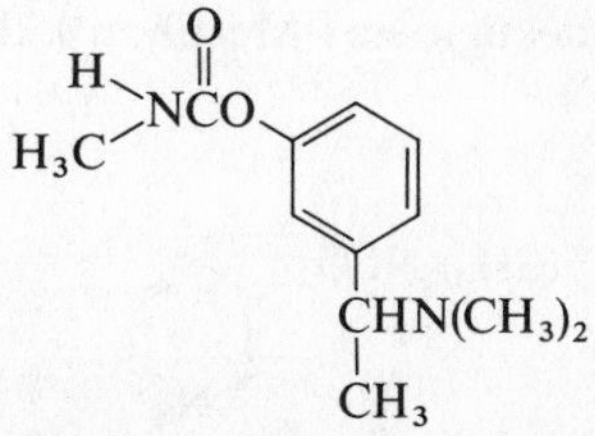

Miotine

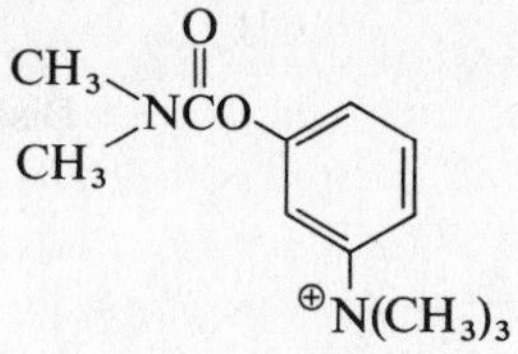

Neostigmine

His early studies on Calabar bean set the pattern for Fraser's future researches into ordeal and arrow poisons. But his main interest in pharmacology lay in its ability to solve medical problems rather than as an experimental science. He mentioned to the Edinburgh ophthalmologist Douglas Robertson that Calabar extract caused constriction of the pupil of the eye. This resulted in the two men carrying out a series of studies that showed physostigmine could antagonize the action of atropine on the pupil. This work attracted considerable attention, but it was not until 1875 that Ludwig Laqueur discovered the value of physostigmine in preventing blindness caused by glaucoma.

The structure of physostigmine was elucidated in 1925 by Edgar Stedman and George Barger of the department of medical chemistry in the University of Edinburgh. Utilizing the miotic action as a bioassay procedure, Edgar and Eleanor Stedman examined analogues of physostigmine, thereby establishing which features of the structure were essential for biological activity. They prepared several simpler compounds that were miotic agents, the most potent being miotine. This was tested clinically in 1931, but was inferior to neostigmine ('Prostigmine'), introduced the same year by John Aeschlimann of the Roche Research Laboratories in Basle. He had followed up the Stedmans' work by exploiting the enhanced stability of dimethyl carbamates, thereby overcoming the extreme sensitivity of physostigmine, a monomethyl carbamate, to hydrolytic cleavage.

In 1934, Mary Walker of St Alphege's Hospital in Woolwich, London, noted a resemblance between myasthenia gravis, a rare form of paralysis, and curare poisoning. Since physostigmine was an antidote to curare, she tried it on her patients. The results were so striking that, after confirmation by others, they made headlines in the press in March of the following year. Neostigmine was also used and subsequently found to be superior. It was not without its disadvantages, including unwanted cholinergic actions. Attempts to avoid these resulted in the introduction of several analogues in the 1950s, including pyridostigmine ('Mestinon'), distigmine ('Ubretid'), and ambenonium ('Mytelase').

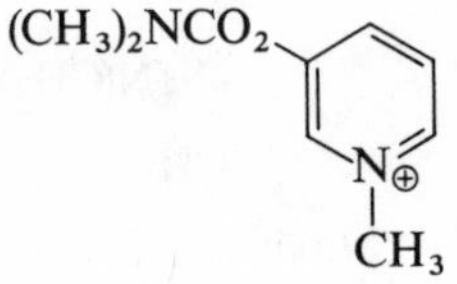

$(CH_3)_2NCO_2$ — $N^{\oplus}$ — CH_3

Pyridostigmine

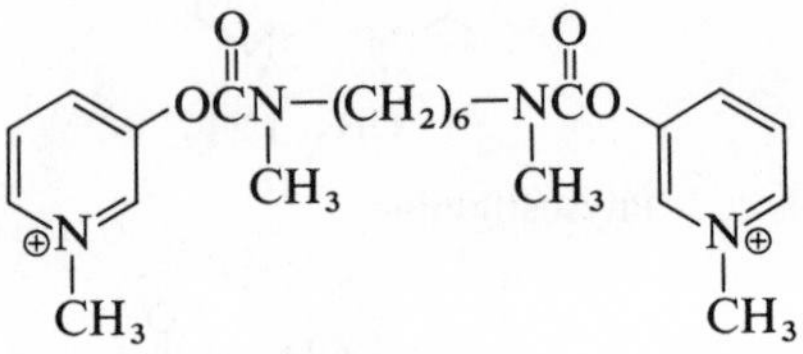

$OCN-(CH_2)_6-NCO$ (CH_3, CH_3; $^{\oplus}N-CH_3$, $N^{\oplus}-CH_3$)

Distigmine

$CH_2-\overset{\oplus}{N}(C_2H_5)_2CH_2CH_2NHCO-CONHCH_2CH_2\overset{\oplus}{N}(C_2H_5)_2-CH_2$ (Cl, Cl)

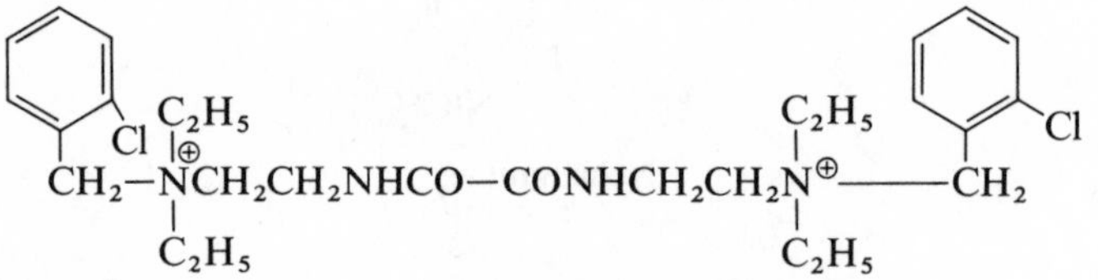

Ambenonium

In 1932, the Stedmans and Easson showed that physostigmine analogues bound to and blocked the active site of the cholinesterase enzyme responsible for the destruction of acetylcholine in the body. The chemical bond eventually broke, permitting the enzymic destruction of acetylcholine to proceed once more. However, certain fluorine-containing organophosphorus compounds form such a strong bond with the enzyme that their action becomes permanent. Under these circumstances, acetylcholine cannot be broken down and nervous transmission is disrupted. This inevitably results in death. The lethal action of such organophosphorus compounds has been ruthlessly exploited in the development of war gases with the terrifying potential to kill entire populations. The initial discovery of these loathesome agents was purely accidental. At the University of Berlin, in 1932, Willy Lange and his student Gerda von Kreuger took advantage of the recent ready availability of fluorine to prepare the first phosphorus-fluorine compounds. In the course of their work they experienced marked pressure in the larynx, followed by breathlessness, clouding of consciousness, and blurring of vision. These effects were mentioned at the end of the purely chemical paper which they published. Lange recognized the potential value of his compounds as insecticides, but left Germany shortly after this. At that time the German agricultural chemical industry was seeking a cheap insecticide as a substitute for imported nicotine, which killed insects by relentlessly mimicking acetylcholine to disrupt nervous transmission. Gerhard Schrader of the Leverkusen laboratories of the Bayer Division of I. G. Farbenindustrie followed up the new lead and prepared a large series of organophosphorus insecticides, more than two thousand compounds ultimately being synthesized in his laboratory. It was shown (although not divulged) by Professor Gross of the company's Elberfield laboratories that these were anticholinesterases, some of which were highly poisonous to laboratory animals. As the war clouds gathered, their potential as chemical warfare agents was fully appreciated by the German authorities, who stockpiled them for military use. The first agent to be so employed was tabun, synthesized by Schrader in 1936. Described as a 'war gas', it was actually a liquid that could be deployed in a fine dispersion. A plant disguised as a soap-making factory was opened near the Polish border in 1942 for production of tabun. By the end of the war, twelve thousand tons had been manufactured, and field forces were equipped with tabun-filled shells. The far more toxic sarin was developed in 1938; a mere one milligram of this was capable of killing an adult within minutes after being absorbed through the skin. In the early fifties, the United States supplied its chemical warfare units around the world with sarin, under the code name of GB. Later in the decade, production began of the even more toxic VX agent.

The British Ministry of Supply arranged for the potential of the fluorophosphonates to be studied at the physiology laboratory in Cambridge University by the Nobel laureate Edgar Adrian and his colleagues. They found, in 1941, dyflos (di-isopropylfluorophosphonate; DFP) to be the most toxic of the compounds originally prepared by Lange and Krueger. Its prolonged miotic action on the eye convinced them that it was an anticholinesterase, and direct evidence for this was obtained. When the war ended, dyflos found occasional clinical application in the treatment of glaucoma.

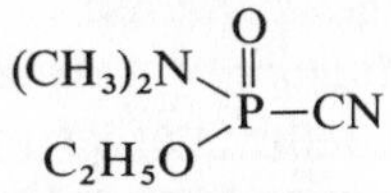

$(CH_3)_2N(C_2H_5O)P(=O){-}CN$

Tabun

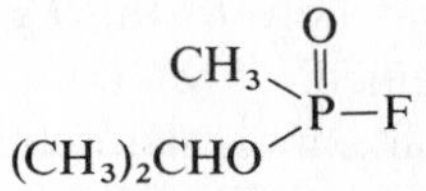

$CH_3((CH_3)_2CHO)P(=O){-}F$

Sarin

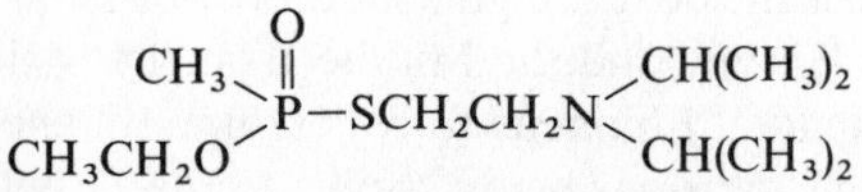

$CH_3(CH_3CH_2O)P(=O){-}SCH_2CH_2N(CH(CH_3)_2)_2$

VX

$((CH_3)_2CHO)_2P(=O){-}F$

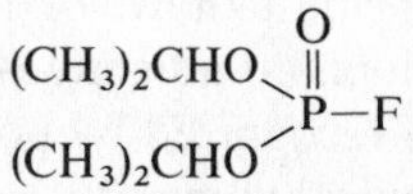

Dyflos

ANTICHOLINERGIC DRUGS

The use of drugs that antagonize the action of acetylcholine long preceded the discovery of the hormone. This should cause little surprise since any substance capable of interfering with the physiological role of acetylcholine must inevitably have such profound effects that its biological activity would be quickly recognized. Several plants of the order Solanaceae contain both hyoscyamine and hyoscine, in varying proportions. Both of these block the access of acetylcholine to the receptor site where it acts after its release from cholinergic nerve endings. Hyoscyamine is present in plants as its laevorotatory stereoisomer, but on isolation it tends to transform into a mixture of equal parts of laevo- and dextrorotatory isomers, i.e. a racemic mixture. This mixture is known as atropine, and is the form in which the alkaloid is used clinically, despite most of the activity residing in the laevoisomer. Atropine was isolated from the roots of belladonna in 1831 by Mein, a German apothecary, but hyoscine was overlooked until after Rudolf Buchheim suspected that the sedative properties of henbane could be due to the presence of a second alkaloid besides hyoscyamine. Eventually, Alfred Ladenburg, who studied the active principles of the solanaceous plants at the University of Kiel during the years 1879–1884, isolated hyoscine and showed that it was chemically closely related to atropine.

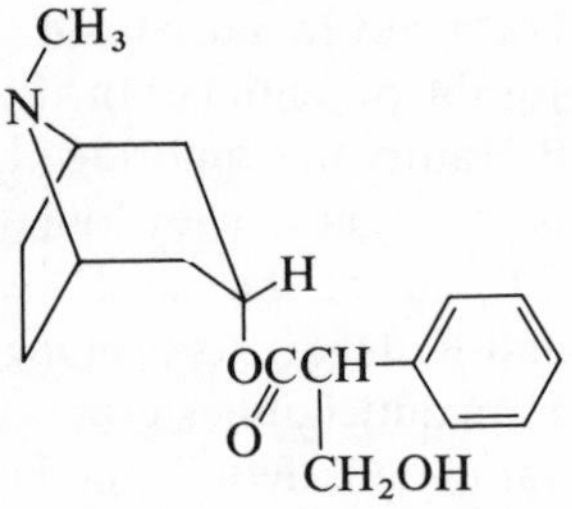

Atropine

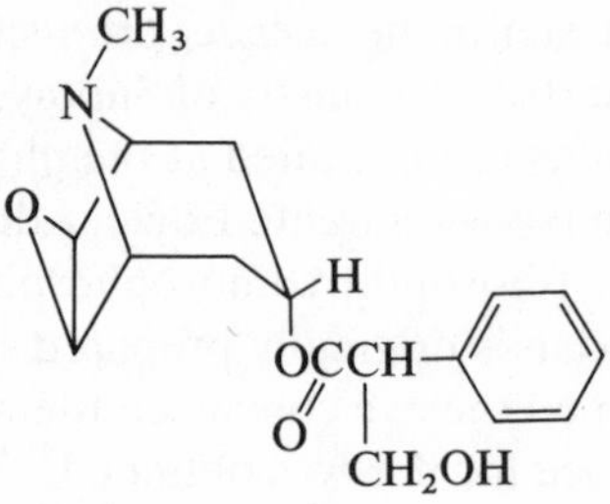

Hyoscine

Biologically active solanaceous plants were known to the ancient Greek and Roman medical writers, and even the Book of Genesis, which in its rejection of all that was holy to the Egyptians normally eschews the use of drugs and magical remedies, refers to the mandrake (*Mandragora officinarum*) for the relief of barrenness. Dioscorides similarly alluded to this, but the practice was based on a fanciful resemblance of its root to the human form, rather than any pharmacological action of its alkaloids.

The best known of the Solanaceae is the deadly nightshade, *Atropa belladonna*, a plant indigenous to southern Europe and the Middle East. Its toxicity, long recognized by poisoners, persuaded Linnaeus to confer upon it the full botanical name of *Atropa belladonna*, Atropos being known to the Greeks as one of the three fates, who could cut the slender thread of life. The mydriatic action of the plant (i.e. its ability to dilate the pupil of the eye) was exploited by the ladies of the Spanish court to beautify their eyes, thereby earning it the popular name of belladonna. The introduction of the herb into orthodox medical practice occurred in the nineteenth century when ophthalmologists began to dilate the pupil with its extracts. In the 1860s the eminent London pharmacist, Peter Squires, recommended a belladonna liniment for the relief of neuralgia. In doing so, he was merely following folk medical tradition. Such had been the faith in the herb, that it was frequently applied to superficial tumours, and occasionally taken internally in desperate attempts to halt the progress of cancer. Although atropine was later shown by Filehne to have a slight local anaesthetic action, it is probable that any pain-relieving value inherent in the liniment or in belladonna plasters was due to a counter-irritant action. One possible exception to this may have been the application of belladonna in situations where pain was caused by spasm of organs that could be affected by local absorption of the anticholinergic alkaloids. Of much greater value was the internal administration of the herb for the relief of colic, a practice long known to countrymen in the areas where the plant grew.

Another rich source of solanaceous alkaloids is thorn apple, *Datura stramonium*, also known as devil's apple, Jamestown weed, or Jimson weed. These latter names recall an incident in Jamestown, Virginia, where fatalities followed the introduction of the plant as a 'pot herb' by early settlers. A poultice or ointment prepared from the leaves of the plant was a popular American domestic remedy for insect bites and stings during the nineteenth century, while asthmatics were encouraged to smoke the dried leaves. Such a practice is not so far removed from the inhalation of the atropine analogue ipratropium ('Atrovent'), which was introduced by C. H. Boehringer Sohn in 1968, having fewer side-effects than atropine. Henbane (*Hyoscyamus niger*), which contains a greater proportion of hyoscine than do the others solanaceous plants, was known to Dioscorides. It has had a long history of folk use in Europe and the Middle East; witness the references to it in the *Arabian Nights*, as well as being an ingredient of witches' brews. It eventually fell into disuse until Baron Antonius Storck of Vienna included it in his *Libellus* of 1760. In the United States, hyoscine is known as scopolamine, reflecting the use of *Scopola carniolica* as a source of this mydriatic alkaloid. This plant was studied by the eighteenth century botanist Johann Scopoli, at the University of Pavia.

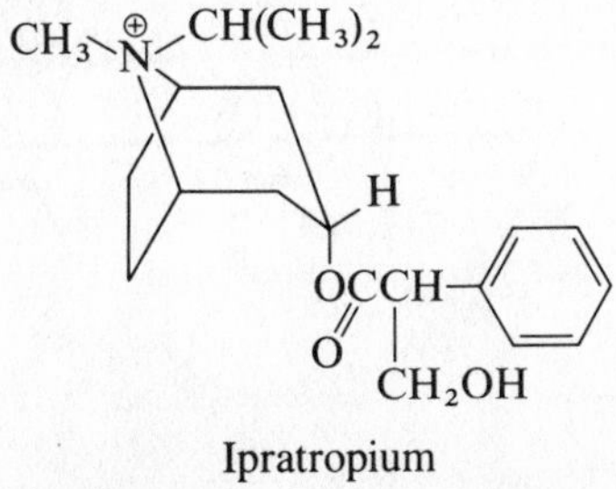

Ipratropium

Workers in the Hoffmann-La Roche laboratories in Switzerland overcame a major impediment to the clinical acceptability of hyoscine preparations when they discovered that racemization of the alkaloid took place in solution, to form equal amounts of laevo and dextro isomers. The dextrorotatory isomer was shown to be a central nervous system stimulant, in contrast to the laevorotatory isomer, which was a sedative. This explained why the racemic combination produced unpredictable effects. It then became possible to use the laevorotatory isomer of hyoscine for pre-anaesthetic medication in formulations such as papaveretum and hyoscine injection ('Omnopon-Scopolamine'), which was introduced in 1910.

When atropine was introduced into medical practice, it was recognized that its clinical effects were very similar to those of belladonna, although faster absorption of the pure alkaloid ensured a quicker onset of action. Furthermore, the sulphate salt of atropine was water-soluble, hence it became possible to inject the alkaloid to dry up salivary and other secretions during anaesthesia, or else to administer it in the form of eye drops to facilitate ophthalmological investigations.

Following the discovery, by Kraut and Lossen, that atropine could be split into two components, viz. tropic acid and a base called tropine, Alfred Ladenburg prepared 'synthetic atropine' by gently heating these two components in hydrochloric acid. Later in 1880, he esterified tropine with a variety of other aromatic acids to produce a series of physiologically active compounds which he called 'tropeines'. The tropeine derived from mandelic acid, homatropine, proved to act more quickly on the eye than did atropine itself, even though it was not as potent. It also had the important added advantage of not paralysing the eye muscles for several days, an annoying feature of atropine which prevented patients from focussing their eyes in order to read. The firm of E. Merck in Darmstadt marketed the new tropeine. It was one of the earliest examples of a synthetic drug that improved upon nature. A quarter of a century later, Albert Jowett and Franck Pyman at the Wellcome Chemical Laboratories in London, investigated the tropeines further, with a view to unravelling the relationship between their chemical structure and pharmacological activity, the latter being investigated by Henry Dale. Thirty compounds were examined during the next few years, but no superior agent was discovered.

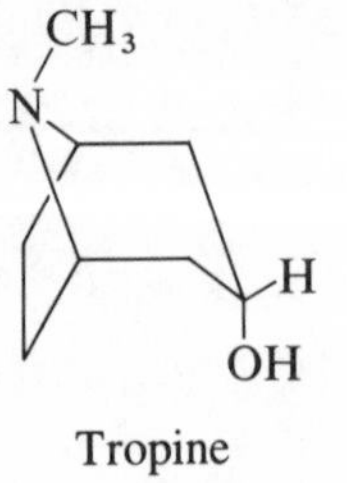

CH_3 N H OH

Tropine

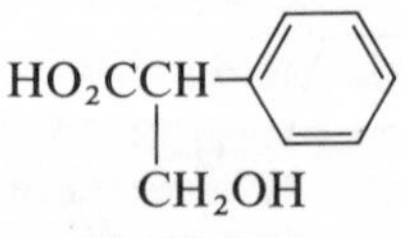

HO_2CCH CH_2OH

Tropic Acid

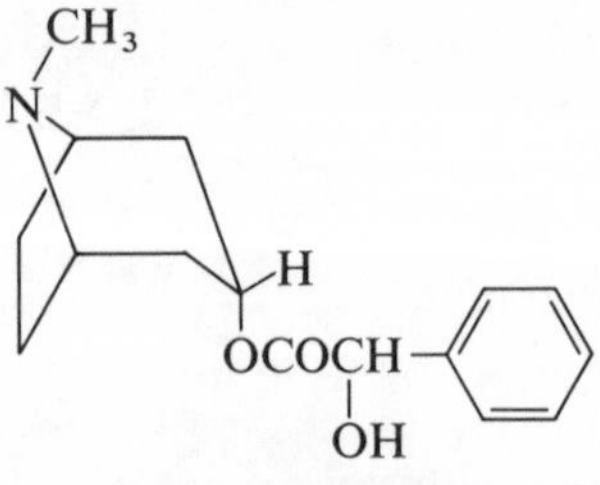

CH_3 N H $OCOCH$ OH

Homatropine

In 1902, the Bayer Company introduced atropine methonitrate ('Eumydrin'), the quaternary salt of atropine, as a mydriatic. Because of its polar nature it penetrated less readily into the central nervous system, thereby being a suitable agent for relieving pyloric spasm in young infants. However, the most important advance in the field of synthetic atropine analogues arose from work

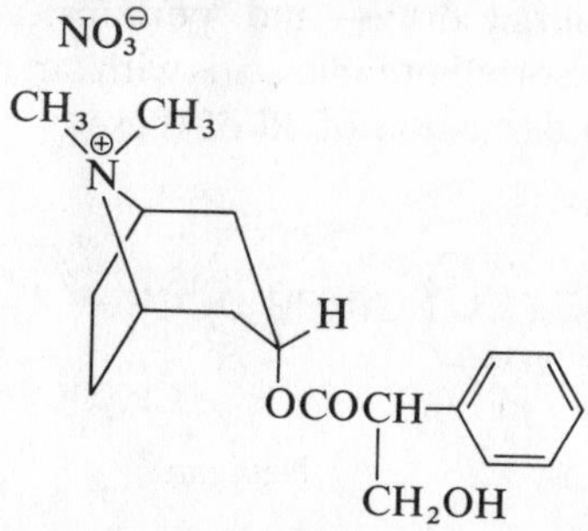

Atropine Methonitrate

conducted at the Landwirtschaftlichen Hochschule (Agricultural College) in Berlin and, later, at the University of Frankfurt by Julius von Braun. He was interested in the effects of transposing functional groups within drug molecules. In the case of atropine, he found that the tropate ester moiety could be moved to a different position on the tropine ring system without loss of mydriatic activity. On further investigation, it became evident that the presence of the intact ring was by no means essential for activity; all that was required was a tertiary amine function situated two or three carbons distant from the tropate ester function. In view of the success achieved in the early years of this century by conducting a similar exercise with the chemically related cocaine molecule to yield novel synthetic local anaesthetics, it is surprising that it took until 1922 before von Braun was able to demonstrate that not only could tropic acid be replaced with retention of mydriatic activity, but so, too, could the tropane ring. This cleared the way for a plethora of compounds to be synthesized in attempts to prepare analogues clinically superior to atropine. However, it was not until later in the decade that the first of these, the acetyl ester of one of the compounds prepared by von Braun, was marketed by the Swiss company Hoffmann-La Roche as a remedy for seasickness, under the name of 'Navigan'. It was claimed this was comparable with hyoscine or atropine. The next development also came from Hoffman-La Roche, and was based on the reasoning that as there was a close structural similarity between atropine and cocaine, it would be worth incorporating into a synthetic analogue of the former an amino alcohol which the company already used in the manufacture of its synthetic cocaine analogue known as dimethocaine ('Larocaine'). This led to the introduction of amprotropine ('Syntropan') in 1933. It had a fraction of the spasmolytic activity of atropine in the gut, but being still less potent with regard to antisecretory activity, it had a somewhat diminished tendency to cause a dry mouth when effective antispasmodic doses were administered. Although amprotropine only attained a modest degree of separation of antisecretory from antispasmodic activity, it was the forerunner of the synthetic selective hormone antagonists (e.g. adrenaline and histamine blockers) developed by medicinal chemists much later. It also stimulated pharmaceutical companies to begin a still unfinished quest that has produced a vast range of anticholinergic drugs—and, perhaps, a few useful therapeutic agents—by combining assorted organic acids with aminoalcohols. Little would be gained by a detailed discussion of all of these.

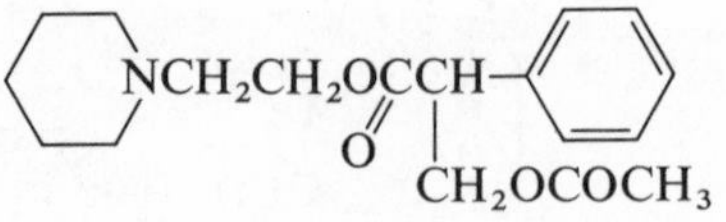

NCH_2CH_2OCCH / O / CH_2OCOCH_3

'Navigan'

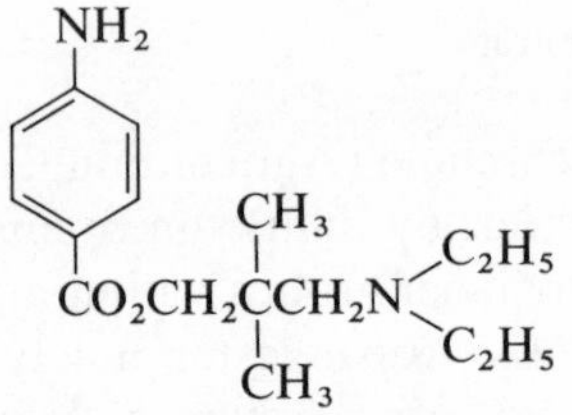

NH_2 / CH_3 / C_2H_5 / $CO_2CH_2CCH_2N$ / C_2H_5 / CH_3

Dimethocaine

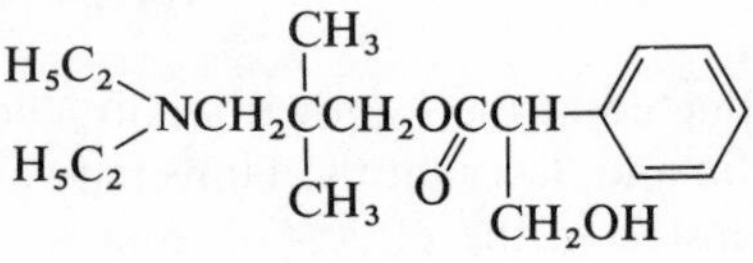

CH_3 / H_5C_2 / NCH_2CCH_2OCCH / H_5C_2 / CH_3 O / CH_2OH

Amprotropine

Although Fromherz of Hoffmann-La Roche also prepared highly potent analogues derived from benzilic acid when he synthesized amprotropine, it was left to Frederick Blicke of the University of Michigan to draw attention to the value of this acid in 1942. His observations were noted by H. R. Ing of the pharmacology department at the University of Oxford, who had been commissioned by the wartime Ministry of Supply to develop synthetic mydriatics, ostensibly for use in the event of supplies of atropine from the Middle East being cut off. At least, that was the explanation given to the public. However, the Ministry had been informed by B. A. Kilby and M. Kilby of Cambridge University that atropine afforded protection against the top secret fluorophosphonate anticholinesterase chemical warfare agents. Against this background, one can understand the priority given to the development of 'mydriatics' at the height of the war effort. To meet the requirements of the Ministry, A. H. Ford-Moore of the Dyson-Perrins Laboratory at Oxford synthesized some fifty benzilic acid esters. These were submitted to Ing's department for testing by Edith Bulbring and Izabella Wajda. Esters made from choline and its analogues proved to be of particular interest, one of these being equipotent with atropine as a mydriatic, although shorter acting. When injected, its toxic effects were less severe than those of atropine, and more like those of hyoscine. After the war ended it was introduced as a mydriatic. Ing gave it the name lachesine, after Lachesis who was the sister of Atropos. Many analogues of lachesine have been developed, such as pipenzolate ('Piptal'), piperidolate

('Dactil'), and mepenzolate ('Cantil') introduced in the fifties by Biel and his colleagues of Lakeside Laboratories, and Blicke's poldine ('Nacton').

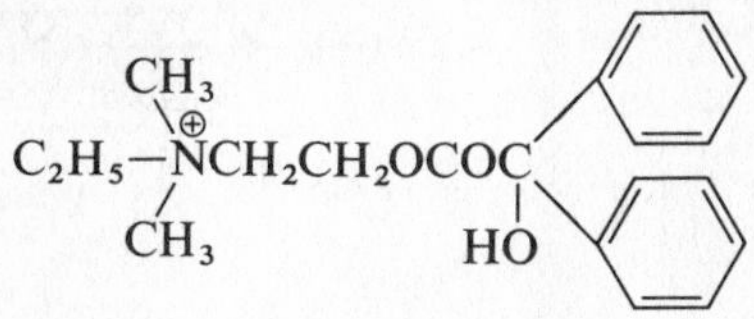

Lachesine

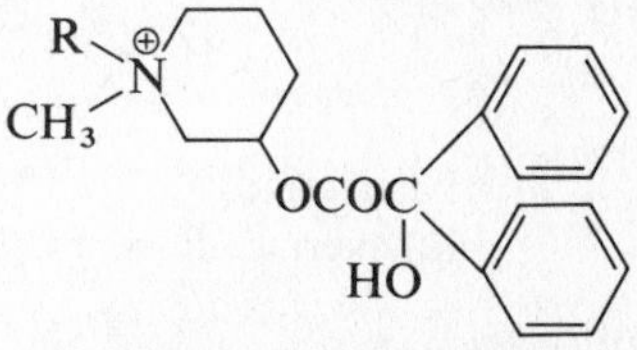

Pipenzolate, R = $-C_2H_5$
Mepenzolate, R = $-CH_3$

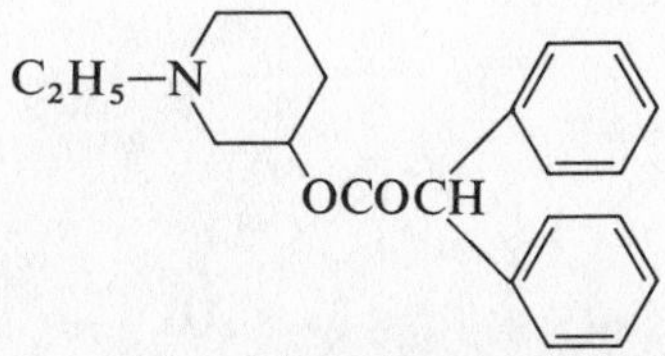

Piperidolate

Not all antispasmodics are derived from atropine. Some are analogues of papaverine, a benzylisoquinoline alkaloid discovered in 1848 by George Merck (son of Emmanuel Merck, the German alkaloid manufacturer), who recovered it from the mother liquors after extraction of morphine from opium. As it had only very slight analgesic activity, it was ignored until David Macht, at Johns Hopkins Medical School, carried out a detailed study of its pharmacology in 1917. This revealed it to act directly on smooth muscle to produce diminution of the movements of the stomach, intestines, and uterus. Macht attributed this to the presence of the benzyl group and introduced benzyl esters such as benzyl benzoate and benzyl phthalimide ('Akinetone') as antispasmodics. These could in no way rival the popularity of papaverine itself in the years prior to the introduction of the synthetic atropine analogues. This popularity created something of a problem in the mid-twenties when international legislation was introduced against the uncontrolled production of opium. The threat of supplies of papaverine being curbed through a shortage of opium resulted in renewed interest in the synthesis of it and its derivatives. Although papaverine had been prepared synthetically in 1909 by Ame Pictet and Alfons Gams at the University of Gens, this was of no commercial value since the overall yield was very low. An efficient synthesis by Carl Mannich at the University of Berlin in 1927 permitted the preparation of papaverine analogues in which either or both of the two pairs of methoxy-ether groups were replaced by methylenedioxy functions. Within two years, Otto Wolfes had obtained a patent on behalf of Merck and Company of Darmstadt for a series of papaverine analogues similar to Mannich's compounds. Merck claimed that one of these, 'Eupaverin' (a name later given to a different benzylisoquinoline alkaloid), was less toxic than

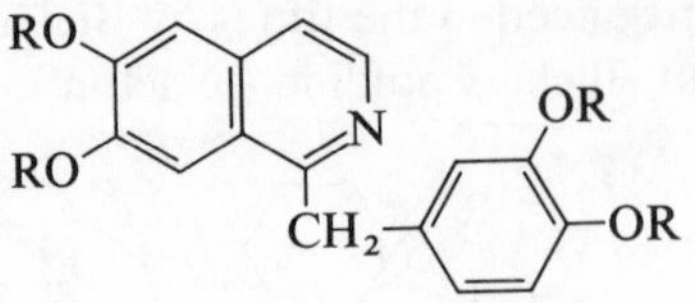

Papaverine, R = $-CH_3$
Ethaverine, R = $-C_2H_5$

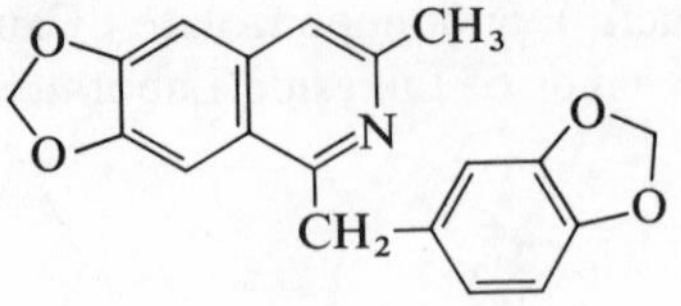

'Eupaverin'

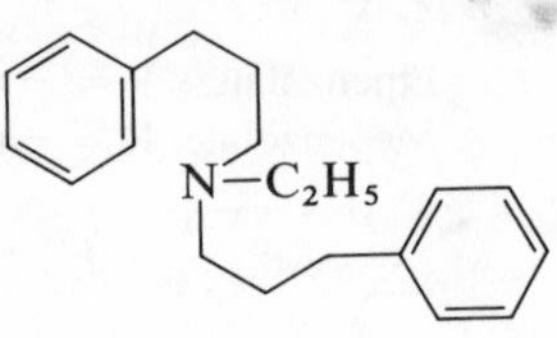

Alverine

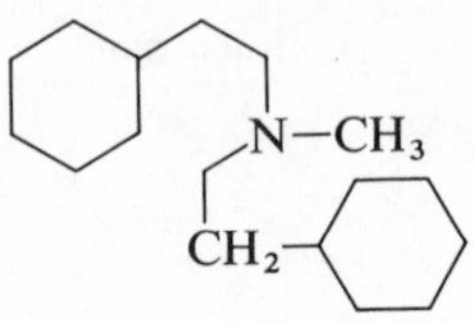

Cyverine

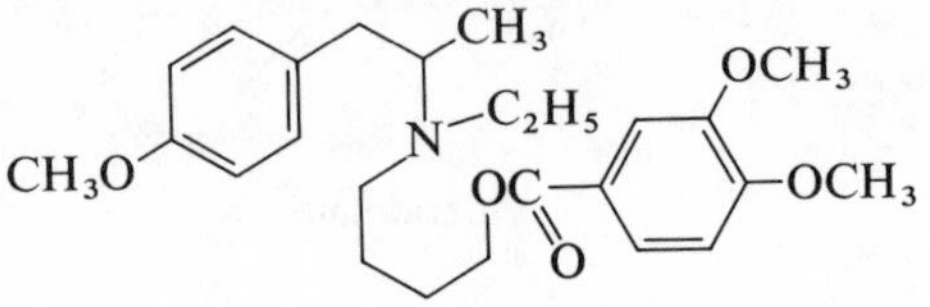

Mebeverine

papaverine. The Chinoin Company of Budapest, which also devised a synthesis of papaverine in 1930, developed several analogues, including ethaverine ('Perparine'), in which all the methoxyl groups of the natural alkaloid were replaced by ethoxyl groups. This was said to be three times as potent as papaverine, with reduced toxicity. It and 'Eupaverin' were widely prescribed as antispasmodics during the 1930s. Open-chain analogues of papaverine were also synthesized, such as alverine ('Sestron') by Karl Rosenmund and Fritz Kulz of the University of Kiel in 1935, cyverine by Frederick Blicke at the University of Michigan four years later, and mebeverine ('Colofac') by Kralt and his colleagues of the Dutch manufacturer Philips Gloielampenfabrieken in 1962. Although these papaverine analogues acting directly on smooth muscle are much less toxic antispasmodics than atropine and its analogues, they tend not to be as effective.

VII

Neuromuscular Blocking Agents

Pieter Martyr d'Anglera, a leading churchman at the court of Queen Isabella of Spain, published *De Orbe Novo* in 1516, this being an account of the experiences related to him by the earliest travellers to return from the New World. In his book, Martyr told of a Spanish soldier being mortally wounded near Santa Cruz by an arrow tipped with poison, further details of which emerged from more than fifty reports written by visitors to South America during the sixteenth and seventeenth centuries. Laurence Keynes, who served with Sir Walter Raleigh on his expedition to the region now known as Venezuela, compiled a list of poisonous herbs, one of which was known by the natives as *ourari*. From this term arose the word 'curare' to describe those South American arrow poisons that killed by causing paralysis, as distinct from those found elsewhere and which acted on the heart. The main use of curare was in hunting wild animals, for which purpose it was eminently suitable since it decomposed during the cooking of the slaughtered animal.

The first scientist who received permission from the Spanish authorities to explore the part of South America where the arrow poison was prepared was Charles Marie de la Condamine. In 1735, he was put in charge of an expedition to Ecuador sponsored by the French Academy of Sciences to settle a geographical dispute concerning the relationship of the equator and the polar regions. During the ten years that he spent in Ecuador, he collected samples of curare which he brought to France on his return. He and others showed these could kill chickens and other animals when injected. A later explorer was the physician and naturalist, Edward Bancroft, who, in 1769, established that there were several different arrow poisons used by various tribes.

The introduction of curare preparations into Europe naturally led to its examination by physiologists who sought to establish how it could paralyse animals. The most celebrated of these early studies was carried out in Paris by Claude Bernard in 1842, being reported eight years later. He found that although curare prevented the ligated limb of a frog from responding to nerve stimulation, both the nerve and muscle retained their ability to function. From this he concluded that the drug acted at the neuromuscular junction to interrupt the stimulation of the muscle by the nerve impulse. However, it was not until 1936, when Henry Dale, Walter Feldberg, and Martha Vogt at the National Institute for Medical Research showed acetylcholine was the chemotransmitter at the neuromuscular junction, that it was realized that curare owed its effects to antagonism of this neurohormone.

During the nineteenth century, Magendie and many other physiologists used curare to render conscious animals immobile during experiments, secure in the knowledge that only the neuromuscular junction was affected. It was necessary to maintain respiration artificially by means of a bellows. Such was the extent of this practice, that public indignation was aroused in the United Kingdom, leading to the setting up of a Royal Commission on Vivisection in 1875 and the subsequent passing of legislation to control experiments on animals.

Sporadic attempts to employ curare in the treatment of rabies, tetanus, chorea, and epilepsy met with little success during the nineteenth century. The extracts were of dubious quality and variable potency due to the uncertainty that surrounded the source of the plant from which they were made. It was not until 1939 that a firm identification of the source of a physiologically active curare preparation was made, thereby removing the main obstacle to the introduction of curare into medicine. The man responsible for this was Richard Gill, a former salesman for an American company in Lima, who moved to Ecuador after the Wall Street crash in 1929. There, he set up a successful plantation to produce coffee and cacao. Troubled by a muscular spasm in his leg, he visited Washington to consult Walter Freeman, a neurologist. During the consultation, Freeman remarked that he would welcome the opportunity of obtaining curare of sufficiently high and consistent quality as to permit experimental work to be conducted. Gill decided to do whatever was possible to help. He contacted the Merck Institute for practical advice, then secured adequate funding from Sayre Merrill to support an expedition into the jungle. This departed for Ecuador in 1938, spending five months there while Gill collected twenty-six varieties of lianas (vines) from which he prepared over fifty kilograms of curare. On returning to the United States in December, Gill was informed that Merck and Company were no longer interested in curare. Shortly after, Walter Freeman met Abram Bennett, a neuropsychiatrist practising in Omaha, Nebraska, who had been trying to obtain samples of curare to prevent the unacceptable high incidence of vertebral fractures during the newly developed technique of convulsive shock therapy. Freeman introduced Gill to Bennett, who explained that he would like to have some curare standardized so as to test its effects in spastic children, as well as in shock therapy. Gill immediately supplied a large quantity of his curare, and it was agreed that this would be standardized by Professor A. R. McIntyre of the department of pharmacology at the University of Nebraska. Bennett then carried out a clinical trial of the drug on twelve spastic children at the Nebraska Orthopedic Hospital in Omaha (funding for this and all subsequent studies in Nebraska was raised from Bennett's own private research foundation). The relaxing action of curare in the spastic children proved to be only of a transient nature, but the results in shock therapy were much more encouraging. It rendered this otherwise dangerous therapy for depressive patients safe enough for routine application. Meanwhile, Gill signed a contract with E. R. Squibb and Sons, who purchased his remaining material. One of the company's pharmacologists, Horace Holaday, simplified the standardization procedure by devising the rabbit head drop assay, after which Squibb made curare available free of charge to clinical researchers as 'Intocostrin', which they described as 'Unauthenticated Extract of Curare'. Subsequently, the plant brought back by Gill which provided the highest yield of curare was identified as *Chondodendron tomentosum* by botanists at the New York Botanical Gardens.

Curare was successfully used in the Nebraska State Mental Hospital to permit pelvic examinations to be conducted on disturbed female patients. This gave Lewis Wright, a medical consultant to Squibb, the idea that it might be possible to administer curare to overcome the lack of adequate muscle relaxation associated with the recently introduced anaesthetics such as ethylene, cyclopropane, and hexobarbitone. Furthermore, it might also permit adequate muscle relaxation with lighter levels of the older general anaesthetics. Wright made arrangements for 'Intocostrin' to be tested at the New York University School of Medicine and at the department of anesthesiology in the University of Iowa. The results when the drug was administered to cats and dogs anaesthetized with ether were most discouraging, the cats all dying and the dogs becoming deeply distressed. No further investigations were carried out. This left Wright with the problem of finding someone to do further tests. Eventually, Harold Griffith, a country doctor who served as a part-time anaesthetist at the Homeopathic Hospital in Montreal, agreed to try the drug on patients anaesthetized with cyclopropane. His wide experience with this anaesthetic had convinced him that patients should be kept on artificial respiration because of the risk of respiratory depression. Unwittingly, therefore, he overcame the two problems that had led the previous investigators to abandon the use of 'Intocostrin' in anaesthesia. Firstly, ether itself had, as was shown later, neuromuscular blocking activity in its own right. This meant it potentiated the effects of the drug. Secondly, the paralysis of the respiratory musculature by 'Intocostrin' necessitated artificial respiration. The outcome of a test of the drug in a twenty-year-old youth undergoing an appendectomy was so encouraging that Griffith went on to inject curare in a further twenty-five patients undergoing anaesthesia. Within eighteen months, this and other clinical trials confirmed the revolutionary role of the drug in making anaesthesia much safer by permitting lower doses of anaesthetic for the attainment of full muscular flaccidity.

The active principle responsible for the efficacy of 'Intocostrin' was actually isolated in crystalline form in 1935 by Harold King at the National Institute for Medical Research, using a preparation of curare obtained from the British Museum. Its botanical source was unknown, but since the material had been packed in the traditional bamboo tube, King named it tubocurarine. It was another eight years before the Squibb chemists Oskar Wintersteiner and James Dutcher isolated it from the curare extracted from *Chondodendron tomentosum*. Only then could adequate supplies of this alkaloid to meet clinical requirements be guaranteed. However, this more or less coincided with the completion of the trials on 'Intocostrin', so that it was the pure alkaloid rather than the crude preparation which anaesthetists began to use.

Tubocurarine, R = —H
(King's original structure, R = —CH_3)

Alcuronium

The work of Heinrich Wieland in Germany and Paul Karrer in Switzerland following on King's isolation of tubocurarine resulted in the identification and characterization of scores of potent curare alkaloids from the bark of varieties of *Strychnos*, especially *Strychnos toxifera*. In 1961, Hoffmann-La Roche introduced a derivative of one of these toxiferines, namely alcuronium (diallylnortoxiferine; 'Alloferin'). This had similar properties to tubocurarine.

In 1946, Daniele Bovet and his colleagues in the department of chemical therapeutics at the Pasteur Institute in Paris initiated a project to synthesize simplified tubocurarine analogues. Since the structure of the alkaloid as proposed by King contained two quinoline rings and two quaternary ammonium functions, Bovet's group synthesized a series of compounds in which two molecules of 8-hydroxyquinoline were linked by phenolic ether formation with an alkyl chain prior to quaternization. One of the series, RP 3381, which was examined by the Rhône-Poulenc Laboratories, proved to be the first synthetic substance whose curare-like action on mammals had a specificity of action comparable with that of tubocurarine. Thus encouraged, Bovet concentrated his efforts on the synthesis of phenolic ethers containing quaternary ammonium functions. This led to the synthesis of gallamine ('Flaxedil') in 1947. Its onset of action was quicker than the three to five minutes required for tubocurarine, whilst recovery was briefer than the thirty minutes or more for the natural alkaloid. Unfortunately, it caused acceleration of the heart and could not be safely used in patients with kidney disease.

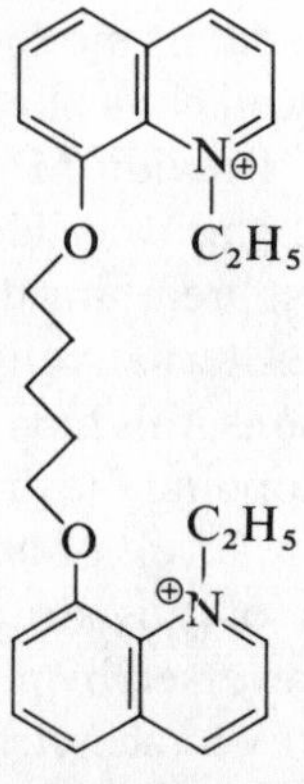

RP 3381

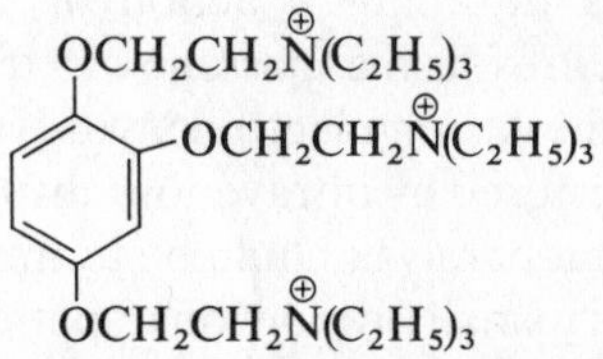

$OCH_2CH_2\overset{\oplus}{N}(C_2H_5)_3$

$OCH_2CH_2\overset{\oplus}{N}(C_2H_5)_3$

$OCH_2CH_2\overset{\oplus}{N}(C_2H_5)_3$

Gallamine

Around the same time that Bovet began his study of tubocurarine analogues in Paris, a similar enquiry was begun at Oxford by R. B. Barlow and H. R Ing. Their approach was to establish whether varying the distance between the two quaternary ammonium functions in King's structure for the alkaloid would increase potency. To this end they examined the simplest possible series of compounds containing two quaternary groups at opposite ends of straight carbon chains, the so-called polymethylene bis-onium compounds. It was found that the compound with a ten carbon chain was three times as potent as tubocurarine when assayed by the rabbit head drop test. By a remarkable coincidence, the identical discovery was made simultaneously by William Paton and Eleanor Zaimis at the National Institute for Medical Research. While investigating the ability of the octamethylene bis-onium compound to liberate histamine in a cat, they discovered that the animal was being asphyxiated, yet was not exhibiting the obvious signs of respiratory distress such as gasping or convulsions. This suggested to them that the drug was acting as a neuromuscular blocking agent. At this point, contact was made with Barlow and Ing, with agreement on simultaneous publications early in 1948. Paton and Zaimis confirmed that the ten-carbon compound was the most potent member of the bis-onium series. They also obtained evidence that it acted by depolarizing the muscle end plate so as to prevent it responding to acetylcholine. Unlike tubocurarine and gallamine, it did not behave as a competitive antagonist of acetylcholine. The compound underwent satisfactory clinical trials and received the approved name of decamethonium. It was marketed for a few years by Allen and Hanburys as 'Eulissin', and by Burroughs Wellcome as 'Syncurine'. Due to its resistance to metabolic deactivation, paralysis was unnecessarily prolonged and could not be reversed with physostigmine. This meant artificial respiration had to be maintained until the effects of decamethonium eventually wore off. To overcome this problem, Bovet examined succinylcholine, a bis-onium analogue of acetylcholine. Reid Hunt had studied this along with acetylcholine forty years earlier, but was unable to observe its neuromuscular blocking activity because his animals had been curarized. Bovet found it to produce very brief neuromuscular blockade; its mode of action was the same as that of decamethonium. It was introduced clinically in 1949, with the approved name of suxamethonium ('Scoline', 'Brevidil M', 'Anectine'). With a duration of action of only five minutes, this was ideal for short procedures such as passage of a tracheal tube to assist breathing during general anaesthesia, or electroconvulsive shock therapy. Alternatively, it could be administered by intravenous drip for longer operations, this having the advantage that paralysis could be terminated merely by stopping the drip. However, in a very small proportion of patients, abnormally low levels of the pseudocholinesterase responsible for the rapid metabolism of this drug resulted in prolonged respiratory paralysis. This could not be reversed by physostigmine. Another disadvantage was that the inherent similarity to acetylcholine caused muscle twitching prior to paralysis, and this often resulted in post-operative cramp.

$$(CH_3)_3\overset{\oplus}{N}-(CH_2)_{10}-\overset{\oplus}{N}(CH_3)_3$$

Decamethonium

$$(CH_3)_3\overset{\oplus}{N}CH_2CH_2O\overset{O}{\overset{\|}{C}}CH_2CH_2\overset{O}{\overset{\|}{C}}OCH_2CH_2\overset{\oplus}{N}(CH_3)_3$$

Suxamethonium

A chance remark by John Stenlake of the department of pharmacy at the Royal College of Science and Technology in Glasgow (now the University of Strathclyde), relating to the preparation of sulphur-containing bis-alkylammonium compounds prepared in his laboratory as potential anti-tubercular agents, was made to John Lewis of the department of materia medica at the University of Glasgow and resulted in a collaborative programme of research aimed at finding an improved analogue of suxamethonium. This objective was ultimately achieved by Stenlake, some twenty years later, but in the shorter term it drew Lewis's attention to the chemical requirements of neuromuscular blockade. When he received some aminosteroids from Christopher Hewett and David Savage of the nearby Organon Laboratories in Newhouse, he and his colleagues Michael Martin-Smith, Thomas Reid, and H. H. Ross decided to synthesize and investigate a series of monoquaternary derivatives of these steroids. One which incorporated an acetylcholine-like residue in ring A of androstane, turned out to have about one-sixteenth of the potency of tubocurarine. This development was reported in 1964. Comparison of this monoquaternary steroid with the naturally occurring bis-quaternary neuromuscular blocking agent known as malouetine, gave Hewett, Savage, and Roger Buckett the idea of duplicating the acetylcholine-like moiety at both ends of a steroid rather than just in ring A. This ploy led to the successful development of pancuronium bromide ('Pavulon'), the pharmacological properties of which were first reported in 1966. Due to its more rapid onset of action and relative freedom from effects on the blood pressure, pancuronium replaced tubocurarine as the relaxant of choice in major surgery. Nevertheless, its onset of action was still slower than that of suxamethonium, whilst its duration was in the order of forty-five minutes.

At the same time as preparing pancuronium, the Organon researchers prepared a variety of closely related compounds. Amongst these was the monoquaternary analogue of pancuronium. At the time of its synthesis, this had not seemed of sufficient interest to warrant detailed study. However, in 1970, A.J. Everett, L. A. Lowe, and S. Wilkinson of the Wellcome Research

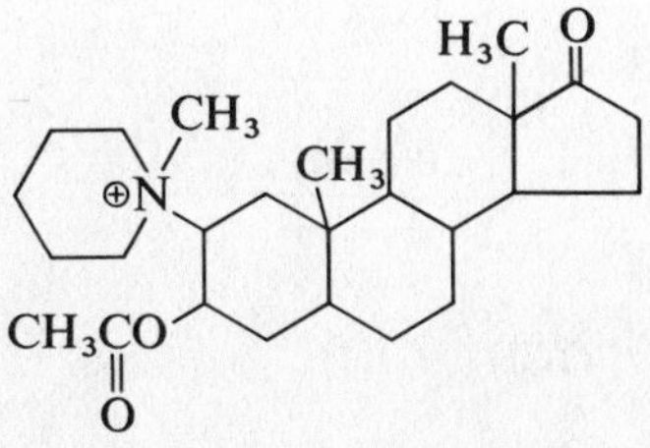

Monoquaternary Aminosteroid

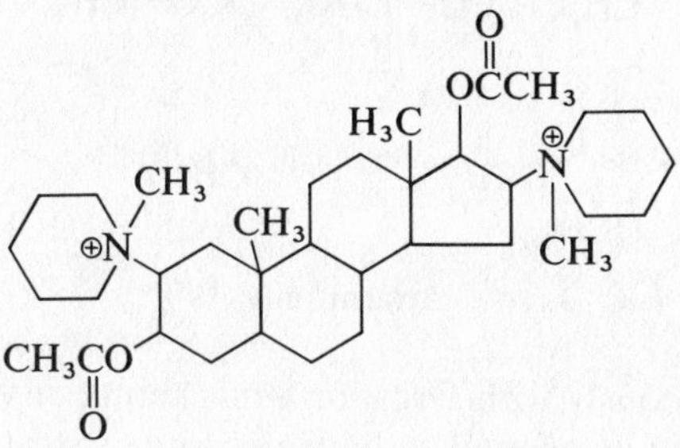

Pancuronium

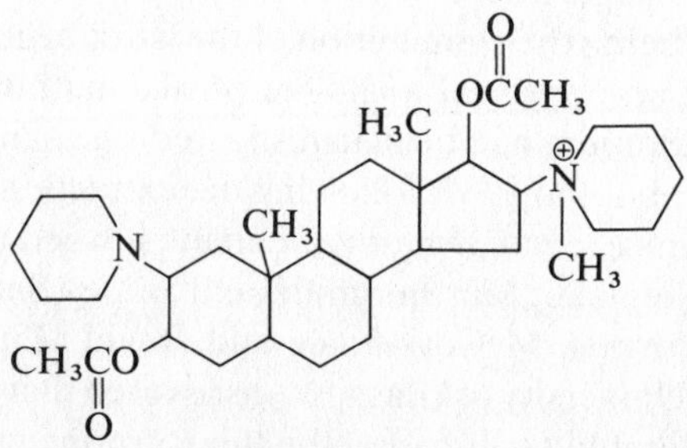

Vecuronium

Laboratories in Beckenham discovered that tubocurarine was not a bisquaternary compound as previously believed, but instead was a monoquaternary alkaloid. One of the amino groups was tertiary. This meant that when Organon submitted more than a hundred compounds to Ian Marshall of the department of pharmacology at the University of Strathclyde for assessment as neuromuscular blocking agents, included amongst these was the monoquaternary analogue of pancuronium. When the Organon compounds were passed through a variety of pharmacological tests by Marshall, this analogue emerged as having the best balance of desirable properties, and was superior to pancuronium itself. It was introduced clinically in the United Kingdom in 1983 with the approved name of vecuronium ('Norcuron').

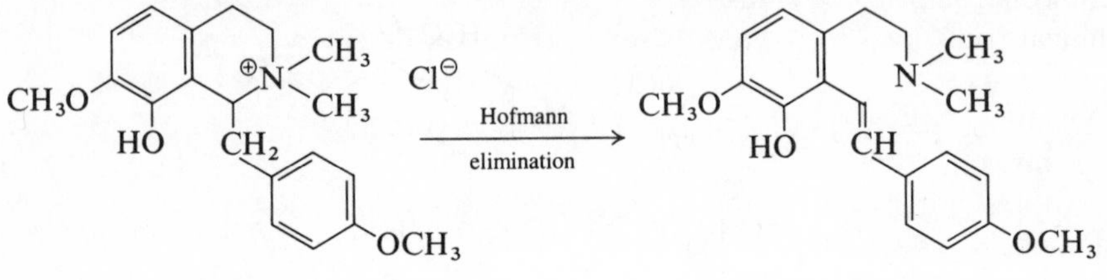

Hofmann elimination

Petaline

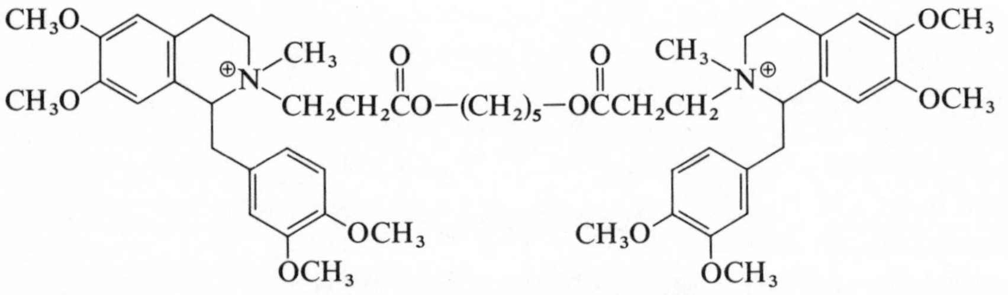

Atracurium

Almost simultaneously with Organon's marketing of vecuronium, another neuromuscular agent developed at the University of Strathclyde was launched by Burroughs Wellcome, namely atracurium ('Tracrium'). This rapidly metabolized tubocurarine substitute was synthesized in the pharmaceutical chemistry division of the department of pharmacy by Professor Stenlake and his research students, being the culmination of the work begun in the late fifties in an attempt to find an improved analogue of suxamethonium. Initially, ester related to suxamethonium had been investigated. A change in the direction of the research came about in 1964, following the extraction of an alkaloid called petaline from *Leontice leontopetalum*, a Lebanese weed used in folk medicine to treat epilepsy. Stenlake, Martin-Smith, and Sidney Smith at Strathclyde, in association with Norman McCorkindale and David Magrill of the chemistry department at the University of Glasgow, discovered that petaline not only had a structure that resembled one-half of the tubocurarine molecule, but was also somewhat unstable. The unusual facility with which petaline underwent the

Hofmann elimination reaction gave Stenlake the idea of synthesizing tubocurarine analogues based on it, with a view to obtaining a rapidly decomposing drug that would persist in the body only a few minutes. With the assistance of his research students, Roger Waigh, John Urwin, George Dewar, and Nirmal Dhar, Stenlake prepared a range of compounds in which the influence of electron-withdrawing properties of substituents in different positions was carefully considered in relationship to the ease of Hofmann elimination. This led to the synthesis of atracurium. It was pharmacologically assessed at the Wellcome Research laboratories by Roy Hughes and Dennis Chapple, who found it to possess outstanding properties. The drug was patented in 1977 by Burroughs Wellcome after the signing of an agreement whereby royalties would be paid to the University of Strathclyde. It was subjected to its first clinical trials by J.P. Payne, research professor of anaesthetics at the Royal College of Surgeons of England, who confirmed in humans its freedom from unwanted cardiac effects, as well as the speedy recovery of all patients, despite the status of their overall health.

Although the Medical Research Council had initially supported Professor Stenlake's research programme with a small grant, it abandoned this after three years. The Wellcome Research Laboratories then showed considerable enthusiasm for the work, as a result of which Burroughs Wellcome has acquired considerable revenue from atracurium, which sells in a market involving well over ten million injections of neuromuscular blocking agents every year. The entire profits of the company are remitted to the Wellcome Foundation, which exists to support medical research. Allen and Hanburys (now Glaxo) was another British manufacturer who exploited a drug synthesized in an educational establishment. On routine screening of a novel cyclic quaternary ammonium salt made in 1972 at the Teeside Polytechnic in Middlesborough by E.E. Glover and M. Yorke, it was found that fazadinium ('Fazadon') had promising activity as a neuromuscular blocking agent. However, clinical experience has shown it to offer little advantage over earlier compounds as it causes acceleration of the heart.

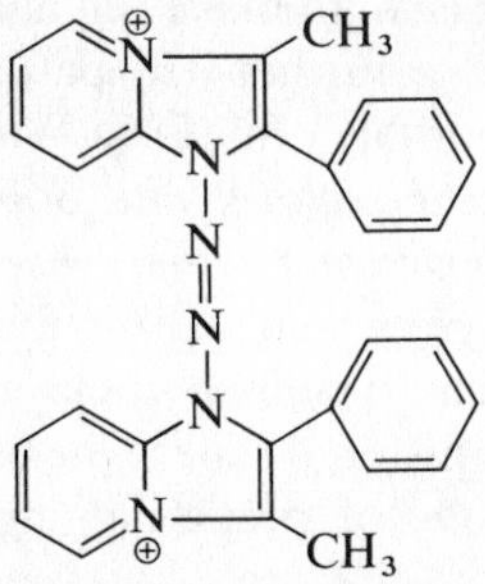

Fazadinium

VIII

Cardiovascular Drugs

CARDIAC STIMULANTS

In May 1775, William Withering moved from Stafford to a post in Birmingham, where a vacancy had arisen following the death of one of the town's leading medical practitioners. It was around this time that his opinion was sought on the merits of a family recipe for the relief of dropsy. This was a herbal tea said to have been concocted by an old woman in Shropshire; it was supposed to have effected cures where conventional treatments had failed. Having rejected the old belief that the external appearance of plants was a pointer to their therapeutic properties, Withering was particularly receptive to reports of the successful administration of herbs to the sick. He believed this was the only way in which the curative powers of herbs could be discovered. As he was in the midst of compiling what was to become the first scientific study of botany in the English language, Withering realised that amongst the twenty or so ingredients in the herbal tea it must have been the foxglove that was causing the violent vomiting and purging that accompanied its use. This plant had a long history of folk use, having been listed amongst herbs used by Edward III (1327–1377). Originally, it was known as the folksglove, apparently on account of the shape of its flowers. This, and their colour, was also indicated by the Latin name *Digitalis purpurea*, given to it in 1542 by the German botanist Leonhard Fuchs, who explained this was an allusion to the word *fingerhut* (thimble). He described digitalis as a violent medicine.

Although Withering treated one of his patients with digitalis in December 1775, followed by a further four during the next twelve months, it was only after he heard that the principal of Brasenose College in Oxford had been relieved of dropsy by digitalis that he began to make regular use of the plant. During the next nine years he treated a further 158 patients with it, of whom about two-thirds responded favourably. In 1785, Withering wrote his celebrated treatise *An Account of the Foxglove, and Some of its Medical Uses: with Practical Remarks on Dropsy, and Other Diseases*. This described how to determine the correct dosage, which was highly relevant since digitalis was a potent poison (once known as 'deadmen's bells' in Scotland!) that was ineffective medicinally unless administered at near the toxic dose level. Withering also discussed the different ways of preparing digitalis, favouring the use of powdered leaves.

Withering was convinced that digitalis acted directly against dropsy since it did not increase urinary flow in people who did not suffer from fluid retention. Although he wrote that the drug '. . . has a power over the motion of the heart, to a degree yet unobserved in any other medicine, and that this power may be converted to salutary ends . . .,' he did not realize that its cardiac stimulant action was responsible for its beneficial role in dropsy. Indeed, it was not until the year of his death, 1799, that anyone suggested this might be the case, when one of his friends, the Manchester physician John Ferriar, argued that the increased urinary output was of secondary importance when compared with the power of digitalis to reduce the pulse rate, whilst Thomas Beddoes wrote that the drug increased the contractile action of cardiac muscle fibres. Unfortunately, these views did not deter Ferriar or Beddoes from recommending digitalis as an excellent remedy for phthisis (pulmonary tuberculosis). Nor was this the sole misuse of the drug, for it soon came to be seen as a panacea for a wide range of ills, including insanity, epilepsy, depression, typhoid, pneumonia, diphtheria, scarlet fever, measles, goitre, haemorrhoids, and habitual abortion, not to mention all conceivable types of heart ailment.

Richard Bright, the leading London physician of his time, was the first to distinguish between dropsy due to kidney disease and that arising from heart failure. In his *Reports of Medical Cases*, published in 1827, he pointed out that if the urine was albuminous, then the dropsy was of renal origin. This should have laid the basis for the correct use of digitalis, yet as late as the 1855 edition of his *Elements of Materia Medica and Therapeutics*, Jonathan Pereira stated that the principal value of digitalis lay in its action against dropsy, without indicating that this should be of cardiac origin. By 1890, however, the sixteenth edition of *The Dispensatory of the United States of America* indicated that the sole use of digitalis should be in heart disease. It is also worthy of note that as recently as 1911, Potter's *Materia Medica and Pharmacy* recommended digitalis as an anaphrodisiac. The writer, at least, had the good grace not to associate this recommendation with his subsequent comment that, 'Digitalis is said by high authority to be particularly adapted to blondes and persons of sanguine and indolent temperament.'

Only in the early years of this century did a proper understanding of the effects of digitalis on the heart emerge. This was principally due to the introduction of the polygraph in 1902 by James Mackenzie in Burnley, followed a year later by the electrocardiograph, invented by William Einthoven at Leyden University. Important investigations were carried out in London by both Arthur Cushny and Mackenzie, and in Holland by Karl Wenckebach, who studied the effects of digitalis on the residents of a home for elderly people. As a result of their work, the correct indications for the use of digitalis became apparent. One of these was atrial fibrillation, in which the coordination of the contraction of individual muscle fibres in the atrium of the heart was so disorganized that the heart beats became weak and irregular. Other indications for the use of digitalis were certain forms of heart failure in sinus rhythm.

Attempts to isolate an active principle from digitalis in the 1820s met with no success. To stimulate research, the Société de Pharmacie in Paris offered a prize of 500 francs for the isolation of a pure principle from the plant, the sum having to be doubled after five years had passed without any claimant coming forward. In 1841, E. Homolle and Theodore Quevenne, the chief pharmacist at the Hôpital de la Charité in Paris, won the award for their isolation of an impure, but active, crystalline material, which probably consisted mainly of digitoxin. They called this digitaline, a name subsequently applied to several different products obtained by other workers (the final 'e' was elided to avoid being mistaken for an alkaloid after it was realized that this substance belonged to a distinct class of compounds known as glycosides). A material called digitaline crystallisée was obtained in 1869 by Claude-Adolphe Nativelle. This may have been very similar to Oswald Schmiedeberg's digitoxin, a crystalline compound of high potency, isolated in 1875 in Strassburg from digitalis leaves by a method similar to that used by Nativelle. This was the principal cardiotonic glycoside present in the leaves of *Digitalis purpurea*. The principal glycoside in the seeds was isolated by Schmiedeberg from 'Digitalin', an extract produced by Henn and Kettler, a local manufacturer. Confusingly, he named this digitalin although the commercial product continued to be available for many years (it was still included in the *British Pharmaceutical Codex* in 1954). It was much less potent than digitoxin. Such was the complexity of the chemistry of these two glycosides that it took almost half a century before Adolf Windaus and his assistants at Göttingen University established the correct structure of digitoxin, in 1928, and digitalin in the following year. In 1933, Arthur Stoll revealed that all the cardiotonic glycosides previously isolated from different varieties of *Digitalis* were artefacts, insofar as decomposition of their sugar component had occurred during extraction.

In the late twenties, it was discovered that the powdered leaves of *Digitalis lanata* ('woolly foxglove') had greater physiological activity than those of *Digitalis purpurea*. This led Sydney Smith of Burroughs Wellcome and Company in Dartford, London, to isolate and separate the glycosides which were present. In the course of this, he obtained a new glycoside which he called digoxin. Nowadays, this is used more widely than either powdered digitalis leaves or digitoxin since it does not bind as strongly to proteins in the tissues and plasma. This means that there is less delay before a therapeutic concentration of unbound, active drug builds up, and its clearance from the body is also faster as only unbound drug is removed by the kidneys. This makes it less cumulative and thus safer to use. If a faster onset of action is required, ouabain can be administered intravenously. This is a substance isolated by Arnaud in 1888 from the roots and bark of the ouabaio tree (*Acocanthera* sp.), from which the Somalis of East Africa prepared their arrow poison. Ouabain is unique amongst the cardiotonic glycosides in the ease with which it can be crystallized, a property which led to its being adopted as the standard against which the activity of all other such substances could be measured. It was also found to be present in the arrow poison prepared from seeds of *Strophanthus gratus* by the Pahouins. Other arrow poisons prepared from infusions of the barks and seeds of different varieties of *Strophanthus* were investigated by Thomas Fraser of Edinburgh University in the last quarter of the nineteenth century. The isolation of cardiac stimulants from these arrow poisons led to renewed interest

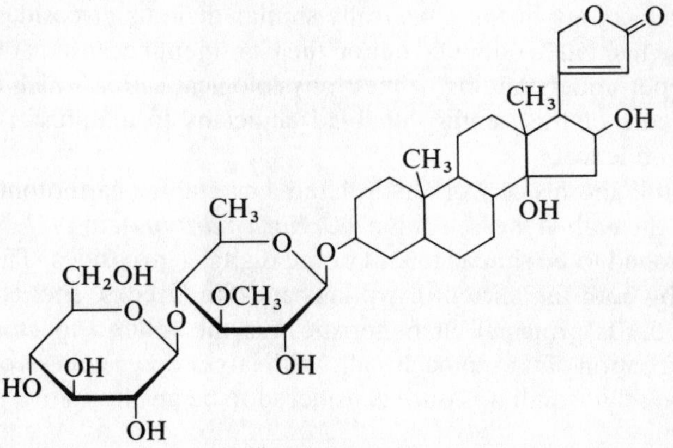

Digitalin

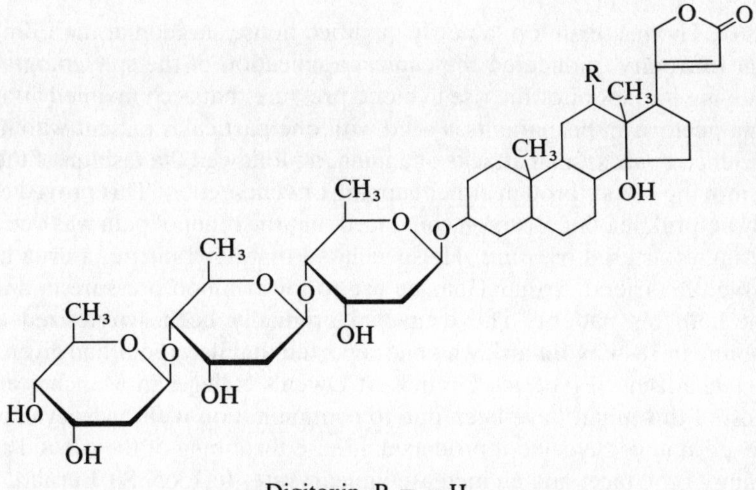

Digitoxin, R = —H
Digoxin, R = —OH

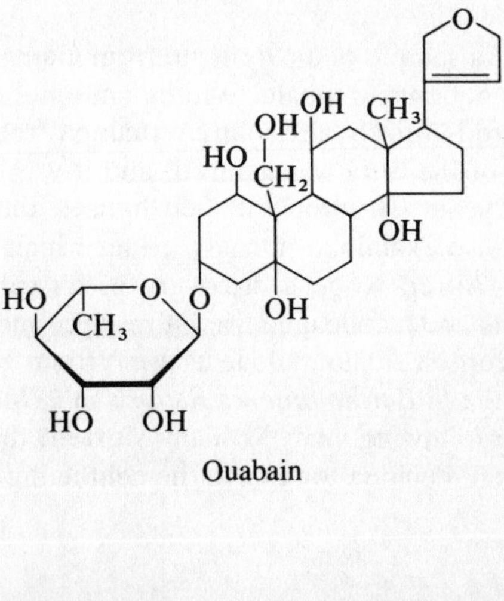

Ouabain

in the pharmacology of the chemically similar digitalis glycosides. To some extent, these had fallen out of favour on the Continent because of their misuse, but the deeper understanding of their physiological action which emerged in the early years of this century enabled clinicians to administer them with renewed confidence.

Arthur Stoll and his co-workers isolated a crystalline cardiotonic glycoside from squill, the bulb of the 'sea onion' (*Urginea maritima*), in 1933. Its chemical structure proved to be similar to that of the digitalis glycosides. The plant had been used by both the ancient Egyptians and the Greeks, and its popularity never waned. Its principal characteristic was the reflex expectorant action caused by irritation of the stomach wall. With larger doses, emesis occurred, so it is surprising that squill was once considered to be an alternative to digitalis.

ANTIANGINAL DRUGS

In 1867, Thomas Brunton, a newly qualified house surgeon at the Edinburgh Royal Infirmary, pioneered the clinical application of the sphygmograph by employing it to monitor the rise in blood pressure that accompanied attacks of angina pectoris in his patients. Faced with one particular patient who nightly experienced paroxysmal attacks of angina, he followed the fashion of the time by removing blood through either cupping or venesection. This proved consistently helpful, leaving Brunton convinced that the relief of pain was due to the lowering of arterial pressure. He speculated that amyl nitrite, a drug he had observed his friend Arthur Gamgee use to lower blood pressure in animals, might help his patient. The drug had originally been synthesized at the Sorbonne in 1844 by Balard, who had reported that its vapour had given him a severe headache. Frederick Guthrie of Owen's college, in Manchester, had suggested this might have been due to contamination with hydrocyanic acid. His experiments revealed it produced intense throbbing of the carotid artery, flushing of the face, and an increase in heart rate. In 1865, Sir Bernard Ward Richardson showed that the flushing was due to dilation of the capillaries, and Gamgee went on to demonstrate that the dilation of blood vessels led to a drop in blood pressure.

Brunton obtained a sample of amyl nitrite from Gamgee and poured some on to a cloth for his patients to inhale. Within a minute, their agonizing chest pains had disappeared entirely and many remained free of pain for several hours. The success of the drug was assured, and it was universally adopted. Over the next few years, Brunton satisfied himself that other nitrites had similar effects. He also examined nitroglycerine, which had become readily available following Alfred Nobel's discovery of its value as an explosive. Finding that when he and a colleague tried nitroglycerine it gave them both a severe headache, Brunton did not pursue its use further. His observations on it were referred to in the *St Bartholomew's Reports* in 1876, but matters did not rest there. Late the following year, William Murrell, then a registrar at the Westminster Hospital, decided to resolve the conflicting reports in the literature over whether or not nitroglycerine caused severe headache. Its discoverer, the French chemist Ascagne Sobrero, had reported in 1847 that he experienced an intense headache after merely tasting a drop of it placed on his finger. Others had a similar experience, including a Brighton dentist, Arthur Field, who had claimed in the *Medical Times and Gazette* that he had alleviated toothache and neuralgia by applying a drop or two of a dilute alcoholic solution of nitroglycerine to the tongue. Were it not for the ensuing headache, so he claimed, nitroglycerine would be a valuable remedy. In the ensuing correspondence, some writers confirmed Field's observations, while others reported no effects whatsoever being produced by nitroglycerine, despite having swallowed relatively large amounts. Murrell suspected that this confusion may have arisen from variation in the susceptibility of different individuals to nitroglycerine. Having decided to experiment on volunteers, Murrell obtained a supply of the drug. In the course of one of his out-patient clinics, he casually licked the moist cork of a bottle of nitroglycerine solution that was in his pocket. Within a few moments he began to experience throbbing in the neck and head, accompanied by pounding of his heart, the very effects that Field had so vividly described. Such was the extent of his discomfiture, that he could not continue with his examination of a patient. After five minutes, Murrell had recovered sufficient to once again attend his patient, although a severe headache lasted all the afternoon.

Murrell tested nitroglycerine on himself on a further thirty or forty occasions before persuading friends and volunteers to take part in a trial of its effects. This brought home to him the similarity between its action and that of amyl nitrite, but there was one important difference. Charting the changes in blood pressure of the volunteers on his sphygmograph, Murrell established that although it took two or three minutes for nitroglycerine to produce its effects, as opposed to only ten seconds or so for amyl nitrite, these persisted for about half an hour. This was in marked contrast to amyl nitrite, the effects of which wore off after five minutes. This persuaded Murrell that nitroglycerine might be superior in the treatment of angina pectoris. He also found that the characteristic headache was due to overdosage. The correct dose turned out to be in the order of half to one milligram when the drug was formulated in tablets that were allowed to dissolve slowly in the mouth, whereas around 100 to 300 milligrams of amyl nitrite had to be inhaled to produce similar effects. If the nitroglycerine was swallowed, a larger dose was required since absorption from the gut was less efficient than from the mouth, or even through the skin (ointments of nitroglycerine have recently been superseded by specially manufactured topical delivery systems that release controlled amounts of the drug over several hours). Murrell treated his first patient with nitroglycerine in January 1878, and two others over the next nine months. The following year, he published his first report in the *Lancet*, this resulting in the general adoption of the drug into clinical practice. British doctors took steps to avoid any unnecessary alarm that might result if patients discovered they were receiving the same explosive as was in dynamite! The drug was instead named glyceryl

$C_5H_{11}ONO$
Amyl Nitrite

CH_2ONO_2–$CHONO_2$–CH_2ONO_2
Nitroglycerine

O_2NOCH_2–$C(CH_2ONO_2)_2$–CH_2ONO_2
Pentaerythritol Tetranitrate

HO, O, O, ONO_2
Isosorbide Mononitrate

trinitrate or trinitrin.

As nitroglycerine was an ester prepared by the nitration of a polyhydroxylic alcohol, analogues of it have been synthesized by the nitration of simple sugars in attempts to obtain longer-acting compounds for prophylactic use. The first of these was pentaerythritol tetranitrate, which was introduced in 1896. There has been much discussion as to whether longer-acting nitrate esters are of any real value in preventing attacks of angina, since the rate of their metabolism in the liver is such that little unchanged drug enters the circulation. Recent years have seen the introduction of isosorbide dinitrate and its active metabolite, isosorbide mononitrate.

After nitroglycerine had been in use for some time, it was realized that it relaxed smooth muscle throughout the body. Today, it is believed that this relaxation of the smooth musculature of the veins reduces the volume of blood returning to the heart, thereby lowering the amount of work done by the heart and thus affording relief from anginal pain. However, until quite recently it was believed that nitroglycerine was effective in angina because it was capable of relaxing the muscle of the coronary artery in the heart, thereby dilating it. Many attempts were made to find alternatives to nitroglycerine by screening compounds for their ability to dilate isolated pieces of coronary artery. Recently, some coronary artery dilators which block the flux of calcium ions into heart muscle and vascular smooth muscle have been shown to be effective agents for preventing anginal attacks. These have the advantage of not being rapidly metabolized like the nitrates. The origin of this type of drug goes back to 1942, when Max Bochmuhl and Gustav Ehrhart of the Hoechst division of I. G. Farbenindustrie synthesized the antispasmodic drug fenpiprane ('Aspasan'), an analogue of methadone (see page 76) which could be used as a papaverine substitute since it was devoid of analgesic activity. Ehrhart continued with this line of investigation after the war ended, eventually synthesizing prenylamine ('Synadrin') in 1958. Because papaverine relaxed smooth muscle and had been tried as an antianginal drug, its analogues were routinely screened for coronary dilator activity. When this was done with prenylamine, it was found to be active. Its superiority over other coronary dilators in preventing angina is now considered to be a consequence of its

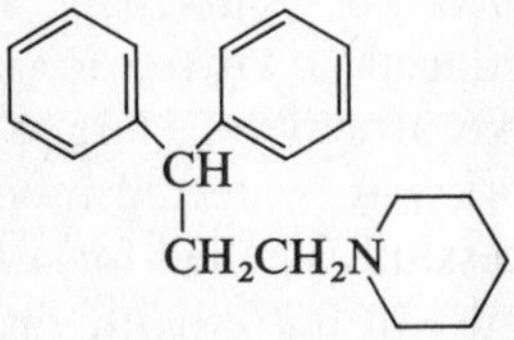

Fenpiprane

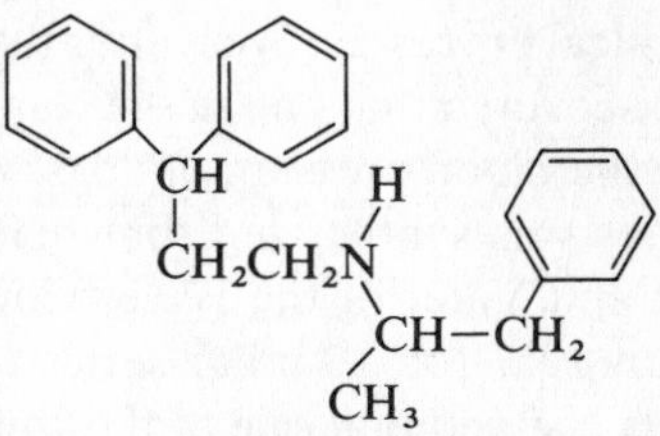

Prenylamine

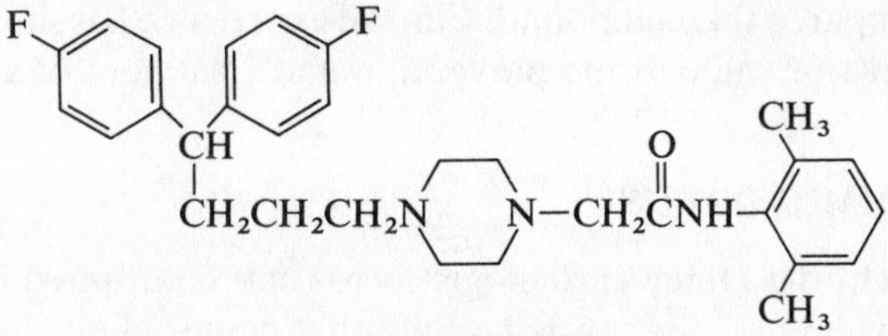

Lidoflazine

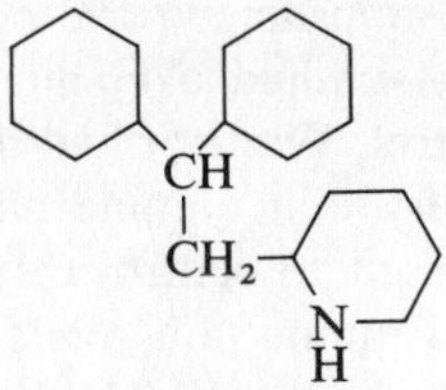

Perhexiline

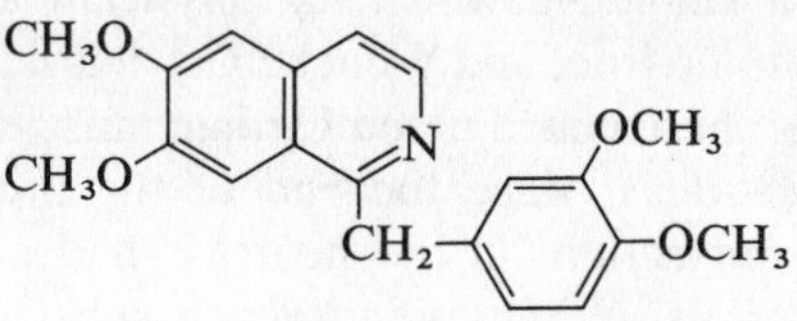

Papaverine

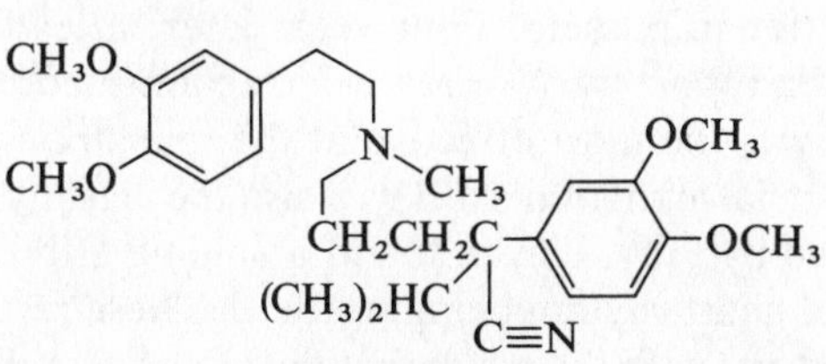

Verapamil

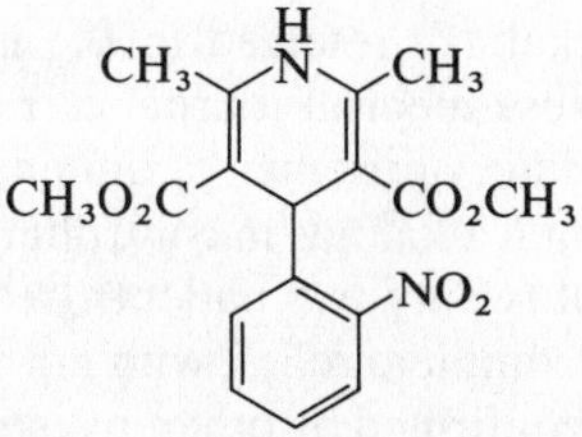

Nifedipine

unsuspected ability to act as a calcium antagonist. Lidoflazine ('Clinium') was also modelled on the fenpiprane system, incorporating a lignocaine residue in an attempt to confer antiarrhythmic activity. It was synthesized by Janssen in 1965, the same year that the Richardson-Merrell Company developed perhexiline ('Pexid'), a fenpiprane analogue in which the diphenhydryl ring system had been chemically reduced. It too is used in the prevention of angina, as is the open-chain papaverine analogue known as verapamil ('Cordilox'), synthesized by Knoll and Company in 1962.

A team of chemists at the Smith Kline and French laboratories in Philadelphia, led by Bernard Loev, undertook to synthesize a series of dihydropyridines since compounds of this type had never been subjected to rigorous pharmacological screening despite their involvement in various biochemical processes. Soon after the publication of Loev's first paper in 1965, it was discovered that one of the compounds produced a relatively long-lasting drop in the blood pressure of dogs. As it lacked oral activity, a large series of analogues was synthesized, but before this work was protected by patents, Bossert and Vater of the Bayer Company successfully applied for a Belgian patent in 1967 for a similar series of compounds. In this patent, the Bayer scientists claimed that one of the compounds, nifedipine ('Adalat') had particularly marked coronary vasodilating activity. It emerged later that the Americans had also prepared this compound! Clinical experience has shown it to be a calcium antagonist of value in the prevention and treatment of angina.

ANTIARRHYTHMIC DRUGS

Karl Wenckebach, the Dutch cardiologist, was once confronted by a patient who disagreed when told there was nothing that could relieve his attacks of atrial fibrillation. The patient offered to return the following day with a normal pulse, a challenge which his physician accepted. Astonishingly, the claim proved to be true, and Wenckebach learned that his patient had taken quinine to bring the incoordinated cardiac muscle under control. Whilst it had been known at least since the end of the previous century that quinine had a depressant action on the heart, it had never been used to control cardiac irregularities. When Wenckebach subsequently tried quinine on other patients, only a few responded. These were all cases in which the arrhythmias were of recent origin. In his book on cardiac arrhythmias, published in 1914, Wenckebach referred to the matter in passing. Four years later, a leading Viennese medical journal carried a report by W. Frey, which confirmed that quinidine, an isomer of quinine, was the most effective of the four principal cinchona alkaloids in controlling atrial arrhythmias. It was also the least toxic. Out of twenty-two patients treated by Frey, eleven found quinidine afforded them complete relief, with normal heart rhythm being restored. These results were confirmed by other investigators, with the result that the use of quinidine became universally established in the early twenties.

Advances in cardiac surgery during the early thirties led Frederick Mautz of Cleveland, Ohio, to investigate several drugs with a view to finding an agent which could be applied directly to the heart to prevent arrhythmias during the operative or post-operative periods. He felt that disturbances of cardiac rhythm might be prevented by local anaesthetics, especially as procaine had already been shown to have some effect on heart muscle. Procaine proved to be highly effective, and it was superior to cocaine or piperocaine in the experiments he carried out on animals. In 1936, he proposed that if a heart proved exceedingly irritable during surgery, or if the heartbeat became irregular, these disturbances could be controlled by the cautious intrapericardial injection of procaine. However, it does not seem to have been until after the war that procaine was used clinically as an alternative to quinidine, the action of which it resembled. Unfortunately, procaine was rapidly metabolized by esterase enzymes in the blood. This severely limited its clinical application, as did a high incidence of central nervous system side-effects. When a procaine analogue resistant to esterases was discovered, namely procainamide, it provided cardiologists with a valuable new drug. It had the added attraction of causing a lower incidence of central nervous system disturbances since it was less lipid-soluble than procaine. In 1951, several reports were published in the United States, all confirming that procainamide was an effective antiarrhythmic agent.

Clinical experience with quinidine showed that it had to be used with caution since it was cumulative in action. Furthermore, a small proportion of patients abreacted severely owing to their hypersensitivity to the drug. With procainamide the problem was one of wide variation in the plasma levels achieved with a fixed dose given by mouth or by injection to different patients. In an attempt to discover a drug free of these disadvantages, pharmacologists working for G. D. Searle and Company screened more than five hundred compounds. One of these, disopyramide ('Norpace', 'Rhythmodan'), was introduced in 1962.

N
C
H_2NOC CH_2CH_2N $CH(CH_3)_2$ $CH(CH_3)_2$

Disopyramide

H_2N–$CONHCH_2CH_2N$ CH_2CH_3 CH_2CH_3

Procainamide

CH_3 –$NHCOCH_2N$ CH_2CH_3 CH_2CH_3 CH_3

Lignocaine

CH_3 –$NHCOCHNH_2$ CH_3 CH_3

Tocainide

CH_3 –OCH_2CHNH_2 CH_3 CH_3

Mexiletine

Lignocaine (see page 56) has proved to be not only one of the most successful local anaesthetics, but it has also turned out to be an important antiarrhythmic drug for use in coronary care units. Unfortunately, it cannot be given to ambulant patients by mouth since a metabolite formed in the liver is somewhat toxic. To avoid this problem, the lignocaine analogues tocainide ('Tonocard'), mexiletine ('Mexitil') and flecainide ('Tambocor') were developed in the mid-seventies by the Astra, Boehringer, and Riker companies respectively. Flecainide was synthesized after Riker researchers in Minnesota had noticed that novel fluorine-substituted benzamides had local anaesthetic and antiarrhythmic activity.

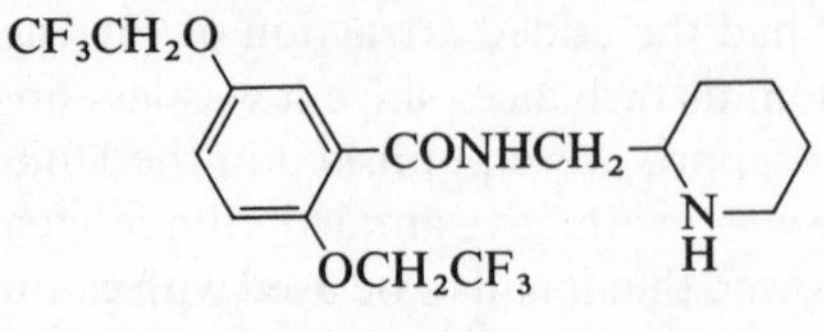

Flecainide

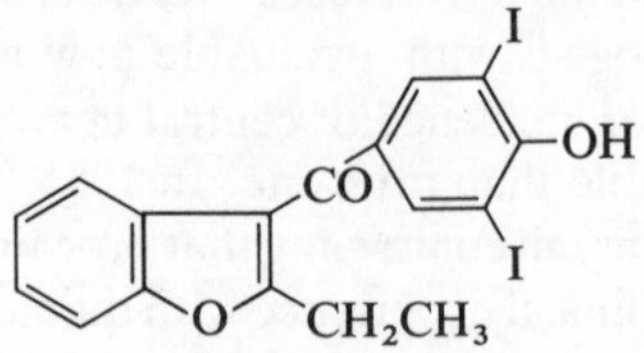

Benziodarone

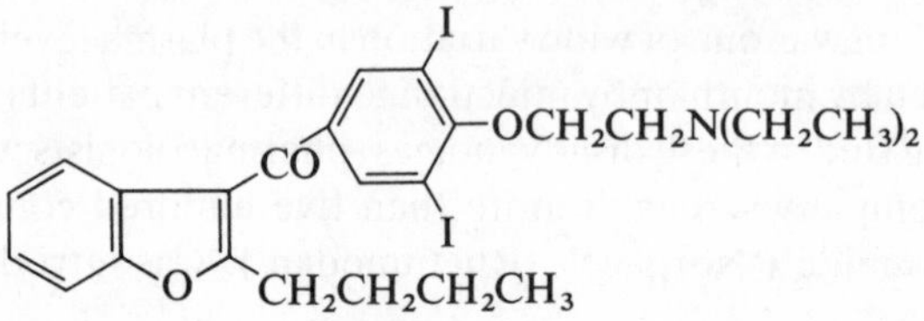

Amiodarone

The antiarrhythmic drugs mentioned so far act by stabilizing the membrane of the heart muscle. This means that the muscle will require a stronger stimulus before it will contract. The development of propranolol in 1964 (see page 112) offered a different means of controlling cardiac irregularities arising from excessive amounts of catecholamines being liberated. An unanticipated membrane-stabilizing direct antiarrhythmic action was also found to be involved when these were used clinically.

In 1957, the Société Belge de l'Azote et des Produits Chimiques du Marly developed benziodarone as a coronary dilator that was marketed on the Continent by their subsidiary company, Labaz. Four years later, an analogue was prepared by reacting benziodarone with a molecule that could be considered as a cross between lignocaine and an alpha-adrenoceptor blocking drug. This was done, presumably, to widen the spectrum of action to include antiarrhythmic activity. This drug, amiodarone ('Cordarone X'), was introduced in Europe as a coronary dilator for angina, but by the late seventies it had become clear that its real value lay in the treatment of a specific type of arrhythmia, the Wolff-Parkinson-White syndrome, and arrhythmias resistant to other drugs. In the United Kingdom, it was licensed specifically for these purposes in 1980.

ANTIHYPERTENSIVE DRUGS

The first effective substance used to lower blood pressure was potassium thiocyanate, which was originally studied at the Sorbonne by Claude Bernard in 1857. It lowered blood pressure by virtue of its effect on smooth muscle, especially that of the vascular system, an action which was not all that dissimilar to that of the organic nitrates. It was to prove to be the forerunner of several chemically unrelated hypotensive vasodilators introduced a century later. However, a high incidence of toxicity, coupled with beneficial effects in less than one in three patients, ultimately rendered it obsolete, although it gained renewed popularity in the late thirties following claims that monitoring of blood levels could control the toxicity. It was replaced by sodium nitroprusside ('Nipride'), introduced by Page as an orally active antihypertensive agent in 1945. The results were rather disappointing, but some years later it was shown to be of value when given by intravenous injection for the control of hypertensive crises. It acts by releasing small amounts of cyanogen and cyanide into the circulation, these substances being rapidly metabolized by a liver enzyme, rhodanase, to form thiocyanate.

In 1949, Ciba patented hydralazine, an orally active hypotensive vasodilator. The company had been interested in hydrazinophthalazines of this type since the end of the war, but reasons for the synthesis of these do not appear to have been disclosed. It must be presumed that the hypotensive activity was unexpectedly revealed during routine phramacological screening. At any rate, hydralazine has, over the years, proved itself to be a useful vasodilator for the long-term relief of high blood pressure.

Diazoxide was a potential thiazide diuretic (see page 158) prepared by a team of chemists from the Schering Corporation after the introduction of chlorothiazide. It was different to other members of this class of compounds insofar as the second sulphonamide function was absent. This straightforward chemical change not only removed diuretic activity, but, unexpectedly, turned the compound into an antihypertensive vasodilator.

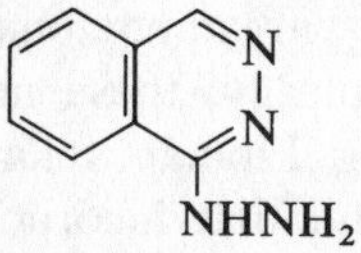

Hydralazine

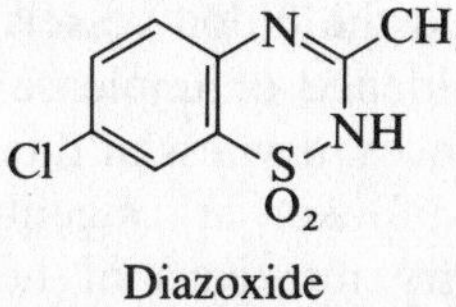

Diazoxide

In 1949, William Paton and Eleanor Zaimis of the National Institute for Medical Research, in London, reported the effect of altering the length of the carbon chain between two positively charged quaternary nitrogen atoms in a series of compounds that they had previously found to have neuromuscular blocking activity (see page 131). The pharmacological properties of these compounds were markedly influenced by the chain length, exhibiting optimal neuromuscular blocking activity if the chain contained ten carbon atoms, as in decamethonium (see page 132). However, when the analogue with a six carbon chain, hexamethonium, was injected into a rabbit, its ears flushed vigorously and became warm, whilst the blood pressure fell. Paton and Zaimis realized these effects were due to vasodilation arising from blockade of the ganglia that mediated chemotransmission of electrical impulses in the autonomic nervous system. The effect of hexamethonium on the blood pressure was reminiscent of surgical sympathectomy, a drastic technique then in vogue for cases of danger-

$(CH_3)_3\overset{\oplus}{N}-(CH_2)_6-\overset{\oplus}{N}(CH_3)_3$

Hexamethonium

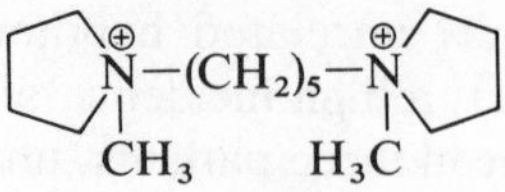

Pentolinium

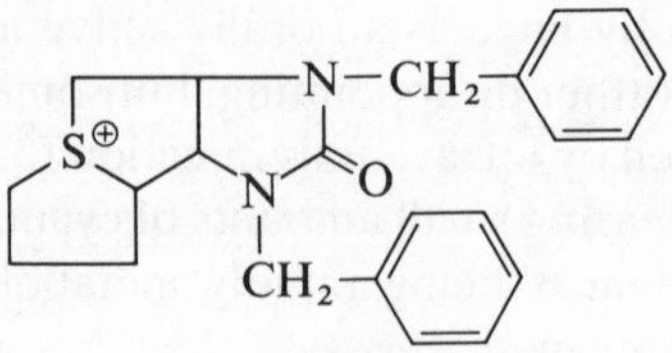

Trimetaphan

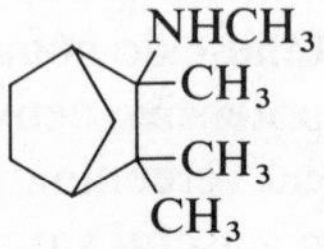

Mecamylamine

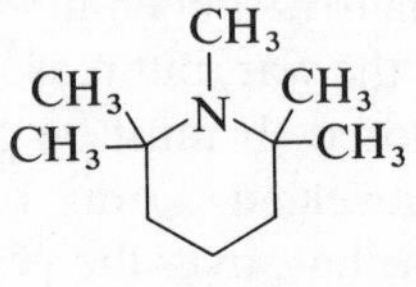

Pempidine

ously high blood pressure. A new approach to the chemical control of high blood pressure was thus apparent.

Clinicians soon began to experiment with hexamethonium as an anti-hypertensive agent. In many respects the drug was disappointing, not least because it was short acting and poorly absorbed by mouth, necessitating frequent injections. As it blocked not only the sympathetic ganglia that controlled the blood vessels, but also the parasympathetic ones, it caused a high incidence of unpleasant side-effects. Typically, these included visual disturbances, dryness of the mouth, and fainting. Libman, Pain, and Slack of May and Baker at Dagenham, London, found that incorporation of the quaternary nitrogens of hexamethonium into pyrrolidine rings enhanced potency and duration of activity, and introduced their pentolinium tartrate ('Ansolysen') in 1952. However, this still suffered from a lack of selectivity. Within four years it was no longer in use, a similar fate also being experienced by trimetaphan, a quaternary sulphonium compound synthesized by Goldberg and Sternbach of Hoffmann-La Roche at Nutley, New Jersey, from an intermediate used in the company's synthesis of biotin.

In 1955, G. A. Stein and his colleagues at the Merck Laboratories in Rahway, New Jersey, obtained an unexpected product on reacting camphene with hydrogen cyanide under acidic conditions. This was sent for routine screening to Karl Beyer at the Sharp and Dohme division of the company. Surprisingly, it had ganglion blocking activity, yet it was not a quaternary ammonium compound, as had been all previous blocking agents. The absence of the polar quaternary function meant that it was well absorbed from the gut and could thus be given by mouth. An additional bonus was that it had a longer duration of action. These features led to its clinical introduction as mecamylamine, and ensured rival manufacturers would try to prepare analogues of it. Indeed, three years later, two British companies, ICI and May and Baker, both announced their synthesis of pempidine, an analogue with a simpler structure. Each had chosen to synthesize it as a consequence of quite different interpretations of what were the probable salient features of the mecamylamine molecular structure that contributed to its activity. Pempidine proved superior to mecamylamine insofar as it was completely excreted, and so did not accumulate in the body to cause problems over the control of the blood pressure. However, it still produced unwelcome side-effects because of its ability to block parasympathetic ganglia.

The discovery of a new class of hypotensive drugs, the adrenergic neuron blocking agents, was made serendipitously in Professor William Bain's department of pharmacology at the University of Leeds. One of Bain's colleagues, Peter Hey, found that a new cholinergic agent he had prepared was also an inhibitor of monoamine oxidase. He synthesized several of its analogues, including one code-named TM 10. When a research student in the department, George Willey, tested Hey's compounds for signs of cholinergic activity, he discovered that TM 10 interfered with the transmission of impulses in adreneregic nerves at dose levels that did not block parasympathetic ganglia. The inhibitory action was confined to the post-ganglionic nerve fibres. This led to the suggestion that the drug might be acting as a local anaesthetic on these fibres. It was then tested and found to be a very long-acting local anaesthetic. Shortly after this, Professor Bain was asked to investigate the action of lignocaine in epilepsy. Finding its effects too brief, he asked another colleague, K. A. Exley, to check whether TM 10 might be more suitable in view of its long-lasting anaesthetic action. Exley's studies soon ruled out any possible suggestion that TM 10 interfered with transmission of electrical impulses along post-ganglionic adrenergic nerve fibres. What it did do, he discovered, was to prevent the release of noradrenaline at the nerve endings, thereby blocking transmission of impulses across the ganglia. This was a unique phenomenon; it opened up the possibility of controlling blood pressure by selective blockade of adrenergic stimuli to the blood vessels. Unfortunately, TM 10 retained pronounced cholinergic properties which prevented it being considered for clinical application.

Exley reported on the novel action of TM 10, now called xylocholine, at a meeting of the British Pharmacological Society in Edinburgh in July 1956. This brought the drug to the attention of A. F. Green and his colleagues from the Wellcome Research Laboratories in Beckenham. He initiated a search for a similar drug without cholinergic activity. It so happened that TM 10 had a close structural resemblance to bephenium, a drug Wellcome had recently introduced for the treatment of hookworm. This resulted in the synthesis of the structurally similar bretylium, in 1957. It was marketed two years later, but soon fell out of favour as a hypotensive drug because of unreliable absorption from the gut. However, it is still administered by injection for the control of cardiac arrhythmias resistant to other drugs.

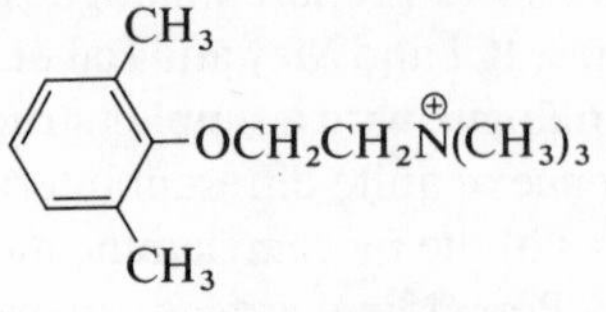

Xylocholine

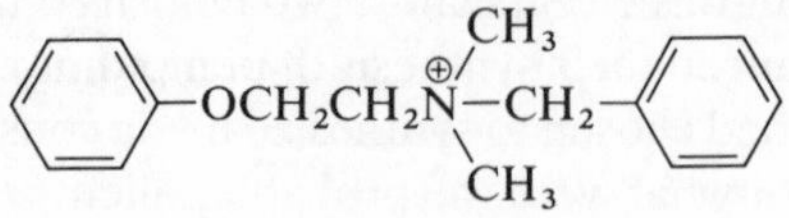

Bephenium

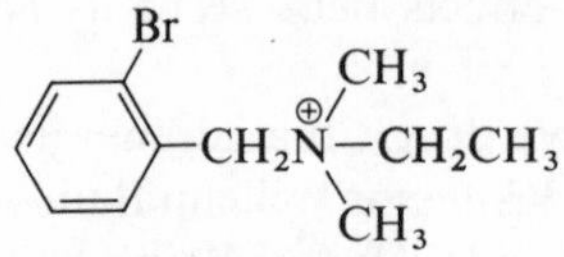

Bretylium

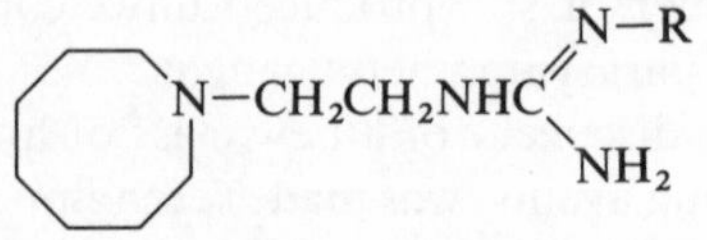

Guanethidine Aldoxime, R = $-OH$
Guanethidine, R = $-H$

Bretylium was quickly challenged by guanethidine ('Ismelin'). This was synthesized after Ciba pharmacologists discovered, in 1957, that one of a series of aldoximes that had trypanocidal, antitubercular and antirickettsial activity also exhibited marked antihypertensive properties. The aldoxime was put on clinical trial, but had to be rejected on account of an unacceptable incidence of side-effects. Guanethidine was then found to be quite acceptable, although it and its analogues often failed to control supine blood pressure without causing dizziness on rising, due to postural hypotension. It is still used, as are its analogues bethanidine ('Esbatal'), debrisoquine ('Declinax'), guanoclor ('Vatensol'), and guanoxan ('Envacor'), all of which were introduced in either 1964 or 1965.

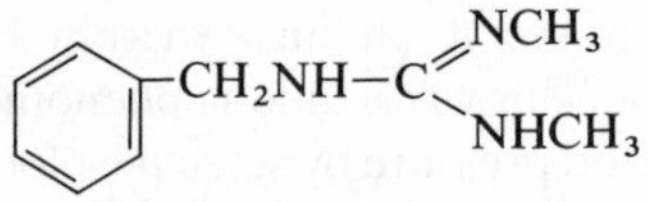

Bethanidine

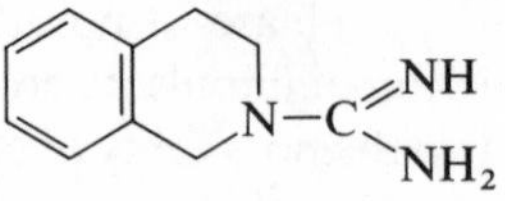

Debrisoquine

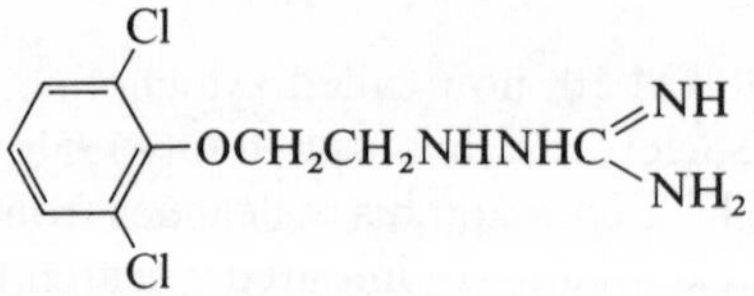

Guanoclor

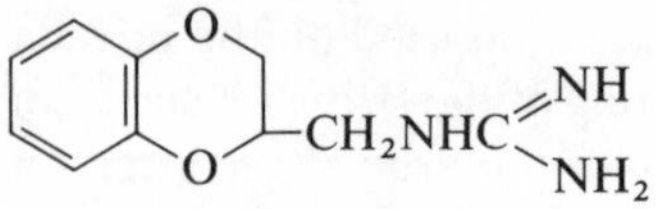

Guanoxan

Attempts to control blood pressure by using alpha-adrenergic blockers were first reported in the early fifties. Studies were carried out with partially hydrogenated ergot alkaloids, but the results were equivocal. Phentolamine (see page 111) was also tried, but its action was transient. Furthermore, it had cholinergic activity, resulting in a variety of side-effects. The long-acting phenoxybenzamine (see page 111) was popular for a while, but it slowed the heart and caused postural hypotension. From these results it appeared that alpha-adrenoceptor blockade was not an effective way to control high blood pressure. However, within the past few years it has become apparent that there are two types of alpha-adrenoceptors, each mediating different physiological responses. It is now recognized that because many adrenergic blockers acted at both receptors, they were of little value in the treatment of hypertension. This has resulted in a complete reassessment of drugs known to act at alpha-adrenoceptors.

Two drugs synthesized in the sixties have been introduced into clinical practice recently as a result of their recognition as specific alpha$_1$-adrenoceptor blockers. In the early sixties, scientists at the Miles Laboratores, in Indiana, discovered that certain piperazine derivatives had hypotensive activity, especially those which contained a cyclic amide function. They prepared a series of quinazoline compounds featuring appropriate piperazine side-chains, the results of this work being published in the *Journal of Medicinal Chemistry* in 1965. Pfizer chemists then synthesized analogues in which the side-chain was attached at a different position in the quinazoline ring. This resulted in the development of prazosin ('Hypovase'), a potent hypotensive for which a patent was secured in 1969. It did not cause any slowing of the heart rate. Recently, it has been found to act specifically as an alpha$_1$-adrenoceptor blocker, as does indoramin ('Baratol'), which was synthesized at the Wyeth research laboratories in Taplow, Berkshire, in 1966. It was marketed in the United Kingdom in 1981.

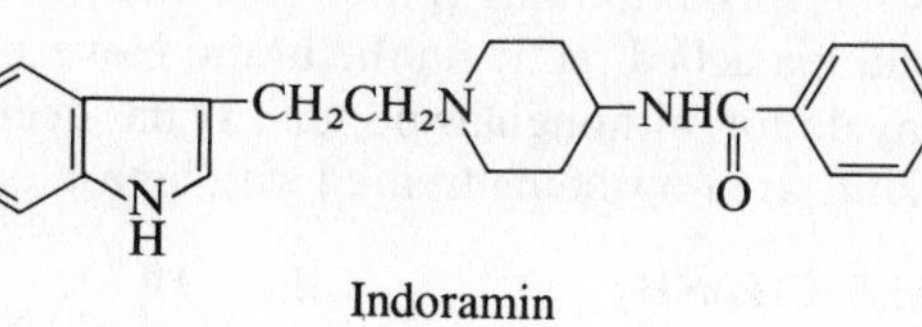

Prazosin

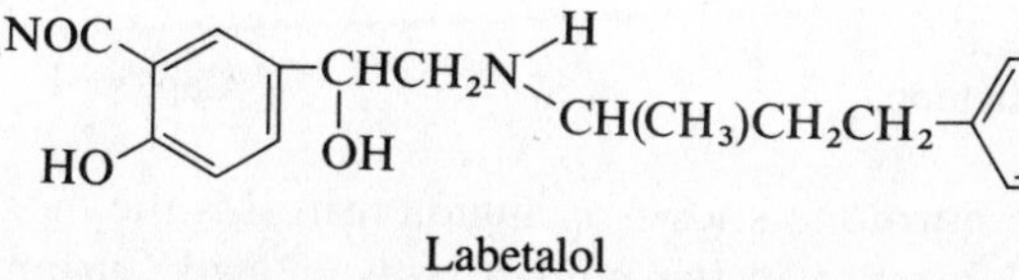

Indoramin

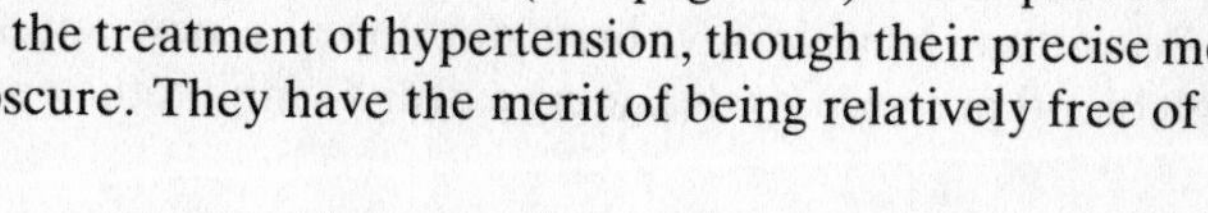

Labetalol

The beta-adrenergic blockers (see page 112) have proved to be highly effective in the treatment of hypertension, though their precise mode of action remains obscure. They have the merit of being relatively free of troublesome side-effects. Although there is little evidence of any significant difference in clinical efficacy between the many products on the market, Allen and Hanburys (part of the Glaxo Group) have made persistent claims that their labetalol ('Trandate') is superior since it blocks both alpha and beta adrenoceptors without causing significant side-effects.

All the antihypertensive drugs so far discussed act by dilating the blood vessels, either directly or through interference with their local innervation. When reserpine was introduced in 1953, it offered a new means of controlling blood pressure, namely through its effect on the brain. It proved disappointing in the clinic since its oral activity could not match that achieved when it was injected. It also caused depression in many patients. Several analogues were introduced, but the problem was never overcome. This, incidentally, gives the lie to repeated claims about *Rauwolfia* being a major antihypertensive used in Indian folk medicine.

Despite the disappointment with the *Rauwolfia* alkaloids, the most widely used hypotensive drug in the world, methyldopa ('Aldomet'), is one that acts directly on the brain. Its hypotensive activity was revealed in 1960 by Sidney Udenfriend and his colleagues at the National Heart Institute in Bethesda, Maryland. They had been provided with the drug by Merck, Sharp and Dohme, in whose laboratories it had been synthesized seven years earlier and shown to be an inhibitor of dopa decarboxylase in test-tube experiments. This was an enzyme that played a crucial role in the biosynthesis of catecholamines such as adrenaline and noradrenaline. After seeing the results of pharmacological studies of methyldopa in animals, Udenfriend was interested in its possible effects on humans since depletion of tissue stores of catecholamines was the basis of the action of the *Rauwolfia* alkaloids. Chronic toxicity studies having been carried out by the manufacturer, Udenfriend and his team administered the drug to hypertensive patients. When it was shown to lower blood pressure it was naturally assumed that the effect was due to a local depletion of adrenergic hormone synthesis leading to dilation of blood vessels. Over the years this interpretation has come to be questioned, and it is now believed that methyldopa acts directly on the brain. Nevertheless, its efficacy has never been in any doubt, although its popularity has declined somewhat as more and more patients are now being treated with beta-blockers or diuretics.

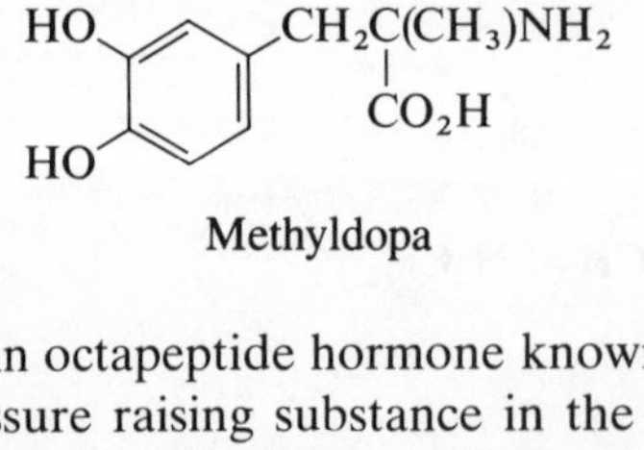

HO, HO, $CH_2C(CH_3)NH_2$, CO_2H

Methyldopa

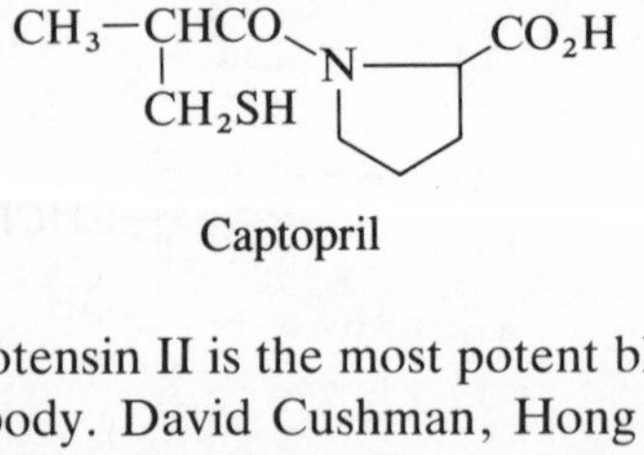

CH_3-CHCO, CH_2SH, N, CO_2H

Captopril

An octapeptide hormone known as angiotensin II is the most potent blood pressure raising substance in the human body. David Cushman, Hong Son Cheung, Emily Sabo, and Miguel Ondetti of the Squibb Institute for Medical Research in Princeton, New Jersey, have developed a new approach to the control of high blood pressure by designing a drug to inhibit the enzyme which catalyses the formation of this hormone. Their work was facilitated by the discovery of a resemblance between the angiotensin-converting enzyme and the carboxypeptidase A enzyme (a pancreatic digestive enzyme), which was known to be inhibited by 2-benzylsuccinic acid. With the assistance of computer graphics, a hypothetical model of the active site on the enzyme was defined, then a large number of potential inhibitors that could fit this were synthesized. The outcome of this highly sophisticated approach to drug design was the introduction of captopril ('Capoten') in 1977. Four years later, it was marketed as an antihypertensive agent in the United Kingdom for use in patients whose raised blood pressure did not respond to other drugs.

DIURETICS

In the past, efforts to reduce oedema (swelling) in various parts of the body relied on the removal of blood, sweat, or urine. Either leeches or the scalpel were used to draw off blood, whilst herbs were concocted to increase either sweating (diaphoresis) or urine flow (diuresis). Nowadays, synthetic diuretics are so effective, that all other remedies for oedema are obsolete. Nevertheless, it is for the relief of hypertension that modern diuretics are principally prescribed.

Numerous herbs that produced a mild diuresis were known to the ancients, but modern investigations of the volatile oils responsible for any effect these might have exhibited have led to their abandonment. The last herbal diuretic appeared in the *British Pharmacopoeia* in 1932, namely Infusum Buchu. At one time, however, the xanthine alkaloids present in tea, cocoa, and coffee, were held to be of clinical value as diuretics. In a doctoral dissertation submitted to the University of Strassburg in 1886, Bronne investigated the diuretic action of caffeine. This led his research supervisor, von Schroeder, to

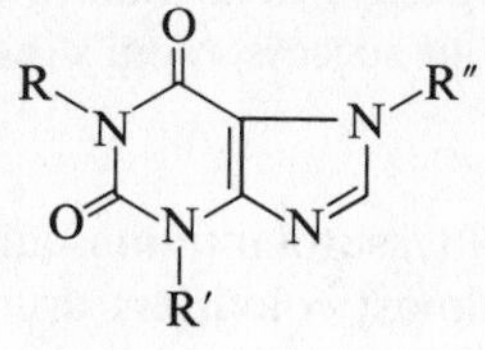

	R	R′	R″
Caffeine	$-CH_3$	$-CH_3$	$-CH_3$
Theobromine	$-H$	$-CH_3$	$-CH_3$
Theophylline	$-CH_3$	$-CH_3$	$-H$

examine the chemically similar cocoa alkaloid, theobromine. This he found to be a more effective diuretic. Shortly after, theophylline was isolated from tea extract, but it was only once it had been synthesized by Wilhelm Traube that it was fully evaluated clinically. In 1902, Minkowsky established that it was around three times as potent a diuretic as caffeine. Although its lack of solubility made it difficult to formulate, this problem was partially overcome five years later on the introduction of its double compound with ethylenediamine. This was named aminophylline ('Euphylline') and was manufactured by Dr H. Byke's chemical factory in Charlottenburg. Other double compounds have been marketed since then. No xanthine derivative, however, has been found to rival any of the more potent diuretics subsequently developed.

In October 1919, Arthur Vogl, then a third year medical student at the Wenckebach Clinic in Vienna's First Medical University, was instructed by the physician in charge to administer mercury salicylate to a severely debilitated syphilitic patient. Not realizing how insoluble mercury salicylate was in water, Vogl blithely ordered a 10% solution of it to be prepared by the hospital pharmacy. When told that such a solution could only be prepared as an oily injection, to save his embarrassment Vogl gladly accepted a colleague's offer of a sample of a mercurial antisyphilitic. This was merbaphen ('Novasurol'), a drug which had been used clinically for about seven years. Vogl then received permission to initiate a course of merbaphen injections.

The nursing staff at the Wenckebach Clinic were meticulous in maintaining records of their patients' progress. This meant that, amongst other things, the amount of urine voided by every patient was entered on a chart. As a consequence, Vogl saw at once that after receiving her first injection of merbaphen, his patient's twenty-four-hour urine output increased to 1200 ml from a mere 200 to 500 ml noted previously. After the third daily injection, this had increased to two litres! On interruption of the treatment for a few days, the urine outflow decreased, only to rise again on resumption of the injections. Convinced that he had discovered a potent diuretic, Vogl decided to administer merbaphen to another syphilitic patient, in whom the disease had damaged the heart, causing advanced congestive heart failure. Conventional diuretics had been tried on this man without success. Vogl's own words describe what now happened:

> 'After the injection of Novasurol intramuscularly, the patient passed a massive amount of almost colourless urine. The flow continued throughout the day and the night and by the next morning he had, to our amazement, eliminated over ten litres. We were convinced that we had witnessed the greatest man-made diuresis in history.
> 'Now everyone became genuinely excited. Under a rather transparent blanket of secrecy, we scurried about in search of more patients on whom to test our discovery. We were repeatedly able to reproduce these miraculous results, causing deluges at will, to the mutual delight of the patients and ourselves.'

Tests with other antisyphilitic mercurials showed they lacked the potent diuretic activity of merbaphen. A thorough clinical evaluation of it was then

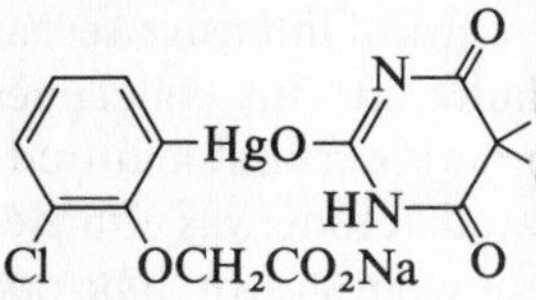

Merbaphen

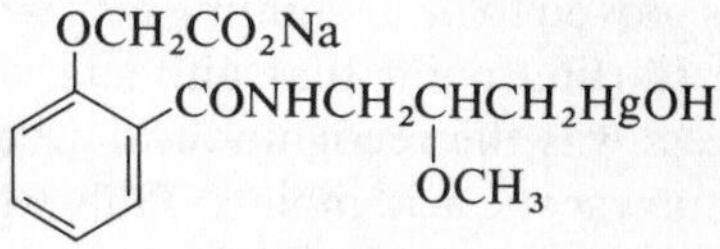

Mersalyl

carried out in the Wenckebach Clinic by Vogl's superior, Paul Saxl. This resulted in what has been described as a revolution in the treatment of the severe oedema of congestive heart failure. By removing fluid, the pressure on the heart was relieved, allowing it to recover normal function. None of the various substances previously used as diuretics had anything like comparable activity. Nevertheless, it was soon recognized that merbaphen injections posed a risk of severe kidney damage or even fatal colitis. In 1924, mersalyl, a mercurial synthesized eighteen years earlier as a chemotherapeutic agent, was found to be somewhat less toxic. It still had to be administered on an intermittent schedule so as to minimize toxicity, but until the development of the thiazide diuretics in the late fifties, it was considered to be a life-saving remedy. Numerous attempts to develop a mercurial compound which could be given by mouth met with failure, principally on account of gastric irritation and unreliable absorption from the gut.

A fresh approach to the problem of designing a non-toxic, potent diuretic was taken by a team of researchers from the Sharp and Dohme company. They had been brought together in 1943 to investigate a major clinical problem posed by the recently introduced antibacterial sulphonamides. This was the tendency of these somewhat insoluble compounds to crystallize out of solution and deposit in the fine tubules of the kidneys. Apart from any damage to the kidneys, the pain was excruciating. In time, it became generally recognized that this problem could be side-stepped simply by using more acidic sulphonamides.

The pharmacological investigations of the Sharp and Dohme team were under the direction of Karl Beyer, Jr. Recent developments in renal physiology convinced him the moment was opportune for a fresh attempt to design an effective diuretic. In 1941, Walker, Bott, and Oliver had employed microanalytical techniques to demonstrate that reabsorption of water from the glomerular filtrate inside the lumen of the kidney tubules depended largely on the efflux of sodium ions across the tubule wall. This transfer of ions was described as active transport, a complex, energy-consuming process that originated within the renal tubular cells. It was believed that mercurial diuretics probably interfered with this movement of sodium (or chloride) ions by inhibiting dehydrogenase enzymes inside the cells of the tubules. The ability of mercury compounds to bind to dehydrogenase enzymes (presumed to involve reaction with thiol groups) was readily demonstrable. This convinced the Sharp and Dohme scientists it would be worth trying to design mercury-free inhibitors of dehydrogenases in order to avoid the toxic effects of mersalyl.

This proved to be a daunting task which took many years of intensive research. The starting point that ultimately led E. M. Schultz and his colleagues to success was the recognition that both merbaphen and mersalyl featured the phenoxyacetic acid moiety. Only when an unsaturated ketone was attached to the 4-position of this did they obtain compounds that were shown to be potent hydrogenase inhibitors. Careful analysis of the influence of further substituents revealed that chlorine or methyl groups attached to the benzene ring enhanced potency. Ultimately, ethacrynic acid ('Edecrin') emerged as a useful diuretic in 1962, five years after a separate group of Sharp and Dohme chemists had announced their discovery of the thiazide diuretics.

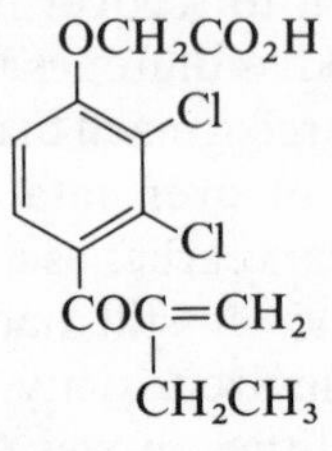

OCH_2CO_2H
Cl
Cl
$COC{=}CH_2$
CH_2CH_3

Ethacrynic Acid

At the University of Cambridge in 1940, T. Mann and D. Keilin carried out an important experiment. They wanted to see whether the fall in the carbon dioxide binding power of the blood caused by some of the recently discovered antibacterial sulphonamides could be accounted for by inhibition of carbonic anhydrase. This enzyme, which they had isolated in a pure state a year before, was known to play an important role in the output of carbon dioxide by the lungs. The experiment confirmed their suspicions. However, only those sulphonamides which retained both hydrogen atoms on the sulphonamide function were enzyme inhibitors. These included 'Prontosil' and sulphanilamide, as well as seven sulphonamides devoid of antibacterial activity. Five drugs then in routine clinical use, including sulphapyridine and sulphathiazole, did not inhibit carbonic anhydrase. Amongst chemically similar compounds, sulphanilic acid and pyridine-3-sulphonic acid were inactive.

Soon after the introduction of sulphanilamide, it had been noticed it caused an alkaline diuresis in patients who had been given massive doses. In 1942, Hober suggested that this could be accounted for by increased excretion of sodium bicarbonate caused by carbonic anhydrase inhibition. It had been shown, shortly before this, that resorption of water from the tubules of the kidney depended principally on the absorption of sodium ions from the lumen. After Pitts and Alexander established, in 1945, that carbonic anhydrase promoted the exchange of sodium for hydrogen ions in the distal portion of the renal tubules, it became clearer what was happening. When the enzyme was inhibited, the sodium ions were excreted in the urine because the process responsible for their reabsorption was blocked. W. B. Schwartz exploited these findings clinically, in 1949, by administering large doses of sulphanilamide by mouth to obtain a diuretic effect in patients with congestive heart failure. He soon abandoned this approach because of the toxic side-effects that occurred.

Schwartz contacted Richard Roblin of the Lederle Division of the American Cyanamid Company and told him about his experiments with sulphanilamide. Five years before this, Roblin had supplied Howard Davenport of the Harvard Medical School with thiophen-2-sulphonamide as a potential potent inhibitor of carbonic anhydrase. A heterocyclic sulphonamide had been chosen in the belief that it would be more acidic than conventional sulphonamides, and that this would enhance its ability to compete with carbon dioxide for the active site on the enzyme. When tested, the compound proved to be about forty times as potent an inhibitor as sulphanilamide, as a consequence of which Davenport recommended it as more reliable for investigating the role of carbonic anhydrase in the tissues. After reading Schwartz's paper, Roblin's former interest in carbonic anhydrase inhibitors was re-awakened. He and James Clapp immediately set about synthesizing some twenty heterocyclic sulphonamides. Within a year, acetazolamide ('Diamox') was found to be ideal for their purposes. It was around 330 times as potent an inhibitor of the isolated enzyme as was sulphanilamide, and it was introduced clinically as a diuretic in 1952. Unfortunately, inhibition of carbonic anhydrase elsewhere in the body besides the kidney led to a variety of complications. The only acceptable way to use acetazolamide as a diuretic was on an intermittent schedule. Happily, the inhibition of carbonic anhydrase in other parts of the body was turned to advantage in the treatment of glaucoma. By acting on the aqueous humour of the eye in much the same way as in the kidney, the build up of pressure from excess fluid was overcome.

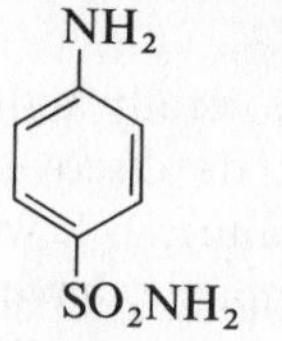

NH_2 SO_2NH_2
Sulphanilamide

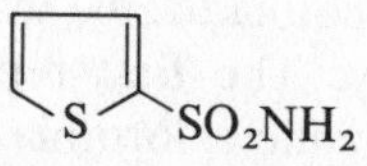

S SO_2NH_2
Thiophen-2-Sulphonamide

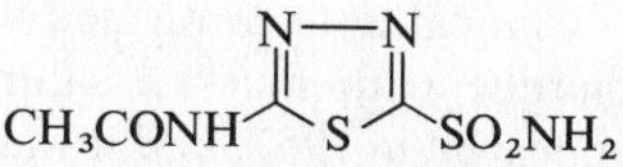

N—N CH_3CONH S SO_2NH_2
Acetazolamide

Schwartz's work with sulphanilamide caught the attention of the Sharp and Dohme renal research group. Karl Beyer felt that the problem with sulphanilamide was essentially that it inhibited carbonic anhydrase at the distal end of the renal tubules, rather than solely at the proximal end. This, he believed, accounted for the increased excretion of bicarbonate. He hoped, therefore, to find a carbonic anhydrase inhibitor that acted in the proximal portion, as indicated by increased excretion of chloride in the form of sodium chloride. Such a drug might, he hoped, have the added bonus of being a useful antihypertensive agent, for clinicians were beginning to believe that low salt diets were an effective means of controlling high blood pressure. To identify a saluretic drug of this type, Beyer carried out salt assays on urines collected from specially trained dogs. These assays were done almost instantaneously by means of flame photometry.

The first carbonic anhydrase inhibitor which the Sharp and Dohme researchers found to increase chloride excretion was 4-sulphonamidobenzoic acid. This was a cheap by-product from the synthesis of saccharin. It had previously been shown to be a carbonic anhydrase inhibitor by Mann and Keilin in their 1940 investigation. It still increased bicarbonate excretion, indicating a lack of specificity of action within the kidneys. Nevertheless, it was the lead compound which spurred James Sprague and Frederick Novello to synthesize more aromatic sulphonamides for Beyer and his associates to test. When two sulphonamido groups were introduced, the ability to increase chloride ion excretion was quite marked. Further enhancement of activity followed the introduction of a chlorine atom into the benzene ring. The outcome of this was the synthesis of dichlorphenamide ('Daranide'), a potent carbonic anhydrase inhibitor. Like acetazolamide, it was a weak diuretic which caused increased bicarbonate excretion, and so was used mainly in the treatment of glaucoma.

When an amino group was attached to the benzene ring of compounds like dichlorphenamide, there was a reduction in the carbonic anhydrase inhibitory potency. Surprisingly, there was no corresponding reduction in chloride ion excretion. This proved to be a major step towards the goal of obtaining a saluretic agent, even if it was unanticipated. Novello now proceeded to synthesize analogues with various substituents on the amino group. In the course of this, he attempted to prepare the *N*-formyl analogue by treatment with formic acid. This resulted in an unplanned ring closure to form a benzothiadiazine derivative. As a matter of routine, this novel compound was included in the screening programme. One can well imagine his delight when John Baer found it to be a potent diuretic which did not increase bicarbonate excretion!

Clinical tests confirmed that this benzothiadiazine was a safe, orally active diuretic with marked saluretic activity. The first reports of its discovery appeared in 1957, and it was given the name chlorothiazide ('Saluric'). It was marketed by Merck, Sharp and Dohme, a merger of the two companies having recently taken place. Chlorothiazide was to be the first of many so-called thiazide diuretics. Literally overnight, these rendered mercurial diuretics obsolete for the treatment of cardiac oedema associated with congestive heart failure. But that was not all. Beyer was correct in his long-held belief that a safe diuretic which could increase sodium chloride excretion would be of value in the treatment of hypertension. Today, the thiazide diuretics and related compounds are universally used for this purpose, either alone or in combination with methyldopa or a beta-blocker.

Many analogues of chlorothiazide have been marketed. The Merck, Sharp and Dohme workers found that its dihydro derivative, prepared by means of a different type of ring closure, was ten times as potent. This was named hydrochlorothiazide ('Hydrosaluric'), and was introduced shortly after chlorothiazide. Over the next year or so, at least four American companies synthesized hydroflumethazide, which was about as potent as hydrochlorothiazide. Of the analogues made since then, the principal differences are in

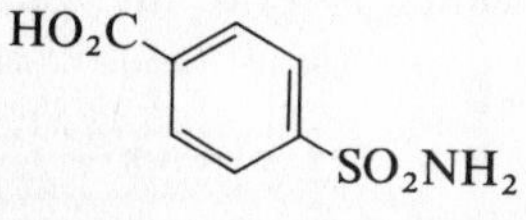

4-Sulphonamidobenzoic Acid

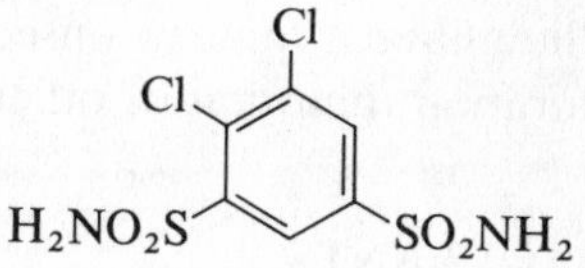

Dichlorphenamide

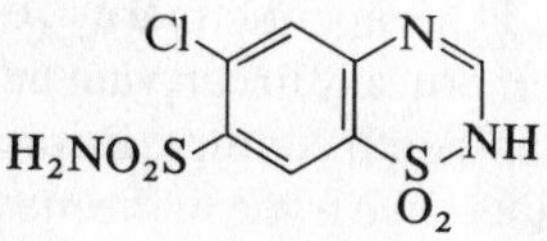

Chlorothiazide

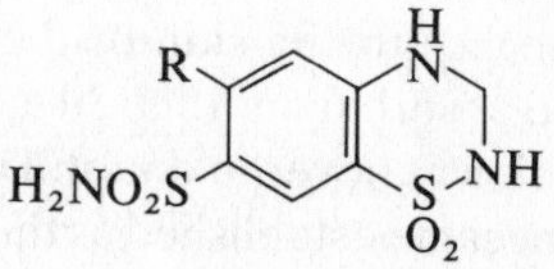

Hydrochlorothiazide, R = —Cl
Hydroflumethazide, R = $—CF_3$

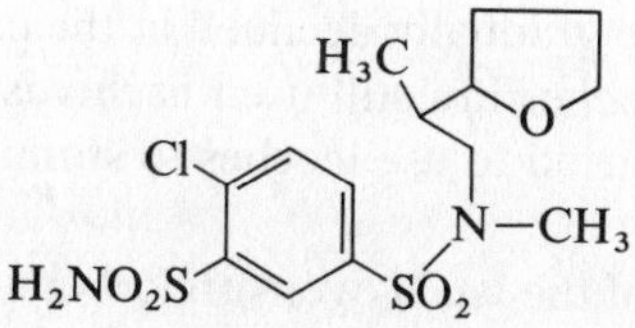

Mefruside

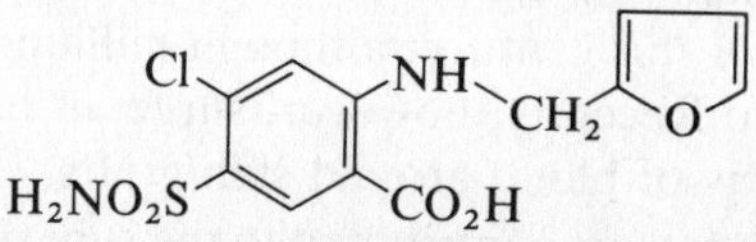

Frusemide

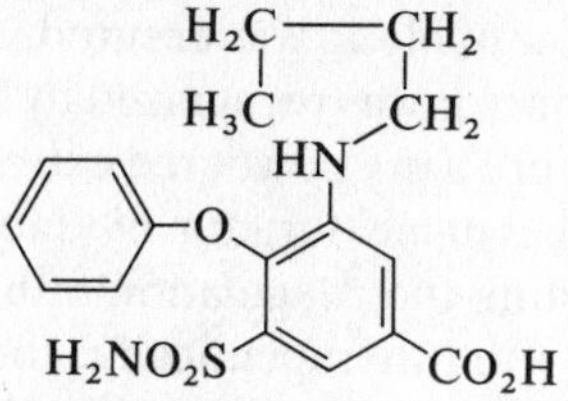

Bumetanide

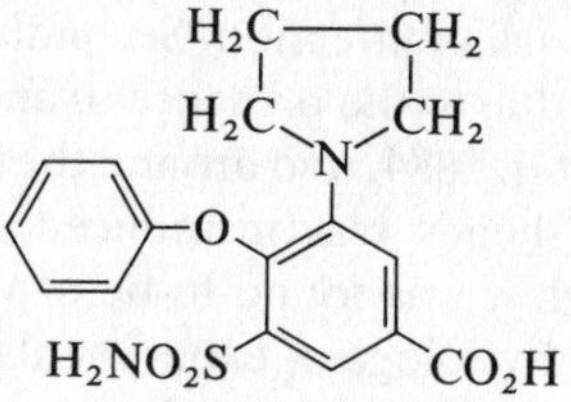

Piretanide

their potency, which can be up to two thousand times that of chlorothiazide, and in their duration of action, which varies from 6 to 24 hours. Whether there is any particular merit in increased potency is a debatable point.

The enhancement of potency of the amino analogues of dichlorphenamide through ring closure may have simply been due to an appropriate substitution pattern rather than the cyclization itself. An example of an open-chain substituted derivative is the Bayer Company's mefruside ('Baycaron'). Similarly, the second acidic group in dichlorphenamide may be replaced with a carboxyl group, so long as an appropriate substituent is present on the amino group. This approach led to the introduction by Hoechst of frusemide (furosemide, 'Lasix') in 1962, and Leo's more potent bumetanide ('Burinex'), ten years later. A subsidiary of Hoechst, Albert Products, introduced a close structural analogue of bumetanide, namely piretanide ('Arelix'), on to the British market in 1984, the patent on frusemide having expired. These three compounds have a different site of action within the kidney tubule, being known as 'loop diuretics' because they act in the region known as the loop of

Henle. They have a quicker onset of activity, which is more intense and of shorter duration than that of other diuretics.

ANTICOAGULANTS

For around a thousand years physicians believed that application of blood-sucking leeches to the skin of their patients had therapeutic merit. As each leech could absorb from 15 to 30 ml of blood, their efficacy in carrying out their allotted task was never in doubt. During the seventeenth century, Bavaria and Bohemia became established as the centres of a lucrative trade in the medicinal leech, *Hirudo medicinalis*, which was carried by the 'leech express' to Paris (no doubt providing a basis for lurid tales of Transylvanian vampires). By the end of the nineteenth century the principal use of leeches in the West lay in the treatment of local inflammation. This practice gradually declined in the early years of this century, although millions of leeches are still used each year in Russia. Recently, however, surgeons have started to use leeches to stimulate the flow of blood around skin grafts.

Two factors contributed to the superiority of the leech over surgical incision as a means of removing blood. Firstly, the small wound was remarkably clean-cut, being inflicted without pain. This enabled rapid healing to proceed without risk of infection. Secondly, a plentiful flow of blood was ensured. That this was due to the presence of an anticoagulant factor was recognized by J. B. Haycraft in 1884, and around the turn of the century a dry powdered extract of leeches' heads was introduced, being given the name hirudin by Jacoby. Although it was found to be of value in preventing the coagulation of blood during physiological experiments, it was rarely used therapeutically, for the best way of administering it remained that provided by nature, i.e. by the leech. At the start of the Great War, direct arm to arm blood transfusion was tried, but clotting of the blood proved hazardous. Hirudin was used for a time in order to prevent this.

Towards the end of the nineteenth century, several European investigators found that the clotting of blood was accelerated by material extracted from body tissues with fat solvents. It was also discovered that once the coagulating factor had been extracted by fat solvents, aqueous extraction of the defatted tissues yielded an anticoagulant. In France during the years 1911–1927, Doyon studied the effects of one such anticoagulant prepared from dog liver. A similar material was obtained in 1922 in Baltimore by William Howell, the professor of physiology at Johns Hopkins University. Howell called this heparin, the same name he had already given to a different, inferior anticoagulant extracted from liver by fat rather than aqueous solvents. The presence in liver of this earlier 'heparin' had been discovered serendipitously in Howell's laboratory in 1916 by a pre-medical student, Jay McLean, after extracts of the fat-soluble coagulating substance had deteriorated during storage. This unmasked the previously undetected anticoagulant. It was a phospholipid. Howell spent some time trying to isolate and purify McLean's anticoagulant before switching to the use of aqueous solvents for liver extraction. During the twenties, he studied the chemistry and physiological activity of the non-phospholipid anticoagulant from dog liver, identifying it as a sulphur-containing polysaccharide. He persuaded the Baltimore firm Hynson, Westcott, and Dunning to sell it for laboratory investigations. This material, however, contained less than 2% of active material. Two or three attempts were made to use it as an anticoagulant while transfusing blood into patients, but side-effects were caused by the impurities. Nevertheless, the introduction of this product brought the anticoagulant to the attention of researchers in Canada and Sweden.

In 1928, Charles Best and his colleagues at the Connaught Laboratories in the University of Toronto joined in the effort to prepare samples of heparin that would be sufficiently pure for clinical investigations. They hoped their earlier experience in preparing insulin would stand then in good stead. Whilst this work was progressing, Best, Gordon Murray, Louis Jaques, and T. S. Perrett began to study the effects of Howell's heparin on experimental thrombosis in dogs. Then, in 1933, David Scott and Arthur Charles completed an investigation into the amounts of heparin in different animal tissues. This pointed to beef lung as a promising source of the anticoagulant. However, the material extracted from this source was different to that originally introduced by Howell. During the next three years, Scott and Charles worked with this new heparin and made a series of advances that finally provided samples that were pure enough for clinical studies. This was shown by Erich Jorpes, in Stockholm, to be an acidic sulphated polysaccharide. It acted mainly by interfering with thrombin formation.

An international standard for the sodium salt of this new heparin was published in 1935, the year that Gordon Murray initiated clinical trials in Toronto with a view to establishing its value in preventing thrombosis after injuries. Around the same time, in Stockholm, Crafoord injected patients with the purified heparin to prevent post-operative thrombosis. The mistaken belief that the new heparin was the substance principally responsible for the prevention of clotting of circulating blood led to its immediate acceptance in hospital practice. Best went on to use heparin in exchange transfusion of blood between patients and donors, just two years before the outbreak of war, when the technique saved many lives.

In 1944, W. J. Kolff and H. Th. J. Berk invented the artificial kidney in Holland, the use of which only became possible as a result of the introduction of pure heparin. The later development of heart–lung by-pass techniques in surgery also depended on the availability of heparin. However, in patients exposed to a high risk of bleeding, the use of heparin in such circumstances may be hazardous. In these cases, epoprostenol ('Cyclo-prostin'; prostacyclin) may serve as a somewhat expensive alternative. This is a prostanoid discovered by John Vane of Burroughs Wellcome in 1976. It occurs in the walls of blood vessels, where it produces vasodilation and prevents clotting. Because it is related to the prostaglandins, it is rapidly metabolized, half of it being destroyed within three minutes of administration to a patient. This is put to advantage in those patients who cannot be given heparin during renal dialysis or heart–lung by-pass surgery.

The prostaglandins were originally discovered in 1934 by Ulf von Euler at the Karolinska Institute, Stockholm. He showed them to reduce blood pressure and stimulate smooth muscle, but they seemed of little interest as potential therapeutic agents until Bergstrom, one of his students, obtained a $100 000 grant from the Upjohn Company in the mid-fifties and then used combined gas chromatographic–mass spectrometric analysis to establish their chemical structures. It was not until 1965 that the first synthetic method was devised by Beal and his colleagues at the Upjohn laboratories, and only then could clinical investigations be considered. The Upjohn Company made a policy decision to supply the highly expensive prostaglandins free of charge to any researcher wishing to obtain these. In 1972, a prostaglandin preparation was marketed in the United Kingdom for obstetric use. Dinoprost ('Prostin F2' alpha) and dinoprostone (Prostin E2') are used for the induction of abortion.

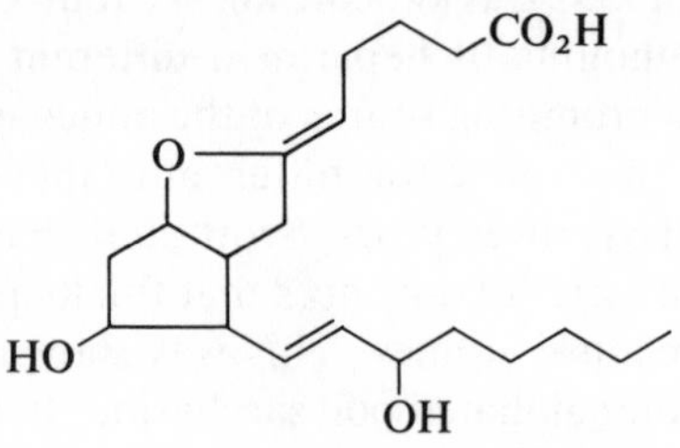

Epoprostenol

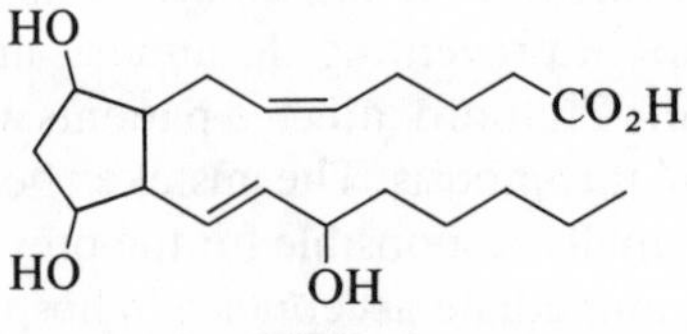

Dinoprost

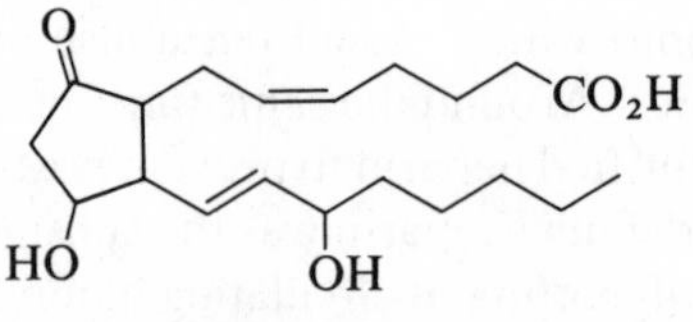

Dinoprostone

The discovery of the oral anticoagulants resulted from investigations in the early 1920s into a strange cattle disease that broke out in North Dakota and Alberta. In 1922, F. W. Schofield, a veterinarian in Alberta, showed that this so-called haemorrhagic septicaemia from which many cattle were bleeding to death, was caused by their eating hay prepared from spoiled sweet clover. By 1931, another veterinarian, L. M. Roderick of North Dakota, had found that the haemorrhagic factor acted by reducing the activity of prothrombin, a precursor of thrombin. However, attempts to isolate it were abortive.

In January 1933, Karl Link, the director of the experimental station run by the University of Wisconsin College of Agriculture, was consulted by a colleague who wanted to introduce a strain of sweet clover free of coumarin, which was responsible for the bitter taste that cattle obviously disliked. When, only a month later, a farmer struggled desperately through a blizzard to reach the experimental station and deposited the carcass of a poisoned heifer at his feet, Link was persuaded that a more urgent problem confronted him. He decided that attempts should be made to isolate the haemorrhagic factor.

Like those who had tried before him, Link found the task of isolating the haemorrhagic factor to be daunting. Its insolubility in common solvents was part of the problem, but even more perplexing was the difficulty of assaying the various extracts to determine how much of it was present. It was not until 1938 that any real progress was made. This was after Armand Quick, then in practice in Milwaukee, had published a method for the quantitative measurement of prothrombin activity in patients afflicted with various blood diseases. H. A. Campbell, whom Link had put in charge of the investigation, then modified the Quick test to make it suitable for assaying haemorrhagic factor.

The lack of solubility of the haemorrhagic factor was overcome by extracting it with dilute alkali. Unfortunately, much of the hay consisted of acidic material, all of which ended up in the alkaline extract! To eliminate these substances required considerable ingenuity, but at dawn on the 28th June 1939, Campbell finally isolated crystals of a potent anticoagulant. Two members of his group, Mark Strahmann and Charles Huebner, spent the next nine months improving the efficiency of the isolation procedure and trying to identify the haemorrhagic factor. On the 1st April 1940, they confirmed that it was 3,3′-methylenebis(4-hydroxycoumarin), a compound that had actually been synthesized in 1903, albeit in impure form.

Link published the first report of the identification of the anticoagulant in the *Journal of Biological Chemistry*. This appeared at the end of February 1941. Within a few days he received a request from doctors at the Mayo Clinic for a sample of it for clinical trials. In the middle of the following June, H. R. Butt, E. V. Allen, and J. L. Bollman presented their preliminary findings to a staff meeting at the Clinic. These were based on studies on dogs and six humans, confirming the oral activity of the anticoagulant. Following on this, numerous studies appeared, many concerning the merits or otherwise of this and related anticoagulants in thromboembolic disease. During the sixties there was widespread controversy in medical circles over the role of oral anticoagulant therapy in patients recovering from myocardial infarction ('heart attack'). Particular concern was expressed over the risk of haemorrhage. Critics of the use of anticoagulants pointed out that only a small proportion of post-infarct patients died as a result of thromboembolic complications. Furthermore, clinical trials failed to reveal a significant reduction in the death rate when anticoagulants were used. Nowadays, anticoagulants are reserved mainly for the treatment of patients considered to be greatly at risk, such as those with deep-vein thrombosis, or those who have just had artificial heart valve prostheses fitted. If treatment is urgently required in such cases, heparin may be injected during the three or four days it takes for the oral anticoagulant to begin to have an effect.

In 1942, the Abbott and Lilly companies marketed the new anticoagulant,

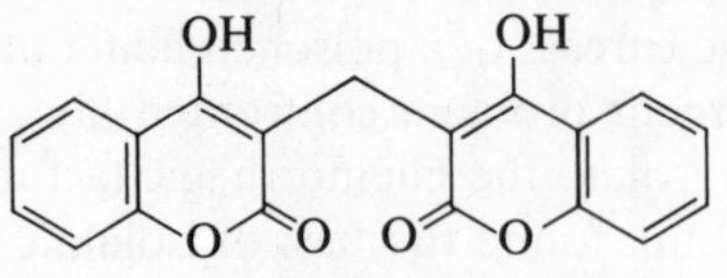

Dicoumarol

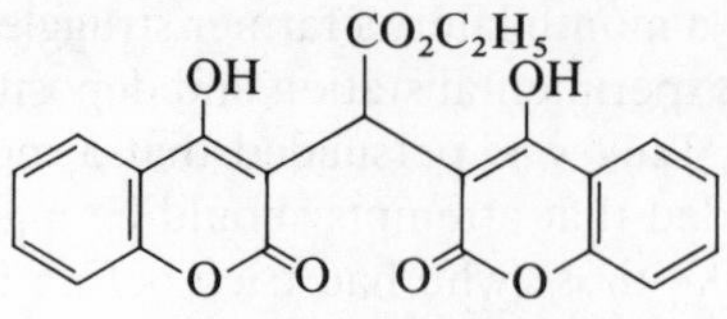

Ethyl Biscoumacetate

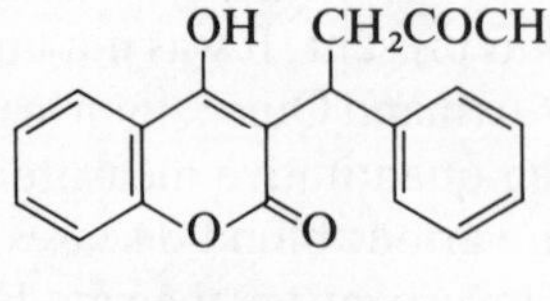

Warfarin

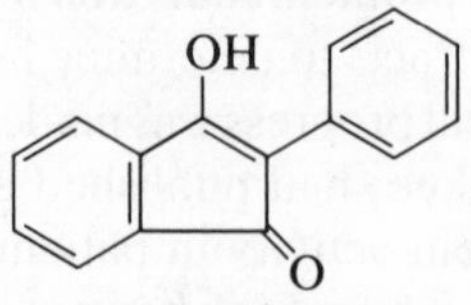

Phenindione

which was given the approved name of dicoumarol. Since then hundreds of coumarins, as well as the structurally related phenindione and its analogues, have been shown to be of clinical value. The principal differences between these drugs concern potency, rate of onset of action, and its duration. Since the mode of action of the oral anticoagulants is to interfere with the role of vitamin K in the liver's biosynthesis of clotting factors, including prothrombin, there is always a time lag before reserves of the existing factors are depleted. Ethyl biscoumacetate, for example, has a quicker onset than dicoumarol, but this is offset by a shorter duration.

The best known of the oral anticoagulants was selected from more than one hundred 3-substituted 4-hydroxycoumarins synthesized in Link's laboratory during the last war. As it was longer acting and also five to ten times more potent than dicoumarol, it was chosen by Link, in 1948, as an anticoagulant for rodenticidal use. It quickly won popularity throughout the world as a rat poison. The handsome royalties from its sales were vested in the Wisconsin Alumni Research Foundation, hence the origin of its names, warfarin. Despite the urging of Link and his associates, it only became the most widely used clinical anticoagulant after its superiority was recognized following an attempt by a US army recruit to commit suicide by taking rat poison. On Link's suggestion, it was then marketed by Endo Laboratories of New York.

IX

Antihistamines and Related Drugs

Whilst attending the International Physiological Congress at Heidelberg in August 1907, Henry Dale watched an experiment conducted by a University of Heidelberg obstetrician, E. Kehrer, in which the effects of an ergot extract were demonstrated on a uterus removed from a cat. Not only was Dale impressed by this new experimental technique, but he was also aware that the ergot extract appeared to contain a potent substance different from those which he had previously encountered insofar as it had an immediate powerful stimulant effect on the uterus. The particular extract used by Kehrer was known as *Egotinum dialysatum*, and it was prepared by a method first used in 1872, which allowed putrefaction to occur. To Dale this was significant, for his colleagues at the Wellcome laboratories were at that time investigating the action on the uterus and blood vessels of tyramine, formed by putrefaction of tissue protein (see page 99). On returning to London, Dale set to work with George Barger to isolate the new uterine stimulant. Shortly before completing their final identification of this material in 1910, D. Ackerman of the University of Wurzburg published an account of its preparation from an aminoacid, histidine, by deliberate putrefaction induced by microorganisms. It was not a new substance, having being synthesized for purely chemical reasons by Windaus and Vogt three years earlier.

Because of its formation from histidine, Dale eventually called the new uterine stimulant histamine. It was quite unsuitable for therapeutic use due to its wide-ranging physiological effects. The Wellcome chemists F.L. Pyman and A. J. Ewins prepared analogues of histamine, but these, too, lacked clinically useful activity, as has been the case with other analogues prepared more recently. However, the recognition by Dale and P. P. Laidlaw, as early as 1910, that the effects of histamine in various species of animals bore a remarkable resemblance to what had recently been described as 'anaphylactic shock', ensured that continuing attention would be paid to it. For many years, however, it appeared that histamine was only present in various tissue extracts as a result of bacterial action.

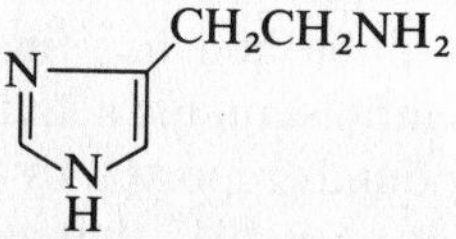

$CH_2CH_2NH_2$
N
N
H

Histamine

In 1926, claims were made in the United States and Canada to the effect that liver extracts could be used clinically to reduce blood pressure. Charles Best, who was visiting Dale to work on the crystallization of insulin, brought Dale a sample of a liver extract that had been used in Toronto. It was examined physiologically by Best and Dale, and chemically by Harold Dudley and W. V. Thorpe at the National Institute for Medical Research. They isolated substantial amounts of histamine and then repeated the extraction process under conditions that ruled out any possibility of formation of the histamine from histidine by putrefaction. This proved that the vasodilator present in liver was histamine. It now became apparent that this also accounted for the activity of gastric and pituitary extracts previously reported to reduce blood pressure by vasodilation. Professor Abel had already isolated histamine from pituitary extract in 1919, believing it was the oxytocic principal. Dale and others, with justification, had rejected this assertion since the possibility of bacterial decomposition could not be ruled out.

Once the presence of histamine in the body had been confirmed, interest in its effects grew. Amongst those who investigated it was a Swiss physiologist, Daniele Bovet, who worked in Fourneau's department of chemical therapeutics in the Pasteur Institute, Paris. Bovet realized that antagonists of acetylcholine and adrenaline, e.g. atropine and ergotamine, had made it possible for physiologists to investigate and understand the actions of these substances more clearly. Since no antagonist of histamine was known, Bovet set about screening possible candidates, assisted by G. Ungar and J.–L. Parrot. His sole lead came from some reports suggesting that the effects of histamine on the guinea pig intestine might be diminished by certain adrenaline analogues and antagonists. Inevitably, Bovet selected the adrenergic blocker, piperoxan (933F), which Fourneau had discovered three years earlier. It was shown to have some antagonistic activity against histamine in isolated intestine, but when injected subcutaneously into live guinea pigs it failed to protect them against the lethal effects of histamine administered by the intrajugular route. Bovet and one of his research students, Anne-Marie Staub, then proceeded to examine a variety of piperoxan analogues that had already been synthesized at the Pasteur Institute. Compound 929 F consistently protected the guinea pigs, counteracting the lethal bronchoconstrictive action of small doses of histamine. It was also the most potent compound in the isolated intestine screen. Several related phenolic ethers also afforded some protection, with 30–60% of animals surviving otherwise lethal exposure to histamine, whilst an analogue in which the ether oxygen was replaced by a nitrogen atom, 1167 F (diethylaminoethylaniline), was also effective. The French workers reported their findings in 1937, and Staub proceeded to investigate several more compounds prepared in Fourneau's laboratory. She published her findings in 1939, indicating the molecular requirements for antihistaminic activity with commendable accuracy. Unfortunately, the compounds reported by her were too toxic for human administration and it was not until 1942 that antihistamines were developed that were safe and potent enough for clinical use.

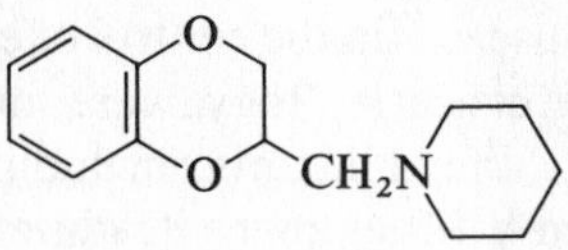

Piperoxan

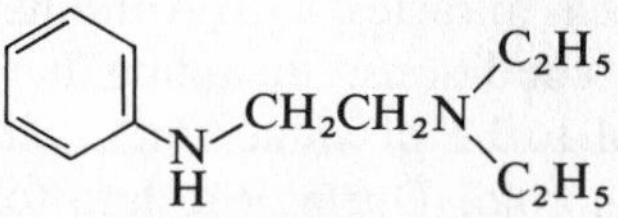

1167F

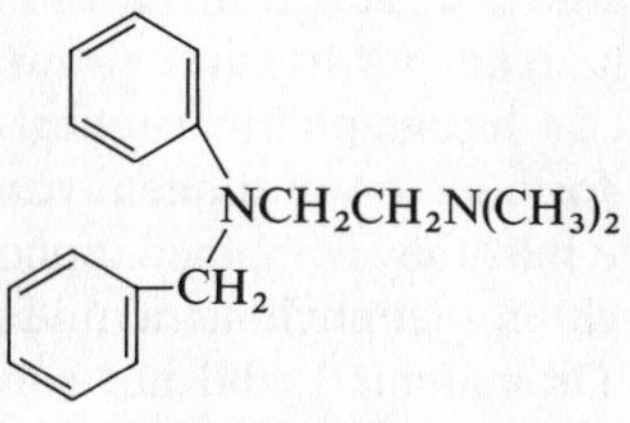

Phenbenzamine

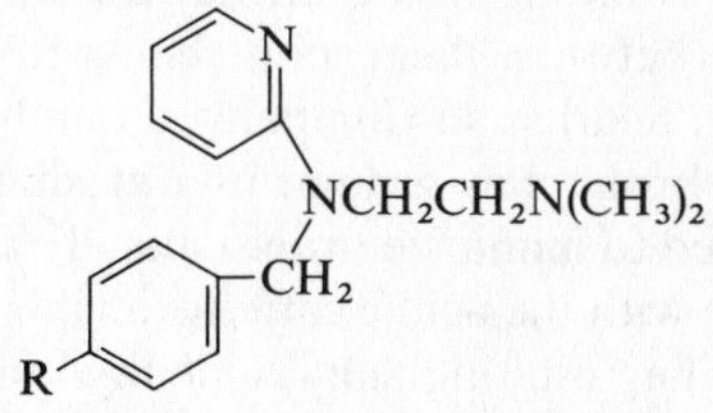

Mepyramine, R = $-OCH_3$
Tripelennamine, R = $-H$

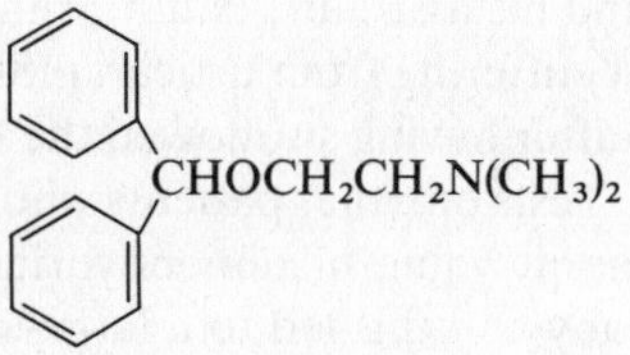

Diphenhydramine

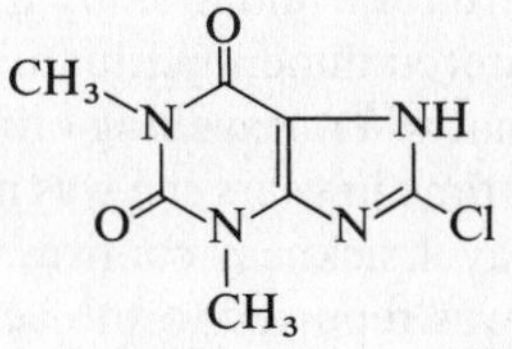

8-Chlorotheophylline

The first antihistamine to be used clinically, phenbenzamine ('Antergan'), was synthesized by Mosnier of the Rhône-Poulene Company. In 1942, B. N. Halpern reported its pharmacological activity, whilst two years later Bovet and his colleagues published their studies on the closely related mepyramine (pyrilamine, 'Neo-antergan') in which one of the benzene rings had been replaced by a pyridine ring. This molecular alteration was probably introduced because of the experience Rhone-Poulenc had acquired from their British subsidiary, May and Baker, who had developed sulphapyridine as a valuable antibacterial agent a few years earlier. Meanwhile, researchers at Ciba Pharmaceuticals in Summit, New Jersey, synthesized a similar antihistamine which differed only in the absence of the methoxyl group attached to the benzene ring. This was known as tripelennamine ('Pyribenzamine'). Also around the same time, diphenhydramine ('Benadryl') was designed by a young lecturer at the University of Cincinnati, George Rieveschl, and synthesized in 1943 by one of his research students, Wilson Huber. Parke, Davis, and Company tested the new compound then bought the patent rights from Rieveschl, granting him royalties on sales of diphenhydramine. These payments were later revealed to be providing Rieveschl with an annual income greater than that of the president of the company! Rieveschl also became director of research at Parke, Davis.

Antihistamines now became immensely popular, being hailed in some quarters as miracle drugs! Although their main use was in the control of certain allergic conditions, including hay fever and urticaria, they were initially believed to be of value for a wide range of other ailments, including the common cold. During the late forties and early fifties many leading pharmaceutical manufacturers competed vigorously to develop new antihistamines, resulting in the introduction of a plethora of new drugs with little to choose between them, none being free of the tendency to cause drowsiness.

G.D. Searle and Company, a family-owned Chicago pharmaceutical distributor, broke new ground by introducing a formulation of diphenhydramine designed to minimize drowsiness. To achieve this, they combined diphenhydramine with the acidic caffeine analogue, 8-chlorotheophylline, a mild stimulant. The resulting salt, dimenhydrinate ('Dramamine'), did not solve the problem of drowsiness, but it became one of the most profitable antihistamines on the market after it was found to have an unexpected therapeutic action! Samples had been sent to the allergy clinic at Johns Hopkins University in Baltimore for evaluation by Leslie Gay and his assistant, Paul Carliner. They administered dimenhydrinate to a patient suffering from urticaria. She discovered that when travelling on a streetcar after having swallowed the drug, for the first time in years she was not car sick! Tests on other patients who suffered from travel sickness confirmed the apparent value of dimenhydrinate. The matter was reported to the Searle Company and this led to a large-scale trial being organized. On 27th November 1947, the 'General Ballou' sailed from New York to Bremerhaven in Germany. The ten-day crossing was particularly rough, yet only 4% of those on board the troopship who received the drug were sick, in contrast to a quarter of a control group who received a placebo (dummy pill). Furthermore, all but seventeen of 389 sick soldiers recovered within a couple of hours of receiving dimenhydrinate. Unlike earlier remedies such as hyoscine, the only significant side-effect was drowsiness. This was quickly exploited by Searle before their rivals began to discover that other antihistamines were also effective against motion sickness.

During the last war, Rhône-Poulenc investigated amine derivatives of phenothiazine as potential antimalarial compounds, an understandable objective in view of Paul Ehrlich's demonstration of such activity with methylene blue. This line of research was abandoned, but Halpern and Ducrot realized that one of the new phenothiazines synthesized by Paul Charpentier was an analogue of phenbenzamine, the two benzene rings being bridged by a sulphur atom. This compound, 3015 RP or fenethazine, was indeed an antihistamine, but a derivative synthesized in 1946, in which there was an extra methyl group on the dimethylaminoethyl side-chain, turned out to be remarkably potent and very long-acting. This was named promethazine ('Phenergan').

Sperber, Papa, and Schwenk of the Schering Corporation realized that all the potent antihistamines contained two aromatic rings joined to either a nitrogen or an oxygen atom. They therefore synthesized a new series in which these rings were instead attached to a carbon atom since this had roughly similar atomic dimensions (i.e. isosteric replacement of an atom). This led, in

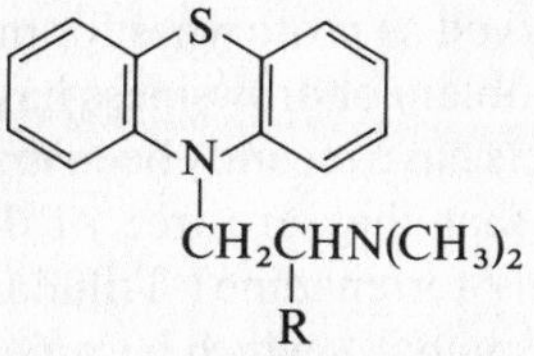

Fenethazine, R = —H
Promethazine, R = —CH_3

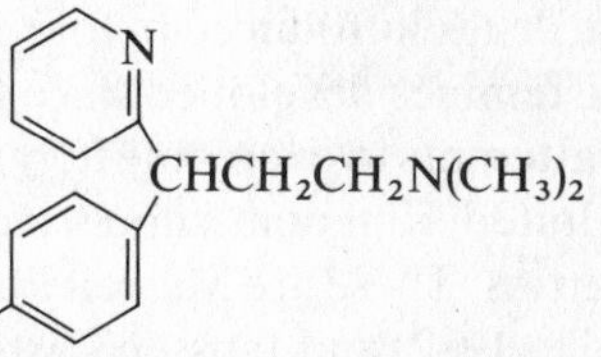

Pheniramine, R = —H
Brompheniramine, R = —Br
Chlorpheniramine, R = —Cl

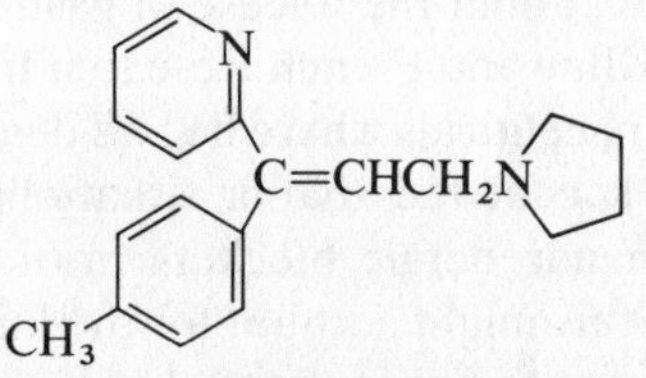

Triprolidine

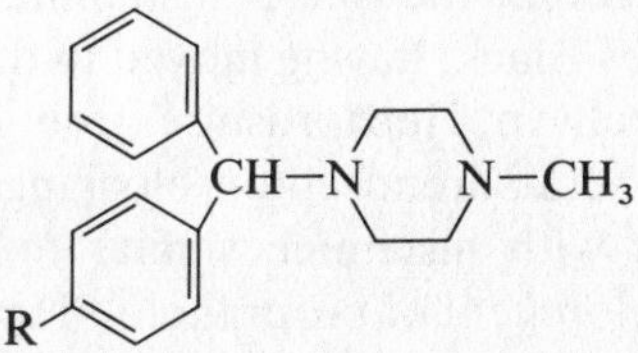

Cyclizine, R = —H
Chlorcyclizine, R = —Cl

1948, to the development of pheniramine ('Trimeton'), as well as its more potent derivatives known as chlorpheniramine and brompheniramine. Around the same time, a similar approach was adopted by D. W. Adamson of the Wellcome Research Laboratories in England, although it was a few years later that this resulted in the introduction of triprolidine ('Actidil'). While this programme of research was continuing, the American division of Burroughs Wellcome developed antihistamines in which the amino group was derived from piperazine instead of dimethylamine. In 1949, Baltzly, DuBreuil, Ide, and Lorz reported their synthesis of cyclizine ('Marzine') and chlorcyclizine, a long-acting antihistamine. Cyclizine achieved the notable distinction of being selected by the US National Aeronautic and Space Agency for use as a space sickness remedy on the first manned flight to the moon.

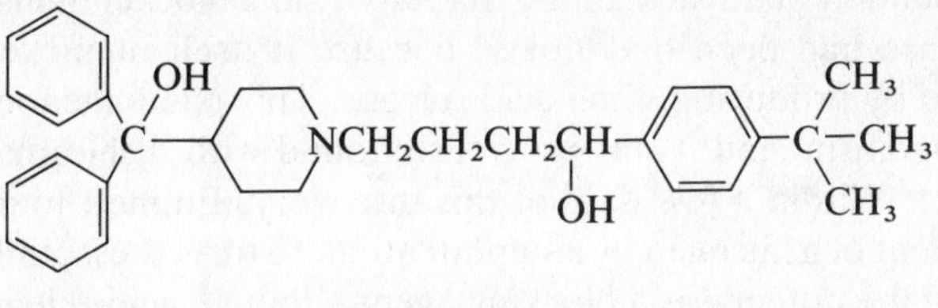

Terfenadine

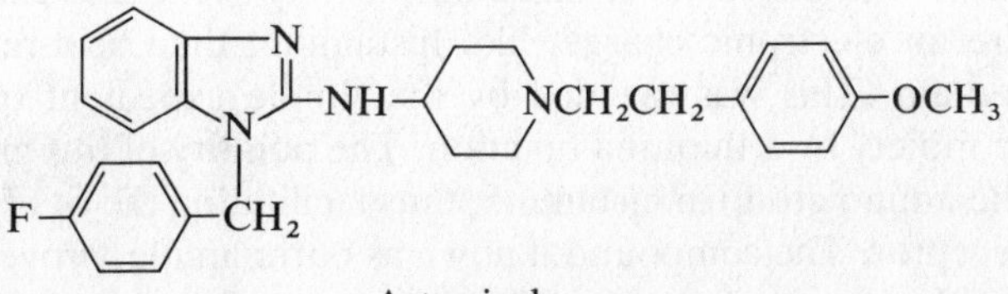

Astemizole

The drugs introduced before 1950 have served as prototypes for most of the antihistamines developed since then. The problem of drowsiness has defeated most attempts to overcome it, but two products have recently been marketed in the United Kingdom supported by claims that they are free of depressant properties. These are Merrell Pharmaceuticals' terfenadine ('Triludan'), introduced in 1982, and Janssen's astemizole ('Hismanal'), which became available the following year. The latter does not enter into the central nervous system.

Remarkable progress has also been made recently in dealing with one major shortcoming of the antihistamines already described, namely their failure to antagonize the role of histamine in bringing about the release of gastric acid. James Black, having moved to the Smith Kline and French Research Institute in Welwyn, Hertfordshire, from ICI Pharmaceuticals where he had developed the beta-adrenoceptor blocking agents, speculated that a situation might exist with histamine similar to that with adrenergic blockers prior to the development of propranolol. He believed this might account for the failure of antihistimines to prevent gastric acid release, and speculated that it might be possible to treat gastric ulcers with a new type of antihistamine. In 1964, he and Michael Parsons set up an assay procedure to detect such an antihistamine, measuring the effect of experimental compounds in preventing the production of acid in the stomach of anaesthetized rats.

The challenge facing the chemists at the Smith Kline and French laboratories was immense, there being no established lead for them to follow. Robin Ganellin was in charge of the project, assisted by Graham Durrant and John Emmett. For four years they examined histamine analogues in the hope that one would resemble histamine enough to fit its receptor in the stomach without triggering acid release; during this period over 200 compounds were synthesized without success. Indeed, the project seemed so hopeless that the company's head office in Philadelphia instructed the British researchers to abandon the project—whereupon no further reports were sent across the Atlantic! By a cruel twist of fate, it turned out that the first breakthrough came with a compound that had actually been synthesized by Durrant at the start of the project, namely guanylhistamine. Its activity as a blocker of histamine-induced acid release had been overlooked because it itself mimicked the action of histamine by producing some acid release. This phenomenon is described as 'partial agonism' and it was previously found with dichloroisoprenaline (see page 112). Within a few days of this discovery, Emmett found that isosteric replacement of a nitrogen by a sulphur atom, to form the isothiourea analogue, enhanced the potency as a blocking agent although acid-releasing activity still remained. Eventually it was realized that so long as the side-chain could ionize to acquire an electronic charge, like histamine, then acid-releasing activity would remain. This was avoided by the simple expedient of replacing the ionizable moiety by a thiourea function. The polarity of this group resembled that of the amino group in histamine, thus facilitating the fit of the drug to the gastric receptor. The compound, known as burimamide, proved to be solely a histamine blocker which, on injection, prevented the release of gastric acid. By now, the gastric receptor was being described as the H_2 receptor. Interestingly, none of the drugs that block this receptor manage to block the H_1 receptor at which the older antihistamines act.

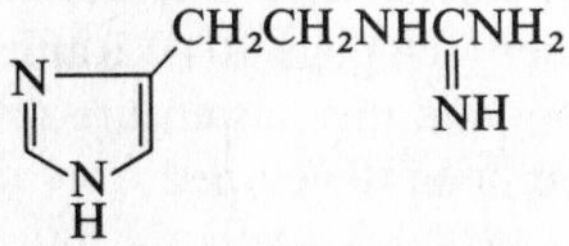

$CH_2CH_2NHCNH_2$ ($=NH$)

Guanylhistamine

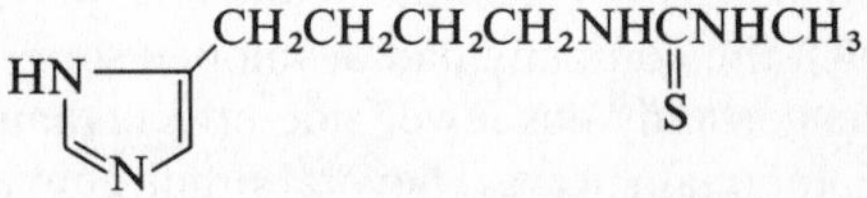

$CH_2CH_2CH_2CH_2NHCNHCH_3$ ($=S$)

Burimamide

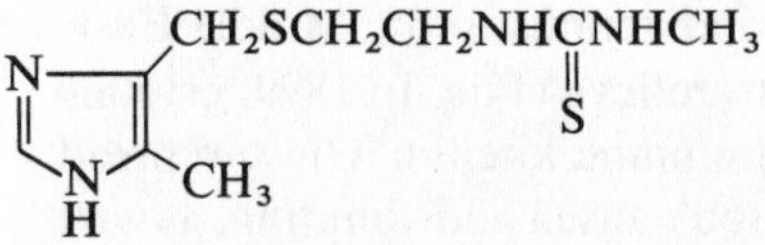

$CH_2SCH_2CH_2NHCNHCH_3$ ($=S$), CH_3

Metiamide

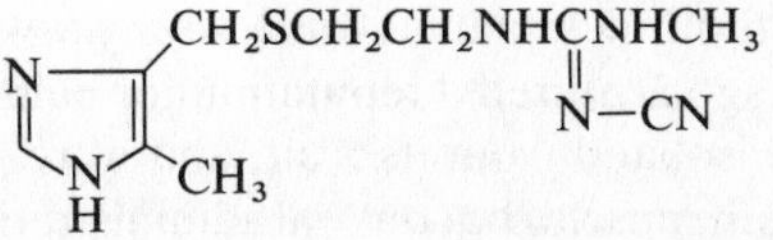

$CH_2SCH_2CH_2NHCNHCH_3$ ($=N-CN$), CH_3

Cimetidine

$(CH_3)_2NCH_2$ — O — $CH_2SCH_2CH_2NHCNHCH_3$ ($=CHNO_2$)

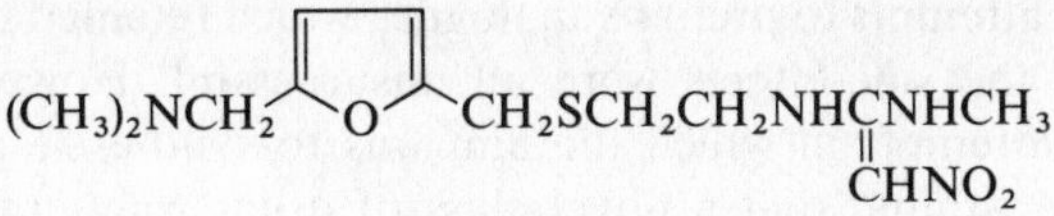

Ranitidine

Burimamide inhibited the release of gastric acid in animals and man, but it lacked sufficient potency for it to be administered by mouth. By careful consideration of electronic effects influencing the disposition of the hydrogen atom attached to one or other of the nitrogens in the imidazole ring, Ganellin obtained a derivative, metiamide, which was ten times as potent and could thus be administered by mouth. Unfortunately, during early clinical trials involving seven hundred patients, a few cases of a blood disorder, granulocytopaenia, were reported and the drug was abandoned even though the condition was reversible. Suspicion fell on the thiourea group, so other polar groups were examined, resulting in the development of cimetidine ('Tagamet'). Finally, it was introduced into clinical practice late in 1976, twelve years after Black's initial experiments. There is no doubt that cimetidine heals peptic ulcers by preventing gastric acid release, although the long-term effects of this type of therapy remain to be fully assessed. There can, however, be little doubt that the development of cimetidine by the Smith Kline and French researchers represents the current state of the art of the medicinal chemist at its best, especially as it was necessary for a lead compound to be generated by rational approaches.

During the early stages of the development of cimetidine, Black and his colleagues had investigated a variety of alternatives to the imidazole ring but none had proved effective. A Glaxo research team led by Roy Brittain, Barry Price, and David Jack discovered that replacement of the imidazole ring by a furan ring was acceptable so long as a basic substituent was present. This resulted in the introduction of ranitidine ('Zantac'), which was designed to be more selective by virtue of an additional moiety that it was hoped would hinder the fit of the molecule to histamine receptors other than those present in the stomach. The reasoning behind this was based on the selectivity of salbutamol, which the company had developed some years earlier (see page 104). Ranitidine certainly has fewer side-effects than cimetidine, and this has ensured its commercial success. Several other new agents have been developed.

ANTIALLERGEN AGENTS

The toothpick plant, *Ammi visnaga*, which is indigenous to the Middle East, long ago acquired a reputation for being able to relieve colic. In 1879, crystals were isolated from its fruit, and were given the name khellin. These relaxed smooth muscle, but on oral administration caused nausea and vomiting, as well as other side–effects. The structure of khellin was determined in 1938 by Ernst Spath and his colleagues at the University of Vienna, who showed it to be a chromone. Several attempts to prepare analogues which retained the relaxant properties without the side-effects were all unsuccessful. However, it was through one such attempt, in which the aim was to synthesize a bronchial relaxant for use in asthma, that a novel class of drugs was serendipitously discovered.

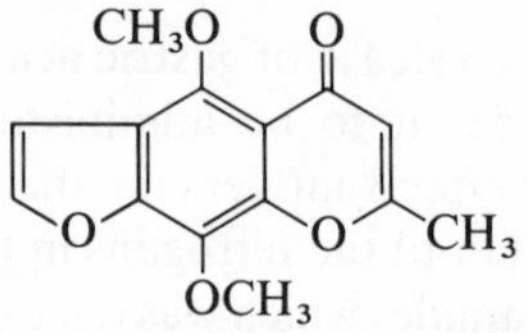

CH_3O O O O CH_3 OCH_3

Khellin

In 1955, a group of chemists at the Benger Laboratories in Cheshire initiated a programme to synthesize khellin analogues as potential bronchodilators. Their compounds were screened for ability to relieve bronchospasm induced in guinea pigs by histamine or carbachol. Active compounds were further examined in sensitized guinea pigs exposed to aerosols of egg white in order to induce a laboratory model of an asthmatic attack. On joining the research team, Roger Altounyan, a general practitioner who had just returned from working in the Middle East, expressed reservations about the validity of such artificial animal tests. He thereupon took the unprecedented step of inhaling a preparation of guinea pig hair, which he knew would induce an asthmatic attack, so as to evaluate compounds on himself! By 1958, Altounyan had found that several members of a series of carboxylated analogues of khellin were bronchodilators. One had unusual properties insofar as it had no bronchodilating activity and afforded animals no protection against histamine or carbachol, yet it somehow protected Altounyan from the effects of guinea pig hair. It only worked when inhaled, but this was itself unpleasant as the compound irritated the lungs. This, plus the fact that it was too polar to be absorbed when taken orally, dampened the company's enthusiasm for the project. In fact, it was soon officially abandoned, which meant the end of funding to support any more investigations. Nevertheless, the individuals working on the project were determined to see it to a successful conclusion even though only two compounds could be tested on Altounyan each week. The long-awaited breakthrough did not come until 1963, when it was found that one of the compounds synthesized by Colin Fitzmaurice had been contaminated with highly active material. Brian Lee believed the most likely contaminant was that formed by a side-reaction between two chromone molecules to form a bischromone. This was put to the test, progress now being faster due to the takeover of Benger by Fisons, and the attitude of the new management being positive. A series of bischromones was prepared for Altounyan to inhale, the six hundred and seventieth compound synthesized since work began more than nine years earlier being prepared by Fitzmaurice in January 1965. Altounyan reported that this bischromone was ideal as a protective inhalant, although the dose required was somewhat inconvenient to administer. While further studies on the safety of the drug were proceeding, Altounyan and a company engineer, Harold Howell, developed a special device that enabled it to be inhaled as a dry powder liberated from a pierced capsule. The first clinical trial was carried out in Manchester in 1967, with favourable results which have been amply confirmed by subsequent clinical experience throughout the world. The new drug, which was to generate handsome profits for Fisons, was given the approved

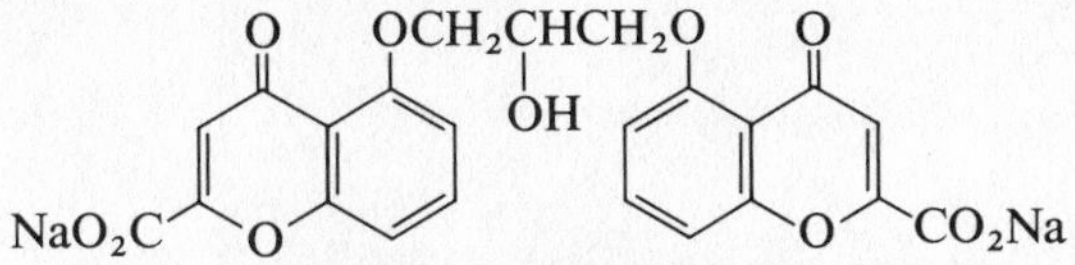

Sodium Cromoglycate

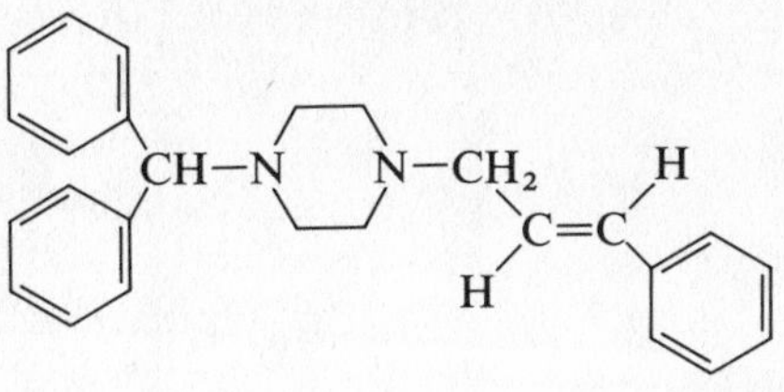

Cinnarizine

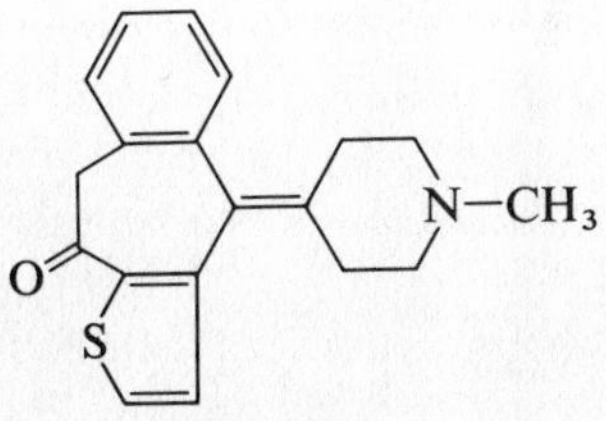

Ketotifen

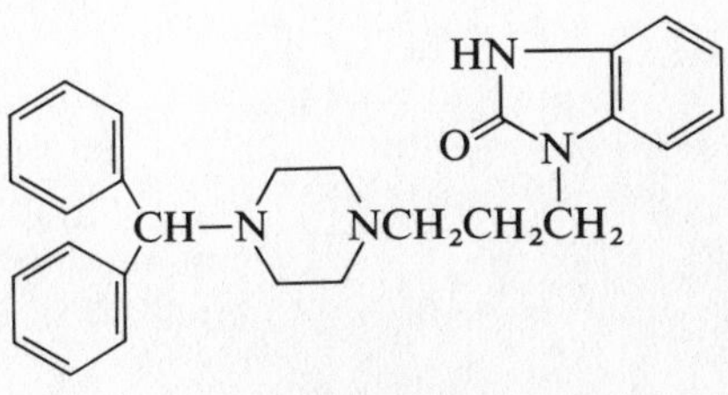

Oxatomide

name of sodium cromoglycate (cromolyn, 'Intal'). It was subsequently shown to act by stabilizing cell membranes in the lung so as to prevent the allergen-induced release of the substances that would otherwise cause bronchoconstriction in allergic asthma patients. Attempts by Fisons and other drug companies to design an orally active analogue met with little success until it was discovered that the antihistamine cinnarizine ('Stugeron'), developed by Janssen in the late fifties, had some activity. Exploitation of this lead resulted in the introduction of two orally active cinnarizine analogues with properties similar to those of sodium cromoglycate, namely ketotifen ('Zaditen'), which Sandoz had previously patented as an antihistamine as long ago as 1971, and Janssen's purpose-designed oxatomide ('Tinset'), these being marketed in the United Kingdom in 1980 and 1982, respectively.

X

Psychopharmacological Agents

TRANQUILLIZERS

In the early years of this century, the British colonial authorities in India grew increasingly concerned about the failure of Western medicine to make significant inroads into the local systems of medicine. Numerous committees of inquiry were established, one result of which was the intensive botanical and chemical investigation of Indian herbal remedies. One of the earliest investigations was into the snakeroot plant, so called on account of its appearance which had, from time immemorial, led to its use as an antidote to snakebite. This is an obvious example of the Doctrine of Signatures, whereby the stars were held to have imprinted their signatures on certain healing plants that could be recognized by characteristic features such as, in this instance, a resemblance to the cause of the malady. The first written mention of snakeroot appeared in the *Charaka Samhita*, a monumental medical treatise written around 600 BC. The plant was described as 'the foremost and most praiseworthy Indian medicine' by Gracia ab Horto in a Portuguese work dealing with Hindu medicine, published in Goa in 1563. The French botanist Charles Plumier in 1703 named the plant *Rauwolfia serpentina* after the German explorer Leonard Rauwolf, who had written a description of it in 1582. Despite the early report of the popularity of the plant in India, where more than thirty applications were known early this century, no serious investigation of it took place until 1931.

Salimuzzaman Siddiqui and Rafat Hussein Siddiqui of the Research Institute for Unani-Ayurvedic Medicine in Delhi purchased some snakeroot in the bazaar at Patna then investigated its constituents. Their first success was the isolation, in 1931, of an alkaloid that they named ajmaline, after the physician-wizard Hakim Azmal Khan who founded the Institute. Another five new alkaloids were later isolated, but when tested on frogs (!) none of them exhibited any properties of interest. Meanwhile, two physicians in Calcutta, Gananath Sen and Kartick Bose, published a report claiming that *Rauwolfia* not only reduced blood pressure if administered for a few weeks, but it also produced sedation, as was well known to many people in Bihar where it was used to put babies to sleep and to calm violent lunatics. They enthused over the plant, urging others to investigate it for themselves. This suggestion was taken up by Professor Ram Nath Chopra and his colleagues at the School of Tropical Medicine in Calcutta, who then spent more than ten years investigating the pharmacology of *Rauwolfia* and its alkaloids, confirming the hypotensive activity (i.e. ability to reduce blood pressure). They also demonstrated that crude extracts of the root had powerful sedative properties, but were unable to isolate any pure alkaloid with such activity.

It has been estimated that around one million Indians received Rauwolfia for their high blood pressure during the 1940s, but there was no interest in the drug in the West until 1949, when Rustom Jal Vakil published in the *British Heart Journal* the findings of a controlled clinical trial of the effects of *Rauwolfia* in fifty patients treated over a five-year period at the King Edward Memorial Hospital in Bombay. His results clearly indicated the beneficial effects of the therapy, and they caught the attention of Robert Wilkins, director of the hypertension clinic at Massachusetts General Hospital, who then asked E. R. Squibb and Sons to obtain a supply of *Rauwolfia* for him to try on patients. This was duly done, and he reported his findings in 1952 to the New England Cardiovascular Society. He confirmed the mild hypotensive activity of the powdered root and drew attention to its unusual sedative effect which caused patients to feel relaxed rather than drowsy. By a remarkable coincidence, the isolation of the long sought after active principle responsible for these therapeutic effects was reported at almost the same time.

The first time an extract of *Rauwolfia* had been found to exhibit sedative activity was in 1944, when J. C. Gupta and his colleagues investigated the action of the so-called resin fraction from the root. Three years later, Emil Schlittler of the Ciba research division in Basle asked his colleague Hugo Bein to test the mother liquors of *Rauwolfia* extracts from which large amounts of ajmaline had already been harvested for Sir Robert Robinson at Oxford, who required the alkaloid for structural studies. The liquors which contained the resinous material were found to have sedative activity, and Schlittler set to work with Johannes Müller to isolate the sedative principle. Their task was complicated by the presence of a wide variety of pharmacologically active compounds, some of which actually antagonized the action of the compound they were hoping to isolate. This confounded efforts to follow the effectiveness of different extraction procedures by assaying the extracts on animals. A further complication was the very slow onset of the sedative-hypotensive action. Bein overcame this difficulty by selecting from the varied physiological effects of the extracts one which seemed to be present in most active fractions, namely the miotic effect on the eye of a hare. This was used to home in on the best method for recovering the sparingly soluble active principle, which was eventually isolated in 1951. The new alkaloid was named reserpine, and it was soon confirmed as being responsible for most of the hypotensive and sedative

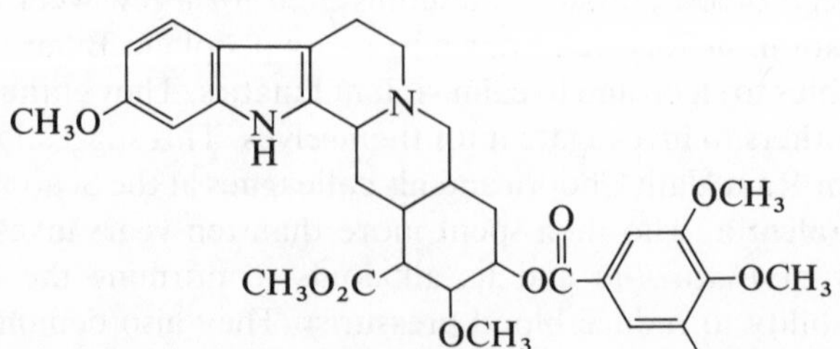

Reserpine

activity of Rauwolfia root. In November 1953, Ciba began to market reserpine ('Serpasil'). It was eventually synthesized in 1956 by Professor Robert Woodward at Harvard.

The term 'tranquillizer' (its spelling depends on which dictionary is consulted!) was first used by F. F. Yonkman in 1953 at a meeting in the Summit, New Jersey, headquarters of Ciba in the United States, when he described the unusual effects of reserpine on the central nervous system. Considerable interest was aroused by the new drug, especially in view of its botanical origin. Drug firms became convinced that a host of medicinal agents were to be found in the remotest corners of the earth, and vast sums of money were spent in obtaining and examining plants of all shapes and sizes. By and large, this effort proved a waste of time and money if viewed from a commercial standpoint, although much important scientific knowledge has been gained. The only major new therapeutic development that arose from it was the introduction of the *Vinca* alkaloids as anticancer drugs.

For several years reserpine was widely prescribed both as a hypotensive agent and as a tranquillizer, but its popularity waned. The effect on blood pressure when given by mouth proved inferior to that of other new drugs, and there was a risk of precipitating a severe depression that could lead to suicide in some patients. The main reason for the eclipse of reserpine as a tranquillizer was the advent of the phenothiazine tranquillizers.

These were developed as a consequence of studies initiated at Bizerte, Tunisia, in April 1949 by Henri Laborit, a French Navy surgeon who wanted to prevent surgical shock by the use of drugs. Whilst examining antihistamines, he found that promethazine was superior to others, but since its mild anti-shock activity was accompanied by side-effects, he tried combining it with various drugs in what he described as a 'lytic cocktail'. This phrase reflected his belief that shock could be prevented by inhibition of the autonomic nervous system to block the cardiovascular response that led to the drastic fall in blood pressure. When it became apparent that promethazine had unusual central actions, he concluded these were somehow responsible for its anti-shock action. The manufacturers of promethazine, the Specia Laboratories of Rhône-Poulenc at Vitry-sur-Seine, near Paris, now became interested in this aspect of Laborit's work. In the autumn of 1950 they began a search for a perfect lytic agent that would prevent surgical shock through its depressant action on the central nervous system. Simone Courvoisier screened the phenothiazines that Paul Charpentier had synthesized since 1944 as potential antihistamines, but instead of drugs with depressant effects being rejected, these were investigated further. When one now known as promazine proved most interesting despite its low level of antihistaminic activity, Charpentier synthesized analogues of it. A chlorinated derivative prepared in December 1950 was passed to Courvoisier, who identified its outstanding activity and low toxicity. In the spring, samples were given to Laborit, now based at the Val de Grâce military hospital in Paris. He confirmed that it was indeed the perfect lytic agent he had long sought. After completing appropriate animal tests, he tried the new drug on patients undergoing surgery. Before long, he observed that not only did they fare better both during and after their operations, due to the anti-shock action, but they also seemed relaxed and unconcerned with what was happening to them during the normally stressful pre-operative period. The significance of this was not lost on Laborit. He persuaded his psychiatric colleagues at the Val de Grâce hospital to test the drug on psychotic patients. On the 19th January 1952, the director of the neuropsychiatric service, Joseph Hamon, with Jean Paraire and Jean Velluz, began to treat a manic patient who was decidedly agitated until he was given his first injection. At once, he became calm and remained so for several hours. It was recognized that the drug was palliative rather than curative, but not only was this patient released from hospital three weeks later, so too were thousands of other psychotics who were later treated with Laborit's drug. Much of the pioneering work was done by Jean Delay and Pierre Deniker at the Hôpital Ste-Anne in Paris.

The new drug was given the approved name of chlorpromazine when it was marketed in France by Rhône-Poulenc in the autumn of 1952. The proprietary name 'Largactil' reflected its diverse range of actions ('Thorazine' in the USA). Clinical trials soon confirmed its value as a tranquillizer for agitated or psychotic patients, with the consequence that psychiatry was transformed overnight.

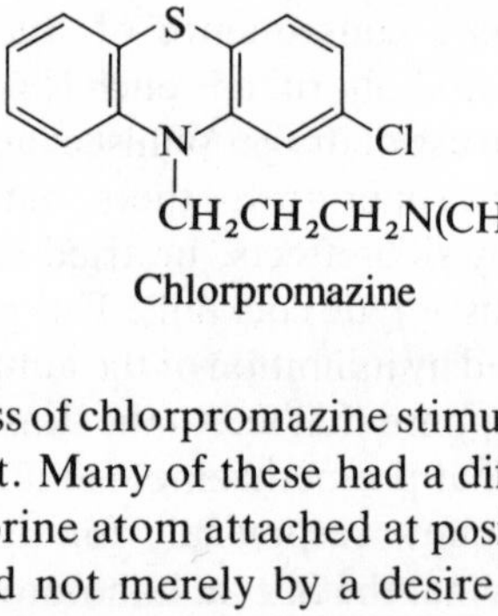

S

N Cl

$CH_2CH_2CH_2N(CH_3)_2$

Chlorpromazine

The remarkable success of chlorpromazine stimulated rival manufacturers to introduce analogues of it. Many of these had a different substituent incorporated in place of the chlorine atom attached at position 2 of the phenothiazine ring. This was motivated not merely by a desire to circumvent the Rhône-Poulenc patents, but also by a genuine conviction that potency was influenced by the electron-withdrawing power of the substituent, a view that is not strictly correct. Typical variants included acetyl, methoxyl, nitrile, trifluoromethyl, thioalkyl, and dialkylsulphonamide groups. More potent analogues were also obtained by replacing the dimethylamine function on the side-chain with a piperazine group, but this increased side-effects involving extrapyramidal pathways in the nervous system, leading to a Parkinson's disease-like tremor in some patients. The differences between the scores of phenothiazine tranquillizers introduced over the past quarter of a century are less significant than the great variability in patient response.

In 1958, P. V. Peterson and his colleagues, who worked for the Danish firm H. Lundbeck and Company, published their first report on a new series of tranquillizers, the thioxanthenes, in which the nitrogen of the phenothiazine ring had been isosterically replaced by a carbon atom. As these tricyclic compounds had strong chemical similarities to the phenothiazine tranquillizers, it is hardly surprising that their therapeutic activity proved to be similar. The first member of the series to be introduced clinically underwent a preliminary trial in 1958. It received the approved name of chlorprothixene ('Taractan'), and several analogues were subsequently developed.

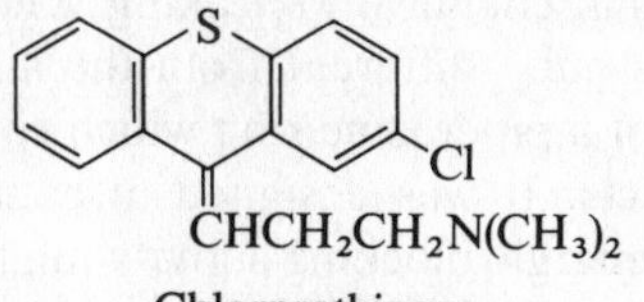

S

Cl

$CHCH_2CH_2N(CH_3)_2$

Chlorprothixene

The only other major group of strong tranquillizers arose not from a search for new tricyclic compounds, but from an investigation into analogues of methadone and pethidine (see page 77) by Janssen Pharmaceutica of Beerse in Belgium. In 1957, Janssen found that replacement of the methyl group attached to the nitrogen atom of pethidine could be achieved by simple chemical methods. He synthesized compound R 951 in which a propiophenone group replaced the methyl. This proved to be a potent analgesic, so it seemed sensible to lengthen the chain to a butyrophenone and test this as well. As anticipated, the new derivative, R 1187, was also an analgesic. However, it was noted that the mice which had been injected with the drug became progressively calm and sedated after initially exhibiting typical pethidine-like excitement, mydriasis, and insensitivity to noxious stimuli. The resemblance of the sedation to that produced by chlorpromazine encouraged Janssen to synthesize analogues of R 1187 in the hope that one might be devoid of analgesic activity whilst retaining tranquillizing activity. This was achieved simply by replacing the ester function, common to pethidine and R 1187, by a hydroxyl group. The next step was to prepare literally hundreds of derivatives with different substituents in the benzene rings as this was known to enhance potency in similar compounds. From this effort there emerged, in 1958, the most potent tranquillizer yet discovered, namely haloperidol ('Haldol'). It was 50–100 times as potent as chlorpromazine, and had fewer side-effects.

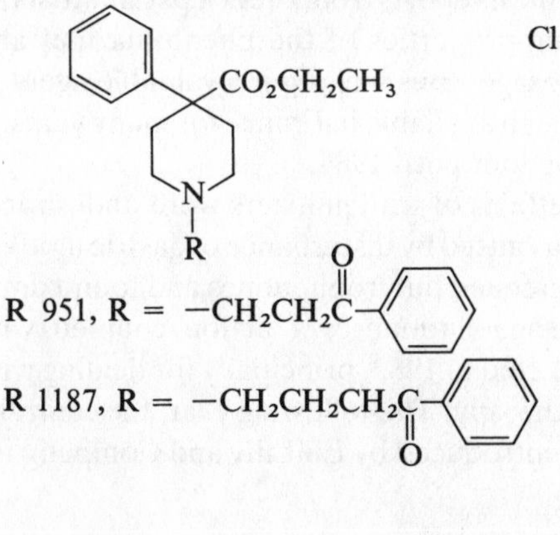

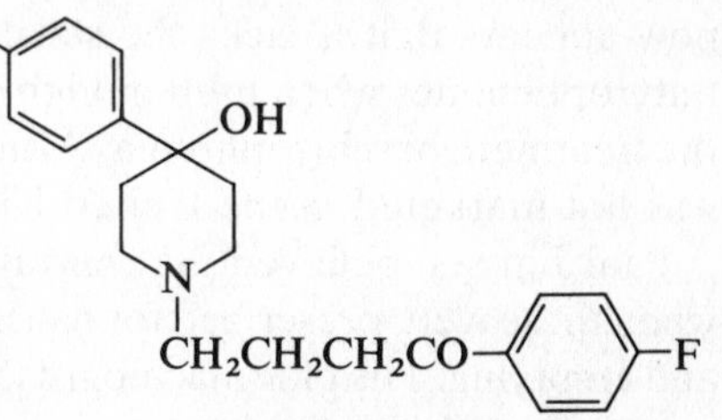

$CO_2CH_2CH_3$

N

R

R 951, R = $-CH_2CH_2C(=O)-C_6H_5$

R 1187, R = $-CH_2CH_2CH_2C(=O)-C_6H_5$

Cl

OH

N

$CH_2CH_2CH_2CO-C_6H_4-F$

Haloperidol

In the decade following the discovery of haloperidol, over 5000 analogues of it were synthesized and tested, of which 4000 were made in the Janssen laboratories. A dozen or so of these compounds were introduced clinically, and new derivatives are still appearing. One of the most important is pimozide ('Orap'), an outstanding drug for the control of apathetic, withdrawn schizophrenic patients, to whom it is given by mouth once daily.

Oxypertine ('Integrin') is, chemically speaking, a novel type of tranquillizer insofar as its structure is quite different from the agents already described, except for the presence of a piperazine ring which could well account for the extrapyramidal side-effects. It was designed at a time (1962) when it was believed that central adrenergic blocking activity might account for the action of chlorpromazine. This persuaded researchers at the Sterling-Winthrop Research Institute in Rensselaer, New York, to combine phenylpiperazine, a weak adrenergic blocker of which they had a plentiful supply, with an indole derivative selected for its resemblance to the neurohormone serotonin. The resulting compound resembled some of the piperazine derived phenothiazines in its clinical effects.

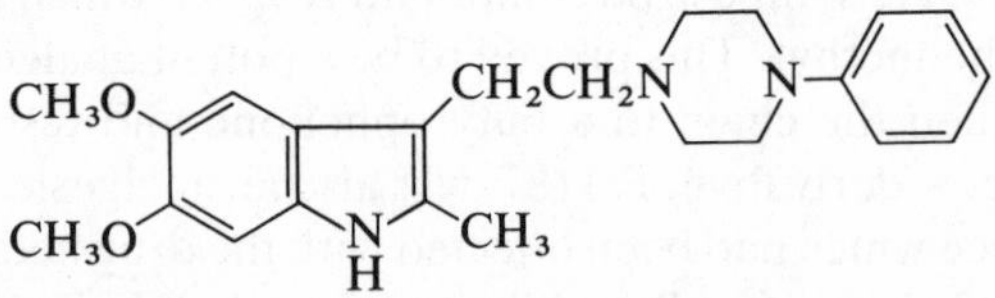

CH_3O CH_2CH_2N N
CH_3O N CH_3
H

Oxypertine

In 1957, F. Besançon of the Delagrange Research Laboratory in France discovered that a procainamide derivative, namely 2-chloroprocainamide, which he was examining as an antiarrhythmic agent, had antiemetic properties. A series of derivatives were synthesized, leading to the marketing of metoclopramide ('Maxolon') as an antiemetic, in 1964. Because of the recent discovery that chlorpromazine was not only an antiemetic but also a tranquillizer, several hundred derivatives of metoclopramide were examined over the next three years, until sulpiride ('Dolmatil') emerged as a potential antipsychotic agent. Animal tests revealed that it did not fit the usual pattern for such compounds, and this was reflected in conflicting clinical reports from French psychiatrists. It now appears that it lacks the sedating properties of the phenothiazines and butyrophenones when used in high dosage, thus making it a valuable agent in the treatment of schizophrenia. Although available in France for many years, it was not marketed in the United Kingdom until 1984.

Paul Janssen believed the central effects of tranquillizers were undesirable when these were prescribed for nausea caused by disturbance of gastric motility and emptying. His pharmacologists screened butyrophenones and found domperidone ('Motilium') to have a strong antinauseant action coupled with minimal central effects. This was marketed in 1982, principally for dealing with nausea associated with cancer chemotherapy. The following year, the centrally acting drug nabilone ('Cesamet') was introduced by Eli Lilly and Company for much the same purpose. It was an analogue of tetrahydrocannabinol, the active principle of *Cannabis sativa*, the euphoriant plant known to the Ancient Greeks, and widely used in the Middle East and India. Analogues of tetrahydrocannabinol were developed following reports that leukaemia patients in the United States were less nauseated if they smoked cannabis before receiving cytotoxic drugs. When tetrahydrocannabinol was administered, a high incidence of cardiovascular and central side-effects created problems for many patients. With nabilone, the central effects were reduced. Unlike cannabis and tetrahydrocannabinol, it was orally active.

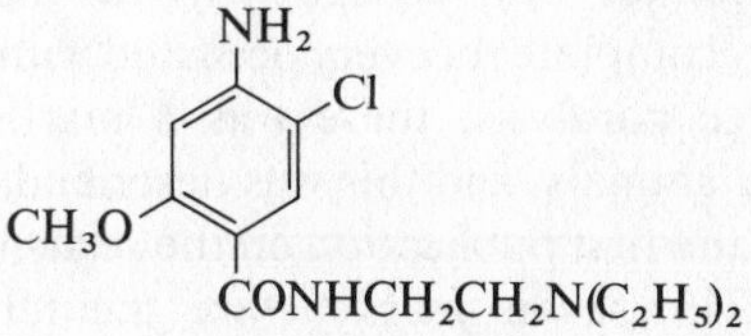

Metoclopramide

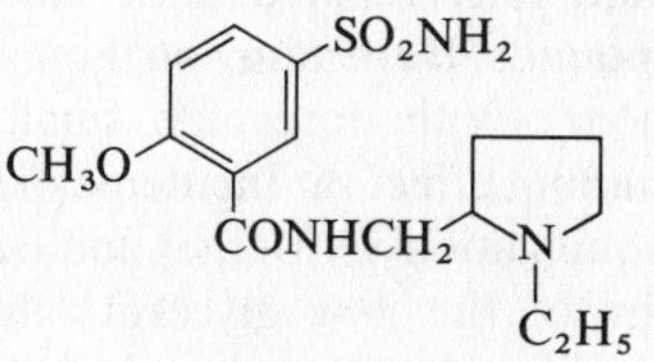

Sulpiride

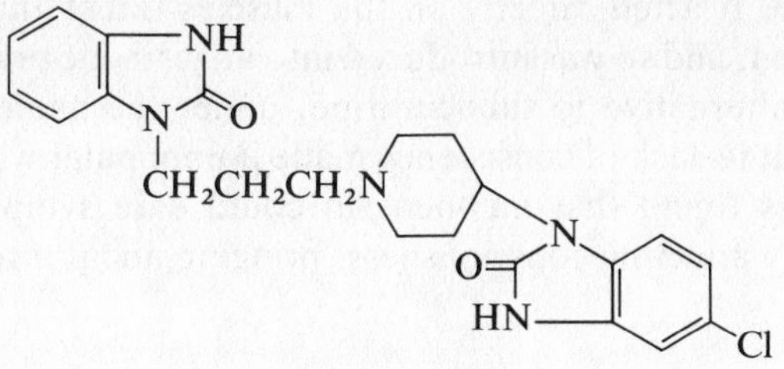

Domperidone

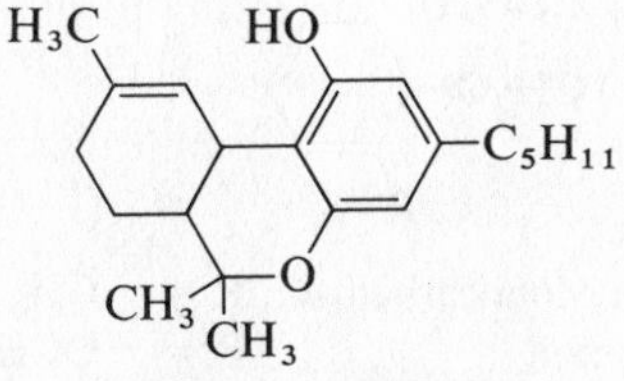

Tetrahydrocannabinol

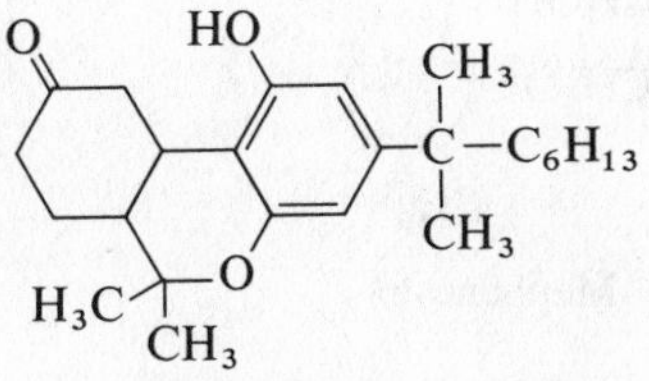

Nabilone

ANTI-ANXIETY DRUGS

Following the introduction of penicillin during the last war, there was concern at its failure to kill certain types of bacteria known as Gram-negative organisms (i.e. those which do not react to the Gram microscopic staining technique). One suggestion for dealing with this problem, especially in the topical treatment of wounds and burns, was to combine penicillin with phenoxyethanol ('Phenoxetol'), an antiseptic known to kill Gram-negative bacteria. As this compound lacked sufficient potency for it to be applied in many situations where it was not possible to achieve the required 2% concentration, William Bradley, chief chemist of the British Drug Houses in London, synthesized several glycerol ethers as analogues of phenoxyethanol, in the hope of finding superior agents. Whilst carrying out toxicity tests on mice with these analogues, an émigré Czechoslovakian pharmacologist, Frank Berger, unexpectedly found they produced flaccid paralysis of the limbs, quite unlike anything he had ever seen before. Walpole's description of serendipity as a discovery made by accident and sagacity, aptly describes Berger's subsequent investigations. He established that conscious animals lost control over their muscles, which became fully relaxed after either injection or oral dosage with the new compounds. Depending on the size of dose, complete recovery occurred within an hour. With doses too small to induce paralysis, there was a marked quietening effect on the demeanour of the animals, and this was described as 'tranquilization' by Berger and Bradley in the first publication on the pharmacology of the new glycerol ethers in 1946, seven years before the term 'tranquillizer' was used to describe reserpine. More compounds were synthesized until it was evident that, out of 143 compounds tested in animals, one had superior activity. It acted directly on the muscles rather than at the neuromuscular junction, and so was introduced into anaesthetic practice the following year as an alternative to tubocurarine, under the name of mephenesin ('Myanesin'), but its lack of consistency made it unpopular with anaesthetists. However, it was found that mephenesin could ease symptoms of anxiety without seriously affecting consciousness, bringing about a relaxed feeling in tense patients.

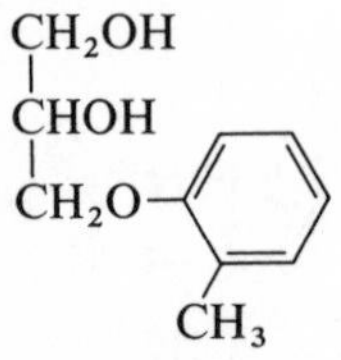

CH_2OH
|
$CHOH$
|
CH_2O–(benzene ring)
CH_3

Mephenesin

CH_2OCONH_2
|
$CHOH$
|
CH_2O–(benzene ring)
R

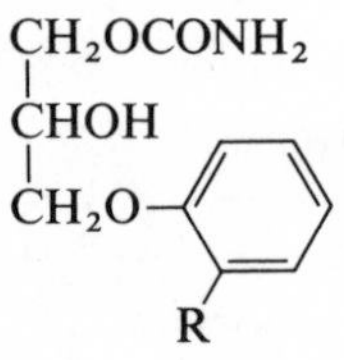

Mephenesin Monocarbamate, R = $-CH_3$
Methocarbamol, R = $-OCH_3$

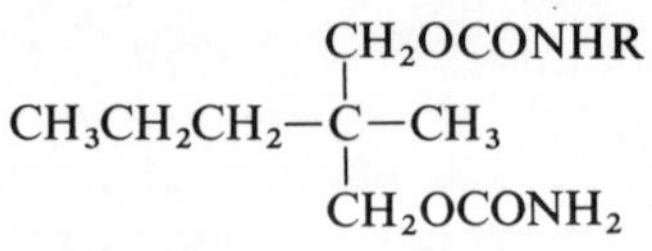

$$CH_3CH_2CH_2-\underset{CH_2OCONH_2}{\overset{CH_2OCONHR}{C}}-CH_3$$

Meprobamate, R = $-H$
Carisoprodol, R = $-CH(CH_3)_2$

Berger crossed the Atlantic to work with Wallace Laboratories in Cranbury, New Jersey, where he eventually became president of the company. His aim was to discover a better anti-anxiety agent, and by the autumn of 1949 he and the chemist B. J. Ludwig had decided the major shortcoming of mephenesin was its short duration of action, which was due to rapid metabolic oxidation of the alcohol function. Whilst this was highly advantageous for a muscle relaxant in anaesthesia, it was quite the opposite when the drug was required to control symptoms of anxiety for several hours between doses. To overcome this problem, all that seemed necessary was to protect the sensitive alcohol function with a carbamate group, as it had long been recognized that this conferred sedative activity. The resulting compound, mephenesin monocarbamate, proved disappointing. Fortunately, Berger had discovered that substitution on the middle carbon atom of 1, 3-propanediol provided a new series of muscle relaxants. These, too, were very short-acting, so it now seemed sensible to prepare their carbamates. In May 1950, Ludwig synthesized meprobamate ('Miltown', 'Equanil') and by the time it was recognized as superior to all others investigated, some 1200 compounds had been synthesized and screened. It was marketed as an anti-anxiety drug in the mid-fifties with unprecedented advertising at a time when doubts were being aired about the safety of barbiturates as sedatives. As a consequence of this, it was an outstanding commercial success, remaining unchallenged until the arrival on the scene of the benzodiazepines, which caused less drowsiness, almost a decade later. Its N-propyl analogue was also developed by Berger and Ludwig, being known as carisoprodol ('Carisoma'), while methocarbamol ('Robaxin'), a close structural analogue of the rejected mephenesin carbamate, was introduced shortly afterwards by A. H. Robbins and Company. Both of these were recommended as centrally acting muscle relaxants for the relief of spasm caused by injury.

Chlormezanone ('Trancopal'), another anti-anxiety agent with central muscle relaxant properties, was also discovered serendipitously, in 1957, at the Sterling-Winthrop Research Institute in Rensselaer, New York. So, too, was Norwich Pharmacal's dantrolene ('Dantrium'), a nitrofurantoin analogue, which was discovered ten years later (see page 70).

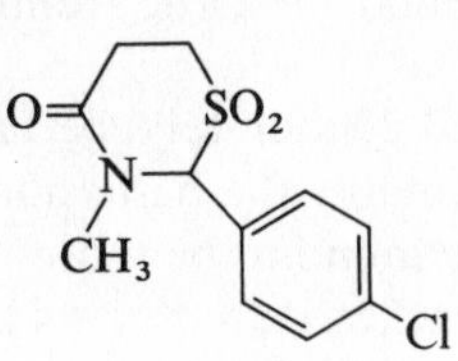

Chlormezanone

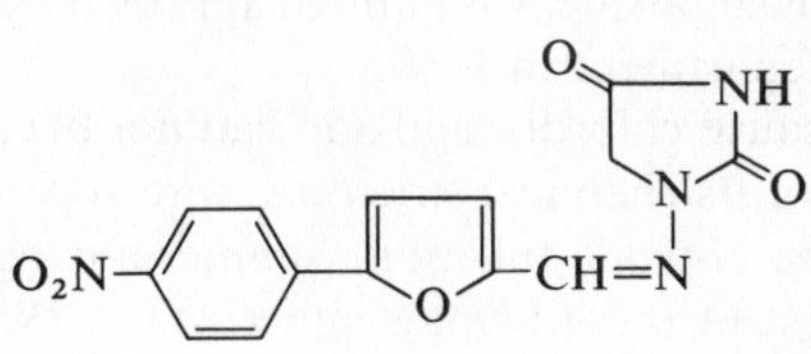

Dantrolene

Serendipity was also to play its part in the discovery of the most successful group of all the anti-anxiety agents, namely the benzodiazepines. Early in 1954, Leo Sternbach of Hoffmann-La Roche in Nutley, New Jersey, decided to reinvestigate some tricyclic compounds he had synthesized about 20 years earlier at the University of Cracow as part of his post-doctoral studies on dyestuffs. He had in mind the tricyclic nature of chlorpromazine, which had just been discovered to be a major tranquillizer, and he believed that the introduction of a basic side-chain into his own compounds might create derivatives with a degree of overall similarity to it. He prepared around forty new compounds by reacting his key intermediate, an alkyl halide, with a variety of secondary amines selected to confer structural analogy with the tricyclics then being patented. When Sternbach's compounds were submitted to Lowell Randall for screening for muscle relaxant, sedative, and anticonvulsant properties, they were all found to be inactive. Renewed chemical studies then revealed that the tricyclic system of the key synethetic intermediate was not that of a so-called benzheptoxdiazine, as had all along been believed, but was instead a quinazoline-3-oxide. This seemed to account for the lack of biological activity in the derivatives synthesized from this intermediate. No further analogues were made by Sternbach, but the last one he had prepared remained untested until a year and a half later, when Earl Reeder, a colleague who was tidying up the cluttered benches in the laboratory, suggested it should be sent for screening. Sternbach agreed, and a few days later Randall informed him that his compound seemed superior to meprobamate in a variety of tests for anti-anxiety and central muscle relaxant activity. It also appeared to approach the activity of chlorpromazine as a tranquillizer. Furthermore, it had a low level of acute toxicity and was free from significant side-effects. This report engendered considerable excitement and raised the question of why only this single compound was active. The answer was soon found when Sternbach reinvestigated its chemistry. It became clear that by using methylamine, a primary amine, in the last stage of the synthesis, the reaction had followed a different pathway (ring enlargement) from that undergone when secondary amines had been employed. The product thus formed was now shown to be a benzodiazepine. Sternbach filed a US patent application for this new tranquillizer, chlordiazepoxide ('Librium') on 15 May 1958, little knowing that it and its derivatives were to become the commercially most successful drugs of all time. The initial clinical studies were conducted on some 16 000 patients before chlordiazepoxide was granted approval by the US Food and Drug Administration department in 1960.

Because chlordiazepoxide had not been designed deliberately, certain features of its chemical structure were superfluous, notably the basic side-chain and the N-oxide function. Simpler analogues were found to be more potent, the first of these being synthesized in 1959 and marketed four years later as diazepam ('Valium'). It had more pronounced muscle relaxant properties.

Literally thousands of benzodiazepines have been synthesized since 1960 by Hoffmann-La Roche and its rivals. About twenty-five are in regular clinical use throughout the world, principally as anti-anxiety agents and hypnotics (see page 34). They are something of a mixed blessing, being of real value in patients whose anxiety interferes with their work, leisure, and personal relationships, but are widely misused in the treatment of the most trivial symptoms of stress. Those benzodiazepines which are slowly cleared from the body are less suitable as hypnotics since patients may awaken feeling drowsy. They are used for treating chronic anxiety states. Some of them, however, form an active metabolite, nordiazepam or something similar, which is responsible for their

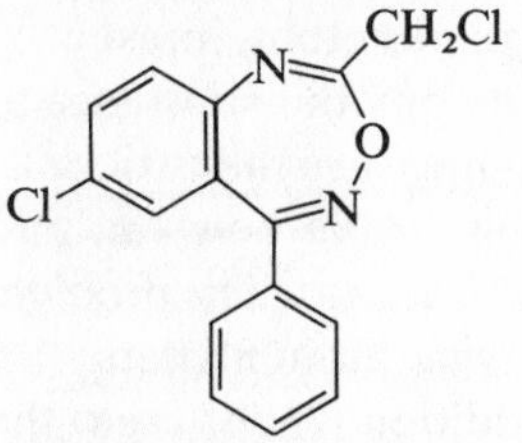

'Benzheptoxdiazine'

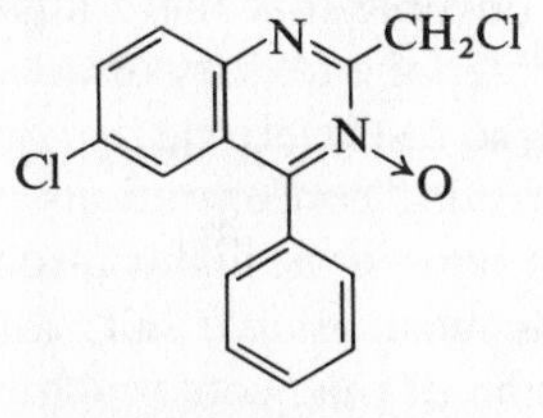

Quinazoline 3-oxide derivative

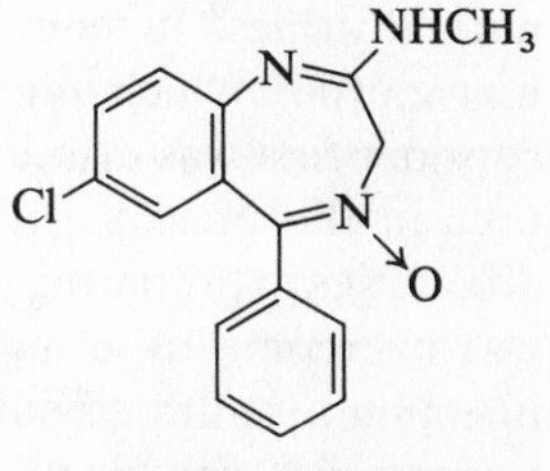

Chlordiazepoxide

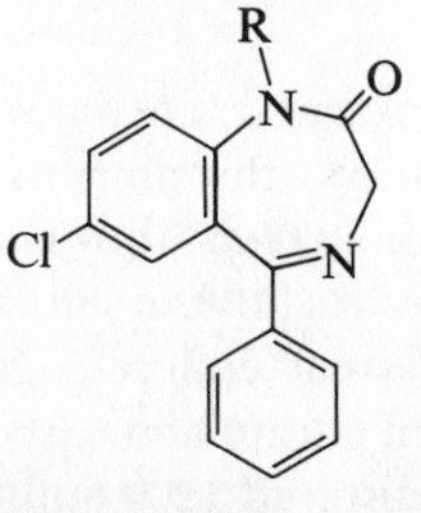

Diazepam, R = $-CH_3$
Nordiazepam, R = $-H$

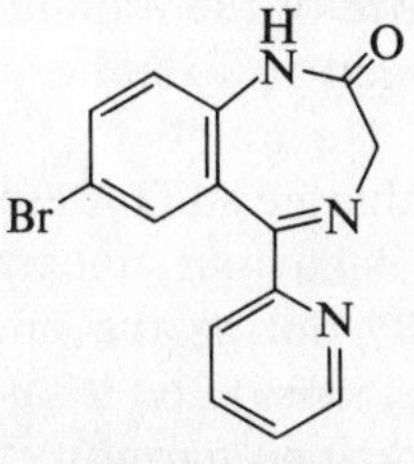

Bromazepam

effects. Unfortunately, nordiazepam takes from two to five days before it is cleared from the body, hence its concentration gradually builds up as more doses are taken. Recognition of this has led to the introduction of benzodiazepines which do not form this type of active metabolite. This is why, in 1982, Roche marketed bromazepam ('Lexotan') in the United Kingdom, nearly twenty years after it was originally synthesized.

ANTIDEPRESSANTS

In 1948, John Cade carried out a simple experiment in an Australian psychiatric hospital, whereby he hoped to obtain evidence that might support his hypothetical conjectures as to the nature of manic-depressive illness. He had observed that patients afflicted with thyroid disorders sometimes behaved in a similar manner to those suffering from manic depression. Extreme hyperactivity of the thyroid seemed to cause a form of mania, whilst a marked absence of thyroid function could be correlated with depression. The question Cade asked himself was whether manic depression could, like thyroid disease, be due to either over- or under-production of a hormone. He collected urine from manic patients, schizophrenics, melancholics, as well as from normal individuals. To test whether an intoxicating substance was present, the urines were promptly concentrated, then injected intraperitoneally into guinea pigs. If sufficient urine was injected, the guinea pigs convulsed, fell unconscious, and died. Cade noted, however, that the urines from manic patients were, in some cases, three times as toxic as any others. He established that the toxicity was caused by urea, but the amount of this in the manic urines did not differ much from that in normal urines. It seemed possible that there might be something present in manic urines which augmented the toxicity of urea. The obvious substance to consider first was uric acid, so tests were carried out to measure the toxicity of urea in the presence of varying concentrations of this. Cade now ran into difficulties in preparing solutions of the highly insoluble uric acid. The problem was overcome by using its most soluble salt, which was lithium urate. On injecting a saturated solution of this, containing in addition 8% of urea, the toxicity was unexpectedly low. Subsequent injections of urea with lithium carbonate confirmed that the protective action against the convulsions was due to the presence of the lithium ion.

Cade decided to examine the effects of lithium carbonate injections on guinea pigs. He found that, after a two-hour delay, the animals became extremely lethargic and unresponsive. Two hours or so later, they reverted to their normal behaviour, unharmed by the injections. As lithium salts had been used clinically during the nineteenth century to treat epilepsy, gout, and cancer, Cade considered it safe to test both lithium citrate and carbonate on himself. Finding no harmful effects, he then administered 1200 milligrams of the citrate thrice daily to a fifty-one-year-old male patient who had been in a state of manic excitement for five years. After five days, there was an improvement in the patient's condition, and by the time three weeks had elapsed he was considered well enough to be transferred to a convalescence ward for the first time. In his fourth month of continuous treatment, he was released from hospital with instructions to take 300 milligrams of lithium carbonate daily, this causing less nausea than the citrate. Such was the improvement that the man returned to his former occupation. Unfortunately, he became negligent about taking his medication, and after a six-week period without treatment, he was temporarily readmitted to hospital on account of his irritability. Cade achieved similar success with nine other cases, the best results being with excited patients.

Cade's pioneering work was put on a firm clinical foundation by Mogens Schou in Denmark. Today, lithium is considered to be a valuable agent for the prophylaxis of manic-depressive illness. Since it is a very toxic drug, careful monitoring of blood levels is required, especially as the amount of lithium ion that enters the circulation varies markedly between different brands containing the same nominal quantity of lithium carbonate.

In 1952, at Sea View Hospital on Statten Island, Drs Selikoff, Robitzek, and Ornstein conducted a trial of a new analogue of isoniazid, an antitubercular drug that had been discovered the previous year (see page 293). The newer drug, iproniazid, had been developed by Herbert Fox and John Gibas of Hoffmann-La Roche Laboratories at Nutley, New Jersey. It seemed just as effective as the earlier one, but it had more side-effects, particularly central nervous system stimulation. Only in New York State did physicians continue to use iproniazid in preference to isoniazid, and it was at a meeting of the American Psychiatric Association in Syracuse, New York, in April 1957 that reports of the value of iproniazid in depression were first presented. George

Crane of the Montefiore Hospital, New York City, told the meeting that the drug had improved the mood of several tuberculous patients who had been depressed. A rheumatic specialist from the Cleveland Clinic and Hospital, Arthur Scherbel, reported a similar experience with his patients. A group of psychiatrists led by Nathan Kline of Rockland State Hospital, Orangeburg, New York, then presented the meeting with results that revealed iproniazid to have been the first drug of value in chronically depressed psychotic patients.

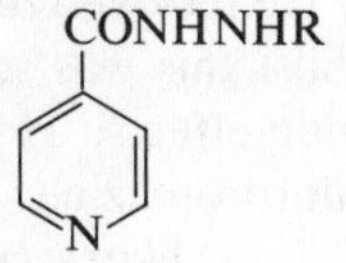

CONHNHR

N

Isoniazid, R = $-H$
Iproniazid, R = $-CH(CH_3)_2$

Kline had begun to study the effects of iproniazid after being shown the results of interesting animal experiments carried out by Charles Scott of the Warner-Lambert Research Laboratories in Morris Plains, New Jersey. At that time it was known that reserpine caused brain cells to liberate two hormones, namely serotonin (5-hydroxytryptamine) and noradrenaline. It was suspected that the tranquillizing effect might be due to the release of the former, so Scott administered iproniazid to inhibit the enzymatic destruction of this substance. That iproniazid could inhibit the enzyme, monoamine oxidase, had been known since 1952. However, to Scott's surprise, pre-medication of animals with iproniazid before administration of reserpine caused stimulation, rather than the expected tranquillization. When Kline saw the results of the animal studies, he carried out similar experiments on humans. He then found that iproniazid on its own could cause stimulation of depressed patients.

Because iproniazid ('Marsilid') was already marketed as an antitubercular drug, psychiatrists were able to obtain supplies as soon as they heard of its antidepressant properties. Within one year of the Syracuse conference, more than 400 000 patients had received the drug for depression. A small number of cases of jaundice were reported, which led to the withdrawal of iproniazid from the American market by the manufacturer in 1961. It was replaced by more potent monoamine oxidase inhibitors such as Hoffman-La Roche's isocarboxazid ('Marplan') and Warner-Lambert's phenelzine ('Nardil').

$CONHNHCH_2-$

CH_3 O N

Isocarboxazid

$NH_2NHCH_2CH_2-$

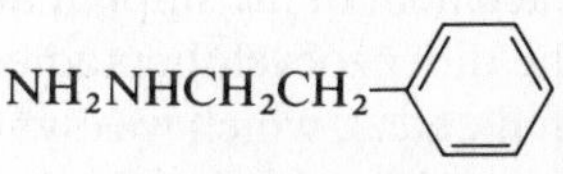

Phenelzine

The recognition of the tranquillizing properties of chlorpromazine in the mid-fifties led psychiatrists to test it and its analogues in a variety of mental disturbances. Roland Kuhn of the Cantonal Psychiatric Clinic, Munsterlingen, Switzerland, noticed that chlorpromazine produced effects which reminded him of those he had observed in 1950 when testing an antihistamine that had been sent to him by the J. R. Geigy Company of Basle for testing as a hypnotic. On that occasion, Kuhn had suggested further studies would be worthwhile, but this suggestion was ignored. Now, however, a long letter he wrote to Geigy was taken seriously, especially as the antihistamine had a striking structural resemblance to chlorpromazine. He received further samples of the antihistamine, code named G22150. Whilst this was soon found to have interesting properties, it had too many side-effects. However, the analogue with a side-chain identical to that of chlorpromazine, G22355, was sent to Kuhn for thorough evaluation in a variety of psychiatric conditions. Early in 1956, it was administered to several patients suffering from endogenous depressions. After only three patients had been treated, it became clear that this new tricyclic compound had unique properties. A letter sent to Geigy at the beginning of February that year referred to the pronounced antidepressant activity of the new drug. Seven months later, at the Second International Congress of Psychiatry, held in Zurich, an audience of only a dozen people heard the first disclosure of this major advance. The new drug was named imipramine, being marketed by Geigy under the proprietary name of 'Tofranil'. Rival manufacturers soon introduced similar compounds, e.g. clomipramine, desipramine, and trimipramine.

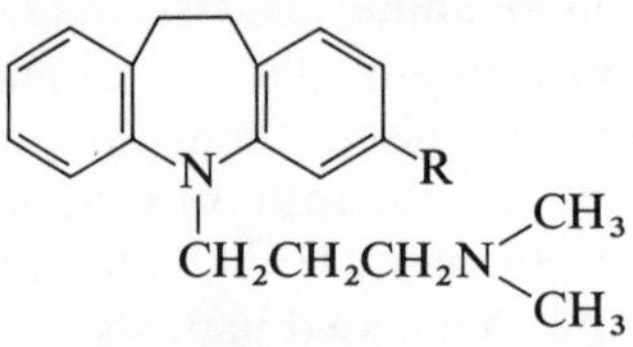

N, R, $CH_2CH_2CH_2N(CH_3)_2$

Imipramine, R = —H
Clomipramine, R = —Cl

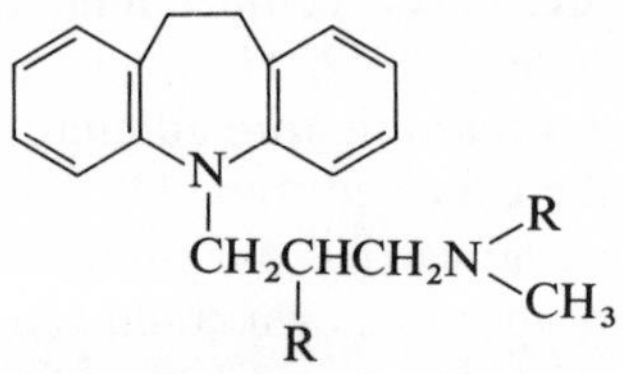

N, $CH_2CH(R)CH_2N(R)CH_3$

Desipramine, R = —H
Trimipramine, R = —CH_3

The recognition of the antidepressant action of imipramine revealed that a minor structural alteration in the central ring of phenothiazine tranquillizers could radically change their pharmacological profile. This stimulated medicinal chemists to synthesize analogous tricyclic compounds. As has been mentioned, replacement of the nitrogen atom in the central ring of chlorpromazine led to the introduction of the thioxanthenes as tranquillizers in 1958. Similarly, replacement of the sulphur atom in the thioxanthene system resulted in the first of the dibenzocycloheptadienes, namely amitriptyline ('Tryptizol', 'Lentizol', 'Elavil', etc.), which was synthesized by several drug companies in 1960. It was found to resemble imipramine insofar as it was an antidepressant rather than a tranquillizer, but it was noticeably less stimulating. This made it more suitable than imipramine for treating agitated, anxious patients who were depressed. Several of its analogues have been introduced, e.g. butriptyline ('Evadyne'), nortriptyline ('Allegron', 'Aventyl'), and protriptyline ('Concordin').

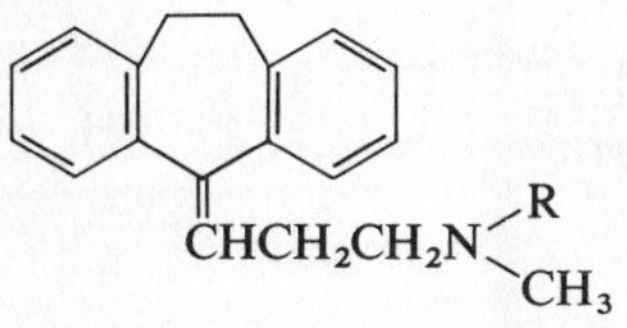

$CHCH_2CH_2N(R)CH_3$

Amitriptyline, R = $-CH_3$
Nortriptyline, R = $-H$

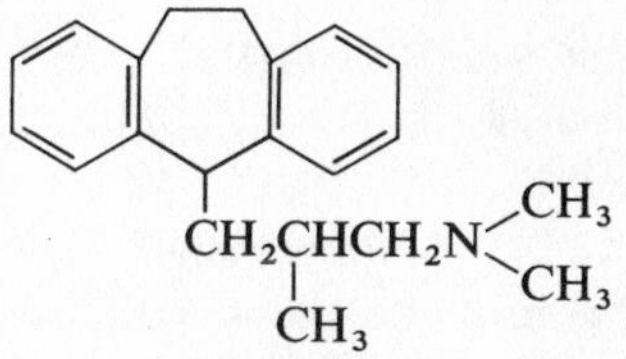

$CH_2CH(CH_3)CH_2N(CH_3)_2$

Butriptyline

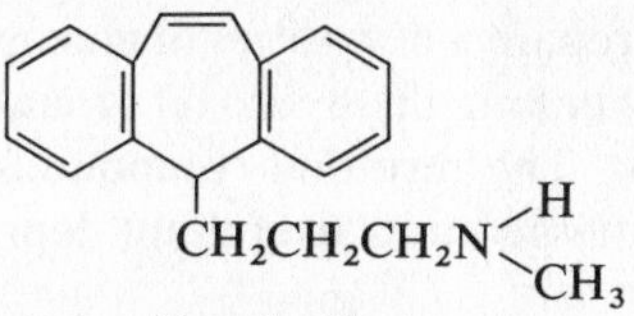

$CH_2CH_2CH_2N(H)CH_3$

Protriptyline

X

$CHCH_2CH_2N(CH_3)_2$

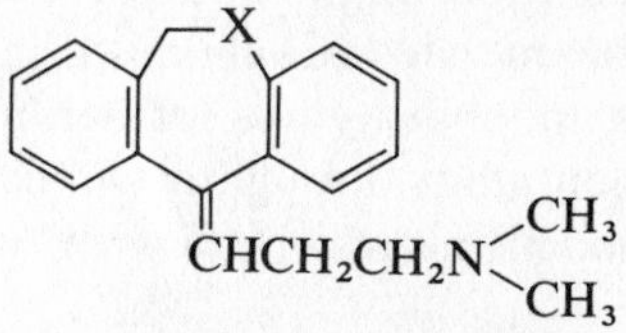

Dothiepin, X = S
Doxepin, X = O

Molecular modification of the central ring of amitriptyline was an inevitable ploy. This resulted in the synthesis of both doxepin ('Sinequan') and dothiepin ('Prothiaden'), in which one of the carbon atoms in the central ring is replaced by an oxygen or sulphur, respectively. It has been claimed that these analogues have fewer side-effects. Another antidepressant resulting from modification of the tricyclic system is iprindole ('Prondol'), which was synthesized in 1963.

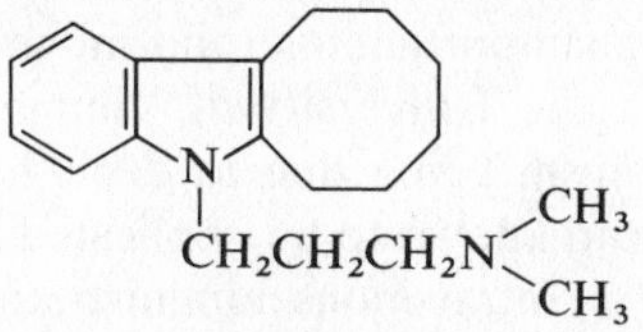

N
$CH_2CH_2CH_2N(CH_3)_2$

Iprindole

N
N
CH_3

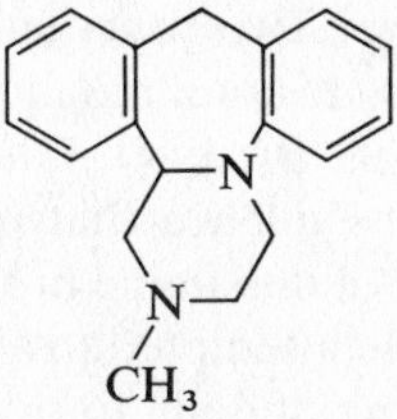

Mianserin

The increase in potency that occurred when the structure of antihistamines was rendered more rigid by the introduction of a sulphur bridge between the aromatic rings, as in promethazine, caught the attention of chemists at the Organon Laboratories in Oss, in The Netherlands. In 1966, they introduced further molecular rigidity by linking the normally flexible aliphatic chain back on to the rigid cyclic system to form a new drug known as mianserin ('Bolvidon', 'Norval'). Not only did this have antihistaminic properties, but it also antagonized the action of serotonin in preliminary animal tests. Although the clinical results in hay fever and other allergic conditions were disappointing, trials in Ireland during 1969 earned the new compound the nickname of 'the good humour pill' because of its mood-elevating action. Further studies eventually confirmed that it had similar activity to amitriptyline.

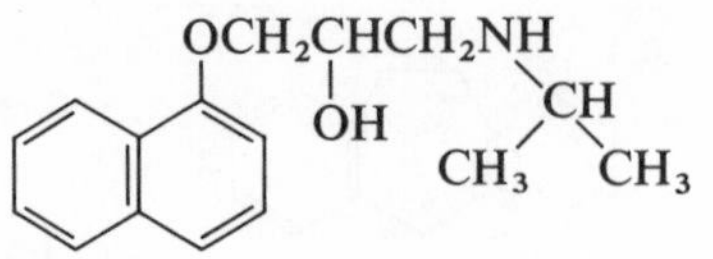

OCH_2CHCH_2NH ... Propranolol

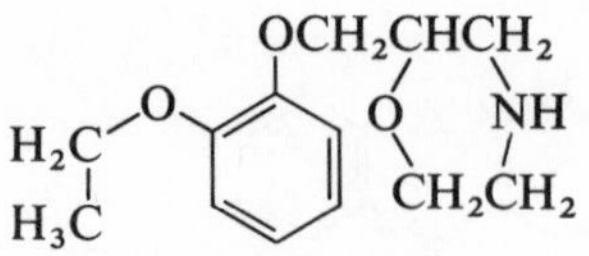

Viloxazine

An annoying side-effect of propanolol and several other lipophilic beta-adrenergic blocking agents is a tendency to produce vivid dreams or even hallucinations in patients receiving large doses. Interest in the nature of this phenomenon led ICI chemists, in 1969, to prepare a new series of analogues in which lipophilicity was further increased by linking the hydroxyl group to the nitrogen atom through a two-carbon chain. The resulting compounds were psychoactive, and viloxazine ('Vivalan') emerged as a useful antidepressant drug.

DRUGS USED IN PARKINSONISM

A dose-dependent side-effect of reserpine was the appearance of signs of Parkinson's disease, characterized by tremors, rigidity of the limbs, and awkward movement. In 1957, Arvid Carlsson and his colleagues at the University of Lund in Sweden demonstrated antagonism between the effects of the catecholamine precursor dihydroxyphenylalanine (dopa) and reserpine in animals. This was important evidence to support the contention that reserpine acted by depleting catecholamine reserves in the brain, especially those of dopamine. Two years later, Carlsson proposed that dopamine was not only converted to noradrenaline, but was actually an important neurohormone in its own right. By 1960, Oleh Hornykiewicz at the University of Vienna had acquired evidence that pointed to parkinsonism being due to depletion of reserves of dopamine in the region of the brain known to be implicated in its cause. He attempted to alleviate the disease by intravenous administration of 50–150 mg of dopa to twenty patients, dopamine itself being unable to cross into the brain from the general circulation. His results were favourable, as were those reported around the same time by Andre Barbeau from the University of Montreal, who gave dopa by mouth to six patients. These early findings were subsequently disputed by other investigators. It was not until 1967 that the treatment protocols were perfected when George Cotzias and his colleagues at the Medical Research Centre in Brookhaven National Laboratory at Upton, New York, demonstrated that oral doses of up to 16 grams each day consistently improved the general clinical condition of at least fifty per cent of patients. This improvement only lasted whilst treatment was continued. Because of the expense involved, the active laevorotatory isomer now known as levodopa was not used in the early trials. Since then, however, it has become the universal treatment for Parkinson's disease.

As soon as the first attempts were made to treat parkinsonism with dopa, it was realised that much of the drug was being metabolized before it could reach the brain. The possibility of finding an inhibitor of the enzyme responsible, dopa decarboxylase, was investigated by Willy Burkard, Karl Gey, and Alfred Pletscher of Hoffmann-La Roche in Switzerland, and Karl Pfister of the Merck Institute in the United States. The Swiss group found that benserazide, a compound synthesized as a potential monoamine oxidase inhibitor and which did not enter the brain, was capable of inhibiting extracerebral dopa decarboxylase to bring about a large reduction in the dose of levodopa required by patients. The combined formulation was introduced under the proprietary name of 'Madopar'. Pfister demonstrated the same effect with carbidopa, which was synthesized specifically as a dopa decarboxylase inhibitor. It was also marketed in a combined formulation with levodopa, known as 'Sinemet'.

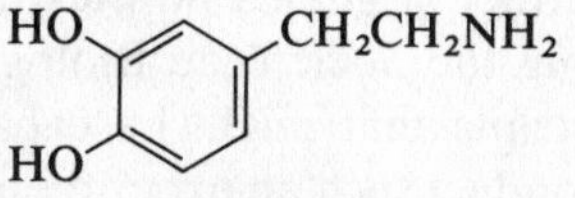

HO, HO, $CH_2CH_2NH_2$

Dopamine

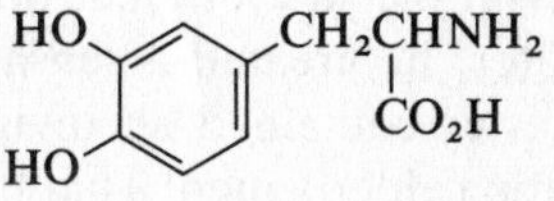

HO, HO, CH_2CHNH_2, CO_2H

Dopa

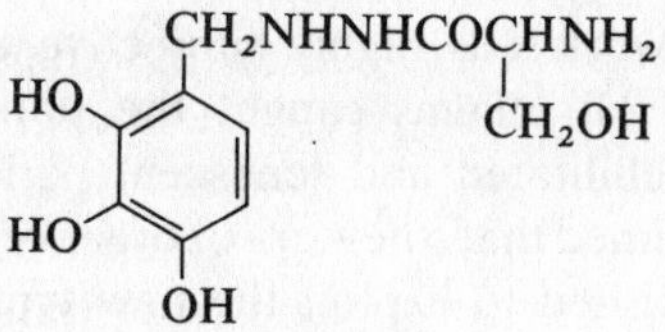

$CH_2NHNHCOCHNH_2$, CH_2OH, HO, HO, OH

Benserazide

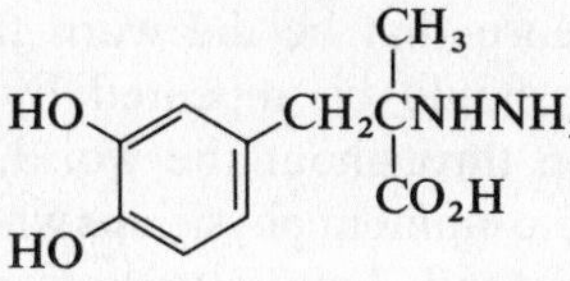

CH_3, HO, HO, CH_2CNHNH_2, CO_2H

Carbidopa

XI

Endocrine Hormones

SEX HORMONES

On the 15th May 1889, Charles-Édouard Brown-Séquard, the highly respected seventy-two-year-old professor of medicine at the College de France, injected himself with the first of eight doses of an extract of guinea pigs testicles. A month later, he created a sensation by telling the Société de Biologie that animal testes contained an invigorating principle that might be capable of rejuvenating elderly men! That he chose to conduct such an experiment upon himself rather than a batch of laboratory animals convinced some of his audience that he was very much in need of rejuvenation, but it should be stated in his defence that he did warn that his work was highly subjective and, therefore, should be repeated by others. His claims caught the popular imagination throughout the world, with debilitated and senescent patients rushing off to eminent physicians who proclaimed that a new era of organotherapy had dawned. Laboratories were established to exploit the new type of therapy, offering the public a variety of glandular extracts. It seemed to many that the Elixir of Life, long sought after by the alchemists, had at last been found.

Brown-Séquard made his extract by grinding either a guinea pig or dog testicle in a small volume of water, then filtering through paper to obtain a clear solution. Since male hormone is insoluble in water, and Ernst Laqueur eventually required more than a ton of bulls' testicles when he became the first scientist to isolate enough testosterone for a single course of injections, it can safely be concluded that the invigoration experienced by Brown-Séquard was due entirely to autosuggestion. It was only a matter of time before Brown-Séquard became totally discredited, but not before a German chemist, Poehl, had marketed a nitrogenous base extracted from animal testes, claiming it to be the active principle. This was appropriately named spermine.

It has been said that Brown-Séquard set back the development of endocrinology by surrounding it with a disreputable aura. Whilst there may be some truth in this so far as investigations into sex hormones were concerned, his concept of organotherapy led directly to the preparation of extracts from other endocrine glands (i.e. those that release their internal secretions directly into the blood, as opposed to exocrine glands whose secretions are carried away in ducts). Thyroid, adrenal, and pancreatic extracts were all investigated during the next decade as a consequence of the old professor's outrageous experiment on himself.

At the turn of the century, Emil Knauer, a Viennese gynaecologist, accelerated the onset of sexual maturity in young animals by transplanting ovaries removed from older animals. His work confirmed the presence of a female hormone in the ovary. It inspired others to administer desiccated ovaries or extracts of these to their patients, in the hope this would remedy menopausal disorders. It also gave Serge Voronoff, a Franco-Russian surgeon, the idea of implanting monkey testicles into his grateful patients, blithely oblivious of the problems of graft rejection! In polite society this treatment, which became popular from about 1912 until 1925, was described euphemistically as 'monkey gland therapy'.

The most potent of the ovarian extracts that became available before the First World War were those prepared by Henri Iscovesco in Paris in 1912, and Otfried Fellner in Vienna the following year. Both used fat solvents (alcohol, ether, and acetone) to obtain extracts that produced sexual changes in castrated animals. The Swiss manufacturer Ciba marketed an ovarian extract in 1913. The war put an end to the European work on sex hormones, but Robert Frank was able to continue it in New York, and he managed to refine it somewhat.

The early attempts to use sex hormones therapeutically failed through lack of a cheap, quick assay that would detect the presence of hormone in the various preparations that were tried. Edgar Allen and Edward Doisy of Washington University in St Louis finally overcame this problem in 1923 by taking advantage of an important discovery made six years earlier at Cornell Medical College by C.R. Stockard and G. M. Papanicolaou. The Cornell researchers had found that changes in the appearance of the cells lining the vaginal wall in rodents closely paralleled the phases of the menstrual cycle. By microscopic examination of such cells in immature mice and rodents, Allen and Doisy were able to observe the effects of ovarian extracts. Such was the sensitivity of this simple technique that it afforded an accurate assay for oestrogenic activity. It enabled Doisy, a biochemist, to make considerable advances towards isolating the pure ovarian hormone, although he faced serious difficulties due to the small amount present in the ovarian follicular fluid from which he extracted it.

In 1927, two Berlin gynaecologists, Selman Ascheim and Bernhard Zondek, used the Allen and Doisy method of detecting ovarian hormone in an attempt to devise a pregnancy test based on changes in the urinary excretion pattern of female hormone. They found that the amount of hormone in urine increased markedly with the onset of pregnancy. With the publication of the Ascheim-Zondek findings, biochemists in America and Europe realized that a rich source of female hormone had at last been found, and a race to isolate the hormone began.

In August 1929, Doisy reported to the 15th International Physiology Congress, held in Boston, that he had at last isolated a crystalline oestrogenic hormone. His work had been supported by Parke, Davis, and Company. Two months later, the isolation of 20 milligrams of this same hormone was reported from Göttingen University by Adolf Butenandt. He had worked on a syrupy extract which the Schering-Kahlbaum Company had prepared from a huge volume of pregnancy urine. Early in the following year, Ernst Laqueur of the University of Amsterdam also isolated the hormone. In 1935, it was agreed by a committee of the League of Nations that the hormone should henceforth be known as oestrone.

The isolation of a second oestrogenic hormone from human pregnancy urine was reported in the summer of 1930 by Guy Marrian, a research student in the department of physiology at University College, London. This hormone was later named oestriol, on account of the presence of three hydroxyl groups in its structure. It was markedly less potent than oestrone, into which it could be chemically converted by dehydration. However, two colleagues of Butenandt, Erwin Schwenk and Fritz Hildebrandt, found that hydrogenation of oestrone formed a new substance, oestradiol, which was around ten times as potent as oestrone. In 1935, Doisy found this to be present in sow's ovaries, four tons of which yielded a mere 12 milligrams of oestradiol. It was also detected in pregnant mare's urine. The outcome of all these investigations was that from 1931 onwards, both Parke, Davis, and Company, and Schering-Kahlbaum, were able to provide endocrinologists with oestrone and related hormones.

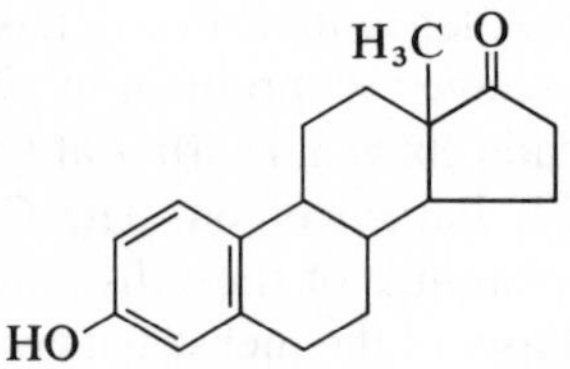

Oestrone

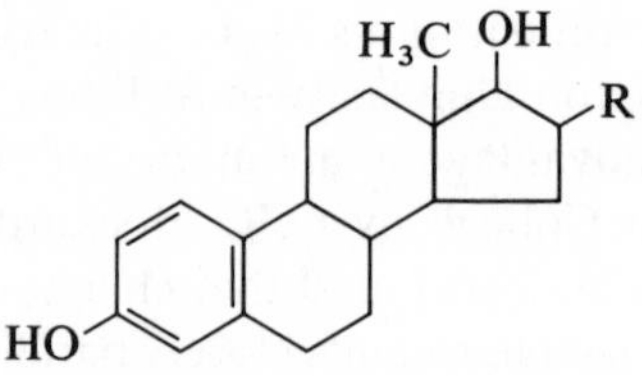

Oestradiol, R = —H
Oestriol, R = —OH

The suspicion that oestrone was a steroid was confirmed by Butenandt. Nor did it take long for its phenolic nature to be recognized, and then exploited in the isolation of other oestrogens, by extraction into alkaline solution. After J. D. Bernal had proved by means of X-ray crystallography that the structure proposed by Windaus and Wieland for the sterol known as ergosterol was wrong, Rosenheim and King deduced the correct one, then went on to propose that the oestrogens may be related to this. By the end of 1932, the present formulae of oestrone and oestriol were considered probable by the University College workers. The final proof was provided four years later in Paris, when Girard developed special reagents that enabled large amounts of oestrone to be isolated. At the Research Institute of the Cancer Hospital, London (now the Chester Beatty Research Institute), James Cook and his colleagues used material supplied by Girard to confirm the proposed structures, in 1938.

The advances made in the field of female hormone research had a direct impact on investigations into male hormones. The introduction of the Allen and Doisy assay for oestrogenic activity encouraged Professor Fred Koch and his research student Lemuel McGee, of the department of physiology at the

University of Chicago, to devise a simple assay for male hormone. They demonstrated that extracts of bulls' testicles could be assayed by their ability to induce growth of the capon's comb. Despite having to pay for the continuation of the research out of his own pocket, Koch tried the methods then being used to isolate female hormone. He thereby obtained a potent extract of male hormone, as little as one milligram of which could induce growth of an upstanding red comb in only five days.

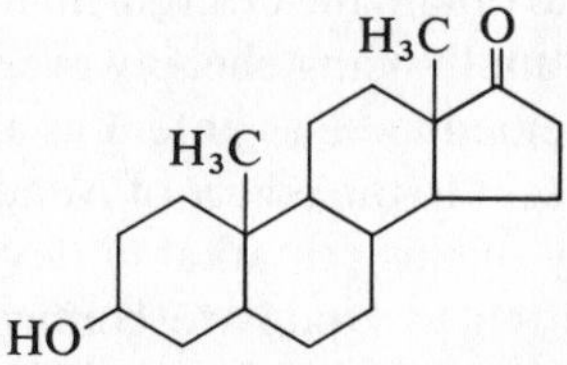

Androsterone

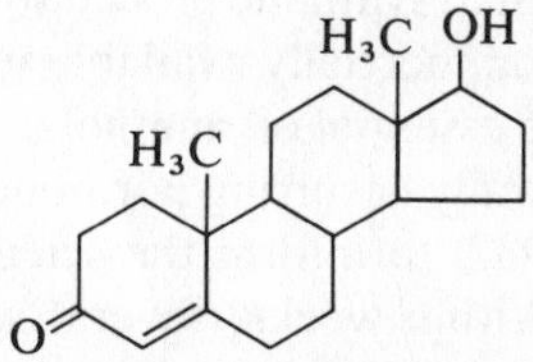

Testosterone

In 1929, S. Loewe and S. E. Voss of Dorpat University in Estonia, detected male hormone in urine. This was seized upon by Adolf Butenandt, and two years later he reported his isolation of 15 milligrams of androsterone from 15 000 litres of male urine. In 1933, Ernst Laqueur isolated 5 mg of another male hormone, testosterone, from nearly one ton of bulls' testicles. The new androgen was at once compared with androsterone. Testosterone was shown to be ten times as active as androsterone in the capon comb test, and about seventy times more potent in the castrated rat seminal vesicle test. It was concluded that androsterone was a urinary metabolite of the true male hormone, testosterone. Androsterone was synthesized from cholesterol in 1934 by the Yugoslavian chemist Leopold Ruzicka, in Switzerland; the following year he synthesized testosterone.

FEMALE HORMONE ANALOGUES

In 1930, it occurred to Charles Dodds of the Courtauld Institute of Biochemistry at the Middlesex Hospital, in London, that since several different substances could induce oestrus in the spayed animal, it might be possible to identify a common feature in their as yet unknown structures. He hoped this might then enable him to synthesize superior analogues that contained this moiety. Two years later, there arose an opportunity to pursue this idea further. Rosenheim and King had proposed that oestrogens had a similar structure to ergosterol, the structure of which had been established. Dodds and C. L. Hewett then joined forces with James Cook, an organic chemist working at the Cancer Hospital in London. They synthesized a tricyclic phenanthrene compound which they believed was an analogue of oestrone. This was found to have some oestrogenic activity, although clinical application was out of the question. As this tricyclic compound was the first substance of known constitution which exhibited oestrogenic activity, its synthesis represented an important landmark. Cook and Dodds then found oestrogenic activity to be present in several other phenanthrenes. In 1934, however, they discovered that the phenanthrene ring was not essential for activity. From this point onwards, activity was revealed in a variety of compounds containing two benzene rings linked together via a short carbon chain, one of the most potent being dihydroxystilbene. To establish whether both rings were required for oestrogenic action, one of the benzene rings in this compound was replaced by a methyl group. No complicated synthetic procedure was necessary to obtain the desired analogue. It was commercially available under the name 'anol', being cheaply prepared from the essential oil anethole. At first, anol seemed to be as potent as any of the naturally occurring hormones. A letter was sent to the editor of *Nature*, in April 1937, to inform the scientific community of this remarkable development. Within weeks, several workers had written to Dodds to confirm his findings, while others wrote to say they had been unable to demonstrate any activity at all, even with high doses of anol. Professor W. Schoeller of the Schering-Kahlbaum laboratories in Berlin sent details of experiments which showed that some batches of anol were contaminated with an impurity formed during its formation from anethole. On crystallization of crude anol from chloroform, this impurity remained in the mother liquor, which proved to be highly oestrogenic. Dodds now decided to collaborate with a rival group at the Dyson Perrins Laboratory in Oxford University, led by Robert Robinson, in an effort to synthesize the as yet unidentified impurity. They decided this could be a stilbene derivative formed by the condensation of two molecules of anol. Several such compounds were then prepared. Early in January 1938, the British researchers reported in *Nature* that one of these, subsequently known in the United Kingdom as stilboestrol (diethylstilbestrol in U.S.A.), was two or three times as potent as oestrone. Subsequently, it was discovered to be almost as potent as oestradiol if injected, but around five times as potent when given by mouth. This was presumably on account of greater resistance to metabolic deactivation in the synthetic drug. As it was cheap to synthesize, and well tolerated in patients, stilboestrol provided gynaecologists, for the first time, with a substance that could be used to deal with oestrogen deficiency, especially in the menopausal patient. Later in 1938, Dodds and his collaborators developed the related drugs dienoestrol and hexoestrol.

In his first publication about stilboestrol, Dodds drew attention to a certain structural resemblance between it and the natural hormones. Three years later, the X-ray crystallographer Dorothy Crowfoot established that the molecular dimensions were, in fact, almost identical to those of oestradiol, especially with regard to the distance between the hydroxyl groups at either end of both molecules.

During the early seventies, stilboestrol was the cause of widespread concern in the United States. This came about after the publication of reports of abnormalities in the genital tracts of young women whose mothers had previously been given stilboestrol during their pregnancies. Mercifully, the fear that these cellular changes may have been pre-cancerous has been allayed,

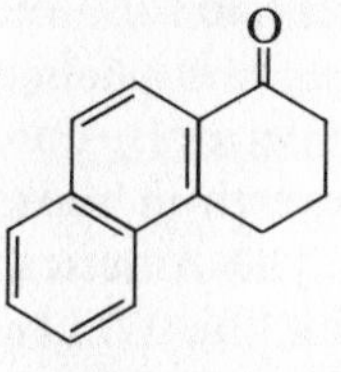

Oestrogenic Phenanthrene

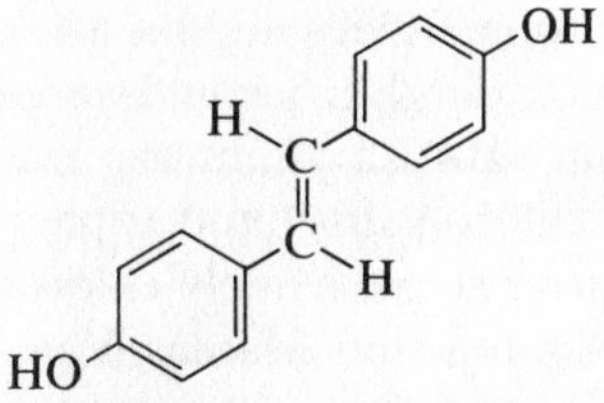

Dihydroxystilbene

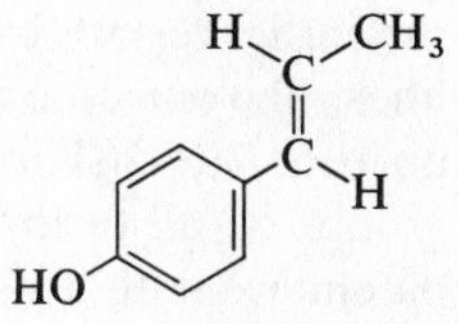

Anol

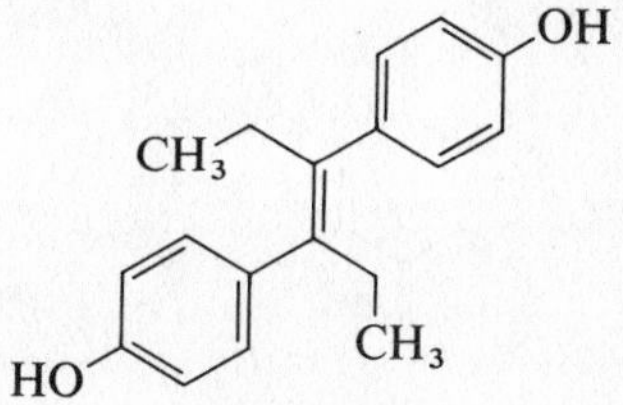

Stilboestrol

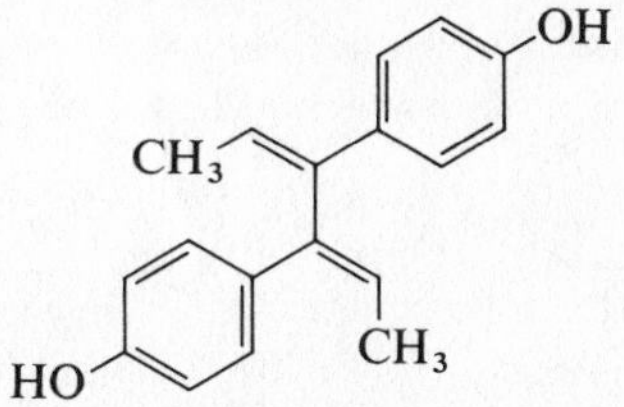

Dienoestrol

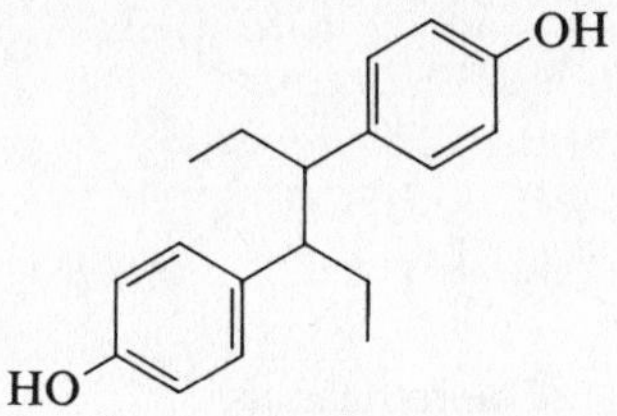

Hexoestrol

but as the drug was administered to somewhere between 500 000 and 2 000 000 American women between the years 1945 and 1955, in order to prevent unwanted abortion, the potential existed for a horrendous human tragedy that would have eclipsed even the thalidomide disaster.

Following Dodd's disclosure of the oestrogenic activity of diphenylethane and diphenylstilbene, John Robson and Alexander Schonberg of the department of pharmacology at the University of Edinburgh found that triphenylethylene had oestrogenic activity, although only about one ten-thousandth that of oestrone. The interesting feature, however, was that when the compound was given by mouth it was just as active as by injection, and the effects of a single dose lasted about one week. This was reported in 1937, and the following year the Edinburgh researchers showed that replacement of the sole ethyleneic hydrogen atom with a chlorine atom increased the potency by a factor of twenty. Robson and Schonberg prepared more analogues, and early in 1942 reported that DBE, a bromine-substituted triphenylethylene with ethoxy groups attached to two of the benzene rings, was a possible alternative to stilboestrol. However, John Davies and his associates at ICI had followed the earlier reports by the Edinburgh group and were now working along similar lines. The outcome was that a British patent was awarded to Frederick Basford on behalf of ICI for the chloro-substituted triphenylethylene with methoxy groups on all of the benzene rings, namely chlorotrianisene. The American patent went to Robert Shelton and Marcus Van Campen for the William Merrell Company of Cincinnati, which marketed the drug with the proprietary name of 'Tace'.

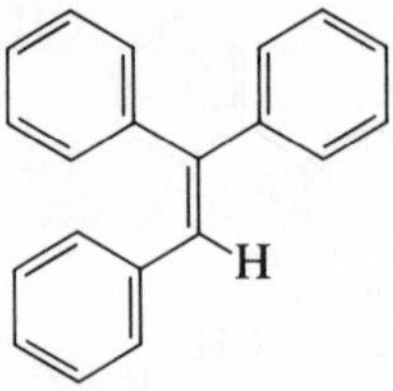

H

Triphenylethylene

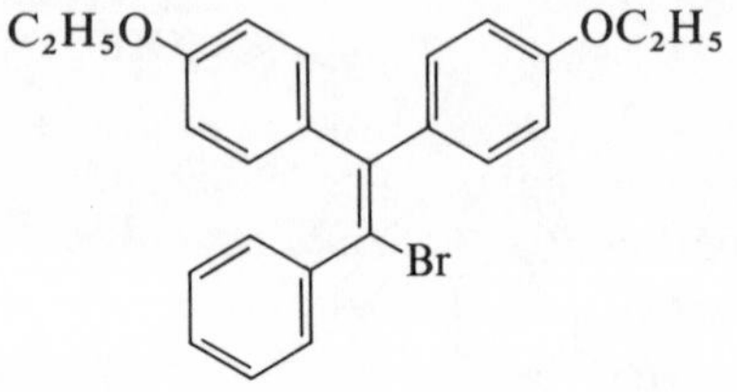

C_2H_5O OC_2H_5 Br

D.B.E.

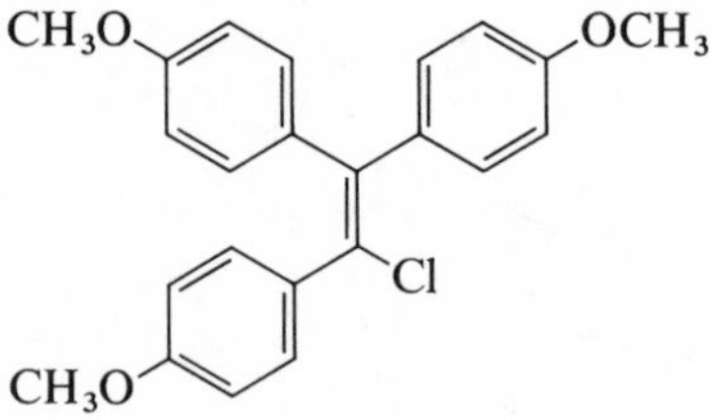

CH_3O OCH_3 Cl CH_3O

Chlorotrianisene

The William Merrell research team continued to work on triphenylethylenes, motivated in part by reports that oestrogens could lower blood cholesterol levels more effectively than could nicotinic acid, massive doses of which were first used in Germany for this purpose in 1955. It was felt that if the cholesterol-lowering action of oestrogens could be separated from their hormonal activity, an alternative drug capable of reducing the risk of heart attacks might be produced. Speculative as this was, it resulted in the patenting of the chlorotrianisene analogue triparanol by Robert Allen and Frank Palopoli in 1959. This was marketed as a hypocholesterolaemic drug for only a short period before it had to be withdrawn, in 1962, because it induced cataract

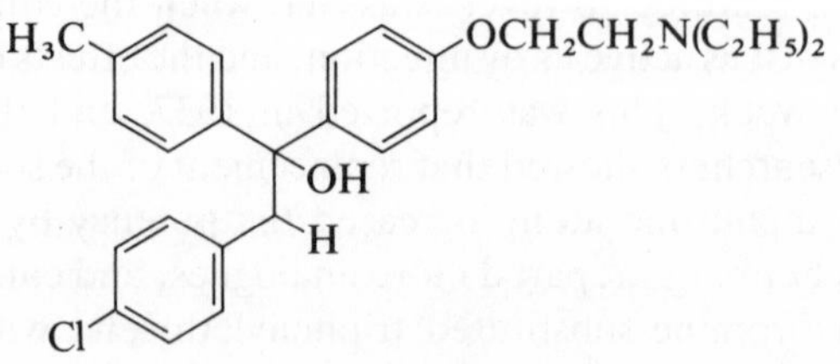

H_3C $OCH_2CH_2N(C_2H_5)_2$ OH H Cl

Triparanol

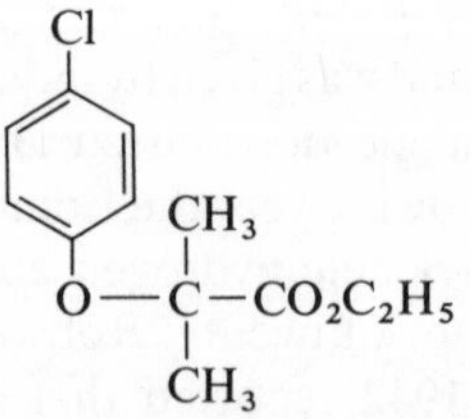

Cl CH_3 $O-C-CO_2C_2H_5$ CH_3

Clofibrate

formation in several patients. An approach that proved more successful was that followed by ICI researchers who, having discovered in 1954 that certain plant hormone analogues lowered cholesterol levels, screened a variety of these and discovered high activity in clofibrate ('Atromid-S') four years later.

In 1962, another chlorotrianisene analogue prepared along with triparanol, namely clomiphene ('Clomid'), was found by R. B. Greenblatt to stimulate ovulation in women with certain types of ovulatory failure that caused infertility. This was not the only serendipitous discovery relating to such compounds, for an analogue known as tamoxifen ('Nolvadex'), patented for ICI in 1964, also had an unanticipated value. When its two isomers were examined by Michael Harper and Arthur Walpole, they found that whilst the *cis* isomer had typical clomiphene-like activity, the *trans* isomer had only weak oestrogenic properties. Further investigation revealed it to act as an anti-oestrogen by virtue of its ability to block oestrogen receptors. This was successfully exploited in the treatment of oestrogen-dependent breast cancer, as demonstrated in 1971 by a clinical trial at the Christie Hospital and Holt Radium Institute in Manchester (see page 333).

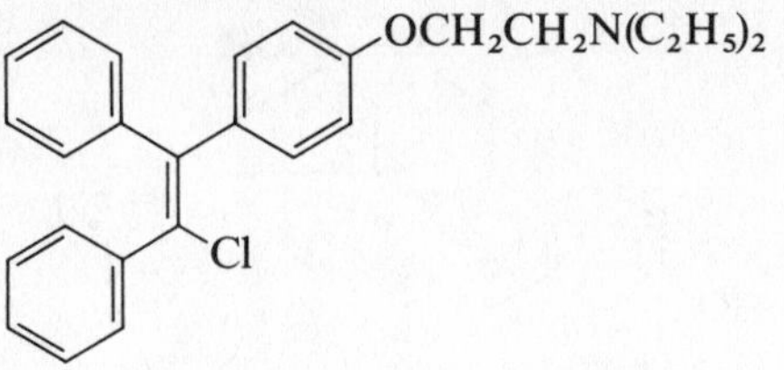

$OCH_2CH_2N(C_2H_5)_2$

Cl

Clomiphene

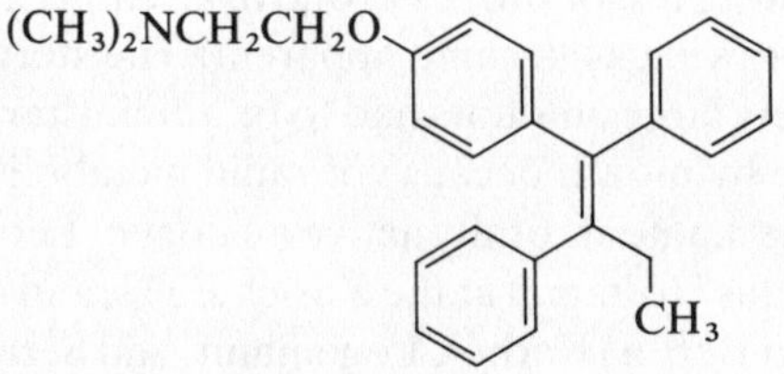

$(CH_3)_2NCH_2CH_2O$

CH_3

Tamoxifen

PROGESTATIONAL AGENTS

In 1903, Ludwig Fraenkel of the University of Breslau discovered that after the cyclical release of ova from the ovary, a yellow substance forms in the ruptured egg sac. The French investigators, Paul Bouin and Albert Ancel, subsequently established that this corpus luteum served to condition the uterus for pregnancy. During the second decade of this century, attempts were made to use extracts of it in menstrual disorders. However, it was not until after the isolation of oestrone that George Corner and his research student, Willard Allen, at the University of Rochester, succeeded in preparing an effective extract. This prevented abortion induced in rabbits by surgical removal of the corpora lutea. Four years later, in 1934, Willard Allen and Oscar Wintersteiner at Columbia University in New York, Alfred Butenandt and K. H. Slotta at Göttingen University, and Adolf Wettstein and Max Hartmann in Switzerland, almost simultaneously announced the isolation of the active principle. Because this was found to be capable of maintaining gestation, it was called progesterone. When injected therapeutically it prevented a high proportion of miscarriages during the early months of pregnancy.

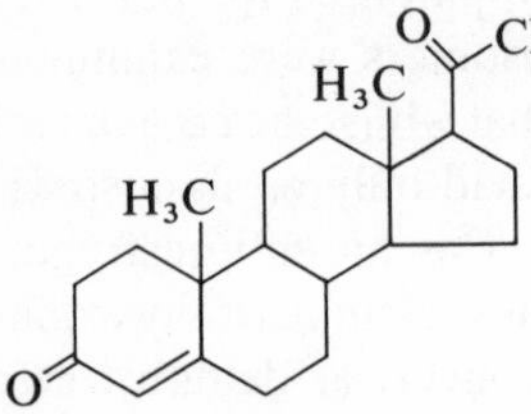

Progesterone

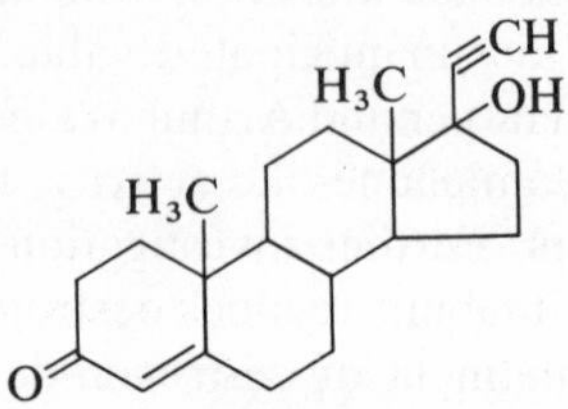

Ethisterone

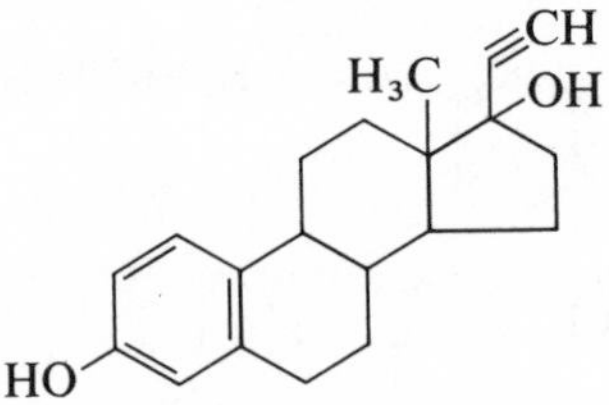

Ethinyloestradiol

Following the increasing use of progesterone in attempts at preventing early miscarriages, certain drawbacks became apparent. The steroid was expensive to produce, yet relatively large amounts had to be administered by injection. It was practically inactive by mouth because of rapid metabolism in the liver. In an attempt to synthesize a potent, orally active analogue, Leopold Ruzicka and his research student Klaus Hofmann at the Zurich ETH, and also a group at the Schering laboratories in Berlin (Kathol, Logemann, and Serini) independently succeeded in adding potassium acetylide to a cheaply available, cholesterol-derived steroid intermediate (dehydroepiandrosterone). The resulting product had a carbon skeleton similar to that of progesterone. The following year, 1938, the Schering group converted this acetylenic steroid into ethisterone, which bore an even closer resemblance to progesterone. Ethisterone proved to be a most successful orally active progestogen. However, its chemical resemblance to testosterone was responsible for the virilizing side-effect of it and its analogues. Nowadays, this contraindicates its use in pregnant women because of risk to the female foetus.

Following their success with ethisterone, the Schering chemists condensed acetylene with oestrone, to form ethinyloestradiol. When this was injected into rats it showed similar potency to oestradiol. Nevertheless, when given by mouth it turned out to be at least twenty times as potent as oral doses of the natural hormone. This was due to prevention of metabolic deactivation in the liver because of the presence of the acetylenic function on the 17-position of the steroid nucleus. Ethinyloestradiol is still widely used for the treatment of menopausal symptoms, although nearly half a century has passed since its synthesis.

During the winter of 1950, Gregory Pincus, co-director of the Worcester Foundation for Experimental Biology, Massachussetts, was awarded £2100 to develop a safe contraceptive. The sum awarded was quite inadequate, but he decided to proceed, raising the necessary funds himself. His experimental approach was to see whether the increase in progesterone output which prevented the release of ova during pregnancy, might serve as a basis for contraception. Since progesterone was known to be extensively metabolized in the liver after oral administration, Pincus and his assistant, Chuey Chang, decided to give very large oral doses of it (more than 5 mg) to female rabbits that were subsequently placed in mating cages. None of them ovulated, and no pregnancies occurred.

Shortly after completing these initial experiments, Pincus happened to meet Professor John Rock, a gynaecologist from Harvard University. Rock had been trying to stimulate the growth of underdeveloped ovarian tubes or uteri in infertile women. To do this, he administered female hormones to mimic the stimulation in the growth of these organs produced by hormonal changes at the onset of pregnancy. In order to achieve the response he required, his patients were sequentially taking large oral doses of stilboestrol followed by progesterone, for three months. This regimen led to a total suppression of menstruation during treatment. Even though many of the women later conceived, this amenorrhoea was considered emotionally distressing. Pincus suggested to Rock that an alternative approach would be to give solely progesterone for three weeks, then withdraw medication so as to induce menstruation. Treatment could restart on the fifth day of the subsequent cycle. When Rock tried this method, it proved acceptable to most of his patients. Several became pregnant after the three-month course was completed. However, one in five of the women in the trial experienced 'breakthrough bleeding'. This was due to the inadequacy of orally administered progesterone. Further, the hormone was unable to maintain total suppression of ovulation for more than around 85% of the time that it was being taken, despite the use of 300 mg daily doses. It was evident that a more potent oral progestogen would be required.

In September 1953, Pincus invited leading pharmaceutical manufacturers to send him supplies of potential progestogens. Nearly two hundred compounds were screened in both rabbits and rats before the end of the year. Fifteen were confirmed as potent ovulation inhibitors suitable for Rock to test in the clinic. These compounds were supplied to fifty infertile women. All were prevented from ovulating with doses in the order of 10–50 mg, a marked improvement over progesterone. In six of the patients, sufficient stimulation of their underdeveloped ovarian tubes or uteri occurred, permitting conception after the course of treatment was completed, the so-called 'Rock rebound'. By now it was clear that Rock and Pincus had developed both oral anti-ovulatory therapy and a remedy for one common type of infertility.

Of the steroids submitted by the drug companies to Pincus, it was clear that the most potent were the 19-norprogestogens. The first of these steroids, in which the methyl group normally found in the 19-position was missing, was prepared in very low yield (0.7%) from the naturally occurring heart stimulant strophanthidin by Maxmillian Ehrenstein, at the University of Pennsylvania in 1944. The small supply of crude 19-norprogesterone was tested on just two rabbits by Willard Allen, who had isolated progesterone ten years earlier when working with Wintersteiner. He found it to be at least as active as the natural hormone, and possibly more so. Further progress was not possible until after Professor A. J. Birch at Oxford University developed a practical synthesis of norsteroids in 1950. Using the Birch procedure, Carl Djerassi and George Rosenkrantz, assisted by L.E. Miramontes, were able to synthesize an isomer of 19-norprogesterone at the Mexico City laboratories of Syntex SA. They submitted a patent in June 1951, claiming it to be more potent than progesterone. Five months later, they patented a still more potent progestogen, norethisterone (norethindrone). The following year, Frank Colton of G. D. Searle and Company in Chicago began to synthesize a series of norsteroids that included norethynodrel which, together with norethisterone from Syntex, was later found by Pincus to be an ideal oral progestogen for clinical use.

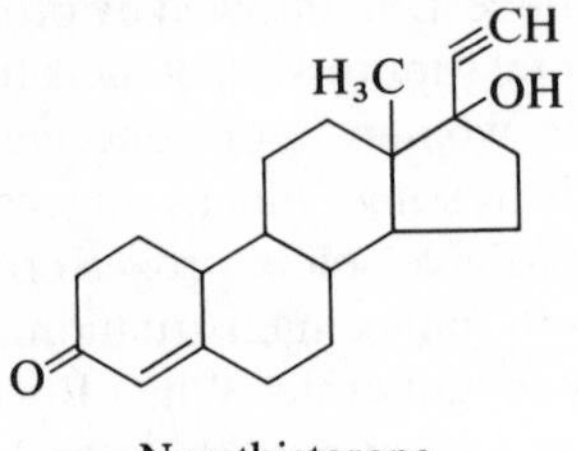

Norethisterone

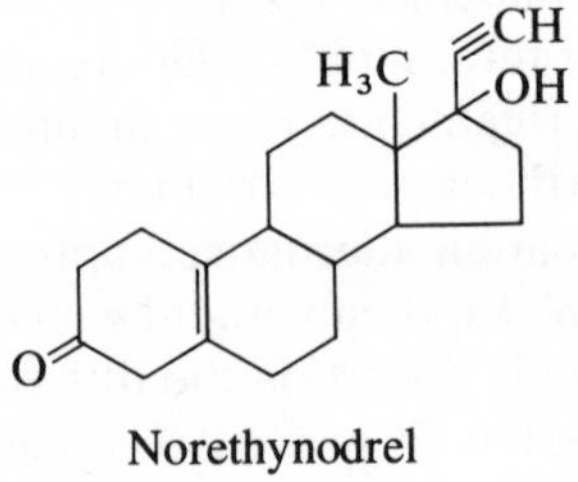

Norethynodrel

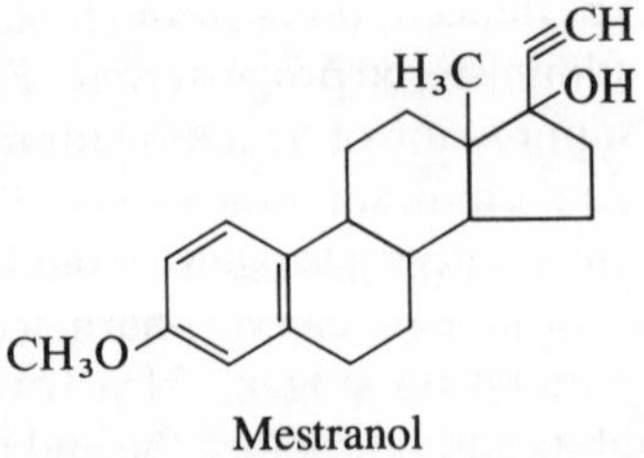

Mestranol

The first large-scale clinical trial of a progestogen as an oral contraceptive began in Puerto Rico in 1956. Norethynodrel was selected as it was the most potent of the anti-ovulatory compounds. It was also free of androgenic activity. During the early phase of the trial, there were few reports of breakthrough bleeding. Later, numerous reports were received of this occurring towards the end of the treatment cycles. Careful investigation revealed that these reports coincided with the introduction of purer batches of norethynodrel from the Searle laboratories. It was soon discovered that the original batches were contaminated with around 1–2% of the 3-methyl ether of ethinyloestradiol, from which the norethynodrel had been synthesized. This was a potent oestrogen, and when small amounts of it were deliberately incorporated in the tablets containing pure norethynodrel ('Enavid tablets'), the problem of breakthrough bleeding was overcome. Later, this oestrogen was given the approved name of mestranol.

After two years, the preliminary results of the Puerto Rico trial were analysed. Of the 221 married women of proven fertility who had taken 'Enavid', none became pregnant, whilst amongst the small number of women who had dropped out of the trial, there were several pregnancies, confirming the return of normal fertility. Finally, after examination of the clinical records of 1600 women who had received 'Enavid', the Food and Drugs Administration in the United States gave G. D. Searle and Company permission, in May 1960, to market their combined progestogen–oestrogen oral contraceptive, the convenience and total reliability of which initiated a social revolution, the full implications of which are not yet known.

HIGH POTENCY PROGESTOGENS

British Drug Houses was a leading United Kingdom manufacturer of steroids during the 1950s, when it saw its ethisterone rendered obsolete by the introduction of the potent oral norprogestogens. One of the BDH chemists, Vladimir Petrow, conceived the idea that the potency of ethisterone might be increased by incorporation of a methyl group on the 6-position of the steroid nucleus. This was based on a report two years earlier, when American researchers discovered that hydrocortisone was metabolically deactivated by oxidation at this position. The 6-methylethisterone was synthesized in 1955, and shown to be more than six times as potent as the parent steroid in animal tests. Eighteen months later, an analogue of it, dimethisterone, was found to be twelve times as potent as ethisterone. It was used as the progestogen in the first sequential oral contraceptive, i.e. where an oestrogen is administered alone for the first part of the cycle, then the progestogen is added near the end of the cycle. This method of contraception is no longer used as there is an increased incidence of side-effects, and also reduced efficacy.

Following the discovery by K. Junkmann of Schering A.G. that the acetylation of the 17-hydroxyl group of ethisterone rendered the molecule soluble enough in oil to be injected intramuscularly for depot medication, there was widespread interest in preparing the acetates, and other esters, of various hydroxysteroids. One such ester, Upjohn's 17-acetoxyprogesterone, proved to be a promising progestogen even though its hydroxy precursor was inactive. Unfortunately, it turned out that no significant prolongation of action was obtained by formulating it in oil. The Upjohn researchers, however, made the unexpected discovery that their acetoxy derivative was orally active, an observation that had been missed by the Schering group, who were primarily interested in the oil solubility of such esters. Several manufacturers then competed to prepare the 6-methyl derivative of 17-acetoxyprogesterone. Priority of discovery for this derivative, subsequently known as medroxyprogesterone acetate, went to Upjohn in the United States for a patent application submitted on 23rd November 1956. Syntex, however, submitted its application on 8th September in France. Medroxyprogesterone acetate proved to be around twenty-five times as potent as ethisterone. When formulated as an oily injection ('Depot-Provera'), it formed a deposit of steroid in the body near the site of intramuscular injection. Small quantities of drug were then slowly released over a period of several months to give prolonged contraceptive cover.

After completion of the British Drug Houses synthesis of medroxyprogesterone acetate, Petrow oxidized the drug to introduce an extra double bond at the 6-position. He did not expect this to have useful progestational activity, since the analogous derivative of progesterone was less potent than progesterone itself. Surprisingly, routine screening in rabbits revealed this new compound, megestrol acetate, to be the most potent ovulation inhibitor discovered up to that time (1958).

An alternative to blocking metabolism at the 6-position of steroids by inserting a methyl substituent, was to insert a chlorine atom. This approach was followed by Syntex, and also by E. Merck in Germany. Both companies

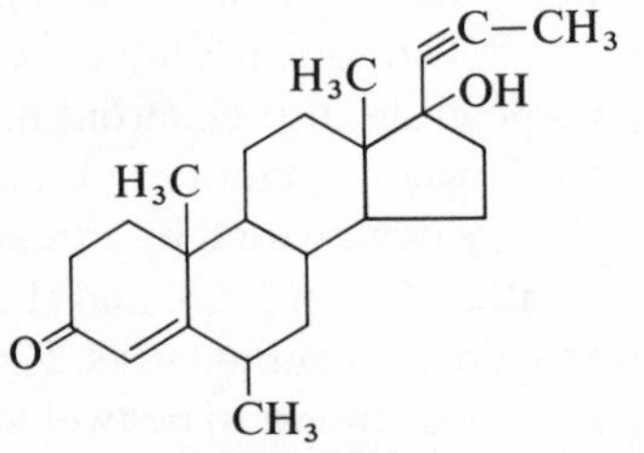

Dimethisterone

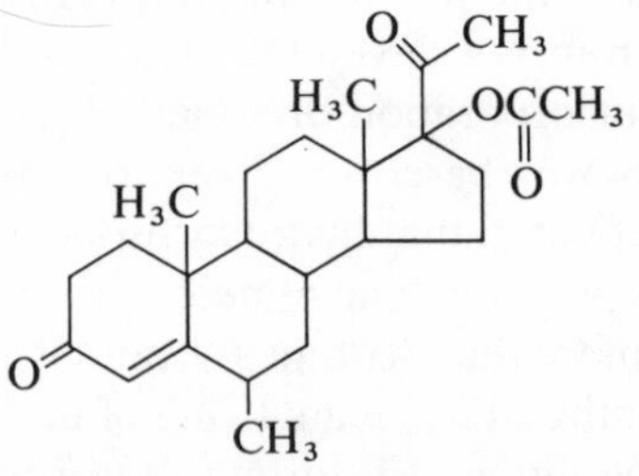

Medroxyprogesterone Acetate

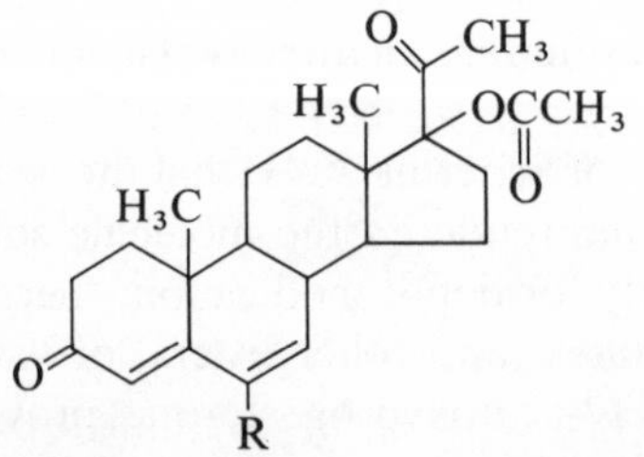

Megestrol Acetate, R = $-CH_3$
Chlormadinone Acetate, R = $-Cl$

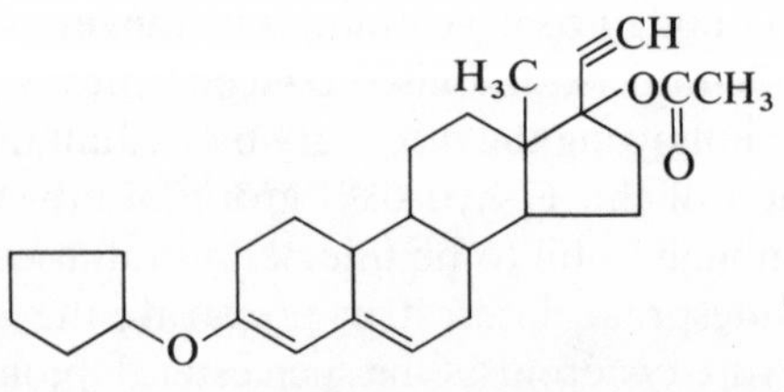

Quingestanol Acetate

independently prepared chlormadinone acetate in the same year that megestrol acetate was synthesized, the two drugs differing only in the nature of the 6-substituent.

In 1959, Alberto Ercoli, who worked in the Vismara Terapeutica laboratories near Milan, reported interesting results from his application of a novel synthetic procedure that he had devised some four years earlier. This enabled him to synthesize ethers of a variety of steroids, such as methyltestosterone, cortisone, and progesterone. Unexpectedly, the resulting ethers, when administered by mouth, were more effective than their parent steroid. This was found to be particularly so with the cyclopentyl ethers. The immediate result of this fortunate discovery was the introduction of the cyclopentyl ether of norethisterone acetate, under the approved name of quingestanol. It was a potent oral progestogen. Later, Ercoli prepared the cyclopentyl ethers of oestradiol and ethinyloestradiol, known respectively as quinestradol and quinestrol.

In 1960, Herschel Smith, a former student of Birch, achieved the total synthesis of an analogue of norethisterone in which the methyl substituent at the 18-position was replaced by an ethyl group. This new steroid, norgestrel, was even more potent, the activity residing in the isomer known as levonorgestrel. Another compound related to norethisterone was synthesized from mestranol by the Dutch company, Organon, around the same time, viz. lynoestrenol. It lacked the carbonyl function at the 3-position of norethisterone, but this was introduced by metabolic oxidation in the liver; hence lynoestrenol is, essentially, a pro-drug of norethisterone. Yet another drug modelled on norethisterone is ethynodiol diacetate, the alcoholic precursor of which was prepared for Searle by Frank Colton in 1954. Ethynodiol itself lacked any promising features, and it was not until the mid-sixties that its diacetate was recognized as a valuable oral contraceptive.

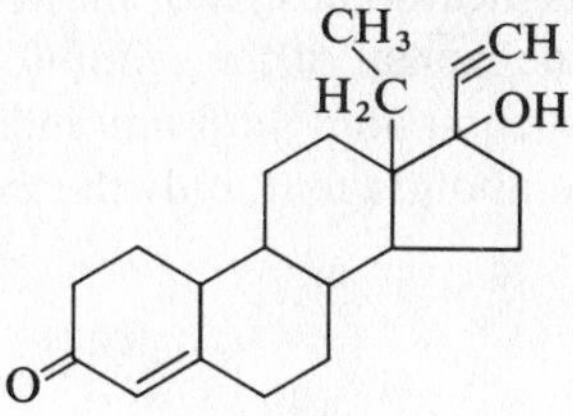

Norgestrel

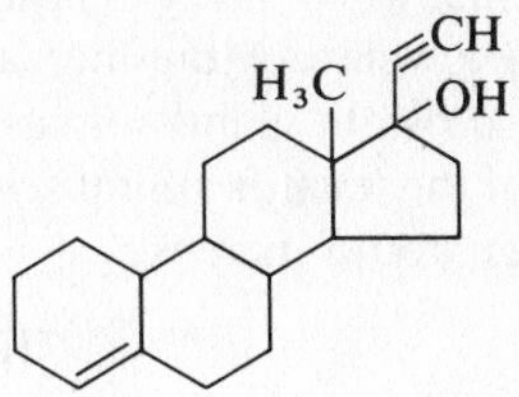

Lynoestrenol

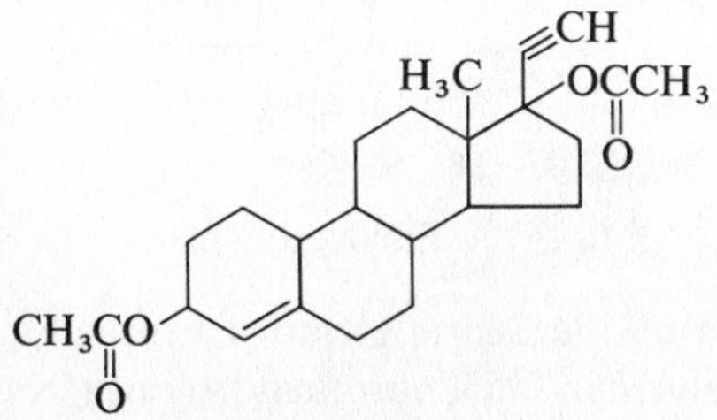

Ethynodiol Diacetate

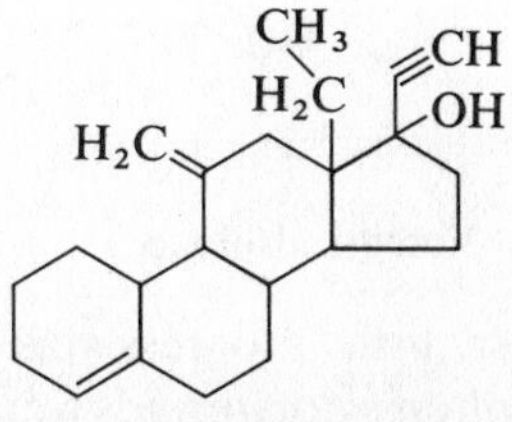

Desogestrel

There was considerable concern during the early seventies that chronic administration of oral contraceptives might cause some deaths as a result of cardiovascular changes. One theory was that residual androgenic activity in synthetic progestogens could be responsible for changes in the blood lipids (fats), leading to arterial disease, the chemical structures of progestogens having certain features in common with testosterone. In attempting to avoid this possibility, Organon introduced desogestrel, a compound originally synthesized in 1975 following reports that introduction of certain substituents in the 11-position of progestogens markedly enhanced potency. It was the first new progestogen to be launched in the United Kingdom for over a decade when it was marketed as a component of 'Marvelon' in 1982. Within a year of its introduction, however, concern was also being expressed about the possible hazard of an increased risk of breast cancer in women who had been taking highly potent progestogens.

ANABOLIC STEROIDS

In 1948, G. D. Searle and Company initiated a long-term project to exploit the muscle-building properties of testosterone. Thirteen years earlier, C. D. Kochakian and J. R. Murlin, at the University of Rochester, had reversed the abnormal excretion of nitrogen (from proteins) in the urine of castrated dogs simply by administering an extract prepared from the urines of their male students. Later, Allen Kenyon at the University of California confirmed a similar effect with testosterone in humans. Due to its hormonal effects, however, testosterone could only be administered to mature male patients. The Searle chemists hoped to overcome this shortcoming. With the assistance of Francis Saunders and Victor Drill, who screened more than one thousand steroids over a seven-year period, the project ultimately succeeded. That it did so was due, in no small measure, to the sophisticated design of the screening procedure, whereby the increase, if any, in the weight of the seminal vesicles and the prostate glands of castrated rats was compared with any increase in weight of the levator ani muscle. In an ideal anabolic agent, only the weight of the latter would increase.

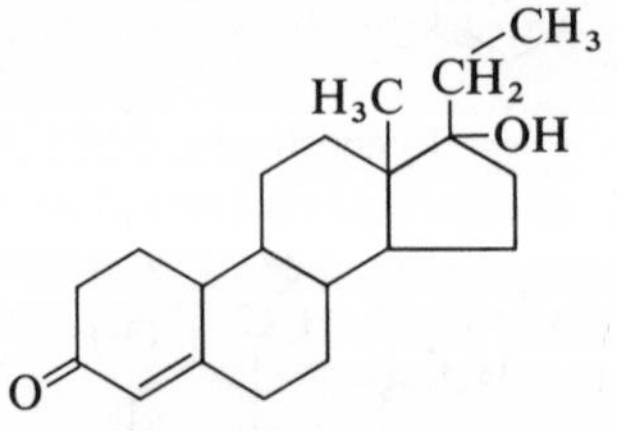

CH_3 H_3C CH_2 OH O

Norethandrolone

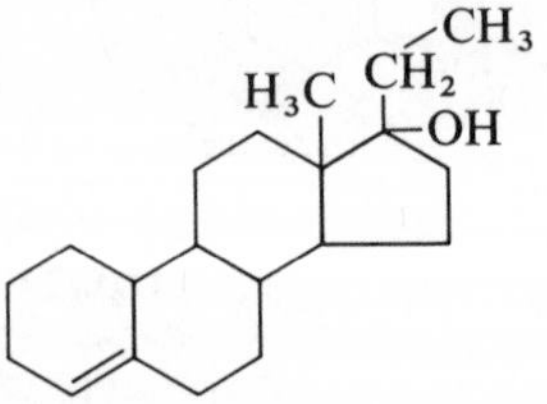

CH_3 H_3C CH_2 OH

Ethyloestrenol

At first, little progress was made by the Searle group. Eventually, when Frank Colton's norsteroids were screened in 1955, norethandrolone ('Nilevar') emerged as having similar anabolic activity to testosterone, with only one-sixteenth the potency of an androgen. Furthermore, it was orally active. After a satisfactory clinical trial, norethandrolone was released for use in debilitated patients in the expectation that enhanced utilization of protein would build up wasting muscles, an expectation that has not been fully realized even with newer drugs. Athletes, especially weight lifters and the like, have seriously abused anabolic steroids in attempts to improve their performance.

Colton had prepared norethandrolone by chemical reduction of the 17-ethinyl group of norethisterone. Organon chemists in Holland similarly reduced this group in their lynoestrenol, to form an anabolic steroid known as ethyloestrenol. It differed from norethandrolone only in the absence of a carbonyl group at the 3-position. Although it may be metabolized to norethandrolone, it is at least ten times as potent, and is actually the most potent anabolic agent in current clinical use.

Several anabolic steroids have been modelled on the structure of methyltestosterone, a compound synthesized by Ruzicka and his colleagues in 1935, where the presence of a methyl group on the 17-position helps to stabilize the existing 17-hydroxyl substituent against oxidation in the liver. Subtle modifications, usually at the opposite end of the molecule, then interfere with the fit of the steroid at receptors mediating the undesired hormonal activity. Examples of this type of anabolic steroid include methandienone (methandrostenolone), oxymetholone, and stanozolol.

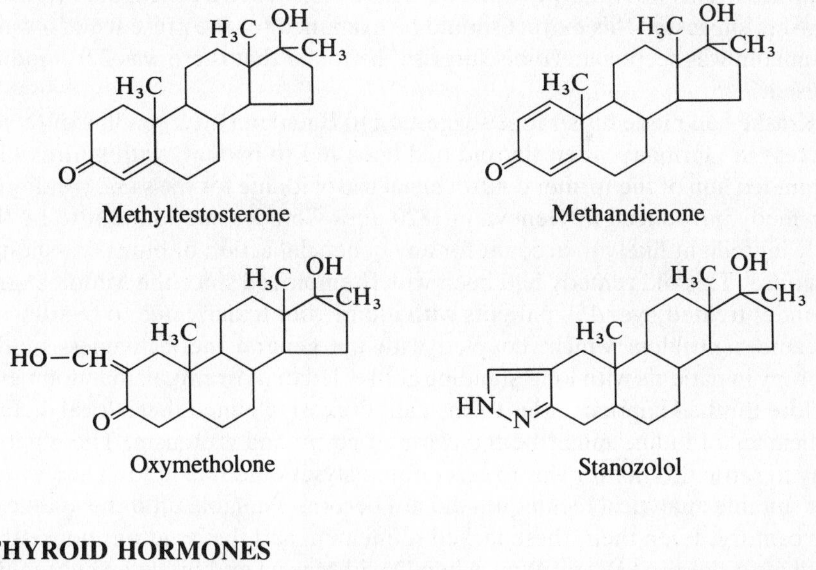

Methyltestosterone

Methandienone

Oxymetholone

Stanozolol

THYROID HORMONES

Sir William Gull of Guy's Hospital was one of London's leading physicians. In 1873, he came to the conclusion that some of his middle-aged female patients were suffering from a form of adult cretinism. This disease, later given the name myxoedema, was manifested by various mental and physical changes. Gull found that atrophy of the thyroid was present in his patients, but later suggestions that this might be the actual cause of the disease were not well received. In 1891, Gley succeeded in extirpating the thyroid without removing the attached parathyroid gland, thereby making it possible to confirm that hypothyroidism was the cause of myxoedema. That same year, no doubt influenced by Brown Séquard's claims for organotherapy, Professor George Murray of Durham University injected a thyroid extract into a patient gravely ill with myxoedema. He made his extract from sheep thyroid by the same procedure then used for extracting pepsin (a digestive enzyme) from the stomach lining of pigs. This involved the use of aqueous glycerol as solvent. Despite being desperately ill at the outset, Murray's patient eventually made a complete recovery. By continuing to receive regular doses of thyroid extract, she remained in good health for twenty-eight years, until her death from other causes. In 1892, several investigators discovered that oral doses of thyroid gland were just as effective as injections. If given in early infancy, dried thyroid could cure cretinism.

Murray's experiments aroused widespread interest. At the University of Freiburg, Professor Eugen Baumann was particularly interested in the biochemistry of the thyroid gland because Freiburg was situated in a part of Germany where there was a high incidence of endemic goitre, a deformity of the neck due to swelling of the thyroid. He prepared his own extracts by boiling the gland in dilute sulphuric acid, cooling, then collecting the flocculent precipitate that was deposited. This was finally extracted with alcohol. When a colleague, Professor Kraske, suggested this extract should be examined for the presence of iodine, Baumann was sceptical. To his surprise, he found that there was 2.9% iodine present!

Kraske had made his strange suggestion to Baumann after reading of recent success in Germany when thyroid had been fed to patients with goitre. This reminded him of the former controversial use of iodine for the same condition, a remedy introduced in Geneva in 1820 after Coindet had found it to be the only ingredient likely to account for any beneficial action of burnt sea-sponge in goitre. This old remedy had been widely employed since the Middle Ages. Coindet treated over 150 patients with iodine, but toxicity due to overdosing became a problem which, coupled with the general ineffectiveness of the therapy in patients with long-standing goitre, led to its eventual abandonment. Before this had happened, his townsman, Prevost, claimed that a local dietary deficiency of iodine might be the cause of goitre and cretinism. The obvious way to settle this matter was to carry out analyses of iodine in food and water, but suitable analytical techniques did not become available until the middle of the century. Even then, these lacked refinement and the issue was not settled until after the First World War, when David Marine and his colleagues in the United States finally confirmed Prevost's suggestion. This led to the introduction of iodine supplements for inhabitants of goitrogenic areas.

Baumann became convinced that there was a correlation between the iodine content of his thyroid extracts and their potency. By 1896, he had isolated a fraction which contained around 10% iodine. When administered to patients, its beneficial effects were similar to those previously obtained with whole gland. Believing he might have obtained the active principle, Baumann named it 'iodothyrin'. However, the isolation of even more potent fractions soon made it evident that this was not the pure, active principle.

When Edward Kendall joined Parke, Davis, and Company in 1910 he was assigned the task of isolating the thyroid hormone. He continued with this after moving first to St Luke's Hospital in New York, one year later, then again in 1914, to the Mayo Clinic in Rochester, Minnesota. By careful attention to details of the initial hydrolysis of the gland, for which he introduced repeated exposure to alkali instead of dilute sulphuric acid, Kendall isolated material containing 23% iodine, double the highest amount previously recorded. On Christmas Day 1914, he finally obtained crystals of thyroid hormone.

When the process was scaled up, Kendall was no longer able to recover active material. It took fourteen months before the cause of this setback was traced to decomposition of the hormone by the galvanized iron vessels in which the hydrolysis of the gland had been conducted. These had to be replaced by enamel vessels. Further problems arose, for thyroid glands collected during the winter months had a low titre of hormone. By 1917, Kendall had amassed about 7 grams of crystals and was thus able to start clinical studies.

Kendall proposed a structure for thyroid hormone in 1917. Believing it was an oxindole, he coined the name 'thyroxin', a term that was later altered to thyroxine (this name has been retained despite the fact that the hormone is not an oxindole). Kendall studiously avoided giving the hormone any name suggestive of the discredited idea that iodine *per se* was responsible for thyroid activity. Thyroxine became available commercially shortly after this when it appeared in the catalogue of E. R. Squibb and Sons of Brooklyn, New York. The company offered it at a price of $350 a gram, which did not even cover the cost of production. At such a high price, it could only be used for biochemical investigations. This meant that when Charles Harrington, at University College Hospital in London, had doubts concerning the proposed chemical structure, he had to devise a more efficient isolation process that would furnish enough material for his own investigations. Supported with funds provided to the hospital by the Rockefeller Foundation, he collaborated with F. H. Carr of British Drug Houses, in 1924, and managed to increase the yield of hormone from dried thyroid twenty-five-fold. This enabled the company to reduce their price from £70 per gram to less than one-tenth of that.

By taking into account prior investigations, Harrington speculated that thyroxine might be formed from two molecules of an amino acid, diiodotyrosine. In 1927, he and George Barger, at the National Institute for Medical Research, synthesized the proposed condensation product and found it to be identical with thyroxine extracted from thyroid gland; both samples were optically inactive. This confirmed the chemical structure as being that proposed by Harrington. However, this structure possessed a centre of asymmetry, which meant that during the alkaline hydrolysis in the first part of the extraction process, thyroxine had probably undergone racemization into a mixture of optical isomers. Harrington and Barger therefore separated thyroxine into its two isomers and established that the laevorotatory one was three times more potent than the dextrorotatory. In 1930, they managed to isolate this isomer from the gland by using enzymes instead of alkaline hydrolysis in a modified extraction procedure.

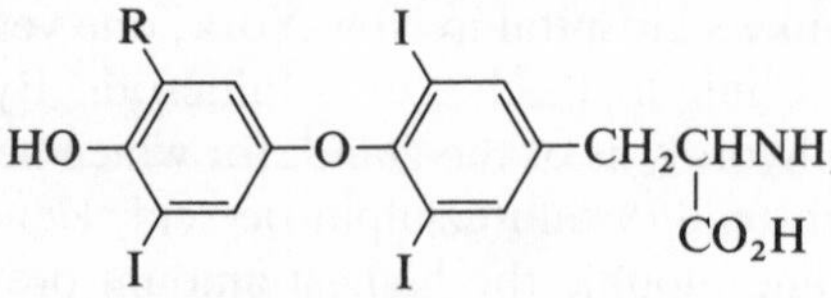

Thyroxine, R = —I
Liothyronine, R = —H

In 1939, Ludwig and Mutzenbecher managed to prepare thyroxine economically by treating casein (a protein from milk) with iodine, followed by careful hydrolysis. Nevertheless, thyroxine did not become available at a price that competed with that of tablets containing dried thyroid gland until B. A. Hems led a team of chemists at the Glaxo Research Laboratories to a successful synthesis of it in the mid-fifties. Although eight stages were involved, the overall yield of optically active hormone ('Eltroxin') was 26%. Over and above the commercial significance of this synthesis, it permitted the preparation of various analogues.

The presence in thyroid gland of a more potent hormone than thyroxine had been held by Kendall for many years, but it was not isolated until 1952. Jack Gross and Rosalind Pitt-Rivers detected its presence whilst carrying out experiments on mice that had been fed with radioactive iodine which became incorporated into their thyroids. By good fortune, it was found to be identical with a trace 'impurity' present in synthetic thyroxine. This contained three iodine atoms instead of the four present in thyroxine. Tests at University College Hospital revealed that this was about twice as potent as thyroxine, and it was suggested that it was the active hormone formed in the body from thyroxine. The new hormone was given the approved name of liothyronine, and it is obtained by the Glaxo synthesis ('Tetroxin').

ANTITHYROID DRUGS

In 1940, the American Cyanamid Company introduced the poorly absorbed sulphonamide known as sulphaguanidine with a view to its being used to treat intestinal infections. Early clinical studies were carried out at the Johns Hopkins Medical School in Baltimore, where Professor Elmer McCollum, a pioneer in vitamin research, was particularly concerned about the effects of the new drug on intestinal bacteria which produced some of the B complex vitamins. With C. G. Mackenzie and Julia Mackenzie, he closely observed the effects of the new sulphonamides on laboratory animals. In 1941, they reported that extensive alterations occurred in the thyroids of rats. The Mackenzies subsequently found some other sulphonamides and thiourea produced a similar deleterious action on the thyroid gland. Simultaneously with the publication of these findings in 1943, there appeared a similar report from Professor E. B. Astwood and his colleagues at Harvard Medical School. Later that year, Astwood reported that 2-thiouracil was the most potent inhibitor of thyroid hormone production amongst 106 sulphonamides and thiourea derivatives that he had examined. He also announced the first clinical use of thiourea and thiouracil in three patients suffering from overactivity of the thyroid, viz. thyrotoxicosis. The metabolic rates of all three were reduced to normal so long as the thiouracil was regularly administered, but the drug had to be withdrawn in one patient because it produced the serious blood condition known as agranulocytosis.

The thiouracil employed by Astwood had been supplied by the American Cyanamid Company. His results interested the company in the possibility of finding a safer antithyroid drug than thiouracil. In 1945, George Anderson, I. F. Halverstadt, Wilbur Miller, and Richard Roblin of Cyanamid's Stanford Research Laboratories introduced propylthiouracil. The next development came four years later when Reuben Jones, Edmund Kornfeld, and Keith McLaughlin of the Eli Lilly Company introduced methimazole as a follow up to Professor Astwood's discovery that mercaptoimidazole had about half the activity of thiouracil in rats. The Lilly researchers happened to have a series of substituted mercaptoimidazoles at hand from another investigation, so they screened these in rats. Although methimazole was less active than propylthiouracil, it was sent to Astwood for clinical trial. It was then found that methimazole was more potent than propylthiouracil in man, but although marketed as 'Tapazole' in the United States, it never achieved the popularity of propylthiouracil. A derivative of it was developed at University College Hospital Medical School in London by Claude Rimington, Alexander Lawson,

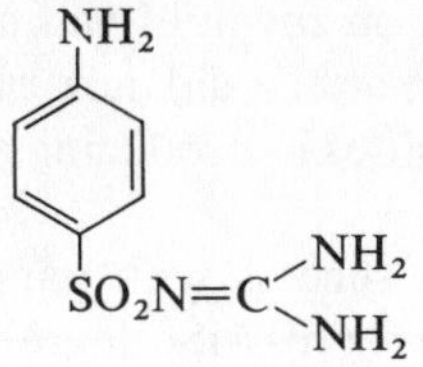

Sulphaguanidine

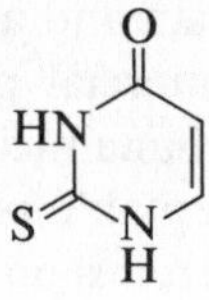

Thiouracil

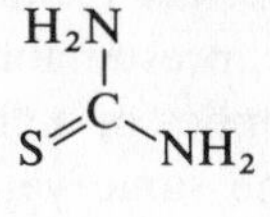

Thiourea

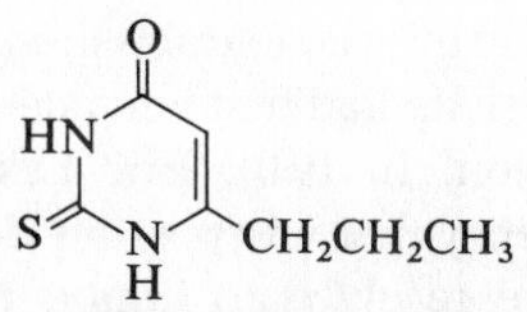

Propylthiouracil

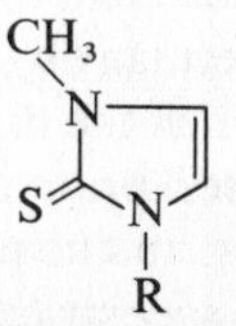

Methimazole, R = —H
Carbimazole, R = —$CO_2C_2H_5$

Charles Searle, and Harold Morley. This was patented by the National Research Development Corporation in 1954 and given the approved name of carbimazole ('Neo-Mercazole'). It is the drug of choice in the United Kingdom.

INSULIN

At Strassburg University in 1889, Josef von Mering and Oscar Minkowski carried out an experiment to establish whether a dog could survive without its pancreas. The outcome of this was that the animal rapidly developed diabetes. Six years later Professor Edward Schäfer proposed that pathological changes in that part of the pancreas described as the islets of Langerhans were responsible for the onset of diabetes. This led to futile attempts to treat patients by feeding them with pancreas, and it was not until 1908 that any progress was made. That year, George Zuelzer, a Berlin physician, found that if a pancreatic extract was intravenously injected into a dog which had had its pancreas removed, the animal's urinary sugar levels could be kept under control so long as daily injections were administered. Once the injections ceased, the sugar rose back to pre-treatment levels. It is of interest to note that the extract was supposedly rendered protein-free by adding alcohol to the expressed juice of the gland until proteins were precipitated. Had the concentration of the alcohol finally exceeded 85% then the hormone, itself a protein, would also have been precipitated and thus lost when the supernatant solution was removed and dried! Zuelzer administered his limited supply of pancreatic extract to eight hospitalized diabetics, achieving an improvement in their diabetes in all cases. He could not persist with the treatment because his supplies ran out and impurities (pyrogens) caused unacceptable fever as a side-effect. His observations were criticized on the grounds that elevation of temperature could itself lower sugar output, so he then spent a further two years purifying his extracts before approaching Hoechst Dyeworks. They were at first reluctant to deal with Zuelzer since his work had been initiated with the support of the Schering Company, but in 1912 a contract and an agreement on royalties was signed. Despite this promise of commercial production, matters did not proceed further, presumably because of unanticipated difficulties in producing adequate supplies of a clinically acceptable product.

In the same year that Zuelzer reported his first success with pancreatic extract, Ernest Lyman Scott decided that his research for his MSc degree in the department of physiology at the University of Chicago should be on pancreatic secretion. He noted at the outset of his investigation that successful biochemical studies in Germany by Cohnheim in 1906 had utilized extracts prepared from pancreas in which the digestive enzymatic activity had first been destroyed by boiling in water prior to alcoholic extraction. In 1910, Eric Leschke had referred to this, and he pointed out that the enzyme could possibly destroy the hormone during extraction. Scott therefore resolved to inhibit pancreatic enzyme at the start of his extraction procedures, albeit by using less drastic methods than boiling the gland in water. The first method he tried was based on earlier studies by Minkowski. This involved tying off the glandular ducts some time before removal of the pancreas, thereby allowing the enzyme-producing cells of the gland to atrophy. Scott abandoned this procedure because it was not possible to ensure total atrophy. Among the alternative approaches he then tried was an extraction technique in which the fresh, moist gland was first treated with alcohol. This effectively served to inactivate the digestive enzyme. In this manner, he obtained material which caused a temporary diminution in the amount of sugar excreted by depancreatized dogs. The lack of a method of measuring the sugar concentration in small amounts of blood prevented Scott from carrying out a more detailed study to eliminate the remaining doubts in his mind that the lowering of sugar output might be due to some other cause. He presented his thesis at the end of the three years allocated to the project, but did not have an opportunity to pursue the problem further and so bring his research to a successful conclusion.

Amongst those who tried to isolate the pancreatic hormone in the 1890s was Nicolas Paulesco, a Rumanian who rose to prominence at that time for his physiological investigations in Paris. After his appointment to the chair of physiology at the University of Bucharest, he turned his attention to the pituitary gland, but he renewed his former interest following the publication of Zuelzer's results on pancreatic extract. In 1916, he prepared an aqueous extract of the gland. This proved active when injected into a diabetic dog. Enemy occupation of Bucharest, however, prevented him from taking the investigation further until 1920. In July the following year, Paulesco reported his successful isolation of the antidiabetic hormone in a preliminary paper published in the prominent French journal *Comptes Rendus de la Société de Biologie*. In this, and in a subsequent paper, he described how the hormone, which he named 'pancreine', dramatically lowered blood sugar levels in normal as well as diabetic dogs when administered intravenously. Nevertheless, Paulesco still had further purification to carry out in order to overcome the problem of irritancy when pancreine was administered subcutaneously, as would be required in future clinical trials. This resulted in a small quantity of a soluble powder being isolated. Before the large-scale production of this material could be organized, a similar material became commercially available as a result of Canadian endeavours.

Enough has been written here to leave no doubt that several early investigators had managed to lower blood sugar levels with their pancreatic extracts. None of them, however, succeeded in returning severely ill diabetics to good health. The first to do so were two young Canadians, Banting and Best, based at the University of Toronto. Their industry and relentless persistence ultimately saved the lives of literally millions afflicted with what had, hitherto, been a fatal condition. Yet, without the knowledge of the painstaking observations made by their predecessors, these two young researchers would probably never have made such rapid progress in the isolation of the hormone. They were fortunate in having support and guidance from one of the world's leading authorities on diabetes, Professor John Macleod. Ironically, Macleod had been approached by Ernest Scott in 1912 in Cleveland, on which occasion he had failed to encourage the young physiologist to pursue his investigations into pancreatic secretion any further!

After returning from service in the armed forces during the First World War, Frederick Grant Banting practised in London, Ontario, as an orthopaedic surgeon. Finding the demands of his practice to be less than enough to keep him fully occupied, he took up a demonstratorship at the university. In October 1920, he read a review article which postulated that a substance secreted by the pancreas might be capable of alleviating diabetes. Soon, he came to the conclusion that the failure to isolate this could be due to destruction of the active principle by the digestive enzymes in the gland; he was unaware that this same suggestion had been made ten years earlier. On the advice of colleagues, Banting approached Professor Macleod, who explained that the proposed investigations would require several months of full-time work. Macleod told Banting that facilities could be provided in his department. This was a generous offer, particularly in view of the fact that Banting was not proposing any radical new approach to a problem which had defeated the efforts of experts in the field for over two decades. Later, Banting was to underestimate the importance of what Macleod did for him, both by way of instruction on how to proceed with the investigation, and by providing the services of a final year biochemistry student, Charles Best. On his own, Banting would probably have made little progress. Best had been instructed in the latest biochemical method which permitted sugar concentrations to be measured using only fractions of a millilitre of blood. Application of this new technique enabled the Canadians to succeed where others had failed.

Banting and Best began their experimental work in the middle of May 1921 in a small laboratory within the university medical building. Working conditions that hot summer were far from ideal, and there was no technical assistance available. However, by late July, the degenerated pancreas (caused by ligation of the duct) had been removed from a dog, and on the 30th of the month 4 ml of a ground-up aqueous suspension of it was injected into a depancreatized dog. Blood sugar levels quickly dropped from 0.20% to 0.12%, subsequently being held at this level by a second injection. This proof that their extraction technique was satisfactory greatly encouraged Banting and Best who, from early August until November, worked night and day injecting dogs with extracts prepared by a variety of methods. Studies of the stability of the hormone were carried out, revealing that it was sensitive to alkali and to the protein-digesting enzyme trypsin. This raised the suspicion that the hormone was a protein. An improvement in the isolation procedure was obtained when Macleod's original suggestion to Best was finally acted upon, and Scott's alcoholic extraction method was employed. The culmination of their endeavours was reached during the Christmas vacation period, when Banting and Best injected each other with their extract. Other than a little redness at the injection site, there were no untoward effects. Tests on patients now began, using the same extract. It was given the name insulin so as to conform with that suggested for the hormone some twelve years earlier. The first to receive insulin, in January 1922, was twelve-year-old Leonard Thompson, a dangerously ill diabetic who was being treated at the Toronto General Hospital by Walter Campbell. He was not expected to live much longer. Unfortunately, the first injections of insulin caused so much local irritation in the sick child that treatment had to be stopped.

Professor J. B. Collip of the University of Edmonton had been working in Macleod's department as a Rockefeller Fellow when he agreed to assist with the purification of insulin. He made the important breakthrough of precipitating insulin from aqueous alcoholic extract of the pancreas by pouring this into several volumes of pure alcohol. This precipitate of insulin was then reconstituted for injection, and was given to Leonard Thompson only a few days after the original injections had been withdrawn. This time the effect of the injections was quite dramatic, for the boy was rapidly restored to excellent health in only a few days. He remained in the best of health for several years, receiving daily injections until his untimely death in a motor cycle accident.

When attempts were made to scale up Collip's manufacturing process to provide insulin for clinical trials, the yield dwindled, and supplies ran out. Tragically, this led to the death of several patients who had been responding well. Facilities to produce insulin had, fortunately, been made available at the Connaught Laboratories. These had been established by the University of Toronto to provide antitoxins and vaccines during the war when supplies from Europe were unavailable. By May 1922, with Best in charge, production was established in the Connaught Laboratories at a satisfactory level, a remarkable achievement when it is realized that there was no previous expertise in this type of production available in Canada. Banting and Best had agreed to apply for patents on the process on the understanding that the University of Toronto would accept and administer these. This it did by setting up an Insulin Committee, which at once put all information it had at the disposal of the Eli Lilly Company of Indianapolis. This public-spirited gesture ensured that mass production of insulin could be rapidly put in hand, thus saving the lives of as

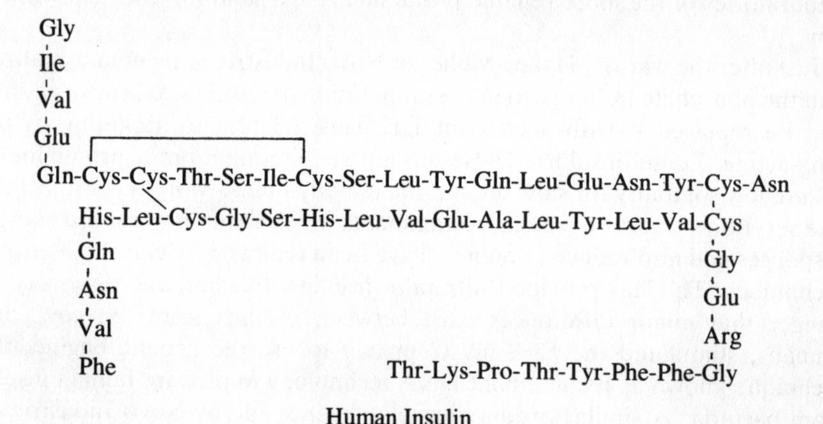

Human Insulin

many diabetics as possible. Similar arrangements were made with other responsible agencies throughout the world. The Lilly group itself consisted of about one hundred workers, headed by George Walden. He introduced the technique of isoelectric fractionation, in which advantage was taken of the insolubility of different proteins at critical levels of acidity. Adjustment of acidity thereby allowed insulin to be separated from protein contaminants. Within a year, the Eli Lilly Company was providing enough pure insulin to meet the need of all diabetics in the United States. This established the company as a leading American pharmaceutical manufacturer.

In 1925, a Nobel Prize was awarded to Banting and Macleod. Banting shared his portion of the prize with Best to express his dismay at his colleague being overlooked. Macleod, in response, shared his award with Collip, whom he considered had been largely responsible for turning the experimental findings into a practical reality. Exactly thirty years later, Frederick Sanger of Cambridge completed a ten-year investigation which won him the 1958 Nobel Prize for chemistry for his elucidation of the chemical structure of insulin. It was the first protein to have its aminoacid sequence determined.

Pure, crystalline insulin did not become available until 1926. It was obtained by the man who had narrowly missed becoming the first person to crystallize adrenaline a quarter of a century earlier, Professor Abel of Johns Hopkins University. The crystalline insulin caused less local irritation than amorphous insulin, but it turned out to be shorter acting. In Copenhagen, H. C. Hagedorn of the Nordisk Insulin Laboratory assumed that the presence of a protein contaminant had accounted for the longer duration of the earlier products. He searched for an acceptable protein that could be added to crystalline insulin, eventually selecting a small protein known as protamine, obtained from fish sperm. Adding this gave a longer-acting insulin preparation that was introduced in 1936. Two years later, D. A. Scott of the Connaught Laboratories showed that the duration of action of protamine insulin was no longer than that of ordinary insulin after removal of zinc, an element he had previously found present in crystalline insulin. This led to the introduction by the Connaught Laboratories of the more reliable protamine zinc insulin for once daily injection.

Just after the war, K. Halles-Moller of Novo Industrie in Denmark realized that the phosphate buffer used in the preparation of insulin was removing zinc, and he replaced it with acetate buffer. This led to the marketing of the long-acting 'Lente' insulins. These did not require addition of protamine to ensure low solubility for slow dissolution in body fluids. More recently, in the late seventies, traces of protein contaminants which sometimes caused allergic responses and also reduced potency, have been removed by chromatographic techniques. This has provided ultrapure insulins. Further, the discovery by Sanger that minor differences exist between insulins from ox, cow, and humans, stimulated the Eli Lilly Company to use the genetic engineering technique known as recombinant DNA technology to prepare human insulin from bacteria. A similar product has also been made by Novo Industrie by substituting one of the aminoacids in pig insulin to convert it into human insulin. As a result of these developments, many patients now routinely receive human insulin.

ORAL ANTIDIABETIC AGENTS

In 1918, C. K. Watanabe reported that the drop in blood sugar levels following removal of the parathyroid gland was due to the presence of abnormal amounts of guanidine in the blood. When guanidine was later tested as a possible antidiabetic agent it proved to be too toxic. During the early twenties, substituted guanidines were shown to be less toxic, but the only one used clinically was galegine, an alkaloid extracted from *Galega officinalis*, the structure of which was elucidated by Barger and White in 1923. K. H. Slotta and R. Tschesche of the Chemistry Institute at the University of Vienna then synthesized a series of compounds with guanidine groups at each end of a long polymethylene chain. These were shown by Professor Hesse of the Pharmacological Institute to be less toxic and more potent than the earlier compounds. Following a clinical trial in 1926, one of these was marketed by Schering AG as an oral hypoglycaemic agent for use in mild cases of diabetes, under the name 'Synthalin'. As a result of adverse reports, another member of the series was introduced as 'Synthalin B' with the claim that it was less toxic. The high incidence of side-effects with both compounds discouraged diabetics from taking these drugs. Following evidence of liver toxicity they were withdrawn in the early forties.

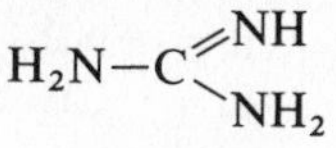

$$H_2N{-}C({=}NH)NH_2$$

Guanidine

$$(CH_3)_2C{=}CHCH_2NHC({=}NH)NH_2$$

Galegine

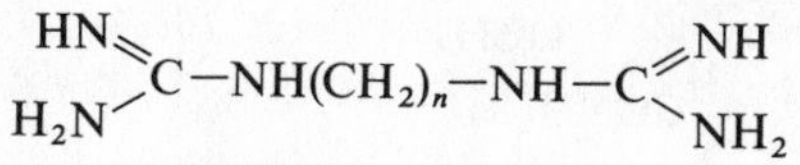

$$(HN{=})(H_2N)C{-}NH(CH_2)_n{-}NH{-}C({=}NH)NH_2$$

Synthalin, $n = 10$
Synthalin B, $n = 8$

In 1942, Professor M. Janbon, head of the medical faculty at Montpellier University, arranged a clinical trial of an isopropylthiadiazole derivative of sulphanilamide, IPTD, which was being considered for use in typhoid cases. On receiving the drug, some of his patients became very ill, and a few died. When intravenous glucose was administered to the sick patients, they made a startling recovery which led to suspicion that the sulphonamide had been producing severe hypoglycaemia. This was confirmed to be the case. Janbon then asked a research student, Auguste Loubatieres, to conduct a full investigation into the effects of the drug on animals, this being an appropriate study for his doctorate. Although Loubatieres concluded in his thesis that IPTD and its derivatives should be of value in diabetes, this suggestion was ignored, presumably because of the recent memory of the problems with the 'Synthalins', as well as the turmoil at the end of the war.

When a sulphonylurea was sent by the C. H. Boehringer company for clinical trial as a long-acting sulphonamide at the August Viktoria Hospital in Berlin early in 1954, it produced severe toxic effects. A young doctor, K. J. Fuchs, tested the drug on himself and discovered it produced the symptoms of severe hypoglycaemia. The chief of the clinic, Professor Hans Franke, arranged for further investigations, these leading to the introduction of the drug as an oral hypoglycaemic agent with the approved name of carbutamide. The Eli Lilly Company in the United States arranged extensive clinical trials of the drug to see whether to conclude a licensing agreement with Boehringer. This revealed an incidence of side-effects which was considered by Lilly to be unacceptable, even though the drug was available for routine use in Europe. A similar trial of Hoechst's closely related tolbutamide, synthesized in 1956 at Elberfield by Gustav Erhart, was arranged by the Upjohn Company. This involved twenty thousand patients and three thousand doctors, and had a happier outcome. The drug received approval from the Food and Drugs Authority for use in

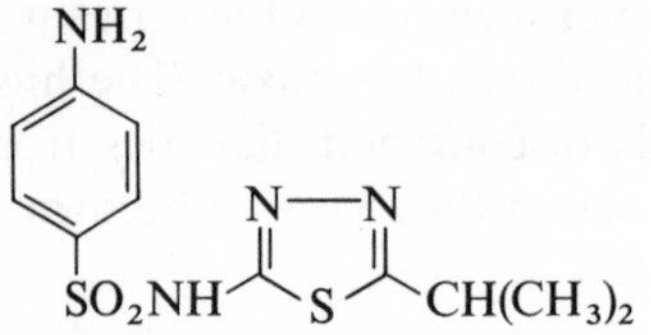

IPTD

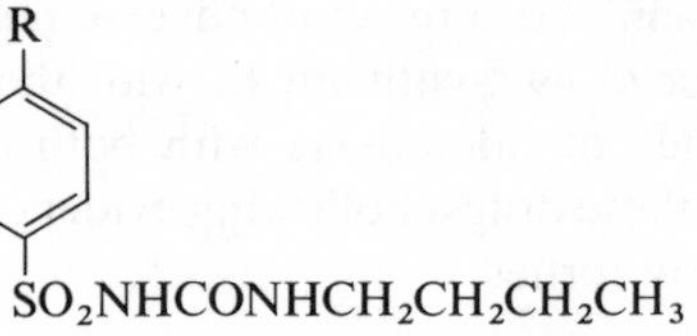

$SO_2NHCONHCH_2CH_2CH_2CH_3$

Carbutamide, R = $-NH_2$
Tolbutamide, R = $-CH_3$

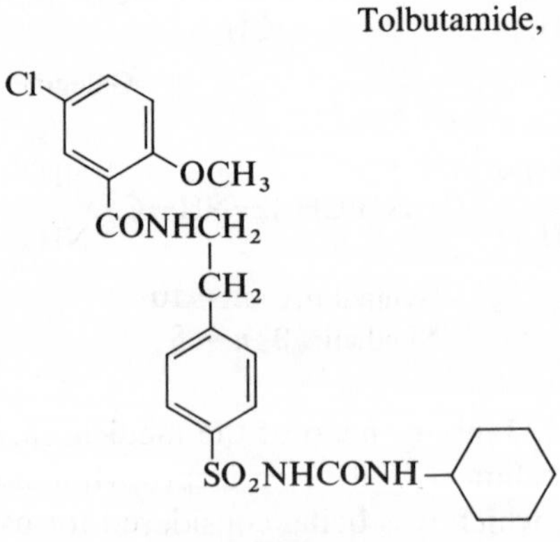

Glibenclamide

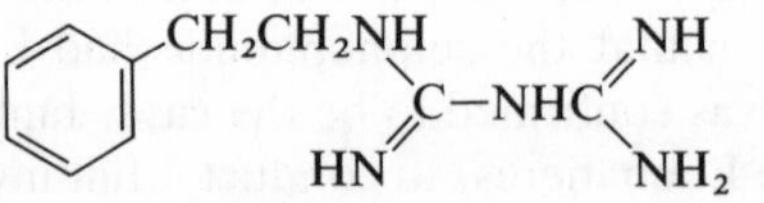

Phenformin

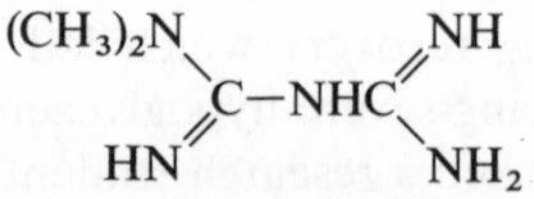

Metformin

maturity onset diabetes. Unlike carbutamide, tolbutamide ('Rastinon') did not possess antibacterial properties, so there was no likelihood of inducing resistant bacteria. Hoechst chemists synthesized thousands of analogues of tolbutamide over the next decade before finding a superior drug in glibenclamide (glyburide, 'Daonil'). Other similar compounds have been marketed.

The introduction of the sulphonylureas as oral antidiabetic agents renewed interest in the guanidines. In 1957, the US Vitamin Corporation introduced two biguanides, phenformin ('Dibotin') and metformin ('Glucophage'), synthesized by Seymour Shapiro, Vincent Parrin, and Louis Freedman at its research centre in Yonkers, New York. They do not act by stimulating the islet cells of the pancreas to produce more insulin, as do the sulphonylureas. Rather, they seem to affect glucose absorption and utilization.

POSTERIOR PITUITARY HORMONES

The pituitary gland, situated at the base of the brain, was one of several glands from which Oliver and Schäfer prepared glycerine extracts in 1894. They found pituitary extract could increase blood pressure, although the effect was much less marked than with adrenal extracts. Four years later, W. H. Howell showed that the activity was only present in extracts prepared from the posterior lobe of the gland. In 1913, van den Velden discovered that the principal physiological effect of pituitary extract was its ability to inhibit urine production. This led to its use in the treatment of diabetes insipidus, in which copious amounts of watery urine are voided.

While working on the pharmacology of ergot extracts at the Wellcome Physiological Laboratories in London, Henry Dale discovered that posterior pituitary extracts had a similar action on the uterus, causing it to contract. By 1909, they were regularly being used in difficult cases of labour. However, it was not until after the introduction of insulin into clinical practice that worthwhile progress was made towards isolation of the active principle. The most significant advance came from the laboratories of Parke, Davis, and Company in Detroit, in 1928, where Oliver Kamm and his colleagues concentrated, and fairly effectively separated, two hormones. These were vasopressin ('Pitressin'), which increased blood pressure, and oxytocin ('Pitocin'). The latter found widespread use in obstetric practice. Further purification of these products was attained at the George Washington University in Washington. The studies there were directed by Vincent du Vigneaud, who had previously worked with Professor Abel on the crystallization of insulin. The task in hand was daunting, since only traces of the active principles were present in the gland. Hundreds of thousands of hog and beef glands were used up during the course of the investigation, which relied mainly on the technique of electrophoretic separation. Progress could only be monitored by testing the purified fractions on animals.

The outbreak of war saw du Vigneaud's attention diverted to work on penicillin. This served to introduce him to Craig's new technique of countercurrent distribution for purifying penicillin. When he renewed his studies on the pituitary after the war, he combined this technique with starch-column chromatography. The outcome was that he purified enough oxytocin for analysis to reveal that it was a peptide (i.e. very small protein), consisting of eight aminoacids. At this point it was realized that the oxytocin could not be more than 50% pure, and another two years passed before a reasonably pure crystalline derivative was obtained. Vasopressin was also found to contain eight aminoacids, two of which differed from those in oxytocin. Du Vigneaud and his colleagues went on to establish how the aminoacids were linked together in each of the hormones, thus completing a herculean task begun two decades earlier. Conclusive confirmation of the proposed structure was obtained in 1953 by the synthesis of oxytocin and vasopressin. The synthetic hormones are now in regular clinical use. For his persistence over quarter of a century, du Vigneaud finally had the satisfaction of being awarded the Nobel Prize, in 1955, for his work on the posterior pituitary hormones.

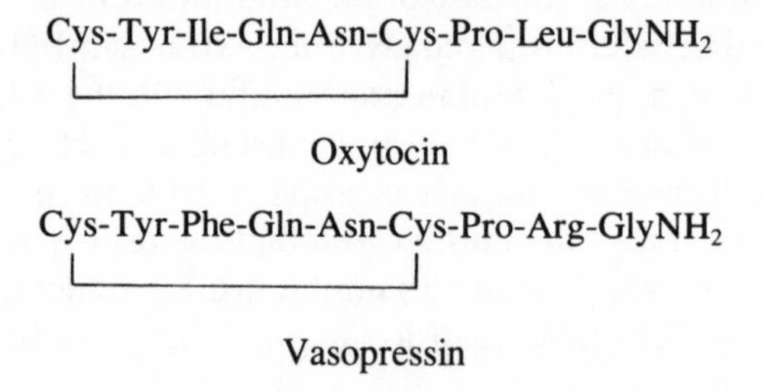

ANTERIOR PITUITARY HORMONES

In 1887, Oscar Minkowski discovered that acromegaly, a disease involving overgrowth of the extremities of the body, was associated with a tumour in the pituitary gland. Over the succeeding years enough evidence accrued to suggest that a growth-promoting hormone was released by the pituitary, so that it was of little surprise when, in 1921, Herbert Evans and Joseph Long at the University of California, Berkeley, injected an extract of anterior lobe into rats and found it to accelerate their growth. What had not been anticipated, however, was the stimulatory effect of this extract on the development of the gonads. A few years later, their colleague, Philip Smith, showed that excision of the pituitary not only caused lack of growth and atrophy of the gonads, but also atrophy of the thyroid and the adrenals. His demonstration that these deficiencies could be corrected by pituitary implants made from the anterior lobe was followed, in 1933, by the preparation of an adrenotropic extract by Professor J. B. Collip and his colleagues at McGill University in Montreal. A purified form of this was used in 1949 by Hensch as an alternative to the then scarce cortisone in the treatment of rheumatoid arthritis. It was also used, for a while, in the chemotherapy of leukaemia, again as an alternative to cortisone. These developments caught the attention of the American Cyanamid Company (Lederle), and Armour and Company, who carried out extensive work on the purification of adrenocorticotrophin (then generally known as ACTH). By 1956, Cyanamid had elucidated the structure of one form of the hormone, a peptide containing thirty-nine aminoacids. Today neither the natural nor the synthetic corticotrophin finds much use in medicine, the direct application of corticosteroids being preferred when steroid therapy is indicated.

Early hopes that pituitary growth hormone would be of value in the treatment of human pituitary dwarfs were not fulfilled, despite the isolation by Evans and Choh Hao Li of pure growth hormone from ox pituitary in 1944. It was not until ten years later, when G. E. Pickford demonstrated that fish growth hormone was active in fish, but not rats, that there was any recognition of species differences. Evidence was soon obtained to show that monkey hormone was active in monkeys, although bovine and porcine varieties were not. In 1957, J. C. Beck and E. H. Venning, in the McGill University Clinic, produced dramatic effects in their patients by injecting pure human growth hormone. This was prepared from cadavers by M. S. Raben of Tufts University in Boston. For many years, supplies of hormone could not satisfy the demand, and the strictest of clinical criteria had to be met before samples would be released from the specialist laboratories which produced it. Recently, recombinant DNA technology has been used to enable genetically modified bacteria to produce this much sought after human hormone.

The only other pituitary hormone occasionally used is the follicle-stimulating hormone. This was obtained pure from sheep pituitary by Li in the early fifties. The human hormone, known as menotrophin, is used in women who, as a consequence of hormone deficiency, are unable to ovulate.

ADRENOCORTICAL STEROIDS

In 1927, Julius Rogoff and George Stewart at Western Reserve University in Cleveland found that dogs from which the outer cortex of the adrenal gland had been removed could be kept alive by repeated intravenous injections of aqueous extracts of canine adrenal cortex. These extracts had only slight and variable activity, hence they were unsuitable for clinical application. Three years later, Professor Wilbur Swingle of Princeton University, assisted by Joseph Pfiffner, prepared a more potent extract by using benzene, a fat solvent. Parke, Davis, and Company then introduced an extract, made in this manner, for the treatment of the hormone deficiency condition known as Addison's disease.

Although the use of benzene as extraction solvent minimized contamination with adrenaline, it was only after Marvin Kuinenga of the Upjohn Company used a form of adsorption chromatography for purification that extracts totally free of adrenaline were introduced. This was in 1935. This advance encouraged four different groups of chemists to try to isolate the active material from adrenal extracts.

Dr Pfinner had by now moved to Columbia University to work with Oscar Wintersteiner, who had recently isolated progesterone. Together with Harold

Vars, they isolated four new steroid hormones before the end of 1936. Three of these were also isolated at the Mayo Clinic in Rochester by Edward Kendall's group. In addition, Kendall isolated two more steroids, Compound A and Compound E. Before the end of the decade, Professor Tadeus Reichstein, head of the department of pharmacy at the University of Basle, had also isolated and identified no less than 25 adrenal steroids, of which all but six were biologically active. This remarkable feat required the extraction of 20 000 cattle adrenal glands! Another success for Reichstein was his conversion of a readily available plant product into desoxycorticosterone, in 1937. This was the first adrenocortical steroid to be used clinically when Ciba marketed it two years later as an oily injection known as 'Percorten'.

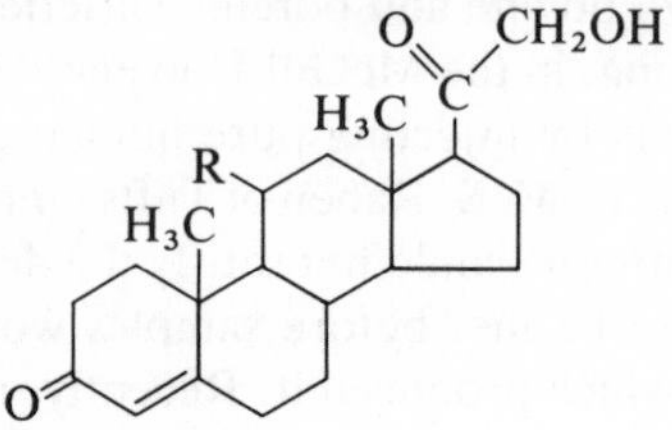

Desoxycorticosterone, R = —H
Corticosterone, R = —OH

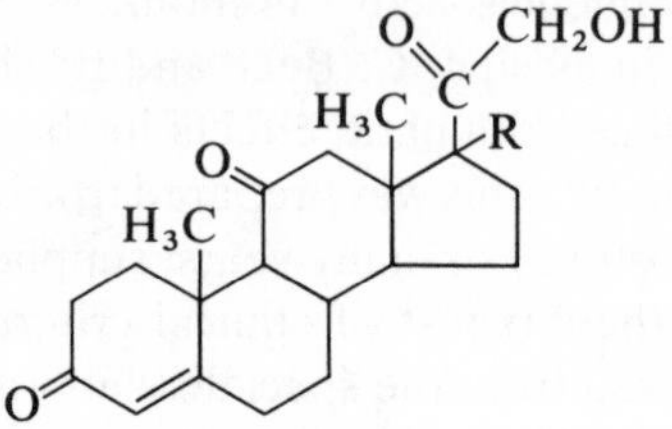

Dehydrocorticosterone, R = —H
Cortisone, R = —OH

A report in the *Lancet*, written by Hans Selye of McGill University, Montreal, in 1940, caught the attention of military medicine experts in the United States. This disclosed that injections of an adrenal steroid known as corticosterone were highly effective in dealing with shock. Desoxycorticosterone lacked this ability. This report was recalled a year later, when American intelligence agents picked up a rumour that Germany was purchasing vast quantities of adrenal glands from slaughter-houses in Argentina. It was believed these were required for the production of extracts that would enable Luftwaffe pilots to withstand the stress of flight at extreme altitudes! Immediately, an approach was made to Kendall by the Office of Scientific Research and Development, inviting him to collaborate with the US drug industry to produce steroids for the air force. Kendall contacted Randolph Major at Merck and Company. The outcome was that, before the war had ended, Kendall devised a simple process for the manufacture of his Compound A (11-dehydrocorticosterone), while twenty-seven-year-old Lewis Sarett of Merck single-handedly synthesized a few milligrams of Compound E. The military interest in the project had, meanwhile, ceased after it was revealed that the rumours from Argentina had no foundation. However, the collaboration between Merck and Kendall continued after the war was over, culminating in Sarett's synthesis of five grams of Compound E in 1948. It was some of this steroid, subsequently renamed cortisone, which Phillip Hensch used so dramatically in the autumn of that year, when he treated a young woman desperately ill with arthritis (see page 91). This success transformed attitudes to the corticosteroids.

Swiss and American chemists realized that biologically active material still remained to be isolated from the solvent mother liquors after the thirty or so known steroids had been recovered. Yet it was left to James Tait, a newly appointed lecturer in medical physics at the Middlesex Hospital in London, to make the next advance. In 1948, he and Sylvia Simpson (later to become Mrs (Tait) used their spare time to test the effect of Allen and Hanbury's 'Eucortone', a brand of adrenal extract, on mineral and water balance. It was known that this was disturbed in patients receiving therapy with 'Eucortone'. Although Tait and Simpson had no grant to support their study, they did have access to radioisotopes of sodium and potassium. This meant they could study mineral balance after administering 'Eucortone' to animals from which the adrenals had been removed. By applying the newly developed technique of paper chromatography, they were soon able to distinguish fractions from 'Eucortone' that were biologically active, but it still took them three years to finally obtain a new hormone that had unprecedented ability to promote salt retention. They gave this the name 'electrocortone', but were persuaded to rename it aldosterone after establishing its full chemical structure, in 1953, in collaboration with Reichstein and Wettstein in Basle. Aldosterone was occasionally used in Addison's disease, or the treatment of shock, but its main importance lay in the recognition that it was a major causative factor in certain types of oedema.

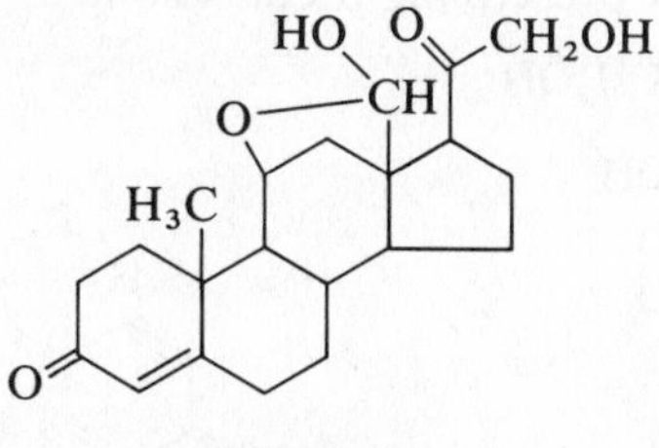

HO O CH_2OH
CH
O
H_3C
O

Aldosterone

ADRENOCORTICAL STEROID ANALOGUES

The first important analogue of the hormones cortisone and hydrocortisone was discovered serendipitously as a consequence of an attempt by a Squibb chemist, Joseph Fried, to synthesize hydrocortisone from its isomer epicortisol. These two compounds differed only in the direction in which the hydroxyl group at the 11-position of the molecule faces. In order to achieve the conversion, Fried first had to introduce a bromine atom at the 9-position. Routine screening revealed the resulting bromine-containing compound had one-third of the activity of cortisone. As this was an analogue of the feebly active epicortisol, Fried prepared several halogenated analogues of cortisone and hydrocortisone. It became apparent that the potency of these compounds was inversely proportional to the size of the halogen substituent, indicating that the small fluorine atom would be an ideal halogen to use in preference to the bulky bromine. In 1954, Fried and Sabo synthesized fludrocortisone ('Florinef'), which was found to be about ten times as potent as cortisone in relieving rheumatoid arthritis. Tantalizingly, not only was the anti-inflammatory (glucocorticoid) activity increased, but so was the aldosterone-like salt and water retaining (mineralocorticoid) activity. This meant that patients became moon-faced and oedematous. Significantly, there was actually a three-hundred- to eight-hundred-fold increase in mineralocorticoid activity, indicating that molecular manipulation could alter the balance of mineralocorticoid and glucocorticoid activity.

At the same time as the Squibb researchers were developing fludrocortisone, Arthur Nobile of the American owned Schering Corporation obtained two novel steroids by fermentation of cortisone in a mould broth. A preliminary clinical trial at the National Institutes of Health suggested both compounds were about five times as potent as hydrocortisone, yet not much more likely to cause salt retention. Encouraged by this, the company took the decision to risk pouring \$100 000 worth of cortisone into a large fermentation tank. The gamble paid off handsomely when prednisone ('Meticorten') notched up more than \$20 000 000 in sales for Squibb within a year of its introduction in 1955! The other new steroid, prednisolone, seemed to have almost identical properties, which is hardly surprising since prednisone is converted into it in the body. At present, it is felt that the unreliability of the metabolic conversion should be avoided by only prescribing prednisolone.

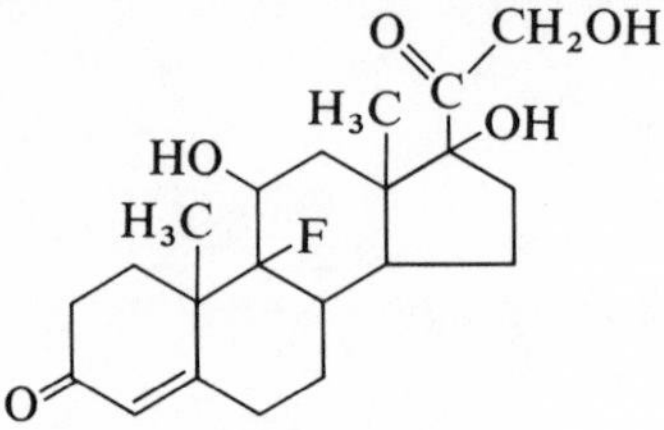

Fludrocortisone

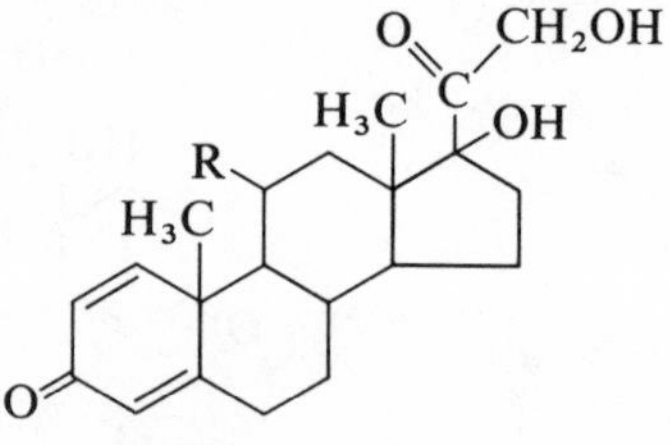

Prednisone, R = =O
Prednisolone, R = —OH

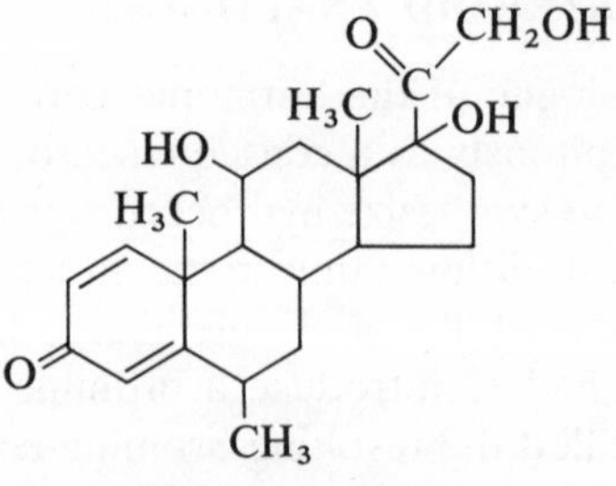

Methylprednisolone

A year before the introduction of prednisolone, Burstein, Dorfman, and Nadel had revealed that hydrocortisone was metabolically deactivated by conversion to its 6-hydroxy derivative. This gave researchers at Upjohn the idea of blocking metabolism of prednisolone by placing an appropriate substituent on the vulnerable position to hinder the fit of the drug at the surface of the oxidative enzymes. The resulting compound, methylprednisolone ('Medrone') proved to be slightly more potent than prednisolone. Schering AG and Syntex subsequently placed chlorine or fluorine atoms at the 6-position of several of their corticosteroids in order to increase potency.

Following their isolation from the urine of a boy with an adrenal tumour, of corticosteroids substituted with a hydroxyl group at the 16-position, a small team of researchers at Lederle Laboratories, led by Seymour Bernstein, synthesized several steroids of this type so that their biological activity could be examined. They found that the 16-hydroxyl group decreased mineralocorticoid activity, an observation that was immediately exploited to overcome the major shortcoming of Squibb's fludrocortisone. The Lederle chemists also took note of the superiority of prednisolone over hydrocortisone. They based their new fluorosteroid on the former structure rather than the latter, as had been the case with fludrocortisone. This resulted in Lederle's introduction of triamcinolone ('Ledercort', 'Aristocort') at the beginning of 1958. Triamcinolone was also marketed by Squibb as 'Adcortyl' as a consequence of a cross-licensing agreement to avoid patent difficulties. Clinicians found this new steroid to be free from mineralocorticoid activity, whilst still as effective an anti-inflammatory agent as prednisolone. Triamcinolone, however, did not live up to its early promise. It had a tendency to cause nausea, dizziness, and other unwelcome effects. Notwithstanding this, Lederle and Squibb were later able to formulate dermatological preparations that were of immense value in the treatment of psoriasis. These contained triamcinolone acetonide, a readily prepared derivative made by dissolving triamcinolone in acetone which contained a small quantity of perchloric acid. Other manufacturers copied this

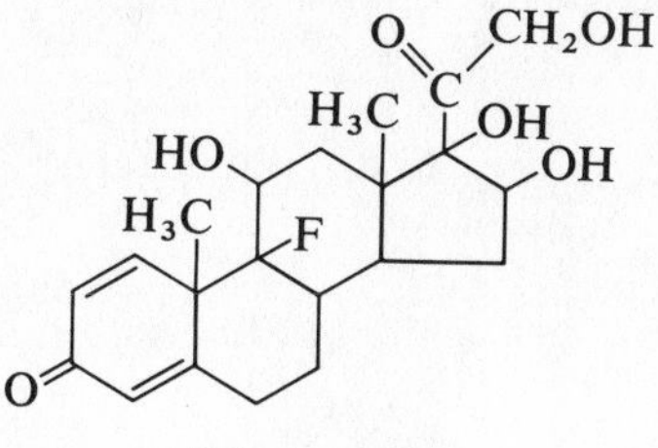

Triamcinolone

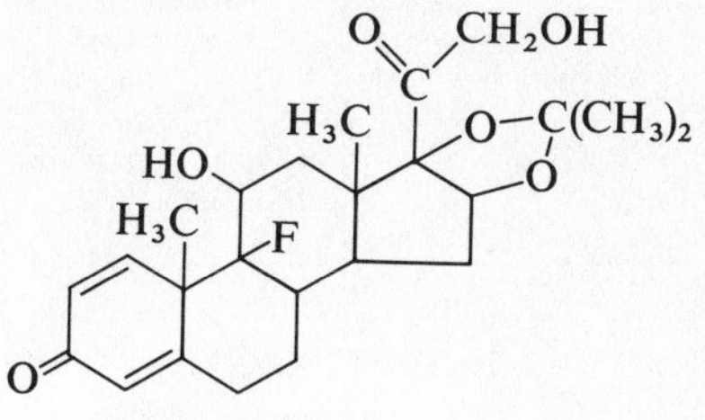

Triamcinolone Acetonide

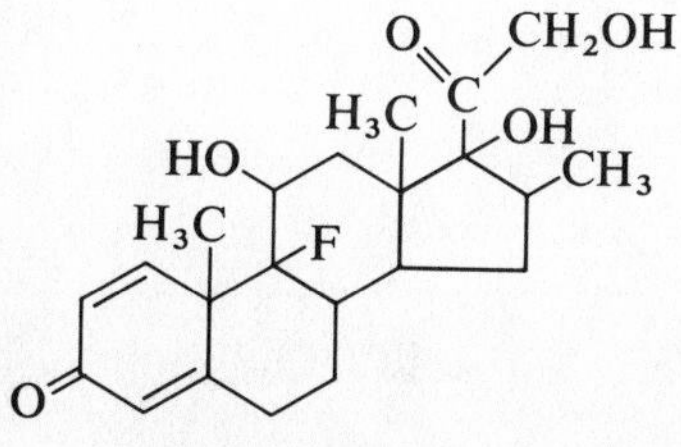

Dexamethasone

approach, utilizing the acetonides of 16-hydroxycorticosteroids in creams and ointments in order to take advantage of improved absorption characteristics arising from the enhanced fat solubility of the acetonides.

In an attempt to hinder metabolism of the side-chain attached to the 17-position of corticosteroids, Merck chemists synthesized analogues with a methyl group in the adjacent 16-position. Whilst this manoeuvre did appear to achieve the desired end, of far greater importance was the apparently unpredicted discovery that the methyl group not only acted like a hydroxyl group to depress mineralocorticoid activity, but it also enhanced the anti-inflammatory activity. The outcome was the successful marketing of dexamethasone ('Decadron') in 1958 (which was simultaneously synthesized by the Schering Corporation), followed shortly after by betamethasone, the isomer in which the methyl group at the 16-position is attached to the other side of the molecule. This stereochemical alteration made little difference to the biological activity, both drugs being six or seven times as potent as prednisolone, with practically no salt-retaining activity at all. Neither of these drugs exhibited the unwelcome side-effects seen with triamcinolone.

Since the discovery of dexamethasone, a variety of molecular permutations has permitted manufacturers to develop a bewildering array of highly potent corticosteroids. So far as systemic medication is concerned, high potency in itself is of little advantage, unlike the situation with topical treatment, where it is possible that the small amounts of drug that enter the circulation will be more readily metabolized than would the greater quantities of a less potent steroid.

XII

Vitamins

During the early years of the nineteenth century the combination of adverse climatic conditions and the Napoleonic wars led to food shortages in parts of Europe. In 1815, the French Academy established a Gelatin Commission to enquire into whether gelatinous extracts prepared by boiling bones in water could be used as a meat substitute for the poor. Magendie, the chairman, carried out a series of studies that revealed dogs were unable to survive on non-nitrogenous foods such as fats and sugars; meat was essential. The Commission then spent several years before concluding that gelatin could be used either as a total or partial substitute for meat. Similar conclusions were reached in Geneva by the Society for the Promotion of the Arts. Matters were taken further in 1841 after Justus von Liebig, the renowned professor of chemistry at the University of Giessen, published his first paper on proteins. This had been inspired by recent work carried out in The Netherlands by G. J. Mulder who had introduced the term *protein*, wrongly believing this to be a single, common component of several albuminous substances. Mulder had claimed that all foods containing protein had nutritional value. Liebig, in turn, mistakenly believed that nearly all of the nitrogen in foodstuffs was present as protein, hence their nutritive value should be proportional to nitrogen content. Magendie, however, found that although gelatin was a protein, dogs could not be kept alive on a diet consisting of this alone or even in combination with bread. A second commission, known as the Magendie Commission, then re-examined the whole issue of human nutrition. It arranged for dogs to be fed solely on raw bones, only to make the surprising discovery that this diet maintained them in good health and ensured normal growth if sufficient bones were provided. All forms of extraction process, such as transformation into gelatin, destroyed the nutritive value of bone. The Commission did not comment on the nutritional value of gelatin, but three years later a Dutch commission confirmed that gelatin was of no nutritive value even when combined with other substances. The Magendie Commission also investigated the nutritional effects of other single substances, including oil, sugar, egg white, and wheat gluten. Some of its conclusions were erroneous, but this was not discovered until the present century.

Liebig proposed that healthy nutrition could be maintained on a diet consisting of protein, fat and sugar (to supply energy), and essential minerals. As he was the most outstanding chemist of his time, his views on nutrition remained unchallenged for many years, but in 1871 the eminent French chemist Jean Baptiste Dumas published his observations made on starving infants while Paris had been under siege by the Prussians. Dumas described the disastrous consequences of feeding these infants on artificial milk prepared by emulsifying fat in a sugared albuminous solution, in accordance with Liebig's teachings. Dumas' observations led to renewed interest in dietary requirements, especially with regard to mineral content.

One of the leading investigators in the field of mineral metabolism was Professor Gustav von Bunge of the University of Dorpat. Being particularly interested in the role of alkaline salts in the diets, he set Nicholas Lunin the task of establishing the effect of a sodium carbonate supplement on mice maintained on an artificial diet consisting of pure protein from milk (casein) and cane sugar. The mice receiving sodium carbonate survived from 12 to 30 days, in contrast to the 11 to 21 days of those denied the supplement. In an attempt to keep his mice alive longer, Lunin fed his mice on casein, milk fat, milk sugar (lactose), and salts, each of these ingredients being carefully purified then incorporated in the same proportion in which it occurred in natural milk. This artificial milk did not improve survival rates. When Lunin gave the mice powdered natural milk they thrived on it. This led him to conclude in his doctoral dissertation of 1880 that milk must contain unknown substances essential for growth and maintenance of good health. Lunin left Dorpat soon after to take up a clinical appointment at St Petersburg. Although his findings had presented a direct challenge to generally accepted views, he did not pursue them any further. Consequently, they were overlooked for many years. A similar fate befell a study by Professor C. A. Pekelharing of the University of Utrecht when it was published in Dutch in 1905. This showed that mice could be kept healthy on an artificial diet to which had been added a small amount of whey (obtained by removal of fat and casein from milk).

Scant attention was paid to the results of feeding animals on artificial diets until Professor Frederick Gowland Hopkins of Cambridge University entered the field. After isolating tryptophan in 1901, he carried out experiments to evaluate the role of this new amino acid in nutrition. By 1906, he had come to the conclusion that '. . . no animal can live upon a mixture of pure protein, fat, and carbohydrate, and even when the necessary inorganic material is carefully supplied the animal still cannot flourish.' That year, Hopkins initiated his classical experiments on feeding rats with purified diets with and without supplements of natural foods. These studies not only confirmed the findings of earlier workers, but also revealed that minute amounts of unknown substances present in normal foods were essential for healthy nutrition. For example, rats could be kept in good health on a diet consisting of casein, fat, starch, sugar, and inorganic salts if a little milk was added. Carefully controlled experiments showed that enough milk to increase the amount of solids in the diet by as little as one per cent was quite sufficient. When Hopkins published his results in 1912, after a delay of five years caused by his ill health, his eminence in the scientific community ensured that the significance of the unknown accessory factors in food would never again be overlooked. By then, new work casting light on diseases caused through dietary deficiencies had given added significance to his conclusions.

WATER-SOLUBLE VITAMINS

Concerned at the high mortality from beriberi amongst soldiers living in barracks in Sumatra and also amongst prisoners in Java, the Dutch government in 1886 sent the bacteriologist Professor Pekelharing and a neurologist, Dr C. Winkler, to Batavia (now Djakarta) in the Dutch East Indies to investigate the cause of the malady. Beriberi was a progressive disease that manifested itself by peripheral neuritis, emaciation, paralysis, then cardiac failure. Its incidence had greatly increased since 1870, especially amongst those living in barracks and prisons for more than three months. Two views prevailed as to its aetiology, namely that it was either of bacterial origin or else caused by a toxin in the rice that formed the staple diet of all those who were afflicted. That rice diets were the key to understanding the cause of the disease should have been evident from the study conducted in 1873 by van Leent, a Dutch naval surgeon. He traced the high mortality from beriberi amongst Indian crews to their rice diet. Simply by putting them on the same diet as their European shipmates, he achieved a dramatic reduction in the morbidity of the disease. Had the influence of Pasteur's work not been so strong at this period, more attention might have been given to the investigation by van Leent.

Pekelharing and Winkler tentatively concluded that beriberi was of bacterial origin, and they recommended the disinfection of barracks and prisons. On returning to The Netherlands they left behind their assistant, Christiaan Eijkman, to isolate the causative organism. He was given a small laboratory in the Dutch military hospital in Batavia where he spent several months studying inoculation of fowls with body fluids from beriberi victims. Although his early studies were inconclusive, after a few months some of the fowls began to stagger around as if intoxicated. Eijkman recognized this as being due to polyneuritis (nerve degeneration) comparable to that seen in humans suffering from beriberi. Surprisingly, it was not confined to fowls that he had inoculated. Six months later, the fowl disease disappeared as mysteriously as it had begun. At this point, Eijkman realized that the diet of the fowls had changed shortly before the period in which they became diseased. When he had begun his investigations, the birds were fed on cheap rice. Later, Eijkman had been able to acquire supplies of the more expensive milled, polished rice fed to patients in the military hospital. When a new director of the hospital was appointed, Eijkman was told to cease this practice. It was the change back to unmilled rice that led to disappearance of the disease in the fowls.

Eijkman carried out an experiment which involved feeding two groups of fowls with milled and unmilled rice respectively. The results proved conclusively that milling removed some unknown ingredient, present in the germ or pericarp, that prevented polyneuritis. Nevertheless, Eijkman was not yet convinced that this simple explanation could account for the occurrence of beriberi in humans. Further investigations would be required before this conclusion could be drawn, but he was unwilling to experiment upon humans. Meanwhile, he published his findings in an obscure Dutch East Indies journal in 1890; it was another seven years before he presented a short account in a prominent German journal.

Eijkman asked his friend Dr A. G. Vorderman, an inspector of health in Java, to ascertain the frequency of beriberi in the 101 prisons on the island, as well as the type of rice being consumed. In the years 1895–1896 Vorderman found that in those prisons where the inmates prepared their own crude rice, the disease was rarely seen. But, in prisons where machine-milled, polished rice was supplied there was a high incidence of beriberi. It was even shown that the degree of milling influenced the incidence of disease. Eijkman then showed that the outer layers of rice, the silverskin, could prevent polyneuritis in fowls. In 1897, Vorderman demonstrated in a prison in the village of Tulung Agung that beriberi could be eliminated by replacing polished rice with unmilled rice.

When Eijkman returned to The Netherlands in 1896 to take up a post at the University of Utrecht, he was convinced that beriberi was caused by gut microbes converting some constituent of the carbohydrate-rich rice into a toxic substance that was then absorbed into the circulation. An antidote to this was presumed present in the outer layers of the rice that were removed during milling.

Eijkman's successor at Batavia, Gerrit Grijns, showed that polyneuritis could be produced in fowls by feeding them on unmilled rice that had been autoclaved for two hours at 120°C. Unmilled rice that had not been exposed to this destructive treatment actually protected the birds against the disease. He also found that feeding fowls on vegetables known to prevent beriberi protected them against the occurrence of polyneuritis. By means of these and other experiments, Grijns finally came to the conclusion, in 1901, that in order to maintain functional integrity of their nervous system, it was essential for the diets of animals and man to contain a protective substance that was present in unmilled rice, meat, and vegetables. If the diet was deficient in this protective substance, beriberi resulted.

In 1906, Eijkman established that the protective substance was soluble in water and had a low molecular weight. Several unsuccessful attempts were made to isolate it in crystalline form. In one by Casimir Funk, a Polish scientist working in the Lister Institute in London, a crystalline compound extracted from rice polishings cured experimentally induced polyneuritis in pigeons. Subsequently, it was established that these crystals had probably consisted of nicotinic acid contaminated with the true antineuritic substance. Believing he had isolated the genuine protective factor and that it was an amine, Funk proposed that it be called 'beriberi vitamine'. His new term was altered to 'vitamin' by his colleague Jack Drummond in 1920, after it was realized that none of the protective substances then isolated were amines. Drummond proposed that different vitamins be distinguished by the use of letters of the alphabet, the antineuritic vitamin being described as vitamin B.

A reliable method of detecting the antineuritic substance was devised by B. C. P. Jansen, who was based in a new laboratory in Batavia, built for the Dutch East Indies Medical Service. With considerable persistence, he established that small birds known as bondols (*Munia maja*) were much more susceptible to polyneuritis than were fowls. When fed on polished rice for only ten days, they developed polyneuritis, instantly detectable by their flying in characteristic circles. By 1920, Jansen had perfected a reliable assay which involved measuring the ability of his rice extracts to prevent, rather than cure, polyneuritis induced by feeding the bondols on polished rice. Due to its instability, years of patient work were required before Jansen and Donath finally isolated crystals of the pure vitamin in 1926. Samples of the crystalline vitamin were sent to Professor Eijkman at Utrecht, where he demonstrated that the addition of two to four milligrams of these to every kilogram of polished rice restored its full antineuritic value. Three years later, Eijkman and Hopkins were jointly awarded the Nobel Prize for physiology or medicine for their work on vitamins.

The Jansen-Donath extraction process was too expensive for commercial exploitation. In 1933, Robert Williams developed a new isolation process for the antineuritic vitamin, by then known as vitamin B_1. He was the chemical director of the Bell Telephone Laboratories in the United States, but had spent more than twenty years researching the vitamin whilst at the Bureau of Chemistry set up in the Philippines by the US Army Medical Commission for the Study of Tropical Disease. His new work on the isolation of the vitamin was carried out privately in his spare time. He contacted Randolph Major, director of research at the Merck Laboratories in Rahway, with the outcome that collaboration with the company began at once. Early in 1936, Williams presented his proposal for the chemical structure of the vitamin. Shortly after, he and Joseph Cline of Merck published a preliminary account of their synthesis which confirmed the validity of the proposed structure. Immediately after this appeared in print, Williams received a copy of a paper by Rudolph Grewe further confirming the structure and containing the information that Andersag and Westphal at the Elberfield laboratories of I. G. Farben had synthesized the vitamin some months earlier. It was also synthesized around this time by Bergel and Todd at the Lister Institute. In 1937, Williams licensed Merck to produce the vitamin commercially by his synthetic process. Much of the profit from the Williams' patents was devoted to a fund which supported nutritional research.

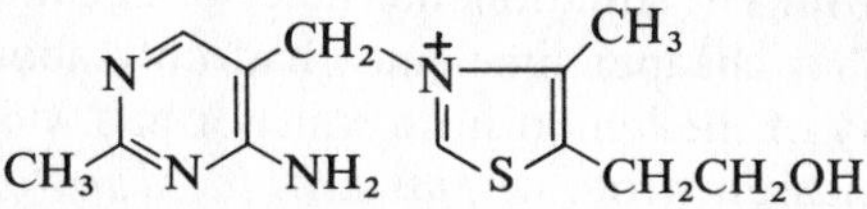

Thiamine

In 1937, several European countries accepted Jansen's proposal that the anti-beriberi vitamin should be known henceforth as aneurine. The US Council on Pharmacy and Chemistry rejected this since their policy was to avoid drug names with any therapeutic connotation. Williams suggested that it be called thiamine, a term which highlighted the presence of the sulphur atom. This was not universally accepted until the International Union of Pure and Applied Chemistry's Commission on the Nomenclature of Biological Chemistry approved the name in 1951.

The investigations into the nature of the antineuritic vitamin were directly responsible for the discovery that absence from the diet of another water-soluble vitamin was the cause of scurvy. In 1536, the French explorer Jacques Cartier vividly described the nature of this disease which afflicted all but ten of the one hundred and ten men aboard his three ships wintering in the frozen St Lawrence river. The victims' weakened limbs became swollen and discoloured, whilst their putrid gums bled profusely. The captain of his ship learned from an Indian how to cure the sailors with a decoction prepared from the leaves of a certain evergreen tree. Miraculously, so it seemed, the remedy proved successful. Nearly thirty years later, the Dutch physician Ronsseus advised that sailors consume oranges to prevent scurvy. In 1639, one of England's leading physicians, John Woodall, recommended lemon juice as an antiscorbutic. Notwithstanding, it is James Lind, a Scottish naval surgeon, who is remembered as the first man in medical history to conduct a controlled clinical trial, whereby he proved that scurvy could be cured by drinking lemon juice. In May 1747, he tested a variety of reputed remedies on twelve scorbutic sailors quartered in the sick bay of the Salisbury. Two of them were restricted to a control diet, but each of the others were additionally given one of the substances under trial. The two seamen who were provided with two oranges and a lemon each day made a speedy recovery, one of them being fit for shipboard duties in only six days. The only other sailors to show any signs of recovery were those who had been given cider. Lind observed no improvement in the condition of those who had been given either oil of vitriol (dilute sulphuric acid), vinegar, sea-water, or only the control diet. He drew the obvious conclusions and these were duly acted upon by Captain James Cook on his second voyage round the world. Although Cook was at sea for three years, not a single member of his crew died from scurvy thanks to adequate provision of lemon juice, as well as fresh fruit and vegetables. Surprisingly, it was not until 1795 that the Admiralty finally agreed to Lind's demands for a regular issue of lemon juice on British ships. The effect of this action was dramatic; in 1780, there had been 1457 cases of scurvy admitted to Haslar naval hospital, but only two admissions took place between 1806 and 1810. The situation then deteriorated for over a century until it was discovered that cheaper lime juice had only about a quarter of the antiscorbutic activity of the lemon juice which it had widely displaced.

In 1899, Stian Erichsen wrote in *Tidsskrift for den Norske Laegeforening* (Journal of the Norwegian Medical Association) that a mysterious illness afflicting sailors on very long voyages was caused by lack of fresh food. Concerned at the growing incidence of this disease which had similarities to both beriberi and scurvy, the Norwegian navy asked Professor Axel Holst and Theodor Frolich of Christiana University in Oslo to investigate the matter. Holst visited Grijns in Batavia before returning to carry out experiments on guinea pigs. Fortunately for him, guinea pigs are exceptionally sensitive to ascorbic acid deficiency, and so he and Frolich readily induced a condition analogous to human scurvy by feeding their animals on polished rice. This was not alleviated by giving the guinea pigs rice polishings, but fresh fruit or vegetables known to cure scurvy restored them to good health. On the basis of these findings, Holst argued, in 1907, that in addition to the antineuritic dietary protective substance postulated by Grijns, there must also exist an antiscorbutic one. The disease amongst Norwegian sailors could be prevented by appropriate dietary measures.

Holst and Frolich went on to demonstrate that the antiscorbutic factor was soluble in water, but insoluble in alcohol unless acidified. They showed that, like the antineuritic substance, it was of low molecular weight. The Norwegians also found that when foods were subjected to drying, the antiscorbutic principle was destroyed. Their pioneering studies were confirmed by work on monkeys carried out at the Lister Institute during the First World War in the wake of outbreaks of scurvy amongst British troops serving in the Middle East; this was despite the provision of lime juice. Only then was it recognized that the juice of West Indian limes had poor antiscorbutic activity in comparison to that of lemons. In 1920, Drummond proposed that the antiscorbutic protective substance be called vitamin C until its chemical structure was established.

Working at the Lister Institute, Sylvester Zilva began to prepare concentrated extracts of the vitamin in 1918. Five years later, he introduced a highly potent concentrate. At the University of Pittsburgh, Charles King later obtained a more stable form of this by removing traces of heavy metals which catalysed oxidation. In the autumn of 1931, after four years of intensive investigations, King finally isolated pure crystals of the vitamin from lemon juice. Tests with these showed that a daily dose of half a milligram could prevent a guinea pig becoming scorbutic on a diet deficient in the vitamin. The crystals turned out to be very similar to an acidic carbohydrate isolated in Professor Hopkins' laboratory at Cambridge in 1928 from adrenal glands, cabbages, and oranges by Albert Szent-Györgyi, a Hungarian biochemist who had been awarded a Rockefeller Fellowship. Szent-Györgyi had been investigating the nature of oxidation-reduction processes in the adrenals. The possibility that his new compound, then thought to be a hexuronic acid, might be the antiscorbutic vitamin had apparently been ruled out by results obtained by Zilva, but King's successful isolation of the vitamin reopened the issue. Assisted by Joseph Sviberly of King's department, visiting the Institute of Medical Chemistry at Szeged on an American-Hungarian Exchange Fellowship, Professor Szent-Györgyi found that one milligram daily of his hexuronic acid protected guinea pigs against scurvy. King gained further confirmation of the identity of his vitamin with Szent-Györgyi's acid. In order to establish the nature of the vitamin, Szent-Györgyi initiated a collaborative programme with the University of Birmingham, a leading centre in the field of carbohydrate chemistry. It soon became evident that the vitamin could not be a hexuronic acid, and Szent-Györgyi and Walter Haworth then proposed that it should be known as ascorbic acid, a term that has met with pharmacopoeial approval. In

1937, Szent-Györgyi received the Nobel Prize for medicine or physiology for his work on the biochemical role of ascorbic acid, whilst Haworth shared the chemistry prize.

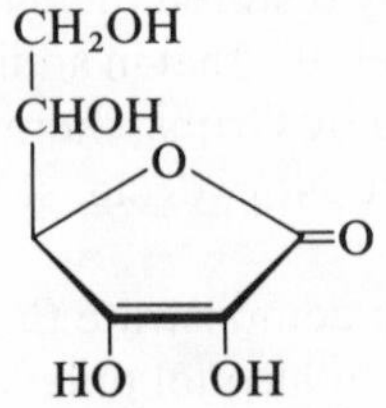

CH_2OH
CHOH
O
=O
HO OH

Ascorbic Acid

In a letter appearing in the *Journal of the Society of Chemistry and Industry* on the 10th March 1933, Edmund Hirst published the correct structure for ascorbic acid as determined by him and his colleagues at Birmingham. A race to synthesize the vitamin began, and on the 11th July Tadeus Reichstein of the Eidgenossische Technische Hochschule (ETH) in Zurich submitted a letter to the editor of the weekly British journal *Nature*, giving a preliminary account of his successful synthesis of ascorbic acid. It did not appear in print until over five weeks later, by which time a preliminary account of a synthesis by Haworth and Edmund Hirst had already appeared in the *Journal of the Society of Chemistry and Industry* on the 4th August, having been submitted only three days earlier! Reichstein, however, amassed a fortune from patent royalties after Hoffmann-La Roche began commercial production of synthetic ascorbic acid in 1934.

Until 1919, it was generally believed that there were only two water-soluble vitamins, namely vitamins B and C. Elmer McCollum at the University of Wisconsin had discovered the existence of a fat-soluble vitamin in 1913 whilst feeding rats on artifical diets (see page 241). Continuing these nutritional studies, he then found that if the commercially supplied lactose (milk sugar) that he had been using was purified by recrystallization, the rats exhibited manifest evidence of a growth disorder due to a dietary deficiency that he could correct merely by supplementing the diet with the aqueous mother liquor from which the lactose had been crystallized. This led McCollum to conclude that a water-soluble vitamin existed, whereupon he then learned of the earlier studies by Eijkman and Grijns. After repeating their work with polished rice, he came to the conclusion that the antineuritic vitamin and his rat growth-promoting vitamin were identical.

In a review article that appeared in the *Journal of Biological Chemistry* in 1919, H.H. Mitchell of the University of Illinois questioned the assertion that the antineuritic vitamin and the rat growth vitamin were identical. All the evidence was circumstantial, so he claimed. During the next few years, experimental results tended to support Mitchell's contention, and researchers began to speak of the vitamin B complex.

In 1927, the British Committee on Accessory Food Factors distinguished between vitamin B_1, the antineuritic vitamin isolated the previous year, and the more heat-stable vitamin B_2. The following year, Harriet Chick and M. H. Roscoe of the Lister Institute developed an assay for vitamin B_2 activity. This involved measuring the effect of test material on young rats fed on a diet deficient in the vitamin B complex, but to which vitamin B_1 had been added. In the absence of any vitamin B_2 supplementation, the rats exhibited loss of hair from the eyelids, sealing of the eyelids by a sticky exudate, dermatitis, blood-stained urine, and stunted growth. Each assay took three or four weeks to complete. Tedious as this must have been, it was sufficient to encourage Professor Richard Kuhn of the Institute of Chemistry in the University of Heidelberg, and Theodore Wagner-Jauregg of the Kaiser-Wilhelm Institute for Medical Research to join with the paediatrician Professor Paul György, an émigré Hungarian, in his attempt to isolate pure vitamin B_2. Wagner-Jauregg noticed that all extracts that proved active by this assay procedure exhibited an intense yellow-green fluorescence, the intensity of which was proportional to potency. When attempts were first made to isolate the fluorescent material, the growth-promoting activity of the extracts deteriorated. It was then realized that other growth-promoting vitamins must have been present prior to refinement of the crude vitamin B_2, thus pointing the way to the discovery of further members of the vitamin B complex. The biological assay procedure had to be modified to allow for this, with the outcome that the yellow-green fluorescent material was isolated from spinach, kidney, and liver, proving identical in each case. The vitamin was crystallized at Heidelberg in 1933 and named riboflavine (American workers at one time referred to it as vitamin G; confusingly, the term vitamin B_2 has been retained despite the fact that this was originally applied to crude preparations containing several members of the vitamin B complex). Two years later, Richard Kuhn at Heidelberg and Paul Karrer of Zurich University almost simultaneously synthesized the vitamin. The latter's process was adapted by Hoffmann-La Roche for commercial production. In 1937, Karrer was awarded a Nobel Prize for chemistry, shared with Walter Haworth, in recognition of his work in vitamin research. The next year, Kuhn was similarly honoured, but the Nazi government of Germany made him decline the award.

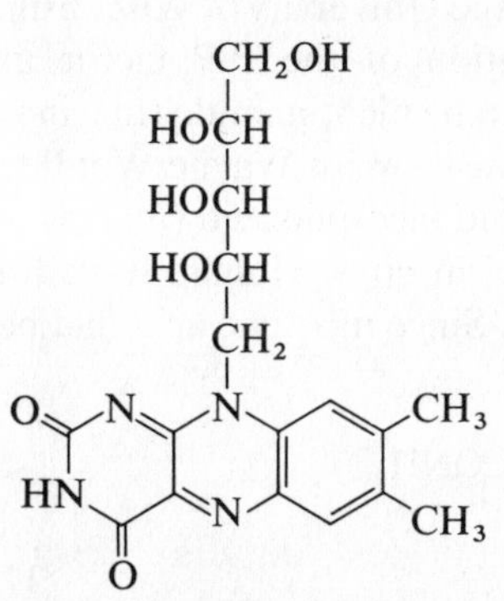

CH_2OH
HOCH
HOCH
HOCH
CH_2
O N N CH_3
HN N CH_3
O

Riboflavine

When the distinction between vitamin B_1 and the crude vitamin B_2 had first been made, it was generally assumed that the latter was the pellagra-preventing vitamin (P-P factor). Pellagra was first described in the early eighteenth century by Gaspar Casal of Spain. He attributed it to consumption of diets rich in maize. The disease was given its name in 1771 by the Italian physician Frapolli on account of the characteristic skin changes (*pelle* = skin, *agra* = dry) that it caused. These were in addition to gastrointestinal disturbances and degenerative changes in the central nervous system that ultimately lead to insanity. It was not until an epidemic swept through the southern United States in the early years of this century that the cause of pellegra was subjected to experimental scrutiny. An extensive series of clinical and epidemiological studies was initiated in 1914 by a team from the US Public Health Service, led by J. Goldberger. The initial conclusion was that diets rich in maize were to blame, this being consistent with the earlier demonstration by Edith Willcock and Gowland Hopkins that young mice failed to grow on diets in which zein from maize was the sole source of protein, zein being deficient in tryptophan. However, in 1920, Professor Carl Voegtlin and his colleagues from the pharmacology division of the Public Health Service discovered that pellagra could be cured by administration of dried yeast or aqueous extracts of yeast, these preparations being known to be a rich source of vitamin B. Further experiments indicated that the active material in the yeast was not destroyed by heating the yeast at 125°C. Since Voegtlin had previously shown that vitamin B_1 had no beneficial value in pellagra, Goldberger and his colleagues concluded that vitamin B_2 must be the pellagra-preventing factor. After vitamin B_2 was found to be a complex mixture, and riboflavine to have no pellagra-preventing activity, the search for the true P-P factor was intensified. This was greatly facilitated through the earlier recognition by T. N. Spencer, a veterinarian from Concord in North Carolina, that a disease of dogs known as 'black tongue' was the canine counterpart of human pellagra. Goldenberger then developed an assay for P-P factor based on prevention of 'black tongue' in dogs. He was able to demonstrate that liver was one of the richest sources of P-P factor. Professor Conrad Elvehjem and his associates in the department of agricultural chemistry at the University of Wisconsin, Madison, used fresh liver in their successful isolation of the P-P factor in 1937. The vitamin was immediately recognised to be nicotinamide (niacinamide), a substance already being studied by biochemists. With Wayne Woolley, Elvehjem demonstrated that both nicotinamide and nicotinic acid (niacin) were capable of preventing and curing 'black tongue' in dogs. Human trials followed at once, and these proved highly successful. Since nicotinic acid had been synthesized by Laden-

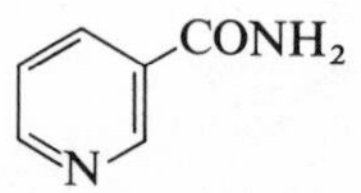

$CONH_2$ N

Nicotinamide

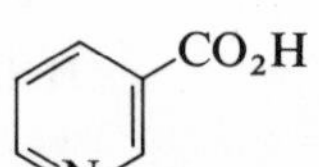

CO_2H N

Nicotinic Acid

burg forty years earlier, there was no problem in producing large amounts for the treatment of pellegra. The amide was readily prepared from the acid.

The isolation of further members of the vitamin B complex rapidly followed that of nicotinamide. The first of these was discovered as a consequence of studies on young animals deliberately deprived of the B group of vitamins other than those already known. The principal difficulty facing the researchers was that of unravelling the complex pathological changes arising from deficiency of unidentified vitamins. It was whilst working at Cambridge University in 1934 that Paul György suggested one such unidentified vitamin could protect rats from a specific type of skin lesion. He proposed the name vitamin B_6 for this rat antidermatitis factor, which he and T.W. Birch did much to characterize chemically. Early in 1938, Samuel Lepkovsky of the College of Agriculture at the University of California, Berkeley, informed György that he and John Keresztesy of Merck and Company were each independently about to submit papers describing the crystallization of the vitamin. This magnanimous gesture enabled György, now at Western Reserve University in Cleveland, to publish his own account of the crystallization shortly after.

Within a year of its isolation, both Karl Folkers of Merck and Company and also Richard Kuhn at Heidelberg had established the chemical structure of vitamin B_6. György proposed that it henceforth be known as pyridoxine. It should, in passing, be mentioned that following American government indications that it favoured the supplementation of foods and cereals with vitamins, Merck and Company had invested heavily in equipment to separate the B vitamins from natural sources such as yeast. The success of their own and rival chemists rapidly rendered this obsolete.

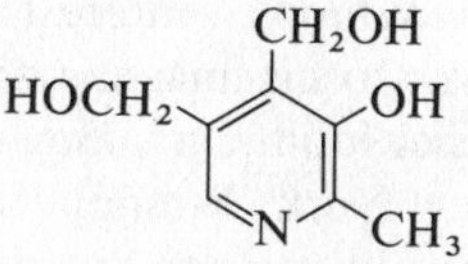

CH_2OH
$HOCH_2$ OH
N CH_3

Pyridoxine

In 1939, Thomas Jukes, a colleague of Lepkovsky at the University of California, and also Woolley, Elvehjem, and Harry Waisman at the University of Wisconsin, simultaneously discovered that pantothenic acid was the hitherto unidentified vitamin whose deficiency had been shown by Jukes' colleague Agnes Morgan to cause dermatitis in chickens. This had been isolated the previous year by Roger Williams at the University of Oregon during investigations into nutrients essential for the growth and replication of cultured yeast cells. It had taken him four years to isolate and purify this new yeast growth factor after differentiating traces of it from other essential nutrients present in food extracts. Williams pioneering studies on yeast growth factors began in 1919 when he tried unsuccessfully to develop a new type of assay for vitamin B_1 activity. Nevertheless, his work stimulated microbiologists to isolate growth factors for other organisms, including bacteria. In 1931, a Canadian scientist, W. L. Miller, detected the presence of two yeast growth factors in malt, namely Bios I and II. He identified the former as inositol, a sugar long known to be present in muscle; five years later, Kogl and Tonnis isolated crystals believed to be Bios II from boiled duck egg yolks. They named these biotin. In 1940, Gyorgy demonstrated that it was identical to a substance he had recently described as vitamin H, whose deficiency caused skin and hair disorders in rats. Vincent du Vigneaud of Cornell University determined the chemical structure of biotin in 1942, and it was synthesized a year later by Folkers and his colleagues of Merck and Company.

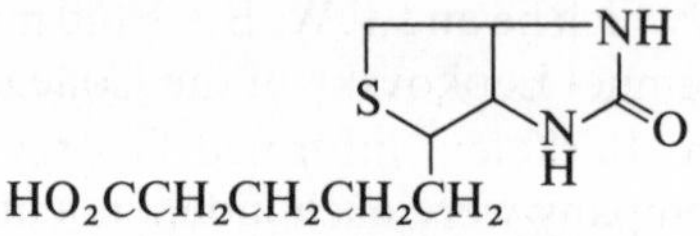

NH
S N O
H
$HO_2CCH_2CH_2CH_2CH_2$

Biotin

The discovery of vitamin B_{12} came about as a consequence of fundamental studies into bile pigment metabolism and its relation to liver disease and damage. These were initiated in 1914 in San Francisco at the University of California Medical School by Professor George Whipple, assisted by C. W. Hooper. They soon found it necessary to extend their investigations to cover the rate of formation of haemoglobin, the pigment in red blood cells, from which the bile pigments were derived. They did this simply by draining blood from dogs and waiting to see how long it took for haemoglobin levels to return to their original level. When this unexpectedly revealed that diet influenced the rate of haemoglobin regeneration, Whipple's interest in liver disease led him to examine the effect of feeding liver to his anaemic dogs. It proved to have a more powerful effect than any other food. On taking up a new appointment at the University of Rochester, New York, Whipple refined his techniques to confirm the earlier studies. His results came to the attention of George Minot at Harvard University, a clinician who, in 1915, had begun to examine the influence of diet on patients with pernicious anaemia because some of its symptoms resembled those of beriberi and pellagra. Pernicious anaemia was at that time an incurable disease characterized by failure of normal red cell formation. Death was inevitable within a period of only a few years at most.

In 1924, Minot began to feed liver to a few patients with pernicious anaemia. Their condition improved, but the results were hardly conclusive. With assistance from William Murphy, a detailed investigation then began, forty-five patients receiving enormous daily doses of liver by mouth in a clinical trial that was completed in May 1926. The results were, to say the least, startling. Many patients showed obvious signs of improvement within a week, their red blood cell count being restored to satisfactory levels within two months. For their outstanding contribution, Whipple, Minot and Murphy received the Nobel Prize for physiology or medicine in 1934.

Eating as much as half a kilogram of liver each day was a daunting prospect for anyone, let alone a sick patient. To overcome this, Edwin Cohn of the Harvard Medical School prepared an extract that was marketed by Eli Lilly and Company in 1928. Two years later, a purer extract suitable for intramuscular injection was introduced by Lederle Laboratories. This was far more satisfactory since the real cause of pernicious anaemia was a defect in gut absorption processes. A single injection once every one to three weeks proved quite adequate. During the early forties it was realized that the loss of activity when liver extracts were decolorized with charcoal was due to adsorption of the active principle. This was ultimately turned to advantage when it was shown that under certain conditions the active material could be eluted from charcoal to give a much purer preparation. This paved the way for eventual isolation of crystals of the active principle in 1948 by Lester Smith of Glaxo Laboratories in the United Kingdom, and also Karl Folkers and his colleagues of Merck and Company in the United States. Later in the year, the Merck researchers also isolated the vitamin from a strain of *Streptomyces griseus* used in streptomycin production. This meant that a cheap source had been found, and commercial production began in 1949. As vitamin B_{12} turned out to be a cobalt-containing molecule it received the approved name of cyanocobalamin. Its chemical structure was elucidated in 1955 through the collaboration of chemists from Cambridge University, led by Professor Alexander Todd, with X-ray crystallographers from Oxford University, led by Dorothy Hodgkin, and a team from Glaxo, led by Lester Smith. Todd was awarded the Nobel Prize for chemistry in 1957, and Hodgkin in 1964.

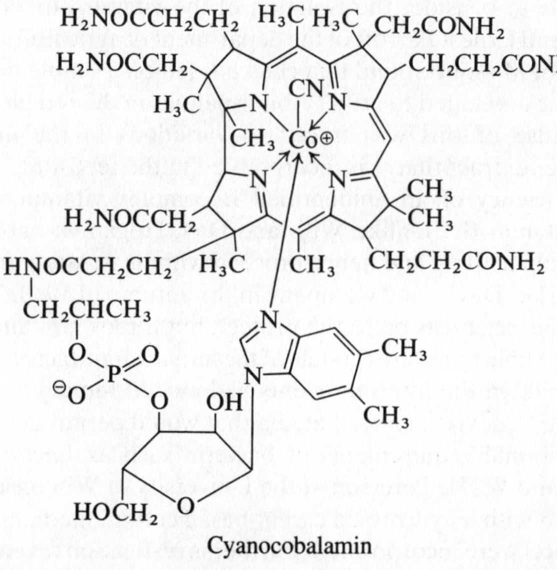

$H_2NOCCH_2CH_2$ H_3C H_3C CH_2CONH_2
H_2NOCCH_2 $CH_2CH_2CONH_2$
CN
H_3C N N
CH_3 $Co^{\oplus}$
N N CH_3
H_2NOCCH_2 CH_3
$HNOCCH_2CH_2$ H_3C CH_3 $CH_2CH_2CONH_2$
CH_2CHCH_3
N
O O CH_3
P N
$^{\ominus}O$ O OH CH_3
$HOCH_2$ O

Cyanocobalamin

In 1930, Lucy Wills and S. N. Talpade of the Haffkine Institute in Bombay reported in the *Indian Journal of Medical Research* that undernourished mothers of premature babies were frequently found to have been consuming diets that were deficient in the vitamin B complex. It was suggested this might even account for the clinical manifestation of pernicious anaemia-like symptoms during pregnancy. Wills went on to study women in Bombay who had developed a form of anaemia resembling pernicious anaemia except for the absence of neurological complications. She described this as 'tropical macrocytic anaemia' because of the presence of many large, immature blood cells. In contrast to pernicious anaemia, this macrocytic anaemia responded positively to treatment with a proprietary brand of yeast extract ('Marmite') that was rich in the vitamin B complex. When monkeys were fed on diets similar to those eaten by the anaemic patients, they too developed the disease. Administration of yeast extract or liver also cured the monkeys, but injections of liver extract normally used in treating pernicious anaemia proved worthless both in monkeys and humans suffering from the macrocytic anaemia. Evidently, the purification of the liver extract had removed a protective factor that was different from vitamin B_{12}.

The significance of Wills' results was not recognized at first. Similar observations on monkeys were reported in 1935 by Paul Day of Little Rock University, Arkansas, in the course of feeding experiments designed to produce cataracts from riboflavine deficiency. He blamed a dietary deficiency when his monkeys developed anaemia and died from complications. After Day had managed to correct the purported deficiency with either yeast supplements or a whole liver preparation, he proposed that the protective factor be called vitamin M. In the absence of a convenient assay system using small animals with a short life span, Day was unable to consider the isolation of the vitamin. In 1939, however, Albert Hogan and Ernest Parrott of the department of agricultural chemistry at the University of Missouri found that chickens fed on a simple diet sometimes became anaemic and failed to grow. Abnormalities in the red blood cells were noted. The cause of this was traced to variations in the quality of the commercial liver extract that was incorporated in the feedstuff. The evidence pointed to deficiency of an unidentified B complex vitamin which Hogan described as vitamin Bc. Unlike Wills and Day, Hogan was able to conduct assays for vitamin activity and hence proceed with its isolation from liver. He approached Parke, Davis, and Company in the autumn of 1940. The company put a team of five scientists on to the project, but it took two and a half years before they were able to isolate crystals of the antianaemic factor, which turned out to be an acid. In the interim, events had moved rapidly.

In attempting to devise artificial media that would permit determination of the exact nutritional requirements of bacteria such as *Lactobacillus casei*, Esmond Snell and W. H. Peterson at the University of Wisconsin found little bacterial growth with a hydrolysed casein-based culture medium, unless plant or animal extracts were incorporated. Further investigation revealed that yeast extract was the richest source of growth-promoting material, and in 1939 an active fraction was separated from this source by means of elution chromatography. Petersen, assisted by Brian Hutchings and Nestor Bohonos, went on to obtain a *Streptococcus lactis* growth-stimulating fraction from liver. Snell transferred from Wisconsin to work with Roger Williams, now at the University of Texas where, with the assistance of Herschell Mitchell, they isolated from spinach a concentrate of a *Lactobacillus casei* growth factor that they named folic acid (L. *folium* = leaf). This was investigated by Professors Elvehjem and Hart and their colleagues at Wisconsin, who found it to be capable of preventing anaemia in chickens. It seemed it must be the same as Hogan's antianaemic factor, vitamin Bc; the physical properties of the two substances were similar. Hogan confirmed that the substances had identical biological properties.

In 1938, Robert Stockstad and P. D. V. Manning of the California-based Western Condensing Company were involved in formulating a diet that would be suitable for assaying riboflavine on chickens, when they came to the conclusion that an unknown dietary growth factor existed. Tentatively, they described it as the U factor. In 1941, Stockstad was recruited by Lederle Laboratories to work on liver extracts at their Pearl River research centre. Two years later, he isolated crystals of a *Lactobacillus casei* growth factor from one and a half tons of liver. These proved to be identical to the vitamin Bc that had just been described by Hogan and the Parke, Davis researchers. In August 1945, the synthesis of folic acid was completed by a team made up of eight scientists from the Lederle Laboratories at Pearl River and a similar number from the laboratories at Bound Brook, New Jersey, of American Cyanamid, the owners of Lederle since 1930.

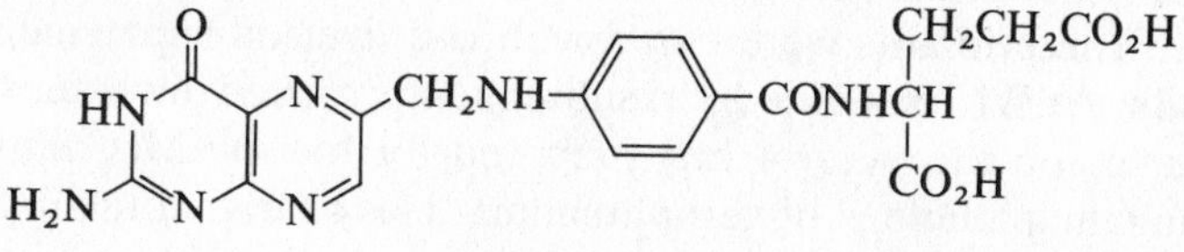

Folic Acid

FAT-SOLUBLE VITAMINS

In the course of examining the effect on rat growth of varying the mineral content of artificial diets, Elmer McCollum and his assistant Marguerite Davis of the department of agricultural chemistry at the University of Wisconsin noted that normal growth patterns could be maintained for only seventy to one hundred and twenty days. However, when natural diets were reintroduced, normal growth was restored. Many experiments had to be carried out before suspicion fell on the nature of the fat content of the artificial diet. To confirm that a fat-soluble accessory factor was present in only certain foods, McCollum and Davis supplemented the deficient diet with ether extracts of fat-containing foods. This proved that the factor was to be found in butter fat and egg yolk, but not in lard. When they reported their findings in 1913, much surprise was engendered in nutritional circles as it had universally been believed that the role of fats in the diet was to produce energy, the qualitative differences between them being of no consequence. Later, when McCollum detected a water-soluble accessory food factor in milk, he named the two factors he had discovered as 'fat soluble A' and 'water soluble B'. These terms were changed in 1920 by Drummond (see page 233) to vitamin A and B respectively, the latter resembling the antineuritic vitamin first discovered by Eijkman.

McCollum's findings were immediately pursued at Yale University by two of the leading American nutritionalists, Thomas Osborne and Lafayette Mendel. They noticed that in animals fed on a diet deficient in the fat-soluble factor, a characteristic eye disease occurred. It had been observed in malnourished animals by Magendie a century before, but neither he nor other investigators had considered it of any particular significance. Once the association with vitamin deficiency became evident, attitudes quickly changed as researchers realized that many clinical reports had associated eye disorders with nutritional factors. Of particular relevance was a report published in 1904 by M. Mori, a Japanese ophthalmologist. This described an eye disease characterized by dryness of the conjunctiva (xerophthalmia), frequently seen amongst infants fed on cereals and beans, but never found amongst the children of fishermen. Mori stated that the disease was due to lack of fat in the diet and could rapidly be cured by administering cod liver oil. Osborne and Mendel were able to demonstrate that both butterfat and cod liver oil could alleviate the ophthalmic disorder in their experimental animals. Not long after, an outbreak of serious eye disease, sometimes blinding, occurred amongst Dutch children fed on fat-free skimmed milk because of wartime measures to ensure increased export of butter. These infants were cured with cod liver oil supplements and full cream milk. At Wisconsin, S. Mori subsequently carried out extensive microscopic studies on the eyes of rats prepared for him by McCollum, thereby elucidating the pathology of xerophthalmia. He found that the dryness of the eyes was due to changes (keratinization) in the cells lining the tear glands. As a result, the tears were unable to exercise their protective role against bacteria, and infection of the inner surfaces of the eyelids ensued. In severe cases, the infection spread into the eye, causing ulceration of the cornea. The interest aroused by the discovery of the relationship between eye disease and vitamin A deficiency drew attention to old reports of night-blindness being cured by the eating of liver. Biochemists eventually established that the vitamin was converted to the pigment in the retina known as visual purple (rhodopsin).

In 1924, Professor Jack Drummond at University College, London, developed a steam distillation process to separate vitamin A from other unchanged fats remaining in cod liver oil after boiling in alcoholic potassium hydroxide (to saponify biologically inactive fats). The following year, he and Otto Rosenheim exploited their discovery that isolation of the vitamin could be greatly facilitated by measuring the intensity of the purplish colour it produced on reacting with arsenic trichloride. In collaboration with Isidor Heilbron and his associates at the University of Liverpool, Drummond made further improvements by developing a high vacuum distillation technique that ultimately yielded almost pure vitamin. In 1929, after it was discovered that the livers of other types of fish were often richer sources of vitamin A than was cod liver, Abbott Laboratories and Parke, Davis, and Company jointly began to process halibut liver oil for its vitamin content. The resulting preparation, though of high potency, was not particularly palatable on account of its strong fishy smell.

In 1931, Paul Karrer at Zurich University introduced adsorption chromatography to isolate a viscous yellow oil consisting of almost pure vitamin A. With this, he determined the chemical structure, reporting it two years later. However, it was not until 1937 that pure vitamin A (retinol) was crystallized by Harry Holmes and Ruth Corbet of Oberlin College, Pennsylvania, using fractional freezing and cold filtration. In 1947, Otto Isler of Hoffmann-La Roche introduced a commercial synthesis of the vitamin, as a consequence of which fish liver oil extraction processes are no longer in use.

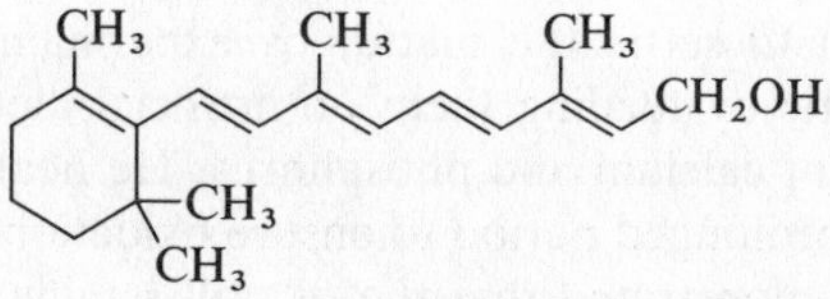

CH_3 CH_3 CH_3
CH_2OH
CH_3
CH_3

Retinol

In 1912, Professor Gowland Hopkins suggested that rickets might be yet another of the diseases caused by deficiency of an accessory food factor. Outwardly, rickets (rachitis) was characterized by deformity of the limbs of infants arising from failure of calcium phosphate to be deposited at the growing ends of their bones. Unchecked, the disease not infrequently involved the central nervous system, which could be fatal. Although known for centuries, rickets reached epidemic proportions early this century in the industrial cities of northern Europe and America. This spurred Gowland Hopkins to recommend to the newly formed Medical Research Committee that it should designate rickets as a subject for special study. He recommended a research enquiry should be undertaken by his former student Edward Mellanby, now lecturing in London University at King's College for Women. The Committee agreed, and Mellanby began work in 1914. Travelling between London and Cambridge, where he had access to a colony of puppies in the Field Laboratories, he painstakingly conducted hundreds of feeding experiments in an attempt to identify the type of diet that induced rickets. In 1918, he was able to inform the Physiological Society that he could produce rickets in puppies by feeding them for three or four months on either a diet of milk, rice, oatmeal, and salt, or on milk and bread. By adding a variety of foods to these rachitic diets, Mellanby was able to confirm that animal fats such as butter, suet, and cod liver oil had antirachitic activity. The latter was a northern European folk remedy that became esteemed as a tonic in the late eighteenth century, since when it was widely employed in the palliation of debilitating diseases such as tuberculosis and rheumatism. Its use in rickets was given prominence in 1861 by the Parisian physician A. Trousseau in his textbook of clinical medicine. However, it was not until Mellanby offered experimental proof of the value of cod liver oil that any significant reduction in the incidence of the disease was recorded. By the early 1930s the disease was no longer seen in London.

Mellanby was inclined to believe that the antirachitic vitamin and vitamin A were identical, although he recognized that the evidence was not altogether conclusive. In an attempt to settle the issue, he took advantage of Gowland Hopkins' new observation that vitamin A activity of hot butterfat was destroyed by bubbling oxygen through it. After treating both butterfat and cod liver oil in this manner, Mellanby found the latter retained antirachitic activity. He was undecided as to whether this proved the existence of a second fat-soluble vitamin, or merely reflected the presence either of a larger initial amount of vitamin A in the cod liver oil, or else of an antioxidant. Elmer McCollum, who had moved from Wisconsin to Johns Hopkins University, immediately set out to settle the matter by experimenting on rats he had already made rachitic by feeding them on artificial diets containing an unfavourable balance of calcium and phosphorus. He heated cod liver oil in a current of air for a prolonged period to ensure oxidation of all the vitamin A present. He then demonstrated that the oil still retained its protective antirachitic action. His results were published in 1922; they proved conclusively the non-identity of vitamin A and the antirachitic factor, described in 1925 as vitamin D since it was the fourth one to have been detected. Shortly after McCollum's announcement of the new vitamin, T. F. Zucker of Columbia University introduced a process which achieved a thousand-fold concentration of it by treatment of cod liver oil. This was used commercially by pharmaceutical manufacturers.

McCollum's demonstration that vitamin D deficiency was the cause of rickets did not settle one outstanding matter. In 1919, a Berlin physician, K. Huldschinsky, had cured rickets in children by exposing them to ultraviolet light emitted from a mercury vapour lamp. His results were corroborated the following year in Vienna by Harriet Chick's group of lady doctors and scientists who, at the end of the war, had been sent from the Lister Institute to assist during a severe epidemic of rickets that affected four out of every five infants in the city. They found the disease did not develop in children exposed to adequate sunlight. Further confirmation came from New York, where A. F. Hess and L. J. Unger of the College of Physicians and Surgeons at Columbia University cured rachitic infants by exposing them to sunlight or untraviolet radiation. Hess suggested that the antirachitic principle might be formed by the action of ultraviolet light on a putative provitamin. He went on to make the surprising discovery, announced in June 1924, that irradiation of certain foods could confer antirachitic properties on them. Before his report appeared in print in October of that year, Harry Steenbock of the University of Wisconsin published similar findings. He took out patents to cover the processing of food by ultraviolet light, assigning these to a body established in 1925 to enable the vast sums earned from license fees to be used in support of research in Wisconsin. This was the Wisconsin Alumni Research Foundation, which earned more than fourteen million dollars from Steenbock's patents during the next twenty years.

Hess and workers in several other laboratories soon established that the substance converted into vitamin D when vegetable oils were irradiated was to be found amongst the plant sterol fraction. He then went to Göttingen to work on the isolation of provitamin D under the guidance of Adolf Windaus. In 1927, Hess and Windaus, with the assistance of the Göttingen physicist R. Pohl, established that the provitamin D was a known substance, namely ergosterol. The following year, Windaus was awarded the Nobel Prize for chemistry in recognition of this and his earlier work on sterols. With the assistance of the Elberfield laboratories of I. G. Farbenindustrie, Windaus isolated the product formed by irradiation of ergosterol in 1932. Windaus named it vitamin D_2, thereby distinguishing it from what he had previously thought was the pure vitamin, namely its complex with lumisterol (a precursor also formed by irradiation of ergosterol). Windaus renamed this complex, calling it vitamin D_1. The chemical structure of vitamin D_2 was established in 1936 by Windaus at Göttingen and by Isidor Heilbron and Frank Spring at the University of Manchester.

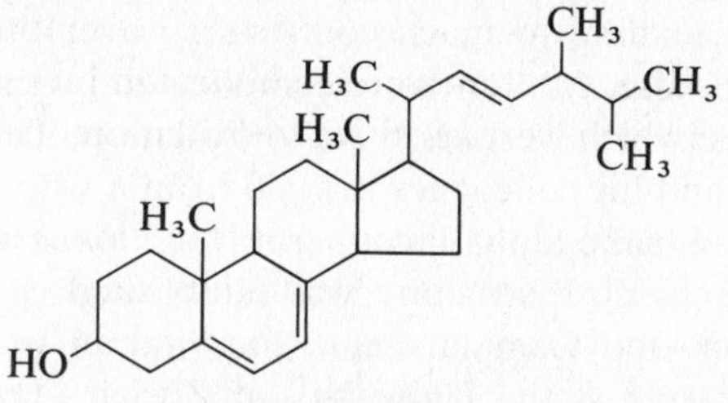

Ergosterol

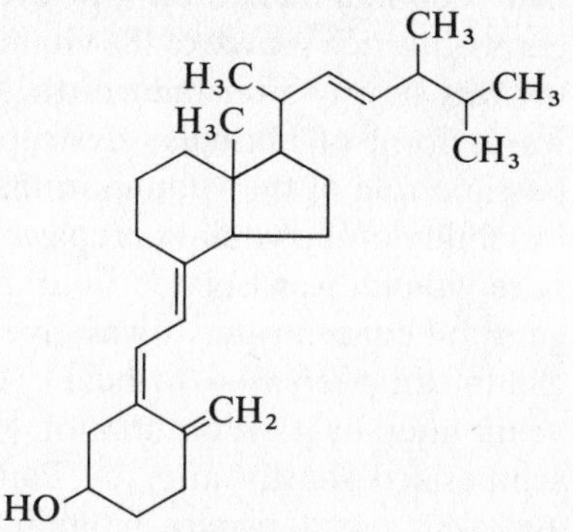

Vitamin D_2

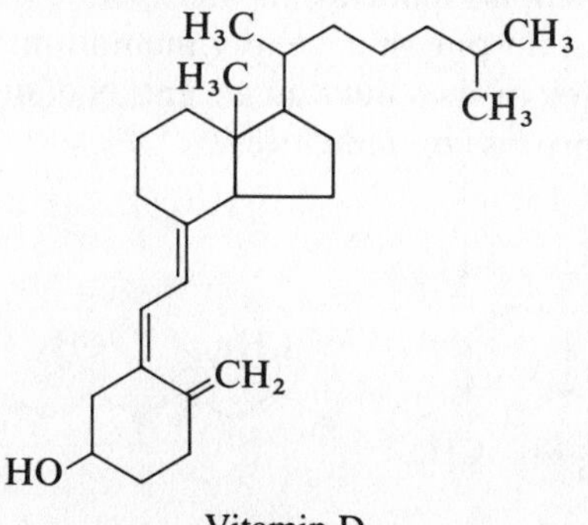

Vitamin D_3

A vitamin D_2 complex was also isolated in 1932 by Askew and his colleagues at the National Institute for Medical Research, London. At that time, this was believed to be homogeneous, and was mistakenly assumed to be identical to Windaus' vitamin D_2. It was given the name calciferol, a term which has been retained in the *British Pharmocopoeia* to describe vitamin D_2, but the *US Pharmacopeia* prefers the term ergocalciferol. Irradiations of other sterols was also found to generate antirachitic products, such as vitamin D_3, cholecalciferol, which was formed from 7-dehydrocholesterol. Windaus and Bock found this sterol present in skin, thereby solving the mystery of how exposure to sunlight could prevent or cure rickets. Vitamin D_3 is the product formed on irradiation of foods from animal sources. 'Vitamin D' is now used as a generic term to describe any substance that can be converted in the body into the active antirachitic metabolite 1,25-dihydroxycholecalciferol.

In 1922, Herbert Evans and K. S. Bishop of the University of California, San Francisco, announced that normal pregnancies did not occur in rats kept for long periods on an artificial diet supplemented with all known vitamins. Few offspring were produced as most foetuses were resorbed a few days after conception. Evans suggested that this arose from deficiency of a substance that became known as vitamin E. Much interest was aroused six years later when Evans and Burr discovered that paralysis occurred in young rats whose mothers had been maintained on low levels of the vitamin during pregnancy. Wheat germ oil, a rich source of the vitamin, could cure the paralysis if administered to the rats shortly after their birth. Other workers later suggested this paralysis was a form of muscular dystrophy, leading to much controversy over the possible role of the vitamin in that disease. Matters were complicated by the instability of the vitamin preparations, which were sensitive to oxidation. The pure vitamin was isolated by Evans and his colleagues in 1936 from a wheat germ oil concentrate. It was given the name alpha-tocopherol (Gr., *tokos* = childbirth, *pherein* = to bear). The chemical structure was established two years later by E. Fernholz of Merck and Company, and the vitamin was synthesized shortly after by Paul Karrer at the University of Zurich. The availability of the pure vitamin from natural or synthetic sources enabled researchers to establish whether it had any role in human nutrition or therapeutics. None of the many claims made for its therapeutic value in diverse diseases has been substantiated, despite much experimentation. The main importance of alpha-tocopherol lies in its safety as an antioxidant for use by the pharmaceutical and food processing industries.

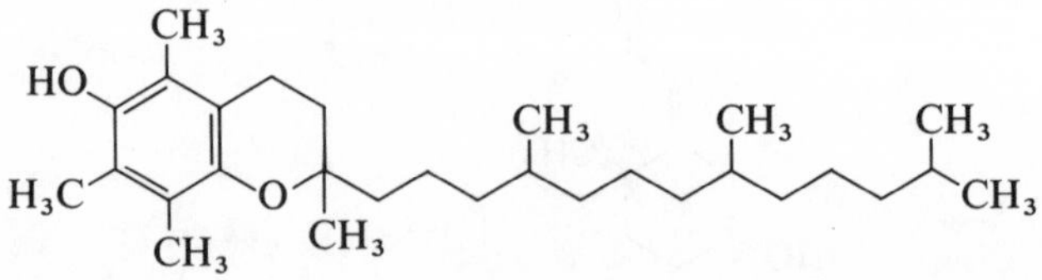

Alpha-Tocopherol

In 1929, Professor Henrik Dam of Copenhagen University carried out a series of experiments to establish whether chickens could synthesize cholesterol, some doubts having existed about this. Dam was able to confirm that cholesterol was indeed synthesized, but in the course of proving this he discovered that his chickens began to haemorrhage after two or three weeks on a fat-free diet supplemented with the known fat-soluble vitamins. Samples of their blood showed delayed coagulation. Dam doubted that this could be a form of scurvy since chickens were already known not to require vitamin C. Nevertheless, he added lemon juice to their diet, but to no avail. Only large amounts of cereals and seeds in the diet afforded protection. In 1934, he reported the existence of a new accessory food factor, then went on to show, in the following year, that this was fat-soluble, but different from vitamins A, D, or E. He described it as vitamin K since it was required for blood coagulation ('Koagulation' in German). A substance with similar activity was reported shortly after by Professor H. J. Almquist and Robert Stockstad of the University of California, Berkeley. They had discovered that alfalfa meal contained a factor which protected chickens against a scurvy-like haemorrhagic disease induced by being fed on diets in which the source of protein was sardine meal. Almquist was able to demonstrate that meat meal was satisfactory because its slower processing allowed bacterial production of an antihaemorrhagic factor. After this had finally been isolated, a report submitted to the American journal *Science* was rejected. It was belatedly sent to *Nature* where it appeared a few weeks after Dam's paper had been published. Dam proceeded to seek the assistance of Professor Karrer in establishing the chemical nature of vitamin K. They isolated the pure vitamin from alfalfa as an oil in 1939 at Zurich. Almost simultaneously, Edward Doisy of the St Louis University School of Medicine, Missouri, also isolated both this and a closely related antihaemorrhagic factor from putrefied fish meal. He named the alfalfa factor vitamin K_1 and that formed by bacteria as vitamin K_2. In 1939, Doisy determined the chemical structure of vitamin K_1, and from it also that of vitamin K_2. This was promptly confirmed by its synthesis at Harvard University by Louis Fieser. Dam and Doisy shared the Nobel Prize for physiology or medicine in 1943 for their work on the vitamin. Vitamin K_1 received the approved name of phytomenadione.

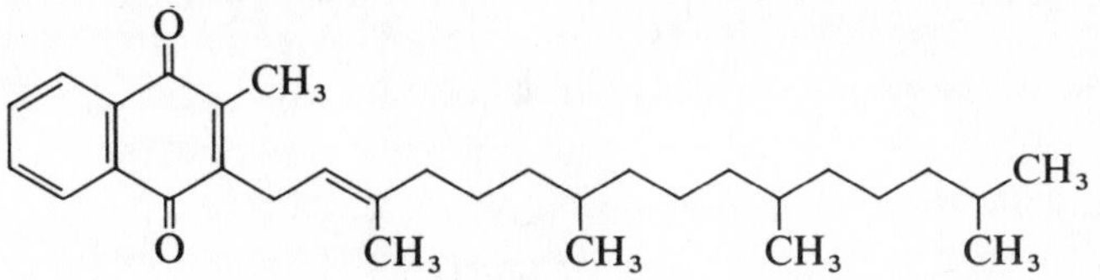

Phytomenadione

XIII

Antiprotozoal Drugs

The first attempt to stain microbes so that they could be readily examined under the microscope was that made in 1869 by Hermann Hoffmann, the professor of botany at Geissen, who used aqueous solutions of two natural dyes, carmine and fuchsin. Carl Weigert, an assistant to Professor Wilhelm Waldeyer at the Institute of Pathology in the University of Breslau, adapted the procedure to meet pathological requirements by employing the botanical microtome to slice infected tissues so thinly that these could be stained and viewed under a microscope. He also devised a system for the differential staining of these tissue slices with dyes that selectively coloured various cellular structures. His first opportunity to exploit his new technique arose when a smallpox epidemic broke out in Breslau in 1871. Weigert had to carry out many post mortem examinations at the All Saints' Hospital, in the course of which he stained serial slices of smallpox pustules obtained from the skin of over two hundred corpses. His method was to use a combination of haematoxylin, picric acid, and carmine. The latter coloured the bacteria selectively, for the first time revealing their presence in association with an infectious disease. By 1875, Weigert was staining various types of bacteria with methyl violet, the first aniline dye to have been manufactured. It had been synthesized ten years earlier and popularized at the Paris Exhibition of 1867, when a 150 kg block was put on display before the public! This application of a synthetic dye by Weigert transformed bacteriological staining into a powerful technique, especially in the hands of Robert Koch, who met Weigert the following year during his visit to Breslau to announce his elucidation of the complete life-cycle of the anthrax bacillus. Meanwhile, Weigert's work had also caught the imagination of his cousin, Paul Ehrlich, nine years his junior.

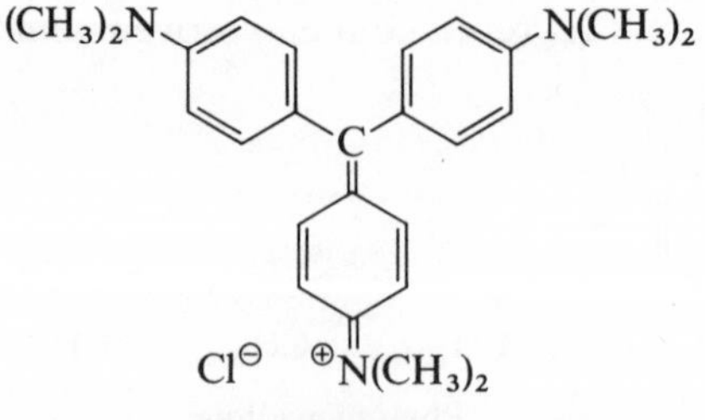

$(CH_3)_2N$ $N(CH_3)_2$
C
$Cl^{\ominus}$ $^{\oplus}N(CH_3)_2$

Methyl Violet

Ehrlich entered the University of Breslau in 1872, at the age of eighteen. He was unsettled in those early days until Weigert persuaded him to study medicine. When Professor Waldeyer moved from Breslau to the new University of Strassburg, Ehrlich went with him. As Ehrlich's tutor, the great anatomist both welcomed him into his home and encouraged him to carry on with the microscopic investigations on which he spent so much time. Ehrlich thrived on this opportunity to conduct original research on his own. Stimulated by reading Huebel's book on lead poisoning, which explained that chemical analysis of various organs had established that lead concentrated in the brain, he went on to examine slides of brain tissue under the microscope in the hope of establishing in which cells the lead was stored. This proved to be a futile exercise that severely disrupted his academic studies. Nevertheless, it forced Ehrlich to change his approach. Instead of lead salts, he injected dyes that could easily be detected microscopically in cellular components where they concentrated. In 1874, having completed his pre-clinical studies, Ehrlich returned to Breslau, where he avoided all unnecessary clinical involvement. He preferred to study in the laboratory of Julius Cohnheim, the pathologist with whom Weigert now worked, and who did so much to encourage both him, Ehrlich and Robert Koch in their endeavours. When Cohnheim first introduced Koch to Ehrlich, he is reputed to have said, 'This is little Ehrlich. He is a very good stainer, but he won't pass his examinations.' That Ehrlich did pass despite missing so many clinical sessions, says much for the flexibility of the examination system at that time! The final part of Ehrlich's medical studies was completed at Leipzig. His doctoral dissertation, submitted in 1878, was entitled, 'Contributions to the Theory of Histological Staining'. It was highly critical of histologists for failing to base their work on a theoretical understanding of how dyes bind to tissue components.

On taking up his first post, in Professor von Friedrichs' medical clinic at the Charitié Hospital in Berlin, Ehrlich was permitted to spend most of his time on histological studies, especially in the field of haematology. He continued to strive for an understanding of the factors influencing the uptake of dyes by cells, and soon came to the conclusion that the size of the dye molecule was a critical factor. He became disillusioned with the limitations imposed by examination of tissues exposed to dyes only after the death of the animal from which they were taken. By 1885, Ehrlich had developed a radical, new approach. This was, simply, to inject dyes into living animals and allow these to diffuse into the tissues before the animals were killed. For the first time, it became possible to examine the disposition of chemical substances in living animals, a process that Ehrlich called 'vital staining'.

The wide diversity of synthetic dyes produced by the German dyestuffs industry over the preceding two decades enabled Ehrlich to draw certain conclusions about the influence of chemical structure on the distribution in live animals of different types of molecules. These conclusions still influence attempts at drug design a complete century later. Foremost amongst them were his findings with regard to penetration of molecules into the brain.

Ehrlich observed the acidic dyes, of which he tested a great number possessing the sulphonic acid function used by dye manufacturers to enhance water solubility, were unable to penetrate into the brain or adipose (fat) tissue. In marked contrast, many basic dyes readily stained tissues. To explain this,

Ehrlich argued that the alkalinity of the blood caused acidic dyes, but not basic ones, to exist as water-soluble salts. Ehrlich saw an analogy between, on the one hand, the transfer of basic dyes from the blood to the brain, and on the other, the extraction of fat-soluble basic alkaloids from alkaline solution into ether. Alkaloids could not be extracted into ether from acid solutions, where they existed in the form of their salts which were insoluble in ether. This remarkably accurate assessment of the situation was promulgated by Ehrlich during the years 1886 and 1887, and it enabled him to explain the cumulative toxicity of the antipyretic drug thalline as being due to its lipophilicity, i.e. its proclivity to dissolve in fat rather than water. Thus, the tendency of thalline to be held in the body for a long time was accounted for by its affinity for fat tissue. Ehrlich was able to demonstrate this by converting thalline residues in these tissues to a coloured derivative by oxidizing pieces of tissue with ferric chloride solution.

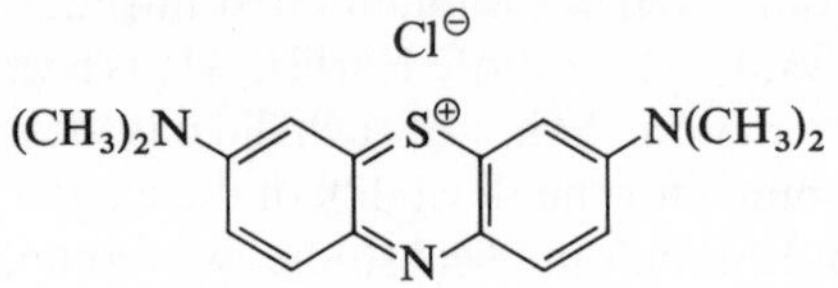

$Cl^{\ominus}$

$(CH_3)_2N$ — $S^{\oplus}$ — $N(CH_3)_2$ — N

Methylene Blue

Ehrlich was particularly interested in a basic dye known as methylene blue, which was synthesized by Baeyer and Caro in 1876. He first used it around 1880 to stain bacteria. By 1885, Ehrlich had found that it had a remarkable affinity for nerve fibres whilst leaving other tissues unaffected. He described it as being neurotropic, for upon injecting it into a living frog, all the nerve fibres were gradually tinted blue. But even more striking was the experiment Ehrlich conducted with it on a frog with a parasitic urinary infection. On excision of the frog's bladder, not only was the parasite still to be seen sucking blood from the frog tissue, but its nerve fibres were stained blue.

In 1888, Ehrlich reasoned that as methylene blue stained nerves it might possibly interfere with nervous transmission and thereby exert an analgesic action. He and Dr A. Leppmann gave it to many patients suffering from a variety of severe neuritic and arthritic conditions. They found the dye did relieve pain in such patients, a view later confirmed independently by at least one other physician. However, the tendency of methylene blue to damage the kidney if used continuously discouraged its use as an analgesic. Examination of Ehrlich's unpublished letters by Sir Henry Dale revealed that Ehrlich wrote to distinguished chemists in the dyestuffs industry asking advice about the possibility of obtaining analogues of methylene blue that would be more potent analgesics. In 1891, Ehrlich carried out further experiments with methylene blue after returning from Egypt, where he had gone to recuperate from tuberculosis. Knowing that this dye stained the plasmodia that caused malaria, and that it would be safe to administer to patients, he and Professor Guttmann administered daily five capsules each containing 100 mg of methylene blue to two patients who had been admitted to the Moabite Hospital in Berlin with malaria. Both recovered as a result of this treatment. Although it was later found to be ineffective against the more severe manifestations of the disease experienced in the tropics, this cure of a mild form of malaria represented the first instance of a synthetic drug being used with success against a specific disease.

Ehrlich did not actively pursue his work with methylene blue any further, for two reasons. Firstly, the inability to infect animals with malaria prevented testing of potential drugs in the laboratory, an unavoidable prerequisite for the development of any chemotherapeutic agent for use in human or veterinary medicine. Secondly, he was then working in Robert Koch's Institute for Infectious Diseases, in Berlin, where his skills were fully deployed in transforming Emil von Behring's diphtheria antitoxin into a clinically effective preparation. This was later to result in his sharing a Nobel Prize for Medicine with Behring. In 1896, he became Director of the Prussian State Institute for the Investigation and Control of Sera. This was established at Steglitz, outside Berlin, to supervise the production of diphtheria antitoxin. Despite his responsibilities in this new post, Ehrlich still managed to continue with important immunological studies, the recognition of which led, three years later, to him being provided with a purpose-built, well-equipped research centre near the hospital complex in the south of Frankfurt. This was known as the Royal Prussian Institute for Experimental Therapy, and it was here that Ehrlich was to lay the foundations of modern chemotherapy.

In 1902, Alphonse Laveran and Felix Mesnil of the Pasteur Institute, announced that they had been able to infect laboratory mice and rats with two varieties of the causative organism of trypanosomiasis. Subcutaneous injections of the sodium salt of arsenious acid into the infected rodents resulted in a rapid disappearance of trypanosomes from their blood, but the parasites reappeared within a few days to cause the death of the animals. This announcement from Paris gave Paul Ehrlich his long-awaited opportunity to return once more to the field of chemotherapy. Trypanosomes were protozoa somewhat larger than red blood corpuscles. Several varieties had been shown to be responsible for what had previously been classified as eighty different tropical diseases affecting both man and animals. The most devastating of these forms of trypanosomiasis in humans was sleeping sickness, caused by *Trypanosoma gambiense*. In one recorded instance, the mortality had risen from 13% to 73% between the years 1896 and 1900. There were even fears that the disease could depopulate the whole of Central Africa, where it was being transmitted by the tse-tse fly. Attempts to prepare vaccines had proved futile.

In the autumn of 1902, Professor Edmond Nocard of the Pasteur Institute visited Frankfurt. He and Ehrlich established a warm friendship, and this resulted in cultures of trypanosomes being sent to Ehrlich from Paris. Ehrlich immediately began his researches on trypanosomiasis, in late December 1902, assisted by Dr Shiga who had been sent to Germany by the head of the Tokyo Institute for Infectious Diseases, Professor Kitasato. The professor had worked with Ehrlich at Robert Koch's Institute, and prior to this at the Pasteur Institute. On his return to Japan, he established the Kitasato Institute which was modelled on the European establishments.

Ehrlich's first experiments were with a new arsenic preparation called 'Atoxyl'. It gave negative results. He and Shiga then examined more than one hundred dyes which were injected into mice infected with either *T. equinum*, the organism that caused mal de Caderas in horses, or *T. brucei*, which caused nagana in cattle. The only dye to exhibit activity was one of the benzopurpurin series. Ehrlich called it Nagana Red. Like arsenious acid, it caused the disappearance of trypanosomes from the blood of the mice for a short time, the mice surviving five or six days instead of the usual three or four.

Ehrlich believed that the low solubility of Nagana Red was resulting in it being poorly absorbed into the circulation from the site of its injection under the skin. He therefore contacted his old friend Arthur Weinberg, who was director of the Cassella Dye Works in Mainkur, just to the east of Frankfurt. Leopold Cassella and Company employed around two hundred workers and was the oldest of several dye manufacturers in the Frankfurt area, having been founded in 1812 (it was soon to become affiliated with the larger Hoechst Dyeworks). Ehrlich asked Weinberg if he could supply a derivative of the benzopurpurin dye with an extra sulphonic acid function to increase water solubility. The chemist whom Weinberg entrusted with the responsibility for meeting Ehrlich's request was Louis Benda. This was the start of a fruitful collaboration between the three scientists, and it continued when Cassella merged with Hoechst in 1908.

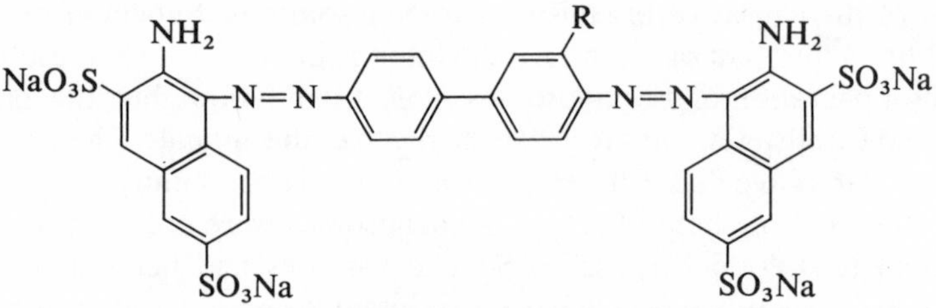

NH_2 R NH_2
NaO_3S —N=N— —N=N— SO_3Na
SO_3Na SO_3Na

Nagana Red, R = —H
Trypan Red, R = $-SO_3Na$

Nagana Red was one of many analogues of Congo Red, the first slightly water-soluble dye that could directly colour cottons. Congo Red was introduced by AGFA in 1884, and a variety of its derivatives were synthesized during the following two decades by the AGFA, Bayer, and Cassella companies. These azo dyes were prepared by condensing aniline derivatives with one of five different naphthalene sulphonic acids. The first of the benzopurpurin dyes was made by Carl Duisberg of Bayer, who synthesized it from a new naphthalene disulphonic acid known commercially as *R* acid. Nagana Red was an analogue of Duisberg's benzopurpurin. Benda was able to supply Ehrlich with another analogue that satisfied his requirement for a third sulphonic acid group to enhance solubility. This dye, first prepared in 1889, was soon to be called Trypan Red.

Ehrlich found that Trypan Red cured mice infected with *Trypanosoma equinum* but not with other strains of trypanosomes. Notwithstanding this, after the publication in 1904 of Ehrlich and Shiga's first paper on Trypan Red, the British Sleeping Sickness Commission requested supplies of the dye for trials in Uganda. The results were disappointing since doses high enough to be effective against the disease were found to be likely to cause blindness, and even death.

Ehrlich obtained about a further fifty derivatives of Trypan Red from Weinberg. The most active of these were more potent than Trypan Red, and were prepared from analogues of *R* acid featuring an extra hydroxyl or amino group on position-7 of its naphthalene ring. A supply of the 7-amino derivative of Trypan Red was tested in Africa during an expedition led by Robert Koch in 1906, but it proved to be no better than Trypan Red.

Further analogues of Trypan Red were examined by Maurice Nicolle and Felix Mesnil at the Pasteur Institute. They, too, obtained their dyes from a dyestuffs manufacturer, in this case Friedrich Bayer and Company. Under its energetic director Carl Duisberg, this company had expanded into pharmaceutical production in 1888, when his phenacetin challenged the phenazone produced by Hoechst. Bayer's 'Aspirin' had subsequently become the best selling medicinal product in the world. As the leading manufacturer of acidic azo dyes, it was hardly surprising that, on learning of Ehrlich's work with dyes supplied by Cassella, Bayer should collaborate with the Pasteur Institute to test their own extensive range of dyes against trypanosome infections. Heymann, who was director of the scientific laboratory at Elberfield, was so impressed by the work done in Paris that he asked Oskar Dressel and Richard Kothe to synthesize new analogues that would be more effective than existing dyes.

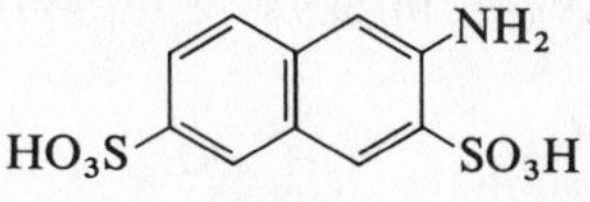

NH_2 HO_3S SO_3H

R Acid

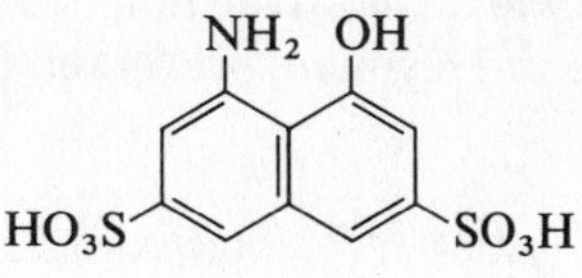

NH_2 OH HO_3S SO_3H

H Acid

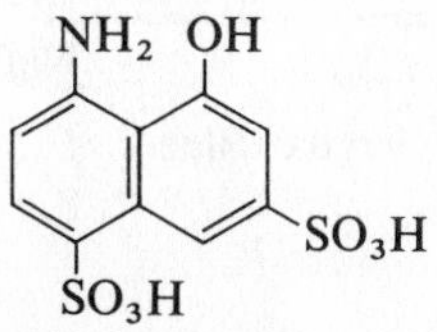

NH_2 OH SO_3H SO_3H

K Acid

Maurice Nicolle had, like Ehrlich, been interested in the physicochemical basis of the staining of tissue components by dyes. After Ehrlich and Shiga had demonstrated the efficacy of Trypan Red against the strains of trypanosomes provided by the Pasteur Institute, Nicolle examined the hundreds of dyes supplied by F. Bayer and Company. These differed from Ehrlich's dyes both in the central part of the molecule, and also in the incorporation of other naphthalene sulphonic acids, such as *K* or *H* acids, in place of *R* acid; Ehrlich had confined his efforts to the introduction of different substituents into the naphthalene ring of *R* acid. In June 1906, the Pasteur workers disclosed that a single injection of Trypan Blue could cause the disappearance of all trypanosomes from the blood. Further trials showed that it had a mild action against several different strains of trypanosomes which caused disease in cattle. On this basis, it was more effective than either 'Atoxyl' or Trypan Red, although still not acceptable for therapeutic use in humans. After the demonstration that it could cure piroplasmosis (babesia), in 1909, it was introduced into veterinary practice.

In 1905, Wilhelm Roehl moved from Ehrlich's laboratory in Frankfurt, where he had been testing the azo dyes provided by the Cassella Company, to join the Bayer research group at Elberfield. He found that none of the dyes prepared by the company for Nicolle and Mesnil was effective in his own infected mice. Nevertheless, he asked for colourless analogues to be made since he doubted that a drug that tinted the skin would ever be acceptable to patients. This provoked Dressel and Kothe to start by modifying the structure of the urea-based Afridol Violet, one of the least strongly coloured Bayer dyes screened by Nicolle and Mesnil. Although it had only feeble activity, Roehl eventually found that several red analogues of it prepared from yet another naphthalene sulphonic acid, *J* acid, exhibited stronger activity against trypanosomes. It was this advance which led to priority being given to the project in 1913. Better analogues were then developed, with the first patents being applied for just before the outbreak of the Great War. At this point there were arguments within the company over the widsom of continuing with a line of research which had still not delivered a truly outstanding drug. Only the insistence of Heymann prevented the dropping of the project. By the autumn

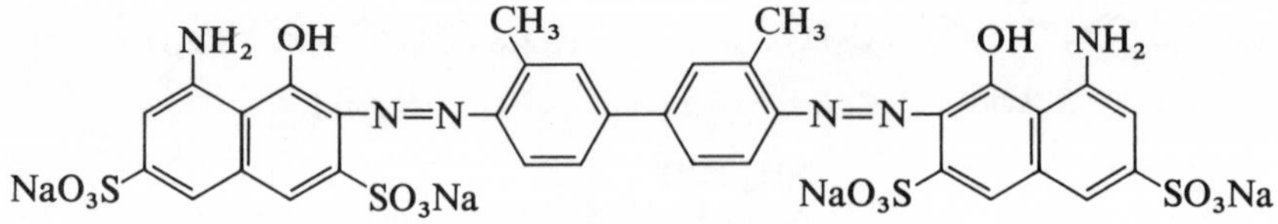

Trypan Blue

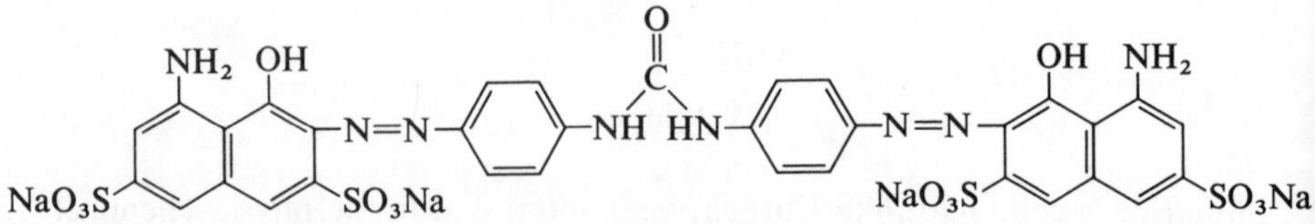

Afridol Violet

of 1917, more than one thousand naphthalene ureas had been synthesized and tested. Only then did the long sought agent emerge. This was a colourless compound that had remarkable antitrypanosomal activity both in experimental animals and in humans. It was later given the approved name of suramin, but became generally known by its code name, 'Bayer 205' (later, it was marketed under the proprietary brand name of 'Germanin'). Its chemical structure was not revealed because the Bayer Company feared this would enable foreign manufacturers to develop similar products.

The first reports of the discovery of the new trypanocide began to circulate outside the Elberfield laboratories towards the end of 1920. For a while the drug was only made available to German doctors and a few foreign investigators who undertook not to allow it to pass into the hands of anyone capable of determining its chemical structure. This irked Ernest Fourneau, head of the medicinal chemistry laboratory at the Pasteur Institute since 1911. Originally trained in France as a pharmacist, Fourneau had gone on to study under leading German chemists, including Emil Fischer and Richard Willstatter. On returning to France, he was anxious to lessen the dependence of his country on drugs imported from Germany. With this conviction, and the on-going commitment of the Pasteur Institute to the chemotherapy of trypanosomiasis, he was determined to establish the chemical structure of 'Bayer 205' by one means or another. To this end, he conducted a critical examination of the seventeen relevant Bayer patents covering trypanocidal ureas derived from naphthalene sulphonic acids. Fourneau's dogged persistence enabled him to narrow the field down until it emerged that 'Bayer 205' must have been one of twenty-five possible structures. Several of these were synthesized by Fourneau, then tested on infected mice. One exhibited antitrypanosomal properties identical to those of 'Bayer 205'. The structure of this was published by Fourneau in 1924, after which it received the approved name of moranyl ('Fourneau 309'). Because the structure had never been previously published, F. Bayer and Company could not claim that its patents were being infringed. For this reason, disclosure of structures thereafter became standard practice in pharmaceutical patents. It was not until 1928 that the German company finally admitted that 'Bayer 205' was identical to 'Fourneau 309'.

The success of suramin can be measured by the fact that sixty-five years after its discovery it remains one of the principal drugs for the prevention and treatment of trypanosomiasis. In keeping with Ehrlich's observation concerning dyes featuring the sulphonic acid function, it cannot enter the central nervous system. This precludes its use in the management of advanced forms of sleeping sickness. However, of even more significance than its undoubted therapeutic value, was the stimulus suramin gave to the subsequent development of chemotherapy. Within twelve years of its discovery, Bayer researchers at Elberfield had developed the first effective synthetic antimalarials, the sulphonamides, and several other chemotherapeutic agents by an extension of the logic that had led to the introduction of suramin. Pentamidine, the other principal drug presently used in trypanosomiasis, was also developed in the

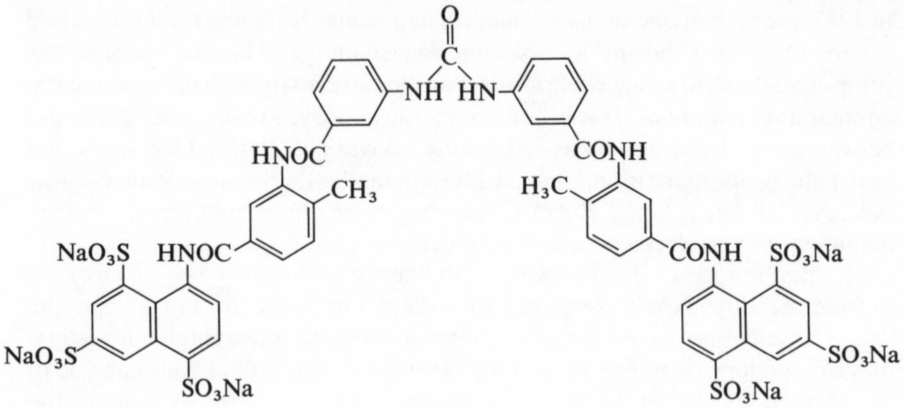

Suramin

wake of suramin.

Evidence obtained by K. Schern in Germany in 1928 indicated that trypanosomes required relatively large amounts of glucose in order to reproduce. Seven years later, Hildrus Poindexter of Howard University School of Medicine in Washington DC demonstrated that survival of animals infected with trypanosomes was prolonged if their blood glucose levels were kept depressed by insulin injections. N. and H. von Jancso likewise exposed mice infected with *Trypanosoma brucei* to the hypoglycaemic drug 'Synthalin' and several of its analogues. These also had a trypanocidal action, which convinced the Jancsos that suramin acted by interfering with glucose metabolism in trypanosomes. Warrington Yorke and E. M. Lourie of the Liverpool School of Tropical Medicine then quickly pointed out that 'Synthalin' killed trypanosomes at dose levels which did not significantly lower blood sugar in mice! The Liverpool researchers demonstrated that 'Synthalin' was even active against trypanosomes growing on culture media rich in glucose. This indicated that the trypanocidal action of the amidines was a direct one, and had nothing to do with lowering blood sugar in the host animal. This was reported to Harold King at the National Institute for Medical Research, and consequently led to the synthesis of a large series of analogues of 'Synthalin' as potential trypanocidal agents. Amongst these were several diamidines that turned out to be powerful trypanocides that cured mice and rabbits infected with *Trypanosoma rhodesiense*. At this point, in 1937, Arthur Ewins of May and Baker was invited to participate. He arranged for the preparation of a new range of diamidines in which the polar amidine groups were separated by an intermediate chain consisting of two benzene rings rather than polymethylene groups, as had previously been the case. The first of his compounds proved to be active, so the series was extended, with a large number of analogues being examined. Many of these were trypanocidal, especially stilbamidine, the 4,4′-diamidinostilbene. However, the related pentamidine had greater water solubility and so was preferred for intramuscular injections. By 1940, these compounds had been tested on over four hundred patients suffering from sleeping sickness or the related tropical disease, leishmaniasis. Because of their polar nature, these diamidines resembled suramin, being unsuitable for treating advanced forms of sleeping sickness in which there was central nervous system involvement. Nevertheless, these diamidines prepared as a result of close collaboration between academia and industry are still widely used against trypanosomiasis.

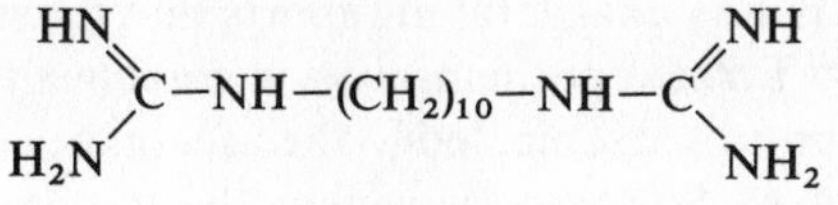

HN NH
$C-NH-(CH_2)_{10}-NH-C$
H_2N NH_2

'Synthalin'

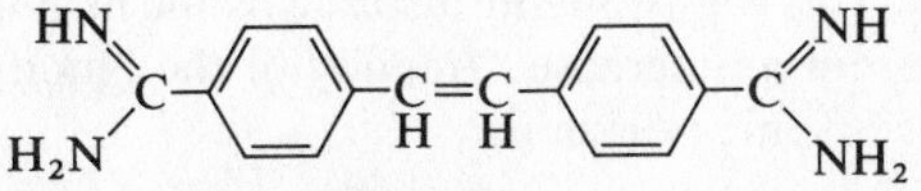

HN NH
C– –C=C– –C
H H
H_2N NH_2

Stilbamidine

HN NH
C– $-O-(CH_2)_5-O-$ –C
H_2N NH_2

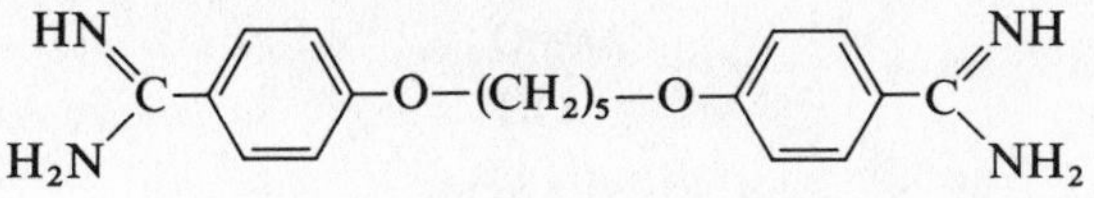

Pentamidine

ARSENICALS

Thomas Fowler, a medical practitioner in Stafford, published his *Medical Report of the Effects of Arsenic in Cases of Agues, Remittent Fevers, and Periodical Headaches* in 1786. He had carried out this investigation into the value of arsenic in the treatment of malaria as a consequence of his favourable results with a patent medicine popular in Lincolnshire, known as 'The Tasteless Ague Drops'. He had been told this was a preparation of arsenic based on products used in Hungary to treat malaria. Fowler introduced his own formulation of potassium arsenite, which was praised by colleagues, including William Withering. Throughout the nineteenth century, Fowler's Solution was generally accepted as a valuable, if less reliable, alternative to quinine in malaria. It is particularly relevant to note, too, that there were also persistent claims attesting to the beneficial effects of arsenic preparations in syphilis, and in 1858 David Livingstone, the Scottish medical missionary who explored much of Central Africa, recommended Fowler's Solution for the alleviation of the symptoms of sleeping sickness.

In 1894, Sir David Bruce demonstrated in Natal that protozoa were present in the blood of animals afflicted with nagana. He demonstrated that these protozoa, later to be described at *Trypanosoma brucei*, could be temporarily eliminated from the blood of infected cattle by administering Fowler's Solution. This was followed up, in 1899, by Professor Lingard at Muktesar in India, who tried unsuccessfully to cure horses infected with the form of trypanosomiasis known as surra. However, in 1902, Professor Alphonse Laveran of the Pasteur Institute, having devised a means of infecting mice with trypanosomes, studies the effects of Fowler's Solution on them under laboratory conditions. It caused a rapid disappearance of the protozoa from the blood of the mice, but within a few days the parasites had reappeared and the animals perished.

Fowler's Solution was widely used for a variety of other purposes during the nineteenth century. Wildly exaggerated reports that the inhabitants of Upper Styria in the Austrian Tyrol kept themselves in good health by sprinkling their food with small amounts of arsenic, led to the acceptance of it as a tonic. Even until the 1940s, Fowler's Solution, as well as the less toxic organic arsenical known as sodium cacodylate, was still being prescribed as a tonic or to treat pernicious anaemia. This was probably inspired by the heightened colouring of the cheeks arising from an increased fragility of the blood capillaries consequent upon chronic arsenic poisoning!

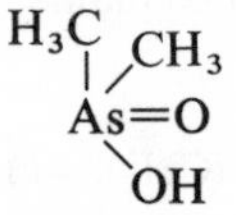

H_3C CH_3
$As{=}O$
OH

Cacodylic Acid

The increasing popularity of cacodylic acid as an alternative to Fowler's Solution encouraged August Michaelis to synthesize a wide range of organic arsenicals during the last quarter of the nineteenth century, first at the Karlesruhe Polytechnic and later at the University of Rostock. His intention had been to develop a drug with lower toxicity than Fowler's Solution. Despite his persistence, it was the Vereinigte Chemische Werke in Charlottenburg that found such a drug with a compound originally synthesized by Bechamp in 1868. On a weight basis, it was twenty or so times less toxic than potassium arsenite, which led to its being marketed with the overly optimistic name of 'Atoxyl'.

The director of the factory at Charlottenburg, Professor Ludwig Darmstadter, had been particularly interested in finding a cure for cancer, and he sponsored some of Ehrlich's work in this field at the Institute for Experimental Therapy. Fowler's Solution had been routinely given to leukaemia patients since 1865 (see page 331), so whether Darmstadter sent Ehrlich 'Atoxyl' with a view to its being evauluated in cancer or in trypanosomiasis is not clear. Ehrlich and Shiga used it as the first substance tested against the trypanosomes sent from the Pasteur Institute, but it proved to be inactive.

After testing 'Atoxyl', Ehrlich began his investigation of dyes. His discovery of the trypanocidal properties of Trypan Red and the research arising from this fully preoccupied him until 1905. In that year, however, he was surprised to read in the *British Medical Journal* a paper by Wolferstan Thomas of the Liverpool School of Hygiene and Tropical Medicine, describing the success of 'Atoxyl' in the treatment of animals experimentally infected with trypanosomiasis. Ehrlich at once reopened his investigation into the activity of 'Atoxyl' and confirmed Thomas' results. He then realized that his earlier study had been misleading because it had been carried out on isolated cultures of trypanosomes, rather than on infected animals. This implied that 'Atoxyl' either stimulated immunity or else it had to be metabolically converted to an active form by the animal before any trypanocidal effect could be exerted.

Robert Koch was asked by the German Sleeping Sickness Commission to evaluate 'Atoxyl' in East Africa, where the mortality from the disease was reaching alarming proportions. Koch established that a 500 mg injection of the drug could cause the disappearance of trypanosomes from the blood for up to eight hours. His final recommendation was that effective treatment would require constant medication for six months, a protocol that would present the risk of blindness through damage to the optic nerve in up to two per cent of patients. This convinced Ehrlich that it would be worth his while examining analogues of 'Atoxyl' to see whether the ratio of efficacy to toxicity could be improved. At first he considered the scope for molecular modification to be limited since 'Atoxyl' was the unreactive anilide of arsenic acid. His unexpected discovery that the drug reacted with nitrous acid to form a diazonium salt led him to the realisation that the structure originally assigned to it by Bechamp was incorrect. 'Atoxyl' was, in fact, the 4-aminophenyl derivative of arsenic acid. As such, it was chemically similar to many of the compounds previously examined by Michaelis, which meant that methods for preparing a wide range of analogues already existed. Ehrlich was now anxious to press ahead with a new programme of research which could be conducted in the new laboratories then being built for him.

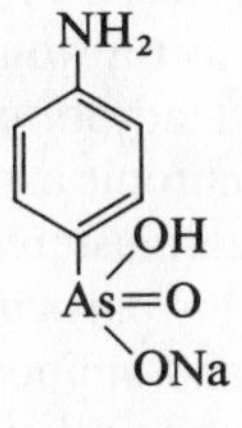

NH_2

OH

As=O

ONa

'Atoxyl'

Thanks to the generosity of the widow of Frankfurt banker George Speyer, who donated a million marks, and also to John D. Rockefeller, who supported Ehrlich's research, a chemotherapy institute was built alongside the Institute for Experimental Therapy. It was called the George Speyer-Haus, and was officially opened in September 1906. There were excellent facilities for chemists to synthesize the arsenicals required by Ehrlich, thus introducing into chemotherapy the same approach that had just borne fruit with the development of procaine as a substitute for cocaine.

Ehrlich was fond of coining Latin aphorisms to express the complex ideas underlying his researchers. To describe the principle guiding the design of new chemotherapeutic agents he proposed to synthesize, he used the phrase *corpora non agunt nisi fixata*. This was his way of explaining that infecting organisms are not killed unless the chemotherapeutic agent had a high affinity for them. Ehrlich termed such drugs as parasitotropic. Their toxic effects he attributed to their also being, to varying extents, organotropic. He believed that the ideal agent would have a high therapeutic index because it exhibited a high parasitotropic activity and a low organotropic activity. Ehrlich measured the therapeutic index simply by comparing both the curative and lethal doses of his dyes and arsenicals in mice. Around the time he moved into the George Speyer-Haus, he began to accept the ideas on the existence of drug receptors newly propounded by the Cambridge physiologist, John Langley. Ehrlich then advanced the hypothesis that the parasitotropic action of arsenicals was due to their binding to arsenoceptors on the surface of the parasites, these being specific types of chemoceptors. It was his fervent hope that he could develop drugs which, like the antibodies discovered by his former colleague Emil Behring, would act like magic bullets insofar as they would bind only to receptors in the parasite and not the patient.

In the spring of 1907, Ehrlich signed a formal contract with the Cassella Dye Works, granting the company exclusive rights to the commercial exploitation of his arsenic compounds in return for their sponsorship of his research. The arrangement continued after the Hoechst Dyeworks took control of Cassella in 1908. Ehrlich's chief chemist, Alfred Bertheim, then synthesized a range of 'Atoxyl' derivatives with substituents on the amino group. One of these, arsacetin, was less toxic to mice than the parent drug, but when administered at the high dose level required to effect a cure, it frequently caused the mice to turn continuously in circles in the manner of Japanese waltzing mice. This damage to the vestibular nerve (associated with balance) was also produced by 'Atoxyl', and it indicated that arsacetin would probably also cause blindness. Remembering his early work on lead poisoning, Ehrlich became convinced that this nerve damage was due to chronic arsenic poisoning caused by the large amount of arsenic that had been administered. In seeking a more potent series of arsenicals he hoped to avoid this particular problem. To this end chloro, hydroxy, cyano, sulphonic acid, and amino groups were introduced into the benzene ring. These variants exhibited their own particular types of toxicity over and above the neurotoxicity associated with large amounts of arsenic. None of them introduced the required enhancement of potency.

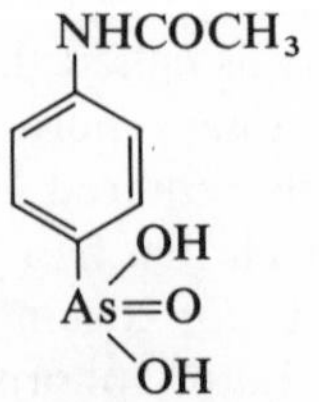

$NHCOCH_3$

OH

As=O

OH

Arsacetin

The discovery that higher concentrations of the phenylarsonic acids were required to kill test tube cultures of trypanosomes than could possibly be achieved when curing infected mice, led to a major breakthrough for Ehrlich. He concluded that the phenylarsonic acids were undergoing metabolic activation, presumably through a process of chemical reduction. To test this hypothesis, he asked Bertheim to prepare the two possible types of reduction products. One type, the arsenoxides, proved highly toxic to trypanosomes, but their general toxicity to the tissues of the host was also high. They were also very irritant when injected, due to the presence of impurities. The other series, the arsenobenzenes, although not as potent, were less toxic to the host whilst still more potent than the phenylarsonic acids. This meant that small doses could be given, thus at last avoiding the problem of neurotoxicity from chronic arsenic poisoning. Ehrlich astutely exploited this and subsequently restricted his investigations to arsenobenzenes. However, he never discovered the true reason for their superiority. He and his contemporaries believed that the arsenobenzenes consisted of two molecules linked by a double bond between their respective arsenic atoms. It was many years later that this was shown to be incorrect; arsenobenzenes are polymers formed from hundreds of molecules joined by single bonds between their arsenic atoms. Since polymers cannot penetrate mammalian cells, the toxicity of arsenobenzenes was much weaker than that of the arsenoxides which were formed from only two molecules, these being joined via arsenic–oxygen–arsenic linkages. The chemotherapeutic action of both arsenobenzenes and arsenoxides was due to the release of arsenite molecules which had only a single arsenic atom that reacted with the sulphydryl group in receptors on the parasite.

Ehrlich soon established that substitution on the amino group of the arsenobenzene analogue of 'Atoxyl' produced what he had been seeking. Arsenophenylglycine (compound 418) emerged as a most promising trypanocidal agent. It was administered to humans in 1907, and for many patients it proved to be safe and effective. Unfortunately, in a small number of patients severe and often fatal hypersensitivity to it was exhibited. Despite this hazard, some physicians deemed it acceptable to continue to use it for those forms of trypanosomiasis with a high mortality rate. For Ehrlich, this was unacceptable.

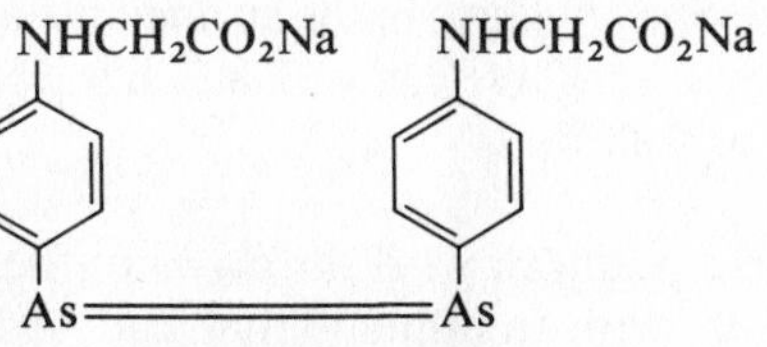

$NHCH_2CO_2Na$ $NHCH_2CO_2Na$

As═As

Ehrich's structure
for Arsenophenylglycine

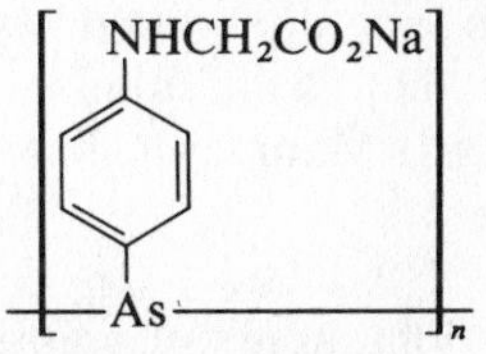

$[NHCH_2CO_2Na]$

—As—$]_n$

Revised structure
of Arsenophenylglycine

Introduction of a hydroxyl group on the 4-position of the benzene ring was readily achieved by warming diazotized 'Atoxyl' in water. Reduction of the product yielded arsenophenol. This was highly effective against trypanosomes, but was prone to oxidation to the severely irritant arsenoxide and exceedingly difficult to purify. However, Ehrlich's experience with dyes such as Trypan Red and Trypan Blue had convinced him that the introduction of a substituent adjacent to a phenolic hydroxyl group enhanced chemotherapeutic activity. Accordingly, an arsenophenol substituted in this manner was prepared in 1907 as potential trypanocidal agent number 606. The assistant who tested it told Ehrlich that it was ineffective, and consequently it remained on a shelf in the laboratory for over a year before being re-examined.

The isolation of the organism that caused syphilis, *Treponema pallidum*, was reported from the Reichsgesundheitsamt in Berlin by Fritz Schaudinn and Erich Hoffmann in 1905. The following year, Hoffmann visited the Speyer-Haus and told Ehrlich that this spirochaete was in many ways similar to the trypanosomes. He asked for samples of Ehrlich's new compounds for testing on syphilitic patients in his clinic at Bonn, where he had just been appointed to the chair of dermatology. Ehrlich agreed to this request, cautioning Professor Hoffmann to exercise great care when using the arsenicals. He then wrote to the Vereinigte Chemische Werke, pointing out that as he had found 'Atoxyl' was an acid, its mercury salt should be produced. Mercurial compounds were the standard treatment for syphilis at that time. Ehrlich also urged Professor Lassar to treat syphilis with arsacetin, and arranged for his friend Albert Neisser, director of the dermatology clinic at Breslau, to take arsacetin and arsenophenylglycine to Java, where he had managed to inoculate apes with the *Treponema pallidum*. The results of Neisser's experiments were published in 1908 and confirmed the efficacy of arsenophenylglycine. This was the only way in which Ehrlich was then able to have his compounds tested on animals. The arrangement lasted until the spring of 1909, when Ehrlich was joined by Sacachiro Hata who, whilst working in the Kitasato Institute in Tokyo, had developed a method of infecting rabbits with syphilis. On his arrival, Hata was asked to test every arsenical that had been synthesized by Bertheim and his colleagues over the preceding three years. Working swiftly and with precision, he discovered that compound 606 had outstanding curative properties in the rabbits infected with syphilis! As a result, Ehrlich arranged for the Farbwerke Hoechst to apply for a patent on '606'; the application was submitted on the 10th June 1909. What happened in the Speyer-Haus in the spring and summer of 1909 was the culmination of a truly remarkable effort. Ehrlich's secretary, Martha Marquardt, has commented as follows:

> 'No outsider can ever realise the amount of work involved in these long hours of animal experiments, with treatments that had to be repeated and repeated for months on end. No one can grasp what meticulous care, what expenditure and amount of time were involved. To get some idea of it we must bear in mind that arsenophenylglycine had the number 418, 'Salvarsan' the number 606. This means that these two substances were the 418th and 606th of the prepara-

> tions which Ehrlich worked out. People often, when writing or speaking about Ehrlich's work, refer to 606 as the 606th experiment that Ehrlich made. This is not correct, for 606 is the number of the substance with which, as with all the previous ones, very numerous animal experiments were made. The amount of detailed work which all this involved is beyond imagination.'

His earlier disappointment over the serious hypersensitivity reactions amongst patients receiving arsenophenylglycine led Ehrlich to exercise great caution over the assessment of '606'. Once the results of animal studies confirmed that it was likely to be safe and effective, samples of '606' were sent to Professor Julius Iversen of the Obuchow Hospital for Men in St Petersburg and Professor Konrad Alt of Uchtspringe (Altmark). Iversen was the first to administer the drug to patients when he successfully treated relapsing fever, a disease transmitted by lice. The causative organism, *Borrelia recurrentis*, was similar to the *Treponema pallidum*. Professor Alt, however, spent three months evaluating the safety of '606' in dogs before he would let two of his assistants volunteer to be injected with the new arsenical. Only after this was the drug administered to several syphilitic patients suffering from general paralysis of the insane. Once its relative safety had thus been confirmed, samples of '606' were given to Professor Schreiber of Magdeburg, where trials were initiated on large numbers of patients with primary syphilis. By January 1910, several other trusted investigators had received '606' for trials. Ehrlich insisted on personally checking the records of every patient who received the drug, and it was not until the following April that he was prepared to make any public accouncement about '606'.

On the 19th April 1910, Paul Ehrlich told the Congress for Internal Medicine at Weisbaden of his work leading up to the discovery of '606'. Dr Hata then described his experiments and Professor Schreiber explained how he had been able to cure syphilis with '606'. These announcements received enthusiastic applause from those present, and in the following days the press throughout the world carried headline reports of the new drug that could cure syphilis. The immediate result of this publicity was that the George Speyer-Haus was besieged by physicians and patients anxious to obtain samples of '606'. Letters also poured in from around the world. Under this unprecedented pressure, Ehrlich modified his policy of only supplying the drug to a small circle of trusted colleagues. He agreed to provide these doctors, at no charge, with a small number of vials of the drug that had been prepared under his supervision, on the condition that he received full reports of every case treated. Between April and December 1910, some sixty-five thousand vials of '606' were freely supplied in this manner while the necessary plant for its commercial production was being installed in the Farbwerke Hoechst. Production began there at the end of the year. '606' was marketed under the proprietary name of 'Salvarsan', the approved name given to it later being arsphenamine. To the general public it remained known as either '606' or 'Ehrlich-Hata 606'. Despite early setbacks because of its careless administration by some doctors, arsphenamine and one or two of its close analogues remained the standard treatment for syphilis until the end of the Second World War, when adequate supplies of penicillin became available. There can be no question whatsoever that '606' represented the first major chemotherapeutic agent and that Ehrlich deserves to be remembered as the founder of modern chemotherapy. His concepts that led to the discovery of arsphenamine still underpin present-day chemotherapeutic research. His concept of combination therapy forms the bedrock of cancer chemotherapy. The concept recognized that it was essential to kill all but the few parasites that could be destroyed by the immune system of the body, otherwise the surviving parasites would reproduce and cause reappearance of the disease. To ensure that parasites resistant to a particular drug were unable to do this, Ehrlich recommended treatment with combinations of drugs that acted in different ways so that the parasites would be destroyed by one of the drugs to which they were sensitive. He feared that if single drug therapy was used, and the first dose did not achieve complete sterilization, surviving parasites would be resistant. Fortunately, this has not proved to be the case in either cancer or antimicrobial chemotherapy, although a drug that could achieve complete destruction of microbes or cancer cells with only a single dose would still be accepted as the target towards which research should ultimately aim.

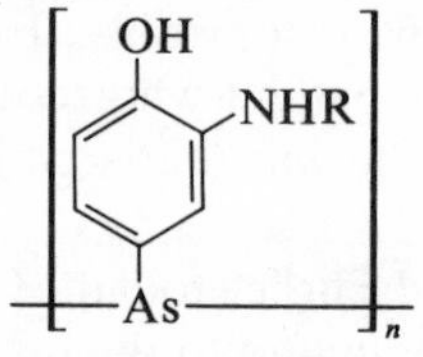

OH
NHR
As
n

Arsphenamine, R = —H
Neoarsphenamine, R = —H or $-CH_2SO_2Na$

Many syphilitic patients were cured by a single 900 mg dose of arsphenamine. This injection was prepared from powdered arsphenamine hydrochloride in a sealed ampoule. Prior to administration, this had to be treated with the correct amount of dilute alkali to form a solution containing the highly unstable sodium salt, a procedure that brought many complaints from physicians. Shortly after starting to produce arsphenamine, the Hoechst Dyeworks patented neoarsphenamine ('Neosalvarsan'), a water-soluble derivative that was made less sensitive to oxidation by condensing some of the amino groups on the polymer with 'Rongalite' (sodium sulphoxylate), an antioxidant widely used in the dyestuffs industry. Trials revealed this to be less potent than arsphenamine, two injections always being required to effect a cure of syphilis. This worried Ehrlich, but he was reluctantly persuaded that it should be made available because of the preference of the medical profession for a preparation that required only the addition of water to the ampoule immediately before injection.

Arsphenamine and neoarsphenamine were exceedingly difficult to manufacture, but this was only fully recognized after the outbreak of the Great War, when attempts were made in Britain and France to synthesize it. Side-reactions occurred during all of the various stages of the synthesis, and there was no way of preventing the formation of impurities or even of removing them. It was this state of affairs that had made Ehrlich insist on every batch being prepared in precisely the same manner before it was biologically tested for both safety and efficacy. To deal principally with the problem of obtaining supplies of arsphenamine, the United Kingdom Medical Research Committee in 1914 set up bacteriology and experimental pathology laboratories at St Mary's Hospital, Paddington, and biochemistry and pharmacology laboratories at the Lister Institute of Preventive Medicine. The director of the Wellcome Research Laboratory, Henry Dale, along with two of his senior colleagues, George Barger and Arthur Ewins, transferred to the Lister Institute to establish the forerunner of the National Institute for Medical Research. Licenses to produce arsphenamine were then granted by the British government to Burroughs Wellcome and Company and also Poulenc Frères of Paris, the latter being required to operate in the United Kingdom through May and Baker, with whom they already had associations. Burroughs Welcome had taken advantage of the fact that 'Atoxyl' could not be patented and had produced their own brand, known as 'Soamin', in 1911, as well as its 3-methyl derivative, which they called 'Orsudan'. Poulenc Frères had been trying to prepare arsphenamine since 1913, unfettered by patent restrictions as French law did not then recognize the right to patent medicines. Throughout the war, Dale and his colleagues freely exchanged information on arsphenamine production with their industrial colleagues. The immediate outcome of this was that by 1916 production of arsphenamine had been successfully achieved, and 94 762 injections had been administered by the French Military Medical Services alone, without any fatalities occurring. This figure, incidentally, provides an insight into the extent of the spread of venereal disease prior to the introduction of arsphenamine therapy. Although the United States had not entered the war until 1917, difficulties in obtaining adequate supplies of arsphenamine had led the Federal Trade Commission to abrogate the patents and issue licenses for its manufacture. American manufacturers then experienced the same difficulties as their British counterparts, and the quality of every batch of the drug had to be checked by the Hygienic Laboratory of the US Public Health Service. These developments on both sides of the Atlantic had far reaching consequences, for they gave a much needed stimulus to the British, French, and American pharmaceutical industries to widen their interests by producing synthetic chemotherapeutic agents.

The war was hardly over before it was announced that Walter Jacobs and Michael Heidelberger of the Rockefeller Institute in New York had synthesized tryparsamide, an arsonic acid analogue of 'Atoxyl' that was more effective than any other arsenical against trypanosomiasis. It is still widely used in the treatment of Gambian sleeping sickness. The Rockefeller Foundation patented this new arsenical, named tryparsamide, but they issued licenses free of charge to responsible manufacturers who wished to produce it. It was truly fitting that the Rockefeller Institute, founded in 1901 with an endowment of $200 000, should succeed in fulfilling Ehrlich's original objective only four years after his death. John D. Rockefeller had been one of the first backers of Ehrlich's work on arsenicals. That the new drug was an analogue of 'Atoxyl' would have surprised Ehrlich, for he had always believed that the arsonic acids caused neurotoxicity because of the large doses that had to be administered. His interpretation was also doubted by Ernest Fourneau, at the Pasteur Institute, in the 1920s. Fourneau was convinced that the blindness caused by 'Atoxyl' was due to impurities, and this led him to synthesize a range of phenylarsonic acids, from which acetarsol emerged as a valuable, safe anti-syphilitic and amoebicide. Acetarsol, in fact, had been used in 1911 as an intermediate in the preparation of some of Ehrlich's arsphenamine analogues, but it was not until 1922 that its value was recognized. Fourneau arranged for acetarsol to be manufactured by Poulenc Frères under the proprietary name of 'Stovarsol', once again a pun on his own name as had been the case with 'Stovaine' (*fourneau* = stove).

$NHCH_2CONH_2$ — C_6H_4 — $As(=O)(OH)ONa$

Tryparsamide

OH, $NHCOCH_3$ — C_6H_3 — $As(=O)(OH)OH$

Acetarsol

Thousands of arsenicals were synthesized in Europe and America during the twenties and thirties. Space does not permit further consideration of them here, but one development is worthy of note. During the First World War, W. Lee Lewis of the US Chemical Warfare Service prepared a highly vesicant arsenical which became known as 'Lewisite'. It produced its destructive effects by penetrating the skin then hydrolysing to form an arsenite. Fears that it would be used by the enemy during the Second World War led the British Ministry of Supply to set up a team of scientists charged with the responsibility of finding an antidote. Based in the department of biochemistry at Oxford, and led by Professor Rudolph Peters, they established that arsenite reacted with two adjacent thiol groups on the vital pyruvate oxidase enzyme system. The discovery that this reaction could be competitively inhibited by the presence of simple sacrificial molecules that also contained two adjacent thiol groups ultimately resulted in the preparation of a highly effective antidote, namely dimercaprol ('BAL', or 'British Anti-Lewisite') by Oxford chemists. It was a viscous liquid that could be applied to the skin in an ointment, or else be injected. This wartime episode constitutes one of the finest examples of rational drug design, and dimercaprol is still used as an antidote to poisoning by arsenic, mercury, and other metallic poisons that react with thiol groups in the pyruvate oxidase system.

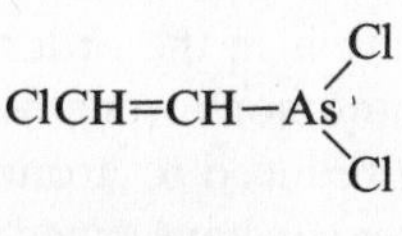

$$ClCH{=}CH{-}AsCl_2$$

Lewisite

$$CH_2SH{-}CHSH{-}CH_2OH$$

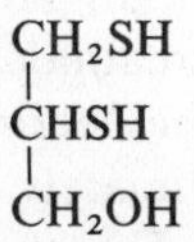

Dimercaprol

ANTIMONIALS

In 1906, Nicolle and Mesnil at the Pasteur Institute followed up their colleague Laveran's demonstration of the value of Fowler's Solution in mice infected with trypanosomiasis, by showing that intravenous injections of tartar emetic (antimony potassium tartrate) could produce similar results. Two years later, cattle infected with trypanosomiasis were injected with tartar emetic, but the doses required to effect a cure were too toxic for human application. Although antimony salts had been used medicinally since biblical times, the parenteral use of tartar emetic was novel. Tartar emetic was prepared by leaving white wine to stand in antimony vessels, and was strongly advocated by Paracelsus, but his followers were criticized for the reckless abandon with which they administered it, resulting in it being banned by the parliament of Paris, in 1566. This is probably the earliest instance of statutory control over a medicinal substance. Notwithstanding the ban, a century later Louis XIV attributed his recovery from typhoid fever to the proscribed drug, as a result of which it not only regained its former popularity, but became even more widely recommended until its demise in the late nineteenth century, again from its persistent misuse.

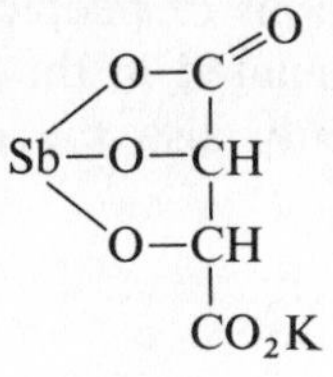

Sb–O–CH / O–C(=O) / O–CH / CO_2K

Antimony Potassium Tartrate
(Tartar Emetic)

The experimental use of tartar emetic in trypanosomiasis led Gaspar de Oliveira Vianna, a young Brazilian physician, to inject the toxic drug in patients afflicted with a disease known locally as espundia. He had been the first person to detect the presence of a hitherto unrecognized organism, *Leishmania braziliensis*, in victims of this disease. The term Leishmaniasis describes a group of tropical diseases caused by species of *Leishmania* protozoa which was first detected in 1900 by a Scot, William Leishman. Leishmaniasis is one of the most widespread of all communicable diseases in the world. A common form is kala-azar, which affects millions of children in Africa, Central and South America, China, India, the Middle East, and parts of Russia. Without treatment the mortality rate is in the order of ninety per cent, but Vianna's pioneering studies with antimonial chemotherapy, first reported in 1912, ultimately led to this being reduced to around the current figure of ten per cent. Few therapeutic measures can have saved so many lives.

Treating a patient in the Civil Hospital at Khartoum in 1918, J. B. Christopherson noticed that tartar emetic also acted against schistosomiasis, a parasitic disease blighting the lives of 200 million people. The parasite, a trematode worm, is transmitted by fresh-water snails.

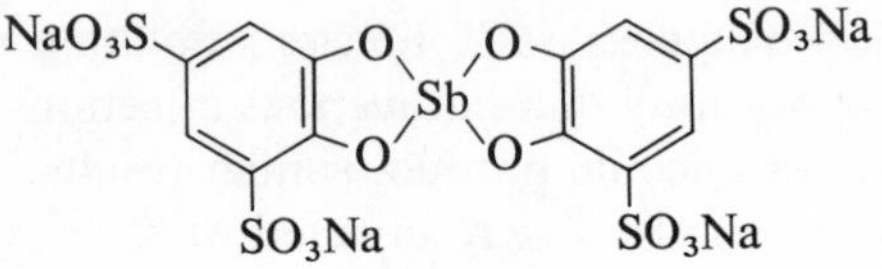

NaO_3S O   O SO_3Na
Sb
O   O
SO_3Na SO_3Na

Stibophen

The undoubted clinical value of tartar emetic was seriously marred by its toxicity and the irritancy it caused when injected. Sudden death after an injection was not unheard of. As for irritancy, the obvious way to reduce this was simply to switch to the more soluble antimony sodium tartrate. Many antimonials, often of complex and uncertain polymeric structure, have been synthesized. Several of the best known ones were prepared in the 1920s by E. Schmidt of the von Heyden Chemical Works in Radebeul, Germany, and tested by Uhlenhuth at the University of Freiburg. One of these was stibophen, astutely marketed under the brand name of 'Fouadin' to impress King Fouad of Egypt, a country where schistosomiasis was, and still remains, the major health problem causing debility in millions of people. Sodium stibogluconate was synthesized by Schmidt and evaluated in the I. G. Farben laboratories at Elberfield in 1937. It is presently also the drug of choice in cutaneous leishmaniasis.

ANTIMALARIALS

Inability to infect small laboratory animals with malaria prevented Paul Ehrlich pursuing the chemotherapeutic studies with methylene blue that he initiated in 1891 in association with Professor Guttmann at the Moabite Hospital in Berlin. Despite the synthesis of many quinoline derivatives in several European laboratories, Ehrlich's arsphenamine was the only synthetic compound other than methylene blue that possessed any clinical value, but this was also markedly inferior to quinine. No worthwhile progress was made until 1924, when Wilhelm Roehl at Bayer's Chemotherapeutic Institute in Elberfield devised a satisfactory technique for screening potential antimalarial drugs in canaries. Compounds that appeared promising were then able to be tested in syphilitic patients suffering from general paralysis of the insane, who were inoculated with malarial parasites to produce fever. This was a new technique, first tried by Professor Julius Wagner-Jauregg of the University Hospital, Vienna, in 1917. He showed that at least one third of his paralysed patients recovered after this form of therapy, for which he was awarded the Nobel Prize for physiology or medicine in 1927.

Professor W. Schulemann and his colleagues at Elberfield, F. Schönhöfer and A. Wingler, took note of the failure of the many quinolines that had been synthesized, and decided to follow up Ehrlich's untried idea of preparing derivatives of methylene blue as potential antimalarials. They started by substituting a diethylaminoethyl side-chain on one of its methyl groups. Roehl found this effective in his infected canaries, with a therapeutic index (i.e. the ratio of the toxic dose to the therapeutic dose) of eight, but he was disturbed by the fact that the compound was a dye. To avoid this problem, Schulemann reverted to the quinoline system, but retained the basic side-chain as he believed this to be essential for strong antimalarial activity. When this side-chain was substituted on to 8-aminoquinoline, which was readily prepared from 8-nitroquinoline, Roehl found the resulting compound cured infected canaries. This was colourless, and it served as the lead compound from which an astonishingly diverse range of analogues were synthesized and tested on canaries. The strategy was to vary the point of attachment of the side-chain to the quinoline nucleus by preparing a full range of aminoquinolines, then to investigate the effect of varying the side-chain in just about every conceivable manner. In order to increase the similarity to quinine, a methoxyl group was introduced at the 6-position of the quinoline ring. As if this were not enough, Schulemann and his colleagues also synthesized a variety of heterocyclic ring systems other than quinoline. Literally hundreds, possibly thousands, of compounds were prepared and tested by the small group of Bayer researchers.

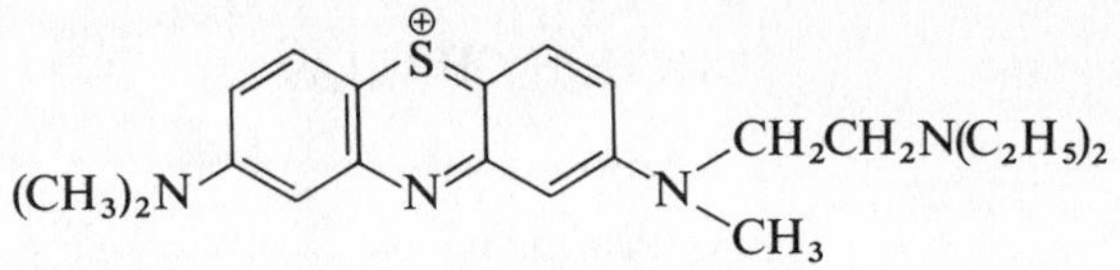

$S^{\oplus}$ $(CH_3)_2N$ N N $CH_2CH_2N(C_2H_5)_2$ CH_3

Diethylaminoethyl derivative of Methylene Blue

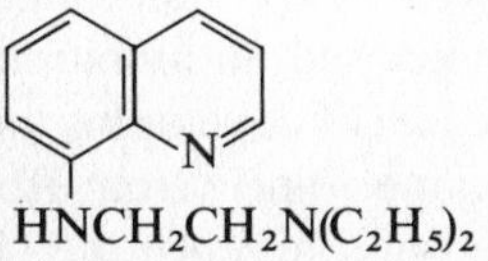

N $HNCH_2CH_2N(C_2H_5)_2$

8-Aminoquinoline Lead Compound

In 1925, a promising 6-methoxyquinoline derivative with a therapeutic index of thirty was selected for clinical evaluation. It was first tried in insane paralysed patients who had been infected with the malarial parasite as part of the Wagner-Jauregg regimen. The new drug was effective, and Roehl himself proceeded to confirm that it was also able to cure patients with naturally acquired malaria. Clinical trials throughout the world followed, and the drug was then marketed as 'Plasmoquine'. It was given the approved name pamaquin, but its full chemical structure was not disclosed until 1928.

The life cycle of the malarial parasites (sporozoites) after they enter the blood of a human bitten by a female Anopheline mosquito is complex. Within an hour, liver cells are invaded, and the sporozoites begin to divide, ultimately causing the cells to rupture. This releases merozoites into the blood, and these penetrate the red cells to initiate the erythrocytic phase of the disease. Inside the blood cells, the merozoites multiply until the cells rupture, causing the patient to experience chills, fever, and sweating. The merozoites then attack further blood cells to renew the cycle, which accounts for the periodicity of malarial attacks. Quinine suppresses this erythrocytic stage of the disease. Roehl, however, discovered that the action of pamaquin was quite different, and it was clearly not simply a substitute for quinine. Large doses of pamaquin greatly lowered the incidence of relapses in patients infected with *Plasmodium vivax*, the commonest type of malaria, and outright cures were even reported amongst those who could tolerate its inevitable side-effects. However, better results were obtained when small doses of pamaquin were administered in conjunction with quinine. It was only some years later that it was established that pamaquin acted by destroying parasites that persisted in the liver. These were responsible for the characteristic relapsing fever associated particularly with *Plasmodium vivax* infection. The combination of pamaquin with quinine eradicated the infection in both the liver and the blood, thus producing outright cures.

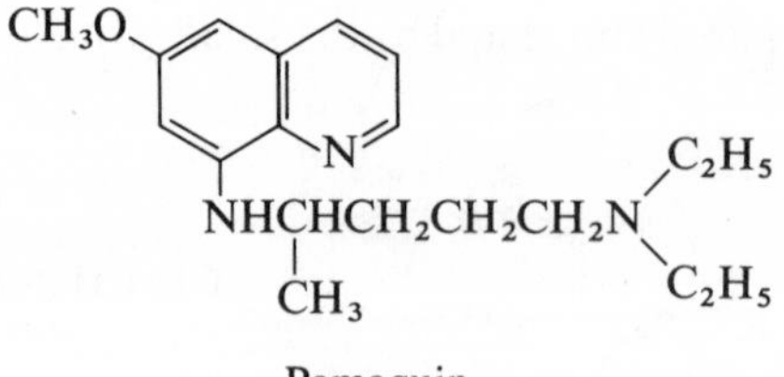

CH_3O (quinoline ring, N)

$NHCH(CH_3)CH_2CH_2CH_2N(C_2H_5)_2$

Pamaquin

Following the clinical introduction of pamaquin, the Joint Chemotherapy Committee of the Medical Research Council and the Department of Scientific and Industrial Research sponsored an ambitious programme of research in British universities, with the aim of developing antimalarial drugs. In 1929, the first of many publications came jointly from Robert Robinson at University College in London, and George Barger at the University of Edinburgh. This described how they had been able to establish the chemical nature of pamaquin before its structure had been disclosed by I. G. Farben, and had proceeded on this basis to make analogues. Although nothing superior to pamaquin was developed out of this work, or from similar investigations by Fourneau in France, Magidson in Russia, Hegner, Shaw, and Manwell in the United States, or Brahmachari in India, the scene was set for the massive wartime effort that was to follow later. The I.G. Farben group at Elberfield, on the other hand, did succeed in discovering more effective quinoline antimalarials in the 1930s, namely sontoquine and chloroquine, but that was after they had introduced mepacrine, the first antimalarial with quinine-like activity.

Working under Schulemann's direction, Mietzsch and Mauss had by 1926 begun synthesizing pamaquin analogues with an extra benzene ring added to the quinoline ring. These were examined by Walter Kikuth, who had taken over responsibility for biological testing after the death of Roehl. Utilizing a new experimental method of his own design, he found optimum activity in these aminoacridines when the pamaquin side-chain was placed on the 9-position and a chlorine atom introduced at the 3-position. No full account of these researches has ever been published, but it is known that more than 12 000 compounds had to be synthesized at Elberfield to achieve success. The activity of this new antimalarial was probably recognized in 1930, and the first reports were published by Kikuth in 1932. The drug was originally called 'Plasmoquine E', but to avoid confusion this was changed first to 'Erion', and later to 'Atebrin'. It was given the approved name of mepacrine (quinacrine in the United States).

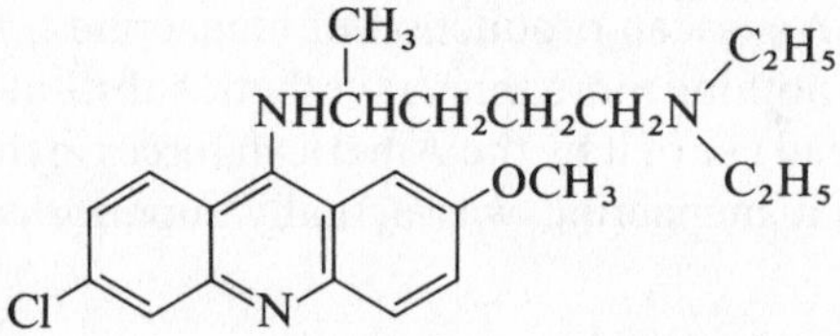

Mepacrine

Kikuth's tests on mepacrine convinced him that its action was similar to quinine insofar as, unlike pamaquin, it could kill merozoites in the erythrocytic phase of malaria. This meant it could suppress the symptoms of malaria and often effect a cure of those types of malaria in which the parasites did not persist in the liver cells. Although it was marketed throughout the world as a substitute for quinine, its full potential was not recognized until after the outbreak of the Second World War.

As the war clouds gathered, the importance of a quinine substitute was recognized. During the First World War, Germany had difficulty obtaining supplies of quinine, but now the tables were likely to be turned if the Japanese gained control of the East Indies, from whence came most of the world's supplies of cinchona bark. Both pamaquin and mepacrine were included in the Association of British Chemical Manufacturers' list of essential drugs that would need to be produced in the United Kingdom if war with Germany broke out. ICI were requested to devise suitable manufacturing processes, and by September 1939, mepacrine was being manufactured in a pilot-plant. Shortly afterwards, full-scale production to meet the requirements of the armed forces had begun. In 1941, the American government also responded to the threat of war. One of its earliest moves involved the Winthrop Chemical Company, which had been set up after the First World War to distribute Bayer pharmaceuticals in the United States following the purchase from the Custodian of Enemy Property, by Sterling Drug Inc., of the Bayer Company of New York. By a subsequent agreement concluded in 1926, the newly constituted I. G. Farben, which had taken over control of Bayer in Germany, became half owners of Winthrop. Three months before the bombing of Pearl Harbor brought the United States into the war, a government anti-trust suit severed the ties between the Winthrop Chemical Company and I. G. Farben, leaving it a wholly American owned company. When the Japanese moved into the East Indies, Winthrop were called upon to supply vast amounts of mepacrine for the army. Prior to this, the company had merely produced about five million of its 'Atabrine' brand tablets annually from six chemical intermediates imported from Germany. Winthrop responded by sublicensing eleven leading American manufacturers on a royalty-free basis, and the outcome was that in 1944 alone some 3500 million mepacrine tablets were produced in the United States. This, and the wartime effort to produce large amounts of penicillin, laid the foundations for the United States to become the biggest producer of pharmaceuticals in the world.

When British and American production of mepacrine first began, the drug was considered to be nothing more than a synthetic substitute for quinine. As a result of the widespread use of it by the American forces in the Far East, it soon became apparent that mepacrine was actually superior to quinine for the

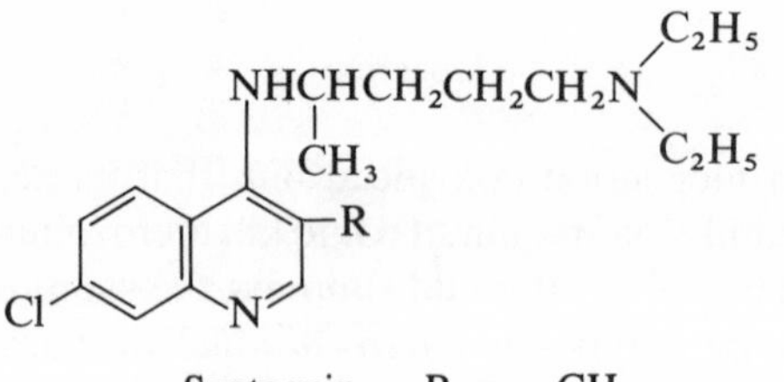

$NHCH(CH_3)CH_2CH_2CH_2N(C_2H_5)_2$, R, Cl, N

Sontoquin, R = $-CH_3$
Chloroquine, R = $-H$

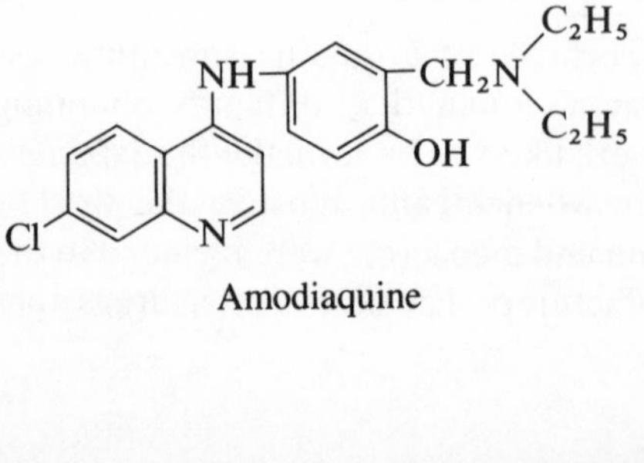

NH, $CH_2N(C_2H_5)_2$, OH, Cl, N

Amodiaquine

CH_3O, N, $NHCH(CH_3)CH_2CH_2CH_2NH_2$

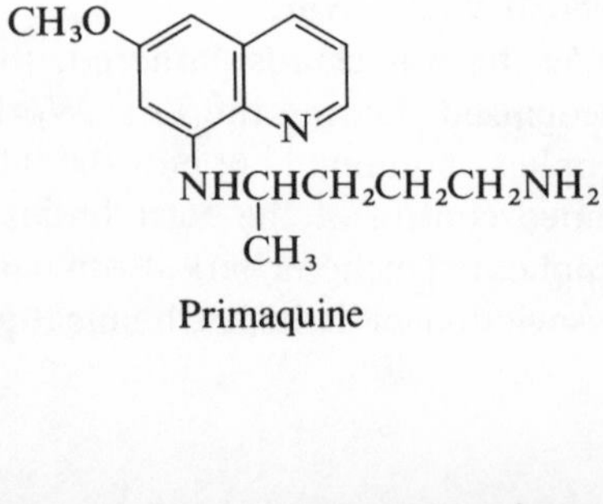

Primaquine

treatment and suppression of malaria. Great care was taken to keep this vital information from the Japanese so that they would continue to use the inferior quinine.

Towards the end of 1939, the British scheme to develop new antimalarials, previously initiated by the Joint Chemotherapy Committee of the Medical Research Council, was given a high priority rating. This resulted in a successful and harmonious collaboration between twenty leading academics and a similar number of industrial chemists, involving the synthesis of about 1700 novel compounds, one-third of which showed significant antimalarial activity against experimental infections. This project was run in conjunction with a similar scheme later established in the United States by the Office of Scientific Research and Development. From 1941 to 1945, the Americans tested over 14 000 compounds, around one-third of these being new substances.

During their North African campaign, German troops were equipped with supplies of sontoquine ('Resochin'). This was an analogue of mepacrine in which one of the benzene rings (i.e. that containing the methoxyl group) was absent. Samples of this were obtained from captured prisoners-of-war, then sent to the United States for analysis. Particular attention was paid to the fact that sontoquine was an aminoquinoline substituted in the 4-position rather than the usual 8-position. Biological screening of a close analogue prepared by Sterling-Winthrop researchers Alexander Surrey and H. F. Hammer, in which the methyl group on the 3-position of sontoquine was absent, revealed outstanding antimalarial activity. This analogue, chloroquine ('Aralen'), had been synthesized at Elberfield in 1934, and a German patent on it was awarded to the I. G. Farben just after the war began. However, Kikuth had dismissed it as being too toxic, preferring sontoquine instead. The Americans found chloroquine to have fewer side-effects than mepacrine, as well as reducing malarial fevers more quickly. Another important advantage over mepacrine was that it did not colour the skin yellow. It was not possible to institute large-scale production of chloroquine before the war ended, but later it superseded mepacrine altogether. It has been more widely used than amodiaquine ('Camoquin'), another substituted 8-aminoquinoline, which was developed near the end of the war by Parke, Davis, and Company.

Robert Elderfield and his colleagues at Columbia University in New York contributed to the wartime programme by noting that amongst the vast range of substituted 8-aminoquinolines related to pamaquin few primary or secondary amines had been reported. A variety of such compounds were then synthesized at Columbia, resulting in the emergence of primaquine as a less toxic analogue of pamaquin, which it immediately superseded as the drug of choice for eradication of benign tertian malaria caused by *Plasmodium vivax*. No superior agent for this purpose has been found during the forty years since then.

The British wartime researchers decided to avoid quinoline or acridine ring systems as a basis for designing novel antimalarials. Their principal success was based on the discovery that newly developed antibacterial sulphonamides containing pyrimidine rings had activity, albeit only weak, against malaria in humans. This was most marked in sulphadimidine ('Sulphamezathine'), which had been synthesized by Francis Rose of ICI. Attributing the antimalarial activity to the pyrimidine ring, Rose and his colleagues synthesized a range of pyrimidines incorporating molecular features that were present in mepacrine. Amongst these pyrimidines was compound '2666', which featured a basic side-chain and a chlorophenyl moiety. It was shown to be active in chickens infected with *Plasmodium gallinaceum*, but was of no clinical value. Nevertheless, using '2666' as the lead compound, several research groups proceeded to synthesize and examine in detail representative compounds of forty distinct chemical classes, as well as isolated examples of numerous other classes. Of these, the most effective proved to be biguanides designed as analogues of '2666' with the pyrimidine ring, as it were, broken open. Biguanides had previously been synthesized by ICI researchers when they were seeking novel analogues of antibacterial sulphonamides possessing pyrimidine rings. The first of these proved to be devoid of antimalarial activity, but fortunately it was realized that this might have been due to the presence of two basic side-chains (the biguanide system being strongly basic), a configuration known to be associated with lack of any biological activity. Replacement of the basic side-chain by simple alkyl groups, of which the isopropyl was consistently the most effective, led to the reappearance of strong antimalarial activity. Around two hundred biguanides were synthesized and tested, together with a similar number of related compounds, before proguanil (chloroguanide in the USA) emerged as being superior to mepacrine. Clinical trials at the Liverpool School of Tropical Medicine confirmed that it was a first-line drug for the treatment of the erythrocytic phase of malaria, although it is now used mainly for the prophylaxis of malaria in those parts of the world where the parasites have not yet developed resistance to it. It is marketed under the brand name of 'Paludrine'.

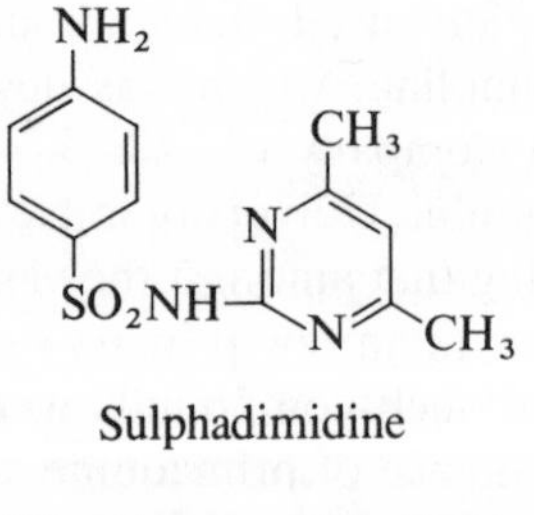

Sulphadimidine

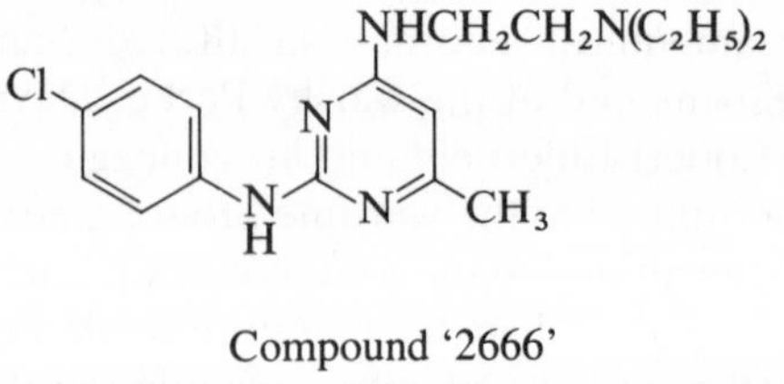

Compound '2666'

In 1949, George Hitchings, associate research director in the Tuckahoe, New York, laboratories of Burroughs Wellcome, with his assistants Elvira Falco and Peter Russell, noticed that one of their potent antifolic acid drugs (see page 291) bore a structural relationship to proguanil, and accordingly tested it as an antimalarial. When encouraging results were obtained, a large series of derivatives was synthesized. This led to the discovery of pyrimethamine ('Daraprim') two years later. It was evaluated in London at the Wellcome

Laboratories of Tropical Medicine, and has been widely used since then as an alternative to proguanil.

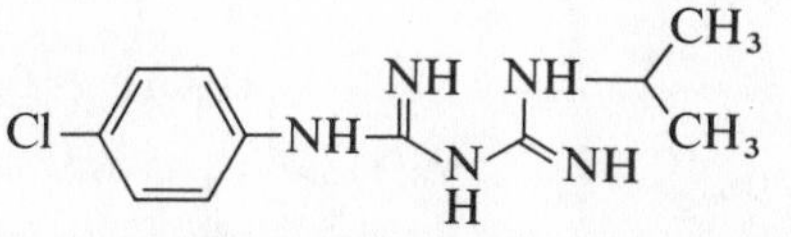

Proguanil

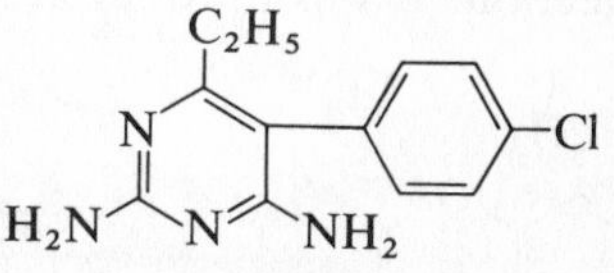

Pyrimethamine

SCHISTOSOMICIDAL DRUGS

In 1936, the I. G. Farben turned its attention to remedies for schistosomiasis, possibly because of German military interest in the Middle East. Kikuth and Goennert, having devised screening methods to reveal active compounds when injected into mice infected with *Schistosoma mansoni*, examined the vast range of heterocyclic compounds prepared in the course of the research programme that resulted in the development of pamaquin and mepacrine. A xanthone compound prepared by Mauss looked promising, and was given the name 'Miracil A' in 1938.Its high toxicity precluded any possibility of clinical application, but the somewhat less toxic analogue 'Miracil D' was acceptable as an orally active remedy for schistosomiasis. It still caused many unpleasant side-effects, such as vomiting, dizziness, and hallucinations. It was employed in Africa during the war, and subsequently received the approved name of lucanthone. Attempts to find a more tolerable analogue were continued at Elberfield after the war. In 1956, Mauss and his colleagues introduced a series of simplified 4-toluidine analogues known as the mirasans. Although active in infected mice, these gave disappointing results in man. Their importance lies in the fact that they have served as the starting point for other developments, notably the discovery of oxamniquine ('Mansil') in 1968 by R. Foster and B.L. Cheetham of the Pfizer laboratories in Sandwich, England. A large-scale trial of this begun in Brazil four years later indicated that a single oral dose cured more than ninety per cent of patients infected with *Schistosoma mansoni*. Unlike earlier drugs, it was well tolerated by patients. Although oxamniquine has become the drug of choice for treatment of the widely encountered *Schistosoma mansoni* infections, it has poor activity against those caused by *Schistosoma haematobium*, which is mainly responsible for the disease in Africa and the Middle East.

There was good reason to believe that lucanthone underwent metabolic conversion to an active metabolite, but all attempts to isolate this failed until 1965, when Archer and his colleagues from the Sterling Winthrop Company in New York found that lucanthone was metabolized by *Aspergillus scleroticum*. Three products were formed, one of which, a hydroxy derivative of lucanthone, was shown to be the long sought after active metabolite. It had similar schistosomicidal activity to the parent compound, but its greater potency permitted sufficient to be dissolved in water for it to be given by intramuscular injection. It was introduced in its own right as hycanthone ('Etrenol').

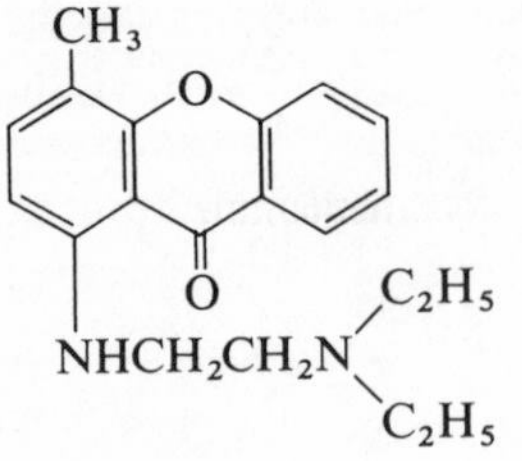

'Miracil A'

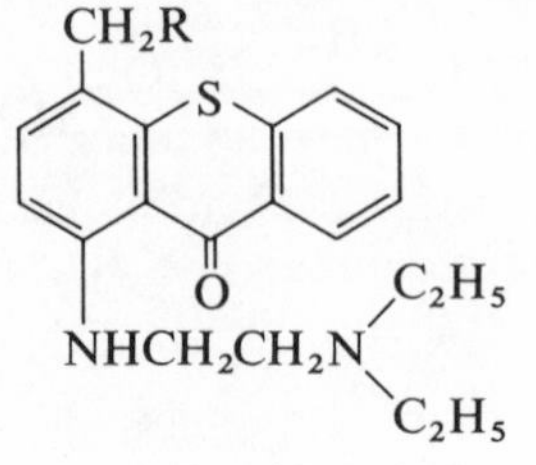

Lucanthone, R = —H
Hycanthone, R = —OH

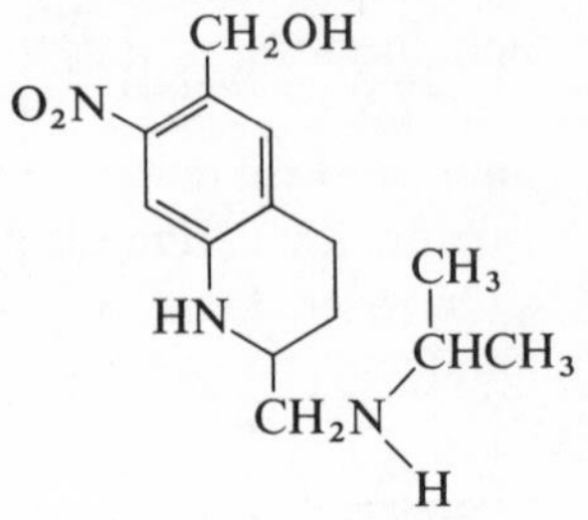

Oxamniquine

NITROIMIDAZOLES AND NITROTHIAZOLES

The introduction in 1949 of chloramphenicol, a nitroaromatic compound, led to the screening of aromatic and heteroaromatic nitrocompounds for chemotherapeutic activity. Since the nitrofurans (see page 70) were useful antiseptics, it was natural that particular attention should be paid to other five-membered heterocyclic rings. Several have proved effective against protozoal infections. Thus, Ciba developed acinitrazole (aminitrazole) for histomoniasis in 1950, while its deacetylated derivative, aminonitrothiazole ('Entromin'), was patented the following year by Monsanto. This was used as an antihistomonad to prevent blackhead disease in turkeys. When Nakamura and Umezawa in Tokyo identified azomycin, an antibiotic isolated in 1953 from a species of *Streptomyces*, and showed it to be 2-nitroimidazole, it was promptly screened. It turned out to have trichomonicidal activity, but was rather toxic. However, in 1957 Jacobs and his colleagues at the Rhône-Poulenc laboratories managed to synthesize a nitroimidazole with antibacterial as well as antiprotozoal activity, and low toxicity. Under the approved name of metronidazole ('Flagyl') this is widely used medicinally in surgical and gynaecological sepsis, notably in the treatment of trichomonal vaginitis. The nitrothiazole known as niridazole ('Ambilhar'), introduced by Ciba in 1963, is a broad-spectrum antiprotozoal agent. Its toxicity limits its range of uses, and it is mainly of value in the elimination of guinea worms, *Dracunculus medinensis*.

Non-nitrated imidazoles have also been found to have chemotherapeutic activity, mainly as anthelmintics for eradication of worms. The active compounds have been discovered through extensive screening of hundreds of compounds, as typified by the introduction of thiabendazole ('Mintezol') by Merck and Company in 1961. Shortly after, the Janssen Research Laboratory at Beerse in Belgium initiated an extensive screening programme in which 2721 novel heterocyclic compounds were tested for anthelmintic activity against three types of parasitic worms before an aminothiazole derivative was found to be effective in chickens and sheep. Its failure in mice and rats pointed to the possibility that it had to undergo metabolic conversion to an active drug that was only formed in some animals. All the metabolites were then isolated and synthesized. The only one which turned out to be active was difficult to produce and, in addition, was unstable in water. A large series of its analogues was then synthesized, one of which met all the requirements for possible clinical application. This was given the approved name of tetramisole, but the laevo-isomer was selected for medicinal use since it was several times more potent, yet no more toxic. It is employed under the name levamisole as an ascaricide to eliminate the common round worm.

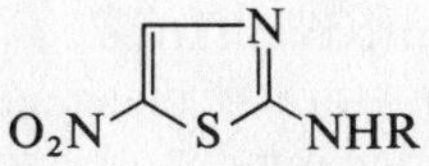

Acinitrazole, R = $-COCH_3$
Aminonitrothiazole, R = $-H$

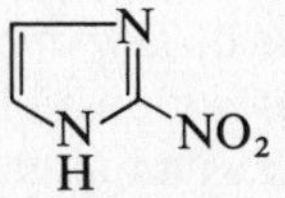

Azomycin

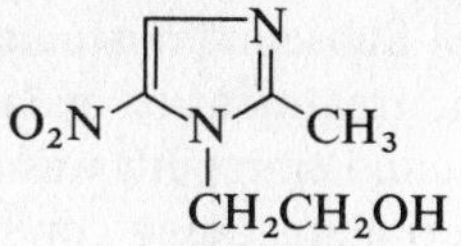

Metronidazole

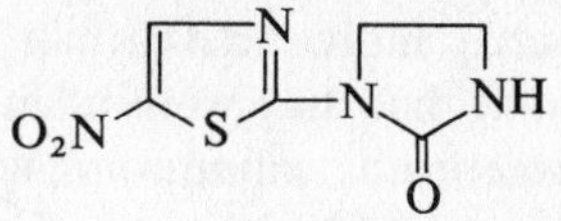

Niridazole

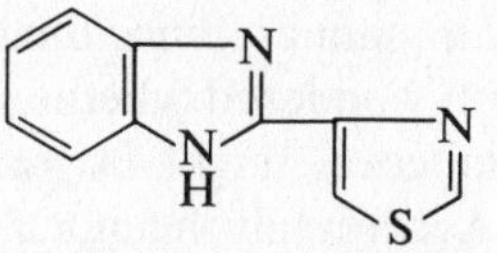

Thiabendazole

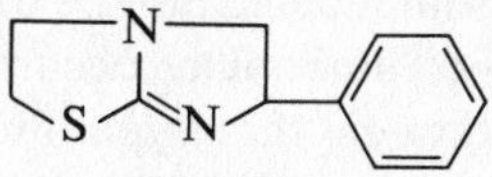

Levamisole

XIV

Synthetic Antibacterial Drugs

The first advances leading to the development of antibacterial chemotherapy were taken by two medically qualified scientists who had both worked with Ehrlich in Frankfurt. In 1911, Julius Morgenroth attempted to devise a biological test that would enable him to screen compounds for antimalarial activity. Working in the Charité Hospital in Berlin, he found that mice infected with a strain of *Trypanosoma brucei* could be cured with large doses of quinine. Since it was not possible to infect mice with any form of plasmodium, he used this system as an alternative. Amongst the compounds to be screened were hydrogenated derivatives of quinine that had been synthesized by G. Giemsa and Josef Halberkann in the chemistry laboratory at the Institute for Maritime and Tropical Diseases, Hamburg. These alkylhydrocupreines proved to be somewhat more active than quinine itself. Subsequent studies, however, indicated that they were of little value in the treatment of malaria.

Morgenroth's laboratory, which was small and cramped, was also used for routine pathological work. One of the experiments being conducted at this time involved dissolving pneumococci (i.e. the bacteria that caused pneumonia) with the aid of bile salts. By accident, an assistant found that Morgenroth's trypanosomes could also be dissolved in a similar manner. Since this was by no means a commonplace occurrence, Morgenroth wondered whether any similarity between the trypanosomes and pneumococci might be reflected in sensitivity towards quinine and its derivatives. Accordingly, he inoculated mice with pneumococci, then tested his compounds by immediately injecting them in the vicinity of the inoculations. Results with quinine were unimpressive, but its hydrogenated analogue, methylhydrocupreine (i.e. dihydroquinine), prevented the spread of infection. The next member of the series, ethylhydrocupreine, was much more effective, and consistently prevented the spread of infection in mice, guinea pigs, and rabbits. However, when injected intravenously or into the abdomen it sometimes damaged the optic and acoustic nerves. Despite this, Zimmer and Company of Frankfurt marketed ethylhydrocupreine ('Optochin') as a remedy for pneumonia. It was the first antibacterial chemotherapeutic agent introduced into medicine, and was prescribed on the Continent and in America until the advent of sulphapyridine in 1938. Some patients were blinded by it, and in an attempt to avoid this it was reformulated as an oral suspension of the almost insoluble free base in order to reduce the rate of absorption from the gut and thus avoid toxic blood levels. Doses of 250 mg were administered five times a day for three days.

Morgenroth also examined several analogues of ethylhydrocupreine and found that as the ethyl ether function was replaced by longer alkyl groups the activity against pneumococci diminished, but that against streptococci and staphylococci increased. During the First World War he introduced isoamylhydrocupreine ('Eucupin') and iso-octylhydrocupreine ('Vuzin') for direct application to deep wounds. These compounds were highly effective even in the presence of serum, but they were too toxic to be of use in the therapy of generalized infections. Similar results were obtained by Walter Jacobs and Michael Heidelberger of the Rockefeller Institute, New York.

Methylhydrocupreine, $R = -CH_3$
Ethylhydrocupreine, $R = -CH_2CH_3$
Isoamylhydrocupreine, $R = -CH_2CH_2CH(CH_3)_2$
Iso-octylhydrocupreine, $R = -(CH_2)_5CH(CH_3)_2$

Shortly after Morgenroth discovered the antipneumococcal activity of ethylhydrocupreine, John Churchman of Johns Hopkins University, Baltimore, published a detailed report on how to kill selectively certain types of bacteria with gentian violet (now known as crystal violet, and similar in composition to methyl violet) in order to permit the isolation of pathogenic bacteria. Reviewing previous work, Churchman mentioned that several dyes had been considered to have bactericidal properties. The earliest to be reported had been a crude form of gentian violet called 'pyoctanin' by J. Stilling in his thesis submitted to the University of Strassburg in 1890. Churchman's citation of this and the subsequent studies caught the attention of Carl Browning, the second of Ehrlich's former associates whose work was to pave the way to the introduction of antibacterial chemotherapy. He was now based at the University of Glasgow medical school in the Western Infirmary, where he proceeded to test a wide variety of dyes for antibacterial activity. Amongst these was 'Trypaflavine', a yellow acridine dye designed by Ehrlich after he discovered that the trypanocidal activity of tryparosan, a triphenylmethane dye related to crystal violet, was due entirely to contamination of it by small amounts of acridines. Louis Benda of the Hoechst Dyeworks had synthesized 'Trypaflavine' for Ehrlich. Browning had recently become interested in it after Ehrlich had claimed it was the most potent trypanocide he had ever worked with, being highly effective against virulent strains of *Trypanosoma brucei* in mice. Unfortunately, it soon proved to be of no value in the therapy of trypanosomiasis in larger animals or man. Nevertheless, Browning found 'Trypaflavine' to be active against a wide range of bacteria. Unexpectedly, he also discovered it to be effective against a variety of pathogenic organisms in the presence of serum. Apart from Morgenroth's ethylhydrocupreine, there was then no other antiseptic known to be active in the presence of body fluids. Ehrlich and Bechhold had demonstrated in 1906 that antiseptics were deactivated by blood serum.

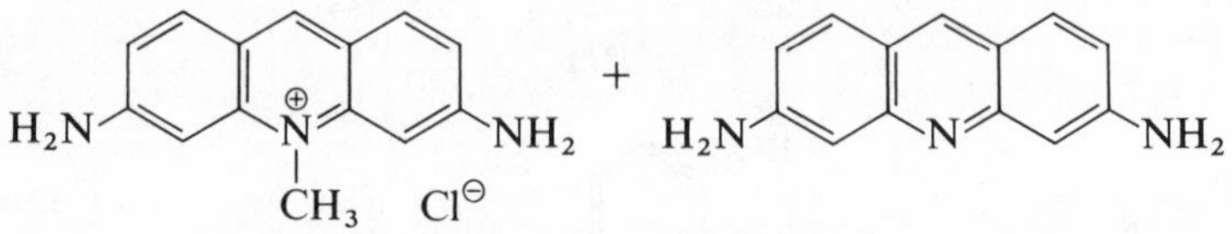

H_2N $N^{\oplus}$ NH_2 CH_3 $Cl^{\ominus}$ + H_2N N NH_2

Acriflavine ('Trypaflavine')

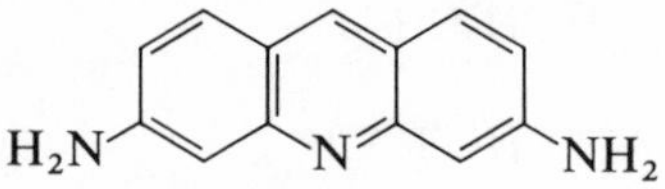

H_2N N NH_2

Proflavine

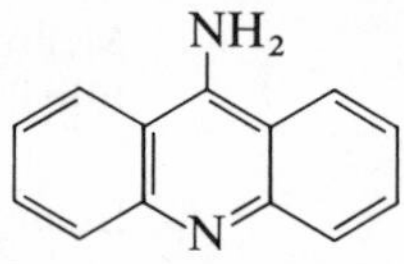

NH_2 N

Aminacrine

The year following Browning's discovery of the antibacterial properties of 'Trypaflavine' saw the outbreak of war. Before long, there was deep concern in military circles over the unprecedented rate of fatal infections amongst the wounded, probably arising from a combination of the nature of the injuries caused by modern weapons and the fact that bacteria thrived in the soil of the well-fertilized battlefields of northern Europe. To meet this crisis, the newly established Medical Research Committee asked Browning to develop his work on 'Trypaflavine' with a view to finding an antiseptic that could be used on deep wounds. He set up a laboratory in the Bland-Sutton Institute of Pathology at the Middlesex Hospital, London, assisted by R. Gulbransen, E. L. Kennaway, and L. H. D. Thornton. After conducting initial studies, they selected two dyes for trial in casualty clearing stations at the front lines and in base hospitals. These were brilliant green (an analogue of crystal violet), and 'Trypaflavine', which, since it had no clinical value in trypanosomiasis, was renamed simply 'Flavine'. The latter proved to be valuable in the prevention and treatment of sepsis in wounds, and was given the approved name of acriflavine. The large amounts of it required for the field trials were synthesized by two chemists recruited from the Burroughs Wellcome company to assist the war effort, namely George Barger and Arthur Ewins, who now worked in the MRC's laboratory at the Lister Institute for Preventive Medicine. It was not until 1934 that it was realized that acriflavine was a mixture of two components, one of which proved superior and was introduced during the Second World War under the approved name of proflavine. A non-staining analogue, aminacrine, was also made available during the war.

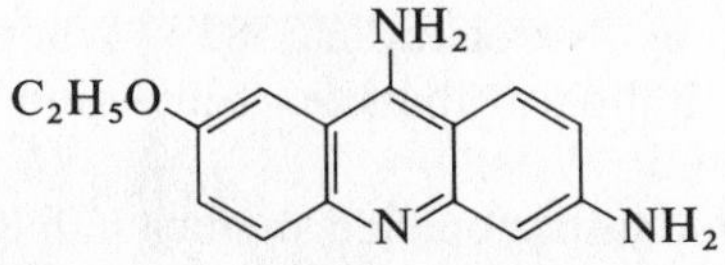

NH_2 C_2H_5O N NH_2

Ethacridine

Morgenroth and Robert Schnitzer asked H. Jensch of the Hoechst Dyeworks to synthesize an acridine derivative that incorporated part of the 'Optochin' structure, hoping that this might be an effective antibacterial when injected. This led to the introduction of ethacridine ('Rivanol') in 1920. An important feature of its molecular structure was the placing of an amino group at the 9-position (sometimes described as the 5-position when a different numbering convention is used) in order to increase the similarity to 'Optochin'. Although this reduced toxicity, clinical experience with 'Rivanol' showed that it was of similar value to acriflavine, being unable to cure generalized infections.

CH_3 $NHCHCH_2CH_2CH_2N(C_2H_5)_2$ O_2N N

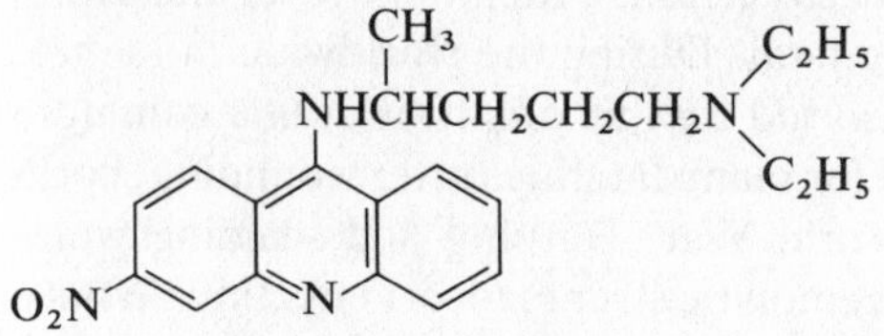

Nitro Analogue of Mepacrine

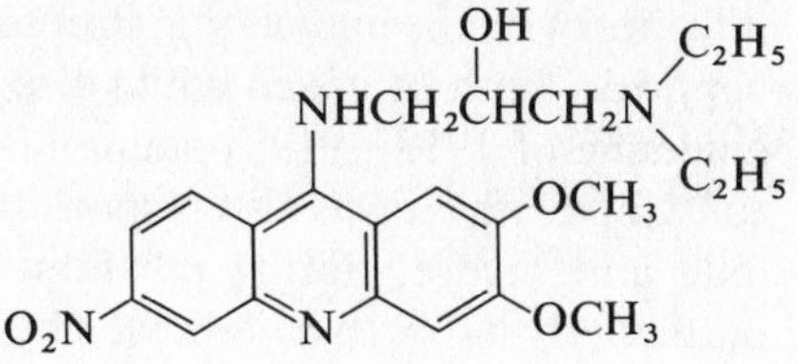

OH $NHCH_2CHCH_2N(C_2H_5)_2$ OCH_3 OCH_3 O_2N N

'Entozon'

After Morgenroth's early death, Schnitzer accepted a post at the Hoechst Dyeworks, where he continued with the attempt to find an acridine that could cure generalized infections. He examined hundreds of compounds and by 1926 was able to control acute streptococcal infections in mice by administering, either orally or by injection, large doses of 9-aminoacridines substituted at the 3-position with a nitro group. One of these contained the basic side-chain that had conferred antimalarial activity on pamaquin, recently discovered at the Elberfield laboratories of I. G. Farbenindustrie, of which the Hoechst Dyeworks was now a constituent part. This compound was actually an analogue of the as yet to be synthesized mepacrine, the only difference being the presence of a nitro group rather than a chlorine atom at the 3-position. This must have been prepared as a potential quinine substitute and then given to Schnitzer for further evaluation, it having by now become company policy to screen promising chemotherapeutic agents for antimalarial, trypanocidal, and antibacterial activity. Whilst effective against both trypanosomes and streptococci in animals, this particular nitroacridine was not potent enough to be considered for clinical application. Nevertheless, it seems to have been the lead compound from which mepacrine was derived. Analogues of it were investigated as potential trypanocides by Schnitzer and Silberstein in 1928, and one these proved to be a promising antibacterial agent. It was marketed as 'Entozon' (also known as 'Nitroakridin 3582'). When tried in the clinic, it caused severe tissue irritation at the site of injection, as well as unpleasant side-effects.

In 1927, I. G. Farbenindustrie opened an outstandingly well-equipped suite of new research laboratories at Elberfield. Gerhard Domagk was appointed as director of the Institute of Experimental Pathology, with the responsibility of continuing the search initiated by Schnitzer at Hoechst for a drug effective against generalized bacterial infection. Recognizing the clinical failure of the nitroacridines despite their promising activity in mice, he introduced more rigorous test conditions that would eliminate all but the most effective compounds. This involved screening dyes on mice inoculated with a highly virulent strain of haemolytic streptococcus, i.e. *Streptococcus pyogenes*. The commonest diseases caused by this organism were tonsillitis and scarlet fever, from which most patients recovered uneventfully. However, the streptococci sometimes invaded the middle ear to cause otitis media, resulting in permanent deafness. Occasionally, fatal meningitis also resulted. Further complications of infection with the haemolytic streptococcus included rheumatic fever and acute nephritis, both of which could also be fatal. During the worldwide influenza epidemic of 1918–1919, pneumonia caused by this organism was a common cause of death. It was also responsible for many fatalities after wounding, both mild and severe, during the First World War. Burning and scalding were particularly likely to be followed by haemolytic streptococcal infection. Whatever the original cause of infection, the appearance of haemolytic streptococci in the blood of a patient, septicaemia, was an ominous sign.

The particular strain of haemolytic streptococci selected by Domagk was isolated from a patient who had died from septicaemia, and its virulence had been increased by repeatedly subculturing it in mice. This resulted in the test system being particularly reliable since one hundred per cent of the mice consistently died within four days of inoculation. Only an exceptional drug could influence this otherwise inevitable outcome.

Domagk began his investigations by testing three classes of substances which had been reported as having clinically useful antibacterial properties, namely gold compounds, acridines, and azo dyes. The first to exhibit activity in the mice were organic gold compounds. Adolf Feldt of the Bacteriology Institute of the Hoechst Dyeworks had been experimenting with gold compounds since 1913, when he established that gold was toxic to the mycobacteria that caused tuberculosis. In 1926, he and Schiemann reported that a single injection of gold sodium thiosulphate ('Sanochrysine'), which had recently been found effective in bovine tuberculosis (see page 90), consistently protected mice against lethal inoculation of streptococci. Domagk confirmed these findings and showed that gold compounds could also cure larger animals that were similarly infected. Unfortunately, kidney damage frequently prevented the administration of the dosage required to cure streptococcal infections in patients.

After this setback with gold therapy, Domagk turned his attention to acridine and azo dyes. One of the latter, phenazopyridine ('Pyridium'), had

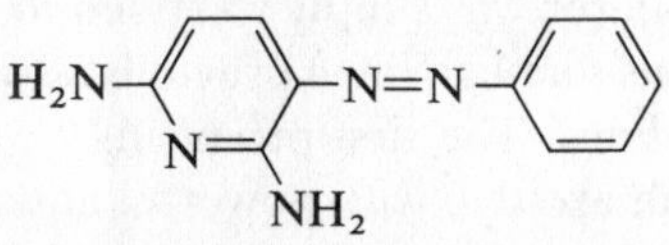

Phenazopyridine

just been introduced as a urinary antiseptic by Ivan Ostromislensky, a New York industrial chemist who had suspected it might have bacteriostatic activity. After confirming this to be the case, he noticed that it imparted a red colour to the urine of animals. This observation was profitably exploited, and the dye is still prescribed to relieve urinary tract pain even though it has little efficacy as an antiseptic. Its main importance, however, lies in the fact that the low toxicity of it and its analogues encouraged Fritz Mietzsch and Josef Klarer to synthesize a range of related azo dyes for Domagk to screen in mice. Such compounds

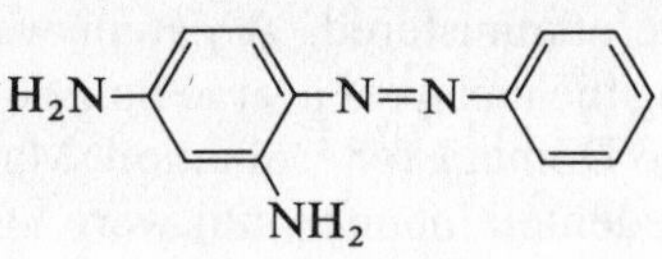

Chrysoidine

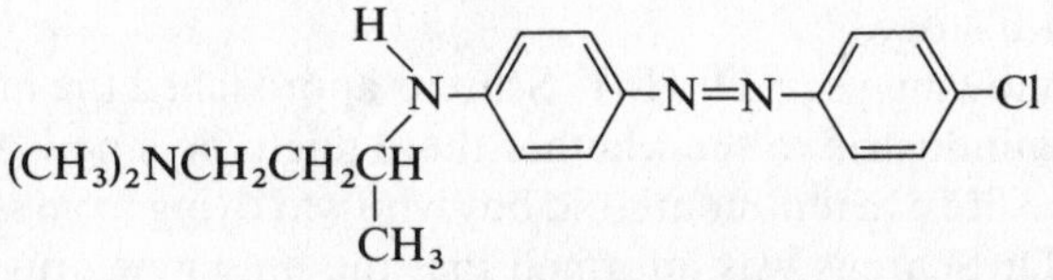

Basic Chrysoidine Derivative

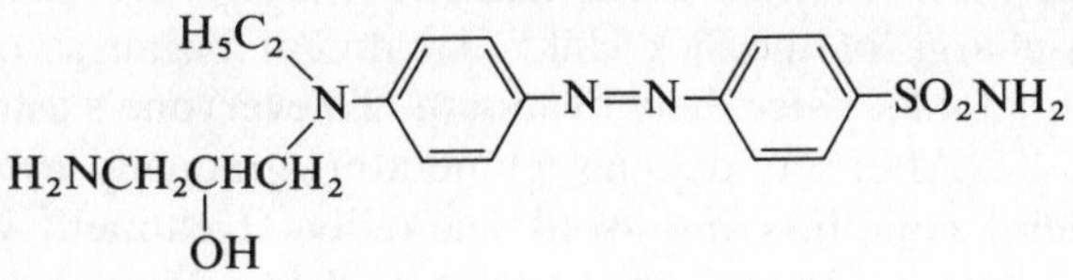

First Antibacterial Sulphonamide

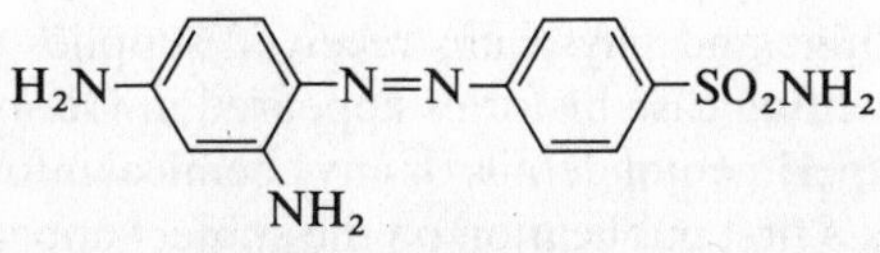

Sulphamidochrysoidine
('Prontosil Rubrum',
'Streptozon')

presented an attractive proposition to the I. G. Farbenindustrie chemists since they were easier to synthesize than the acridines which, in any case, were affording only negative results in Domagk's rigorous test system. The approach taken by Mietzsch and Klarer was simply to attach to an appropriate azo dye molecule those side-chains that had been previously found to confer antistreptococcal activity on acridines. The first promising result came when this was done with chrysoidine, an azo dye with powerful antiseptic properties, which Philipp Eisenberg had evaluated at the University of Breslau in 1913. Domagk found that the basic side-chain attached to this molecule greatly enhanced activity against cultures of streptococci, but negative results were still obtained in infected mice. Mietzsch and Klarer next took up an old idea developed in 1909 by Heinrich Horlein, now director of the medical division of I. G. Farbenindustrie, but then merely a young chemist. This involved introducing a sulphonamido function into azo dyes in order to enhance their ability to bind to wool. When Domagk screened the sulphonamide derivatives of the most promising of the dyes already examined they turned out to be almost ineffective against cultures of streptococci. Undaunted by this, Domagk tested the sulphonamide dyes on infected mice. For the first time since the screening programme had begun, a genuine protective effect was apparent when high doses of the dyes were administered. A patent was applied for on the 7th November 1931. During the next year a large range of sulphonamide dyes were synthesized and given to Domagk for evaluation. Many of these not only cured the mice in an unprecedented manner, but were also remarkably non-toxic. Finally, on Christmas Day 1932, a patent application for another batch of sulphonamide dyes was submitted. Amongst these was a red dye that was to make medical history.

Early the following year, Dr H. T. Schreus approached the medical division of I. G. Farbenindustrie to see whether there might be a new drug that could help to save the life of a ten-month-old boy who was dying from straphylococcal septicaemia. Dr Schreus was informed that the only new drug available was intended for use in streptococcal infections, and then he was promptly supplied with tablets of 'Streptozon'. Schreus handed the tablets over to Dr Foerster, his colleague in charge of the sick child. Treatment began at once, the baby receiving half a tablet twice daily by mouth. To everyone's astonishment, the baby did not die. After four days his temperature gradually lowered to normal and his general condition improved markedly. Treatment was eventually stopped after three weeks and the child was discharged from hospital. The case was reported by Dr Foerster to a meeting of the Düsseldorf Dermatological Society on the 17th May 1933.

Several other Rhineland physicians received supplies of 'Streptozon', and three brief reports citing case histories appeared in German medical journals during 1934. No experimental details or any chemical information appeared in print until Domagk's first publication on the subject appeared in the *Deutsche Medizinische Wochenschrift* of 15th February 1935. In this he described how small, non-toxic doses of the brick-red sulphonamide dye called 'Prontosil

Rubrum' prevented every single mouse which received it by stomach tube from succumbing to an otherwise lethal inoculation of haemolytic streptococci. Thirteen out of fourteen untreated mice died within three to four days. Domagk also explained that in rabbits 'Prontosil Rubrum' had been able to cure chronic streptococcal infections and alleviated those caused by staphylococci. 'Prontosil Rubrum' was a new name for 'Streptozon' and later it was given the approved name of sulphamidochrysoidine, although this was rarely used. Accompanying Domagk's paper were three others with clinical reports of two years of investigations into the remarkable antistreptococcal action of 'Prontosil Rubrum'. Although the bacteriological work supporting these and the earlier clinical studies left much to be desired, they did testify to both the efficacy and safety of the dye, which made it the first truly effective chemotherapeutic agent for any generalized bacterial infection.

Surprisingly, Domagk's paper on 'Prontosil Rubrum' did not create the sensation that might have been expected. There was an understandable conviction in medical circles that chemotherapeutic agents could have little effect against generalized infections. Only clear-cut clinical results could be expected to allay any doubts. That such results quickly became available was largely due to the efforts of Leonard Colebrook of Queen Charlotte's Maternity Hospital in London. A well-equipped unit supported by the Medical Research Council, the Rockefeller Foundation, and the Bernhard Baron Trustees, had been opened there in 1931 to seek a solution to the apparently intractable problem of puerperal fever. This was then the cause of death in two or more out of every thousand women a few days after giving birth. Colebrook was the bacteriologist in charge of the unit; he was the leading British expert on the chemotherapy of streptococcal infection, the cause of about half of the cases of puerperal fever. A particularly worrying form of this disease was that caused by haemolytic streptococci. Typically this had a mortality rate of at least twenty-five per cent. In 1931, Dr John Smith of Aberdeen demonstrated that patients were infected by streptococci from the nose and throats of their medical attendants. This led Colebrook to insist on the wearing of masks by his staff. He also arranged with his cousin, a chemist who worked for the Reckitt company in Hull, for the development of a non-irritant antiseptic that could kill streptococci on the skin of the midwives' hands. He experimented on this by smearing his own hands with virulent bacterial cultures, a bold procedure that led to the introduction of chloroxylenol solution ('Dettol').

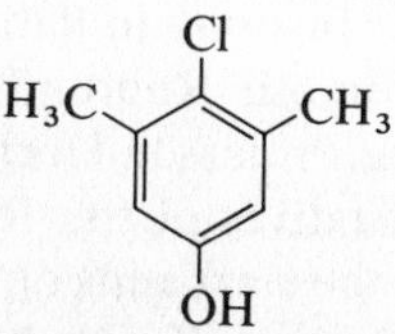

Cl

H_3C CH_3

OH

Chloroxylenol

After Professor Horlein had delivered a lecture on 'Prontosil Rubrum' to the Royal Society of Medicine in London on 3rd October 1935, Colebrook tried to obtain samples of it. With material eventually supplied from France, he confirmed Domagk's results in infected mice, although only after selecting a particularly virulent strain of haemolytic streptococci. In January 1936 he began to use the drug in patients. Shortly after this, he received supplies of 'Prontosil Rubrum' and 'Prontosil Soluble' from Germany for use in a clinical trial that was to involve 38 dangerously ill women. The following June, Colebrook and Meave Kenny reported in the *Lancet* that only three of these patients died. This publication had considerable impact, especially amongst pharmaceutical companies. Following it, Colebrook and his colleagues went on to treat a further 26 seriously ill women, none of whom died.

Domagk was awarded the Nobel Prize for Medicine in 1939 for his discovery of the antibacterial properties of 'Prontosil Rubrum'. After acknowledging notification of the award he was detained by the Gestapo and persuaded to reject it. This strange state of affairs was a direct consequence of Hitler's rage over the award of the 1936 Nobel Peace Prize to Carl von Ossietsky. It was not until two years after the war ended that Domagk was able to travel to Stockholm to receive his medal, by which time the prize money had reverted to the Nobel Foundation. For Domagk, however, the greatest reward must surely have been when, in February 1935, the life of his own daughter, Hildegarde, was saved by 'Prontosil Rubrum' after she developed a severe septicaemia caused by pricking her finger with a needle.

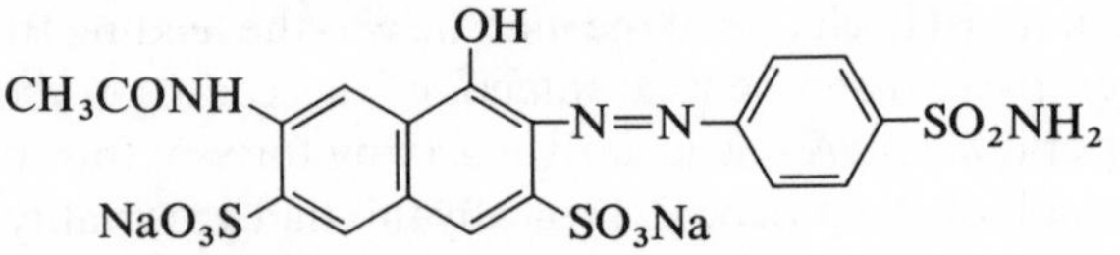

OH

CH_3CONH — $N{=}N$ — SO_2NH_2

NaO_3S SO_3Na

'Prontosil Soluble'

Apart from the introduction of 'Prontosil Soluble', an injectable sulphonamide azo dye, in July 1935, no other major development in this field came from the I. G. Farbenindustrie laboratories. The initiative passed to French, British, American, and Swiss workers despite over one thousand sulphonamides having been synthesized at Elberfield during the five years following the first recognition of antibacterial activity amongst such compounds.

At the beginning of April 1935, Rumanian-born Professor Constantin Levaditi of the Alfred Fournier Institute in Paris asked Louis Benda of I. G. Farbenindustrie for some 'Prontosil Rubrum'. This immediately led to a high-level conference involving Professor Horlein, Ernest Fourneau of the Pasteur Institute, and representatives of the Rhône Poulenc Company. No agreement could be reached on the marketing of 'Prontosil Rubrum' in France, so Fourneau instructed one of his staff, Dr A. Girard, to synthesize it for Levaditi by the method specified in the French patent. Since French law allowed 'Prontosil Rubrum' only to be patented as a dye and not as a medicine, Girard was free to synthesize it and arrange for its manufacture by Roussel Laboratories under the name 'Rubiazol' (strictly speaking, this compound which was identical to 'Prontosil Rubrum' should be described as 'Rubiazol I'; three slightly different analogues were subsequently introduced). By using this, Levaditi and his assistant, A. Vaisman, were able to tell a meeting of the Académie des Sciences on 6th May that Domagk's claims were correct. Shortly after, a group at the Pasteur Institute showed that Domagk had been fortunate in having worked with a strain of virulent streptococci that were atypically sensitive to 'Prontosil Rubrum'. However, by far the most important announcement from Fourneau's laboratory in 1935 was that by the Tréfouëls, Nitti, and Bovet, who suggested that the azo linkage of the red dye was cleaved in the patient's body to form 4-aminobenzene sulphonamide, a colourless compound. This was promptly synthesized by Fourneau as 1162 F, using the method described in the literature in 1908 by Paul Gelmo of Vienna, who had prepared it for his doctoral thesis. Tests quickly revealed that it retained the activity of 'Prontosil Rubrum', hence it was called, for a while at any rate, 'Prontosil Album'. The approved name given to this non-patentable compound was sulphanilamide. A few months later, it was isolated from the urine of patients by Colebrook's assistant, Albert Fuller. Professor Horlein subsequently admitted that the fact that this was actually the active form of 'Prontosil Rubrum' had already been discovered by his company. There has been speculation that the two-year delay by I. G. Farbenindustrie in bringing the sulphonamides on to the market could have been caused by efforts to find some way of protecting their discovery from exploitation by rival manufacturers. The company's reply to this was that the clinical results were unprecedented and thus required careful validation.

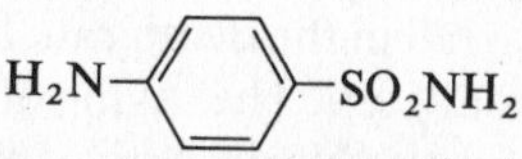

$$H_2N-C_6H_4-SO_2NH_2$$

Sulphanilamide

Since sulphanilamide could not be patented, any manufacturer was free to incorporate it in his own formulation. In 1937 an Elixir of Sulphanilamide was put on sale in the United States. This contained ten per cent of diethylene glycol as a solvent to render the sulphonamide soluble. During the two months that this preparation was on sale, at least 76 people died from severe damage to the liver and kidneys caused by the solvent. As a consequence of this episode, the US Congress enacted a Food and Drug Act in an attempt to prevent such a state of affairs ever arising again. That the United States was spared the thalidomide tragedy twenty years later was one direct consequence of this legislation.

Diedrich and Dohrn of the Schering-Kahlbaum company of Berlin introduced the acetyl derivative of sulphanilamide in 1938. Known as sulphacetamide ('Albucid'), this was rapidly excreted by the kidneys, resulting in high urinary concentrations that favoured its use in urinary tract infections. Since solutions of its sodium salt were not as alkaline as those of other sulphonamides this was ideal for application to the eye, a purpose for which it is still widely employed.

$H_2N-C_6H_4-SO_2NHCOCH_3$

Sulphacetamide

$H_2N-C_6H_4-SO_2NH-C_5H_4N$

Sulphapyridine

The most important development following the introduction of sulphanilamide was the synthesis of sulphapyridine ('M & B 693'). This was made at the suggestion of the director of research at the May and Baker laboratories, Arthur Ewins, who wanted to study the action of sulphonamides substituted with a heterocyclic ring on the sulphonamide nitrogen. The actual synthesis was carried out by Dr M. A. Phillips in March 1937. The compound proved to be not only more potent on a weight basis than sulphanilamide, but it also had a wider spectrum of antibacterial activity. It was effective against pneumococci, meningococci, gonococci, and other organisms. As a result, it was supplied to Dr L. E. Whitby at the Middlesex Hospital in London. He established that it had an unprecedented efficacy against mice inoculated with pneumococci, despite being less toxic than sulphanilamide. After tests on volunteers amongst the staff of May and Baker, the first patient received the drug in March 1938. In the first clinical trials sulphapyridine fully lived up to its early promise by reducing the mortality rate amongst patients with lobar pneumonia from 1 in 4 to only 1 in 25, thereby ensuring that this dread disease was no longer one of the commonest causes of death amongst otherwise healthy adults. Epidemiological tables clearly depict a sharp fall in the death rate from pneumonia amongst the population at large in the year 1939. The saving of the life of Winston Churchill by 'M & B 693' during his visit to North Africa in December 1943 had as great an impact on the popular imagination as had the anaesthetizing of Queen Victoria with chloroform some ninety years earlier.

Other heterocyclic sulphonamides quickly followed sulphapyridine on to the market. Sulphathiazole, first reported in April 1939 by Russel Fosbinder and L. A. Walter of the Maltbie Chemical Company of Newark, New Jersey, proved to be more potent, whilst sulphadiazine, prepared the following year by Richard Roblin and his colleagues at the Stamford Research Laboratories of the American Cyanamid Company, was not only more potent than sulphpyridine, but also less toxic. Furthermore, it had a wider spectrum of activity than any previous sulphonamide, hence it was used extensively during the war. Sulphamerazine, its methyl derivative, was described in the same original publication. It was more soluble than previously known sulphonamides (due to its stronger acidity) and since it was slowly excreted by the kidneys it did not have the disturbing tendency of the earlier sulphonamides to crystallize in the fine tubules of the kidneys (crystalluria), a phenomenon that caused much pain. Because the analogous methyl derivative of sulphathiazole had caused peripheral neuritis amongst 1 in 50 patients, doctors in the United States were at first reluctant to use this sulphonamide. Their prejudice against it was finally overcome through the widespread use by British physicians of the related sulphadimidine (known as sulphamethazine in the USA, this being the brand name used by ICI in the UK), a drug first synthesized at Temple University in Philadelphia by William Caldwell and two of his students, Edmund Kornfeld and Conrad Donnell, as part of their master's degree requirements. It had even greater solubility though it proved somewhat less potent.

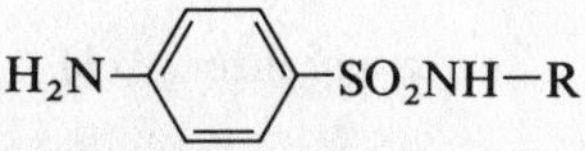

$H_2N-C_6H_4-SO_2NH-R$

	R
Sulphathiazole	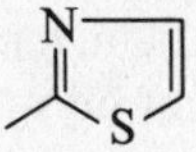
Sulphadiazine	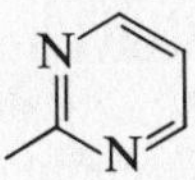
Sulphamerazine	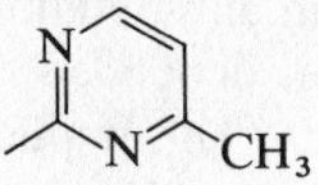 CH_3
Sulphadimidine	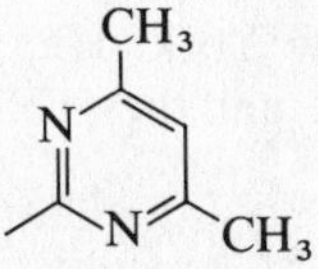 CH_3, CH_3
Sulphamethoxypyridazine	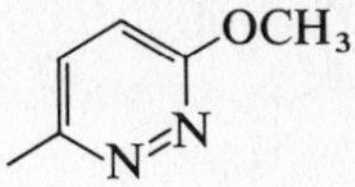 OCH_3
Sulphamethoxazole	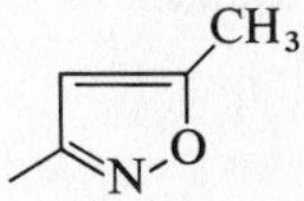 CH_3

During the early forties a variety of heterocyclic sulphonamides were simultaneously discovered in different laboratories. This led to conflicting patent claims, with several manufacturers involved. A further spate of new products appeared on the scene after 1957, when researchers at the Pearl River

Laboratories of the American Cyanamid Company discovered that introduction of a methoxyl group into the heterocyclic ring, to form a compound such as sulphamethoxypyridazine, slowed down elimination by the kidneys. This meant less frequent administration was required. The methoxyl group somehow increased binding to plasma proteins, the bound drug not being filtered through the kidneys. Rival manufacturers then marketed methoxy derivatives of their own sulphonamides. These and other sulphonamides besides those mentioned here have been used clinically. They are dealt with in standard texts.

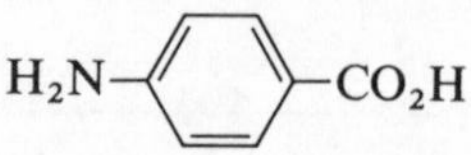

$H_2N-C_6H_4-CO_2H$

4-Aminobenzoic Acid

Soon after the introduction of sulphanilamide into clinical practice, it was discovered that its antibacterial activity was antagonized by pus, as well as by tissue or yeast extracts. In 1940, Donald Woods of the department of biochemistry at Oxford University had the idea that the substance responsible for this antagonism might be structurally similar to sulphanilamide, thereby acting in much the same manner as physostigmine and tubocurarine did in antagonizing the action of acetylcholine, i.e. by competing with it for an unidentified receptor site. Woods showed that 4-aminobenzoic acid was a very effective antagonist of sulphanilamide, which led him to propose that sulphanilamide acted as an antimetabolite of 4-aminobenzoic acid. He emphasized that the antagonism depended on the close structural relationship between the two compounds. The implication of this was recognized by many investigators, notably George Hitchings of the Wellcome Research Laboratories in Tuckahoe, New York. He, Gertrude Elion, Elvira Falco, Peter Russell, M. B. Sherwood, and their colleagues, prepared potential antimetabolites of purines

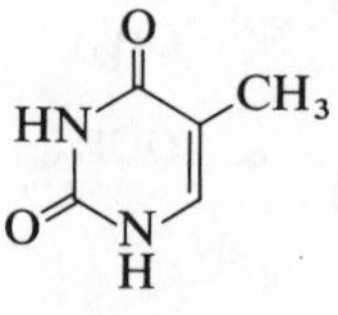

Thymine

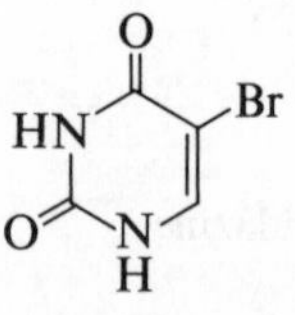

5-Bromouracil

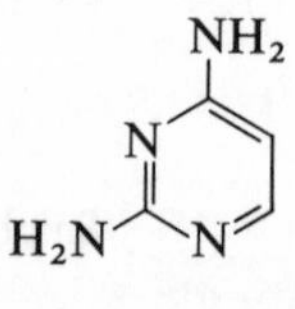

2,4-Diaminopyrimidine

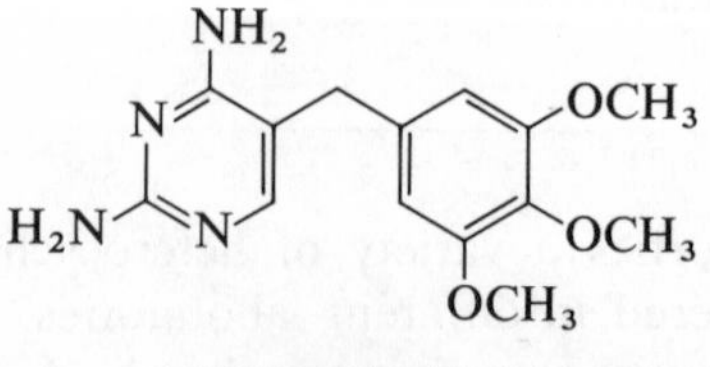

Trimethoprim

and pyrimidines which were required for the synthesis of nucleic acids. This was somewhat inspired, considering that the investigation began in the early forties, long before Watson and Crick elucidated the central role of DNA in controlling cellular activity and reproduction. Their motivation had been the recognition that the rate of nucleic acid synthesis in parasites, be these bacteria, protozoa, viruses, or neoplasms, was apparently more rapid than that of surrounding host tissues as a consequence of the rapid reproduction upon which the parasite depended for its survival. Initially, more than one hundred pyrimidine analogues of thymine were examined for their ability to inhibit the growth of *Lactobacillus casei*. Although replacement of the methyl group in thymine by amino, hydroxyl, chlorine, or iodine atoms resulted in partial inhibition of growth, only the bromo-analogue, i.e. 5-bromouracil, had potent inhibitory activity. The work at Tuckahoe continued with the development of purine antimetabolites, such as mercaptopurine (see page 347), that had considerable value in the treatment of leukaemia.

In 1948, Hitchings and his colleagues demonstrated that nearly all 2,4-diaminopyrimidines and related compounds inhibited the growth of *Lactobacillus casei* by virtue of the fact that they were folic acid antagonists. Four years later, the precise mode of action was shown to involve inhibition of the enzyme known as dihydrofolate reductase. This was a key enzyme which catalysed the transformation of folic acid into a form that could be utilized in the process leading to thymine formation prior to its incorporation into DNA. It also became evident that the degree of inhibition of the enzyme by different antimetabolites varied according to the species from which the enzyme was isolated. This resulted in the discovery of species-specific dihydrofolate reductase inhibitors amongst a series of 5-benzyl derivatives of 2,4-diaminopyrimidine, such as the antimalarial drug pyrimethamine (see page 274). By varying the substituents on the benzene and pyrimidine rings, it was observed that methoxyl groups enhanced inhibitory potency against cultures of *Proteus vulgaris*, whilst not markedly increasing inhibition of enzyme derived from mammalian sources. This was fully exploited in trimethoprim, the trimethoxyderivative, which was 50 000 times as potent an inhibitor of the bacterial enzyme as it was of the human liver dihydrofolate reductase. It was marketed by Burroughs Wellcome in combination with sulphamethoxazole, a preparation known as co-trimoxazolre ('Septrin', 'Bactrim'). This was a form of combination chemotherapy based on the knowledge that sulphonamides acted by preventing the bacterial synthesis of dihydrofolic acid. If a bacterium was resistant to the sulphonamide, it was hoped that the dihydrofolic acid reduction process would still be inhibited by the trimethoprim, or vice versa. The choice of sulphonamide was based purely on the duration of its action, thereby permitting the two drugs to be formulated as a single dose. Co-trimoxazole has been widely prescribed in bronchial and urinary tract infections, as well as some serious infections caused by certain Gram-negative bacteria.

ANTITUBERCULAR DRUGS

In 1938, A. R. Rich and R. H. Follis of the Johns Hopkins Hospital reported that sulphanilamide had weak activity in animals infected with *Mycobacterium tuberculosis*, the bacterium responsible for the death of seven million people each year from tuberculosis. Late the following year, Domagk discovered that sulphathiazole and the related sulphadithiazole were much more effective. The practice at Elberfield of screening all novel intermediates resulted in Domagk being given by Behnisch, in 1941, the thiosemicarbazides from which 2-aminothiadiazoles were prepared prior to their incorporation into the sulphonamide. To his surprise, these were more active than the sulphonamides, benzaldehyde thiosemicarbazone being particularly so. A series of thiosemicarbazones were prepared for Domagk to test, from which emerged as a potential therapeutic agent thiacetazone (originally called tibione, a name that reflected its being the first tuberculostatic compound in the series; the proprietary name was 'Conteben'). The clinical trials in Germany were inadequately organized during an epidemic of tuberculosis at the end of the war and led to unrealistic claims being made. By 1950, American physicians had discovered that thiacetazone was too toxic to the liver for routine therapeutic application in tuberculosis.

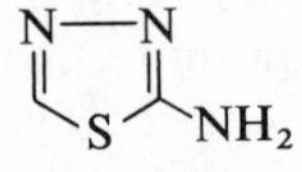

2-Aminothiadiazole

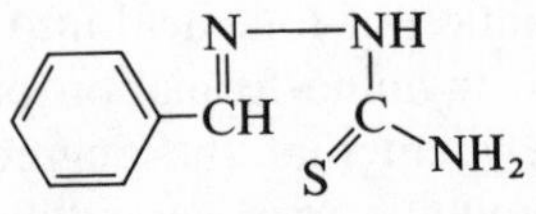

Benzaldehyde Thiosemicarbazone

$CH_3CONH-C_6H_4-CH{=}NNHC(=S)NH_2$

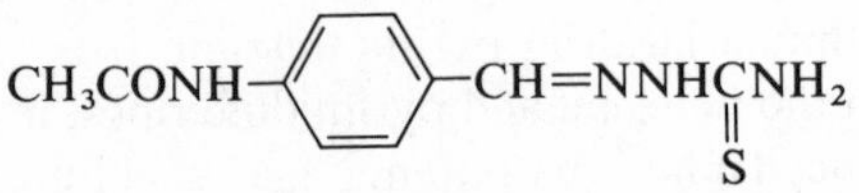

Thiacetazone

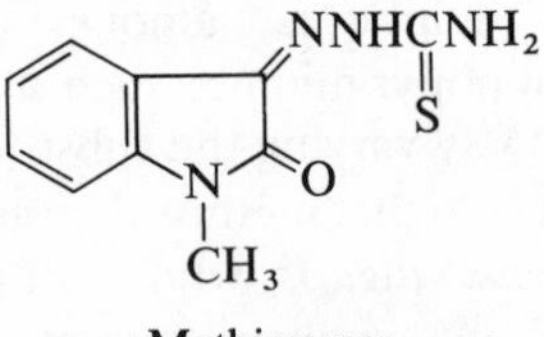

Methisazone

In 1950, it was reported by Hamre that thiacetazone and its desacetyl derivative could afford some protection to mice infected with vaccinia virus. This led to the screening by Thompson, Price, and Minton at the department of microbiology in Indiana University, of a variety of related thiosemicarbazones prepared by drug companies as potential tuberculostatic agents. An isatin thiosemicarbazone obtained from the Wellcome Research Laboratories at Tuckahoe turned out to have enough antiviral activity to justify the launching in London of a collaborative programme between the Wellcome Laboratories for Tropical Medicine and the Courtauld Institute of Biochemistry at the Middlesex Hospital. This resulted in the discovery of high antivaccinial activity in the *N*-methyl derivative of the original isatin thiosemicarbazone. This compound was introduced in 1960, with the approved name of methisazone ('Marboran'), for the prophylaxis of smallpox. It greatly reduced the incidence and severity of the disease amongst a large control group of contacts in an epidemic of smallpox in Madras in 1963. The disease itself was eradicated during the next two decades through the international vaccination campaign organized by the World Health Organization.

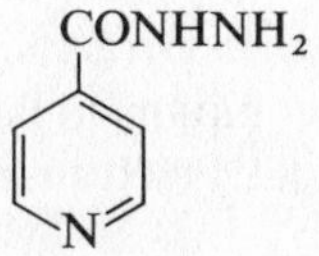

$CONHNH_2$

Isoniazid

Some of the thiosemicarbazones screened for antiviral activity at Indiana University had been supplied by the Squibb Institute in New Brunswick, New Jersey. A research programme had begun there in 1946 under the direction of Frederick Wiseloge. In all, some five thousand compounds were synthesized by Jack Bernstein, Harry Yale, Kathryn Losie, and their colleagues. These were assayed, at a rate of about twenty compounds each week, on a uniform strain of mice infected with tuberculosis. In 1951, Yale synthesized isonicotinaldehyde thiosemicarbazone from a hydrazide intermediate, namely isonicotinic acid hydrazide. Yet again, the screening of a synthetic intermediate resulted in the unexpected discovery that this was much more active than the sought after product! In fact, the hydrazide proved to be at least fifteen times as potent as streptomycin, the antitubercular antibiotic discovered by Selman Waksman at the end of the war (see page 323). It was tested in New York hospitals and quickly became established as far and away the most valuable antitubercular drug ever discovered, being known as isoniazid. Even more remarkable than its serendipitous discovery, was the fact that two weeks or so before the publication of the Squibb breakthrough on the 10th January 1952, a team from the Hoffmann-La Roche Laboratories in Nutley, New Jersey, announced that isonicotinic acid hydrazide was a highly effective antitubercular drug. This was also a serendipitous discovery!

The Roche team was led by Robert Schnitzer, who had formerly worked with Morgenroth and then I. G. Farbenindustrie until his imprisonment in Buchenwald concentration camp. He fled to North America in 1939, with the aid of the Society for Protection of Scientists. At Nutley, he examined the activity of pyridine derivatives synthesized by Hyman Fox, who had been following up a report by Kushner at the Lederle Laboratories, published in 1948, to the effect that the B group vitamin nicotinamide had mild tuberculostatic activity. Fox found that analogues of nicotinamide lacking vitamin activity were still active, thereby disproving the Lederle team's contention that tuberculostatic action was related to vitamin activity. Schnitzer and Fox, as well as Domagk in Germany, at once recognized the possibility of combining a pyridine moiety with a thiosemicarbazone. Both Schnitzer and Domagk then went on to discover that the chemical intermediate used in the synthesis of isonicotinaldehyde thiosemicarbazone was a highly active antitubercular drug.

Domagk's announcement of this appeared shortly after those of the two American groups had appeared.

Shortly after the announcement of the discovery of the activity of isoniazid, both Kushner of Lederle Laboratories and Solotorovsky of Merck and Company simultaneously reported the antitubercular properties of the nicotinamide analogue subsequently known as pyrazinamide ('Zinamide'). Four years later, Libermann of the Theraplix company in Paris introduced ethionamide as yet another tuberculostatic analogue of nicotinamide, but it is now rarely used.

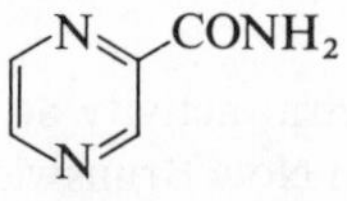

$CONH_2$

Pyrazinamide

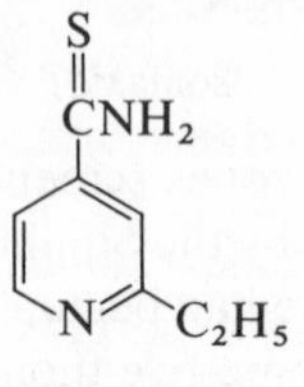

$\overset{S}{\overset{\|}{C}}NH_2$ C_2H_5

Ethionamide

OH CO_2H H_2N

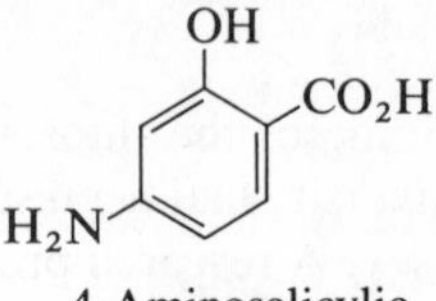

4-Aminosalicylic Acid

F. Bertheim, a biochemist at Duke University, North Carolina, demonstrated in 1940 that benzoic and salicylic acids increased oxygen utilization in *Mycobacterium tuberculosis*. The following year he used iodinated benzoic and salicylic acids as antimetabolites to antagonize this stimulation of oxygen consumption which was a measure of bacterial activity. In conjunction with Alfred Burger and others, Bertheim examined a diverse range of halogenated aromatic acids and phenolic ethers as potential antimetabolites. Some of the latter were active, but they were unsuitable for clinical application because of effects on the central nervous system. This problem did not occur when J. Lehmann of the Gothenburg Hospital tested 4-aminosalicylic acid, prepared for him by the Ferrosan Company of Malmo. In 1946, he announced that this was strongly tuberculostatic. For many years it was widely prescribed in combination with streptomycin and isoniazid.

In the course of an extensive screening programme, researchers at Lederle Laboratories discovered N,N'-diisopropylethylenediamine had antitubercular activity comparable with that of isoniazid. The sole drawback was its somewhat greater toxicity. An extensive series of analogues was prepared, culminating in the development of ethambutol ('Myambutol'), which was reported in 1961 to be a particularly promising drug. The early promise seems to have been fulfilled, although visual side-effects do occur and require monitoring.

By the proper use of combinations of the drugs that have been described here, deaths from tuberculosis—once the commonest cause of death outside of the tropical zones—are unlikely to occur. A further benefit arising from the

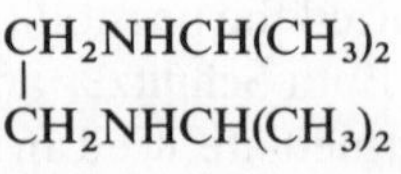

$$\begin{array}{l} CH_2NHCH(CH_3)_2 \\ | \\ CH_2NHCH(CH_3)_2 \end{array}$$

Diisopropylethylenediamine

$$\begin{array}{l} \quad\quad\quad\; CH_2OH \\ \quad\quad\quad\quad | \\ CH_2NHCHCH_2CH_3 \\ | \\ CH_2NHCHCH_2CH_3 \\ \quad\quad\quad\quad | \\ \quad\quad\quad\; CH_2OH \end{array}$$

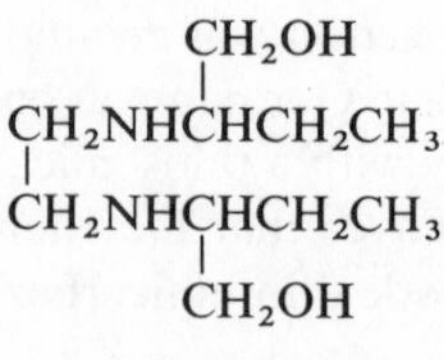

Ethambutol

investigations into antitubercular chemotherapy has been the discovery of valuable antileprotic activity in sulphone analogues of sulphanilamide. Dapsone, a sulphone first synthesized in 1908 by Fromm and Wittmann at the University of Freiburg, was shown by Buttle and his colleagues at the Wellcome Research Laboratories in London, as well as by Fourneau at the Pasteur Institute, to be a more potent antibacterial than sulphanilamide. They considered it to be too toxic for clinical application when compared with other sulphonamides, and this led to several derivatives being made in different laboratories with a view to finding safer sulphones. One of these, prepared in the laboratories of Parke, Davis, and Company in Detroit, was known as glucosulphone ('Promin'). Like dapsone, it had slight activity against *Mycobacterium tuberculosis*. This led E.V. Cowdry and C. Ruangsiri to test it in rats infected with *Mycobacterium leprae*, the causative organism of leprosy. Two years later, in 1943, researchers at the US National Institutes of Health confirmed that another disappointing antitubercular drug, Abbott Laboratories' sulphoxone sodium ('Diasone'), also had antileprotic activity. This was followed up by a study of the drug at the US National Leprosarium in Carville, Louisiana. Clinical studies were then carried out in the Uzuakoli Leprosy Settlement in Nigeria by Davey, who used the Burroughs Wellcome compound known as solapsone (solasulphone, 'Sulphetrone'). The results were published in 1948. Subsequent reports from Cochrane in Madras confirmed the value of dapsone and glucosulphone.

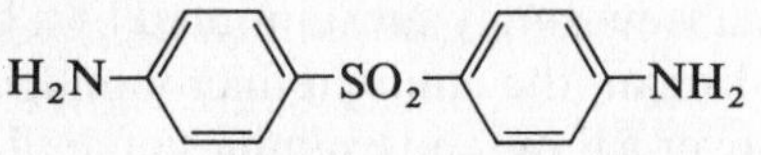

H_2N–C_6H_4–SO_2–C_6H_4–NH_2

Dapsone

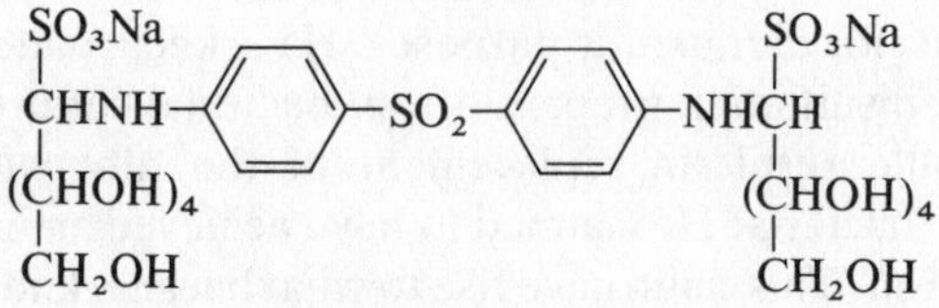

SO_3Na / $CHNH$ / $(CHOH)_4$ / CH_2OH — C_6H_4–SO_2–C_6H_4 — $NHCH$ / SO_3Na / $(CHOH)_4$ / CH_2OH

Glucosulphone Sodium

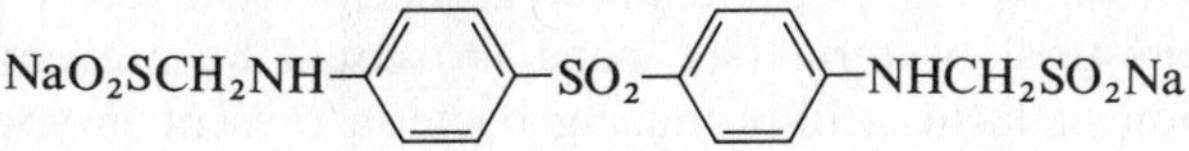

NaO_2SCH_2NH–C_6H_4–SO_2–C_6H_4–$NHCH_2SO_2Na$

Sulphoxone Sodium

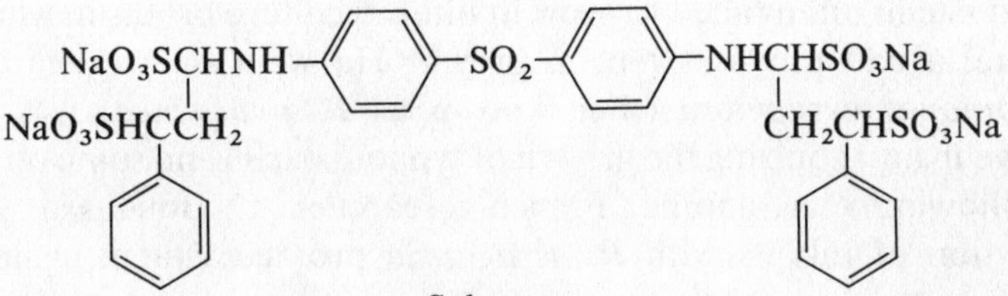

NaO_3SCHNH / NaO_3SHCCH_2 / C_6H_5 — C_6H_4–SO_2–C_6H_4 — $NHCHSO_3Na$ / CH_2CHSO_3Na / C_6H_5

Solapsone

XV

Antibiotics

The concept of antibiosis was described by Louis Pasteur in 1877, when he wrote that amongst the lower organisms '. . . la vie empêché la vie' (life hinders life), a reference to his observation that fermentation of a nutrient liquid induced by one class of microbes inhibited the multiplication of other organisms present in this liquid. As an example, Pasteur described the inhibition of anthrax bacilli in urine by simultaneous inoculation of it with common microbes. In the absence of these, the anthrax bacteria grew profusely. Pasteur then showed that animals inoculated with a mixture of anthrax bacilli and common bacteria did not contract anthrax. He even suggested it might become possible to exploit this therapeutically.

Twelve years later, Paul Vuillemin employed the term 'antibiose', the French equivalent of 'antibiosis', in a paper he presented to the French Association for the Advancement of Science. He contrasted this with the more familiar phenomenon of symbiosis, where two different species thrive upon their mutual association. Vuillemin implied that antibiosis was by no means a rare state of affairs, a view amply substantiated by a bibliography compiled in 1945 by Selman Waksman, the eminent microbiologist who discovered streptomycin. This cited over a thousand examples of antibiosis, one-third of which preceded Fleming's much celebrated discovery of the action of the *Penicillium* mould on bacteria.

In 1885, an Italian physician, A. Cantani, made a somewhat crude attempt to exploit antibiosis for therapeutic purposes. He asked a patient with tuberculosis to inhale an insufflation prepared from bacterial cultures and gelatin. He hoped this would result in replacement of the tuberculosis microbes by non-pathogenic bacteria. He claimed to have achieved an improvement in the patient's condition. This claim gave rise to what became known as replacement therapy, an idea that was later enthusiastically advanced by Elie Metchnikoff, Pasteur's Russian protégé. Metchnikoff argued persuasively that autointoxication by intestinal bacteria (see page 66) could be prevented by regular consumption of lactic acid-producing bacteria present in yoghurt and sour cream. An entire industry was developed as a result of this.

In 1888, E. Freudenreich reported from the Pasteur Institute in Paris that typhoid bacilli often failed to grow in filtered culture broths in which a variety of bacteria had previously been grown. He went on to demonstrate that *Pseudomonas aeruginosa* (then known as *P. pyocyanea*) was particularly effective in antagonizing the growth of typhoid bacilli and some other bacteria. The following year, another French researcher, C. Bouchard, showed that inoculation of rabbits with *P. aeruginosa* protected them against anthrax.

Much information on this phenomenon was gleaned by various workers during the following decade. In 1899, R. Emmerich and O. Low introduced pyocyanase, a product they wrongly believed to be an enzyme. It was probably obtained by allowing cultures of *P. aeruginosa* to incubate for six weeks, during which time the bacteria decomposed and liberated the antibiotic principle into solution. Emmerich and Low never disclosed how they extracted their powdered material from this, so most subsequent workers had to purchase commercially prepared material. The original pyocyanase undoubtedly destroyed a variety of pathogenic bacteria, including the causative organisms of diphtheria, anthrax, plague, and typhoid. This led to the therapeutic use of pyocyanase in a variety of ailments until 1913, when the commercial product suddenly ceased to have any activity. Even then, investigations did not stop altogether, and pyocyanase was one of three antibiotics considered worthy of investigation by Ernst Chain at Oxford in the summer of 1938. Obviously, its activity could never have been outstanding, otherwise it would have remained in use.

The early studies on pyocyanase stimulated investigations on other potential bacterial antibiotics. At the Pasteur Institute in Paris in 1907, Maurice Nicolle followed up Metchnikoff's claim, made ten years earlier, that *Bacillus subtilis* destroyed toxins. Nicolle confirmed that diphtheria and tetanus toxins were indeed destroyed, and in addition established that filtrates of cultures of *B. subtilis* were bactericidal towards several important pathogens, including those causing typhoid, cholera, and pneumonia. These observations convinced Nicolle that the lytic action of the *B. subtilis* filtrate was due to enzymes. He also confirmed that two strains of *B. mesenteroides* had similar activity.

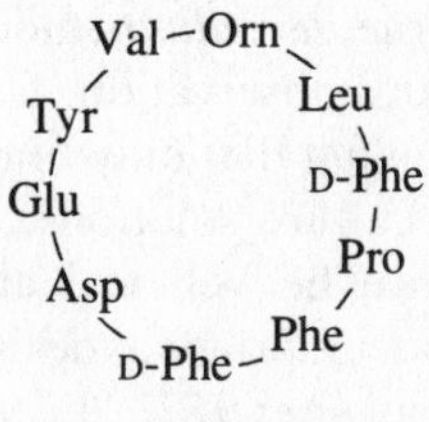

```
      Val—Orn
     /       \
  Tyr         Leu
   |            \
  Glu          D-Phe
    \            |
    Asp         Pro
      \          /
       D-Phe—Phe

    Tyrocidine A
```

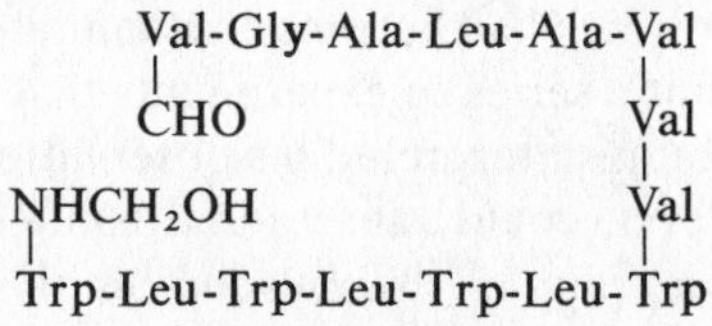

```
Val-Gly-Ala-Leu-Ala-Val
 |                   |
CHO                 Val
                     |
NHCH2OH             Val
 |                   |
Trp-Leu-Trp-Leu-Trp-Leu-Trp

        Gramicidin
 (i.e. Valine Gramicidin A)
```

Nicolle's careful bacteriological studies paved the way for other investigators. When Ernst Chain reviewed the situation in 1938, it was clear that many research workers were in agreement that several strains of bacteria of the genus *Bacillus* produced bactericidal substances. Amongst these, Chain considered *B. subtilis* to be worthy of further investigation, along with pyocyanase and penicillin. The following year, Rene Dubos, a former student of Selman Waksman now at the Rockefeller Institute in New York, published a paper on the isolation of a bactericidal, protein-free extract from *B. brevis*, an organism that had not been hitherto known to have antibiotic activity. This material, tyrothricin, was soon shown to consist mainly of a cyclic peptide called tyrocidine, together with another peptide called gramacidin, which was fifty times as potent. Although both components were able to protect mice against pneumococci, they were too toxic for general use. However, tyrothricin was suitable for topical application and was the first modern antibiotic to be marketed when it was introduced in the United States by Sharp and Dohme in 1942 for treatment of Gram-positive infections. It was widely used in throat lozenges as a non-prescription antibiotic (e.g. 'Tyrozets'), but the commercial success of this type of preparation was in no small measure due to the incorporation of a local anaesthetic, benzocaine, which soothed sore throats.

PENICILLIN

The first scientific report of a fungus exhibiting antimicrobial activity involved the familiar green *Pencillium* mould sometimes found on oranges or jam. The observation was made at St Mary's Hospital in Paddington, London. The investigator concerned was John Burdon-Sanderson. A graduate of the University of Edinburgh, he had studied physiology in Paris with Claude Bernard before his appointment to the lectureship in botany at St Mary's shortly after the opening of its medical school in 1854. As Medical Officer of Health for Paddington, he recognized that the cholera epidemic of 1853 was caused by contamination of drinking water with sewage. He instituted a variety of successful measures to remedy this, including draining the Serpentine and installing a filtration plant to keep its waters clean. Later, as one of the first British physicians to accept Pasteur's ideas on the germ theory of disease, he became an enthusiastic supporter of Lister. In October 1870, Burdon-Smith noted that bacteria did not grow and produce turbidity in sterilized culture solutions which had become contaminated by air-borne *Penicillium* mould. He carried out a series of experiments that confirmed this observation. Unfortunately, he misinterpreted his observations by concluding that only fungi, and not bacteria, could cause aerial contamination of culture solutions. Joseph Lister was puzzled by Burdon-Smith's results since he believed airborne bacteria to be a cause of disease. Indeed, his carbolic acid spray was designed to kill airborne bacteria in operating theatres. In November 1871, Lister confirmed that bacteria did not grow in glasses containing his own urine and which were contamined by a mould. His brother identified this mould as *Penicillium glaucum*. Lister next demonstrated that bacteria could grow in urine that was only lightly contamined by the mould, but no growth occurred if there was heavy contamination. This clear recognition of the inhibitory power of the mould on bacteria led Lister to experiment with it as an antiseptic. He did not publish details, but Dr J. Fraser-Moodie has written an account of how, in 1940, he was told by a former nurse at King's College Hospital, that Lister cured a badly infected wound she had sustained in a street accident by application of 'Pencillium' after other antiseptics had proved ineffective.

In January 1895, Vincenzo Tiberio of the Navy Hospital in Naples published an account of work on the antibacterial properties of moulds, which he had carried out during the previous year. He reported that an extract of a mould, which he identified as *P. glaucum*, had prevented the growth of pathogenic bacteria, and he gave a detailed description of its action on infected rabbits, guinea pigs, and rats. That he did not proceed further with this line of investigation suggests he was far from convinced by his results. The following year, however, his countryman, B. Gosio, isolated a crystalline product from a *Penicillium* which he thought was *P. glaucum*. There was no connection whatsoever between these two investigations, for Gosio had been examining fungal growth on spoiled maize in an attempt to identify the cause of pellagra. Nevertheless, since the chemical properties of the crystalline compound indicated it was a phenol, Gosio tested its antiseptic properties on cultures of the anthrax organism. This showed it to be an inhibitor of their growth. Lack of material prevented Gosio carrying out animal experiments, but during the following year E. Duchesne, a French army doctor, submitted a thesis to the University of Lyon in which he stated that certain green *Penicillia* were able to prevent or attenuate the growth of a variety of bacteria. He subsequently established that extracts of *P. glaucum* protected animals against virulent bacteria, but he died from tuberculosis before being able to conduct further studies. Reports of antibacterial activity associated with *P. glaucum* were published by Tartakovskii in 1904, Adriano Sturli in 1908, and the Belgians Andre Gratia and Sara Dath in 1925.

The phenolic crystalline antibiotic isolated by Gosio in 1896 was named mycophenolic acid in 1913. It seems probable that the early claims of antibacterial activity amongst varieties of *Penicillum* moulds relate to this compound, the species described as *P. glaucum* most likely really being *P. brevicompactum*. Mycophenolic acid was reinvestigated in the 1930s by Harold Raistrick and his colleagues in the department of biochemistry at the London School of Hygiene and Tropical Medicine. Its structure was elucidated in 1952, and was shown to be chemically unrelated to any of the penicillins. Although mycophenolic acid was too toxic for clinical application as an antibiotic, during the 1960s animal tests indicated that it had promising activity as an anticancer agent. Unfortunately, results on humans were unimpressive.

There is no evidence to support the contention that early attempts, in folk medicine or otherwise, to exploit the bacterial antagonism of *Penicillia* involved penicillin-producing strains. Alexander Fleming, the director of the Inoculation Department at St Mary's Hospital, Paddington, was extremely fortunate when a remarkable combination of events occurred in the summer of 1928 and permitted him to discover the antibacterial activity of penicillin. The probable sequence of these events has been reconstructed in an account written by his former assistant, Professor Ronald Hare. Firstly, on the floor beneath Fleming's laboratory a colleague, C. J. La Touche, worked with moulds required for the production of vaccines to treat allergies, and it seems likely that one of these wafted through the air into Fleming's laboratory to settle on a petri dish plated with a layer of agar impregnated with staphylococci. Secondly, this mould was a rare strain of *Penicillium notatum* which produced relatively large amounts of penicillin. Thirdly, Fleming left his culture plate on his work bench instead of placing it in an incubator at body temperature to ensure bacterial growth. Fourthly, an exceptionally cool spell followed when Fleming went on holiday at the end of July, and this favoured growth of the mould in preference to that of the staphylococci. Fifthly, the climatic conditions changed later in the month, by which time the mould had produced sufficient penicillin to kill bacteria in its vicinity. This rise in temperature allowed colonies of staphylococci to grow elsewhere on the culture plate, thus enabling Fleming to observe a zone of inhibition of staphylococcal growth when he returned to the laboratory on 3rd September.

In this plausible reconstruction of the sequence of events leading to the discovery of penicillin, Ronald Hare concluded by describing what happened during a visit to Fleming's laboratory on 3rd September by Dr D. M. Pryce, a former colleague who had assisted with experiments on staphylococci. Fleming complained to him about his increased work load since Pryce had left the department. To illustrate his point, he picked up several petri dishes from a stack piled in a tray of 'Lysol' for disinfecting before being washed for reuse. The dishes at the top of the pile had not yet been soaked in the disinfectant, and it was these that Fleming showed Pryce to illustrate the work he had done with staphylococci before departing on holiday. Most of them were contaminated by yeasts and moulds, which was hardly surprising since they had been lying in the pile for several weeks. Fleming, however, spotted something unusual on one of them, namely a zone of inhibition of staphylococcal growth around a mould. Surely this was the moment at which Fleming discovered penicillin? If this were the case, the remarkable sequence of events already described seem even more so because the famous culture plate was already destined for destruction when Fleming examined it with Dr Pryce! The original plate is kept in the British Museum, Fleming having treated it with formaldehyde vapour to preserve it.

During the First World War, the staff of the Inoculation Department of St Mary's had served in a special unit at Boulogne, where they conducted valuable studies on wound disinfection. When the war ended, Fleming and Leonard Colebrook returned to St Mary's and continued with research on antiseptics and chemotherapy, both men becoming leading authorities on the subject. Fleming made a notable contribution by his discovery of lysozyme, a bactericidal enzyme produced by most living tissues. He was, therefore, well-equipped to make a realistic assessment of the significance of the effect of the *Penicillium* mould on bacterial growth.

Fleming prepared subcultures of the mould and gave the name 'penicillin' to the filtrate of broth in which these had been grown for one or two weeks at room temperature. Two assistants, Frederick Ridley and Stuart Craddock, were given the task of preparing sufficient quantities of penicillin for bacteriological studies. They were also asked to obtain information about the chemical nature of the active substance in the mould juice, this initially being assumed to be an enzyme since it apparently lysed cells in a similar manner to lysozyme. Facilities for this work were hopelessly inadequate, space having to be found in a corridor where they set up their equipment on tables. Despite this, they devised an efficient process for growing the mould in large, flat-sided bottles from which the juice below the surface growth of mould was drained and filtered. Since it was found that boiling the slightly alkaline mould juice destroyed the activity of penicillin, Ridley and Craddock resorted to the use of evaporation of acidified juice at 40°C under reduced atmospheric pressure. The resulting sticky concentrate was taken up in alcohol, in the process of which much of it was precipitated. It was expected that the precipitate would contain the active substance since this was assumed to be an enzyme. When, instead, the activity was found in the alcohol, the probability of penicillin being an enzyme was discounted. The alcohol extract was very potent, but its activity gradually disappeared over a period of weeks. This instability, coupled with the complexity of the isolation process, later discouraged Fleming from pursuing the therapeutic possibilities of penicillin.

With penicillin supplied by Ridley and Craddock, Fleming carried out tests to find which types of bacteria were sensitive to it. He established that it could lyse cultures of major pathogens, including staphylococci, streptococci, meningococci, gonococci, and diphtheria bacilli. Pathogenic organisms unaffected by penicillin included the bacilli causing pneumonia after attacks of influenza, anthrax, and those responsible for intestinal diseases such as typhoid and dysentery. The marked insensitivity of *Haemophilus influenzae* particularly interested Fleming. This had proved to be a difficult organism to isolate from infected patients since cultures of it were readily overgrown by other bacteria. This had prevented Fleming's colleagues from preparing vaccines, which was the prime purpose of the Inoculation Department at St Mary's. There were those who believed (wrongly) that it was not merely associated with influenza, but was actually the cause of it and other respiratory conditions. In an attempt to kill off the bacteria that interfered with the growth of *H. influenzae*, Fleming incorporated penicillin in cultures prepared from throat swabs. This ploy worked admirably, and for the next ten years or so was the principal purpose for which penicillin was used. Indeed, the title of Fleming's first paper on penicillin was 'On the antibacterial action of cultures of a *Penicillium*, with special reference to their use in the isolation of *B. influenzae*'.

As a leading authority on antiseptics Fleming could not overlook the therapeutic potential of penicillin. After completing the tests for antibacterial activity, he established that leucocytes were not harmed by penicillin. The head of the bacteriology department at St Mary's, Sir Almroth Wright, had long since drawn attention to the importance of this, the leucocytes in the blood being one of the body's major defences against pathogenic bacteria. Then, in December 1928, Fleming examined the effect of penicillin on slides of bacteria growing in the presence of blood or serum, the obstacle that had rendered worthless practically every other antiseptic, except for superficial application where high concentrations could be achieved. Fleming had developed this technique for evaluating other antiseptics, and when it indicated that the activity of penicillin was diminished by blood or serum he no longer had any reason to expect it to be active against generalized infections or those in deep wounds. He also learned later from Craddock that penicillin injected into a rabbit disappeared from its blood within thirty minutes. This was most discouraging since Fleming had established that penicillin required around four hours to act. Had he placed less reliance on his own methodical approach to the study of antiseptic action, and instead followed the approach favoured by Morgenroth, Schnitzer, and Domagk, Fleming might have discovered the true value of penicillin in 1929. Unfortunately, he did not administer penicillin to infected mice, and it was left to others to discover its outstanding therapeutic efficacy. He did, however, inject penicillin into both a rabbit and a mouse in order to establish its lack of toxicity. Craddock also ate some *Penicillium* mould without ill effect, commenting that it tasted like Stilton cheese!

Fleming investigated the potential of penicillin as an antiseptic for topical application. On the 9th January 1929, he treated Craddock's infected nasal antrum with penicillin, but to no avail. A second disappointment followed when he failed to cure an infected amputation stump and the patient died from septicaemia. Fleming did have one success with penicillin when he applied it to the eye of another of his assistants, Dr K. B. Rogers, who contracted a pneumococcal conjunctivitis. This time the infection rapidly cleared.

At the beginning of April 1929, Fleming and Craddock carried out an experiment which they believed might explain the failure of penicillin to cure infections. They incubated the organs of a newly killed rabbit in a liquid culture of staphylococci for twenty-four hours, then transferred the organs to a penicillin solution for a similar period. Examination of slices of the organs revealed that staphylococci had penetrated deep within the organs and had survived. It was concluded that penicillin could not penetrate beyond the surface of organs. This was a misleading interpretation since it failed to consider the possibility of circulating blood being able to carry penicillin into the tissues. It cast further doubt on the clinical value of penicillin. Little wonder, then, that when Fleming wrote up his first paper on penicillin just after completing this experiment, he emphasized its importance as a reagent for facilitating the isolation of *Haemophilus influenzae* rather than its therapeutic potential. Having submitted his paper to the *British Journal of Experimental Pathology*, Fleming abandoned further research on the clinical potential of penicillin, although he occasionally applied it locally to treat infections. His subsequent researches with it were mainly as a reagent in bacteriological investigations. As a recent biographer of Fleming, Gwyn Macfarlane, has pointed out, this was just as well since it ensured that cultures of his original rare strain of *P. notatum* were requested by various bacteriology laboratories, including that of the School of Pathology at Oxford.

In the same year that Fleming published his first paper on penicillin, Harold Raistrick was appointed to the new chair of biochemistry at the London School of Hygiene and Tropical Medicine. As one of the foremost authorities on the chemistry of moulds, he became interested in penicillin from a strictly scientific point of view, his medical colleagues having assured him that it was of little clinical importance. Raistrick obtained a culture of Fleming's mould and showed it to a mycologist, J. H. V. Charles, who expressed doubts as to whether it was *P. rubrum*, as Fleming had been told by La Touche. A sample was then sent to the United States, where the acknowledged expert on such matters, Chalres Thom, identified the mould as a rare strain of *P. notatum*. Meanwhile, Raistrick, had assigned one of his assistants, Percival Clutterbuck, to the isolation of penicillin. The bacteriological assays required for monitoring this were done by a colleague, Reginald Lovell, from the department of bacteriology. Lowell spoke to Fleming on the telephone several times about the work he was doing on penicillin, but apparently Fleming did not mention Ridley and Craddock's isolation procedure. This meant that their researches had to be unnecessarily duplicated before further progress could be made. Eventually, an ether extract of penicillin was obtained, but evaporation of the solvent resulted in the loss of its activity. Baffled by this unparalleled behaviour of the extract, Raistrick abandoned the project since there was no good reason to deploy the extensive resources that would be required to pursue the investigation further. Neither Fleming nor anyone else had suggested that the therapeutic potential of penicillin justified this. Even the distinguished fellows of the Royal Society did not attach much importance to penicillin, for they declined to accept Fleming as a member on the basis of his work with it. In any case, penicillin was only one of several mould products in which Raistrick's department was interested. He and his colleagues published a brief report on their work in 1932.

The next attempt to purify penicillin came from Fleming's laboratory in 1934. By now Ridley and Craddock had both left the Inoculation Department, but a chemist, Lewis Holt, had just joined the staff. Fleming asked him to see what he could do with penicillin, and referred him to Raistrick's paper on it. Once again, he omitted to mention the earlier work done at St Mary's by Ridley and Craddock. Holt made quick progress, thanks to Clutterbuck's discovery that penicillin could be extracted into organic solvents from slightly acid solutions. He extracted mould juice directly with amyl acetate, thus avoiding the tedium and inefficiency of the original vacuum evaporation procedure. Holt was able to recover penicillin from the amyl acetate solution by back-extracting it with very slightly alkaline solution, a procedure that had not been tried by anyone else. However, he was dismayed by the instability of penicillin in this solution and abandoned his work on it after only a few weeks without publishing his findings.

In the summer of 1938 at the Sir William Dunn School of Pathology in Oxford, Ernst Chain began investigating penicillin with the assistance of L. A. Epstein, an American Rhodes Scholar who was researching for his D.Phil (he later changed his name to Falk). Chain, a young Jewish refugee from Hitler's Germany, had been invited to Oxford two years earlier on the recommendation of Professor Gowland Hopkins, the eminent Cambridge biochemist in whose department he had obtained his doctorate. At Oxford, Chain made an important contribution towards an understanding of the mode of action of

Fleming's lysozyme. In the course of this work he was made aware of the range of naturally occurring antibacterial substances. In particular, he became interested in pyocyanase and penicillin as these seemed likely to be enzymes, as was lysozyme. Chain believed a detailed study of how they lysed the bacterial cell wall would afford much useful information on the structure of this wall, just as his studies on lysozyme had done. At this time, he had no reason to believe that penicillin would be of any particular therapeutic value. In 1938, he found himself with the opportunity to begin such a study, and he was given every encouragement to pursue this by Howard Florey, the Australian who had moved from Sheffield University in 1935 to take up the chair of pathology at Oxford. Florey had previously encountered penicillin in 1932 when a colleague at Sheffield had employed it in three cases of skin infections.

Chain and Epstein used a culture of Fleming's original mould to prepare penicillin. This was provided by a colleague who had propagated it in the department for several years after it had been obtained from Fleming for use as a bacteriological reagent. Unaware of the unpublished work done by Lewis Holt in Fleming's laboratory in 1934, the Oxford workers rediscovered the solvent extraction procedure. Whilst doing this, Chain realized that the physical properties of penicillin indicated that it could not possibly be a protein enzyme as he had anticipated. It seemed to be a small molecule. Nevertheless he found the challenge of isolating such an unusual substance to be irresistible.

The work on the protective action of tyrothricin in mice infected with pneumococci, reported from the Rockefeller Institute in 1939, together with the onset of the war, gave an added dimension to the research on penicillin. With Chain's research grant due to expire, Florey opportunistically sought financial support for the penicillin project on the grounds that, unlike lysozyme, penicillin was active against pathogenic bacteria. He argued that since previous workers had used only crude penicillin, every effort should be made to obtain it in pure form so that the effects of injecting it intravenously could be properly assessed. In October 1939, the Medical Research Council awarded Chain an annual grant of £300 plus £100 for materials, this to be for each of three years. Shortly after this, the Rockefeller Foundation, which had sponsored Ehrlich's research on 'Salvarsan' many years before, awarded Florey a magnificent annual grant of $5000 for five years, plus an initial sum to purchase equipment. This large sum guaranteed the viability of the project, and must surely rank as the finest investment ever made by a charitable foundation. The Foundation also awarded a fellowship to Norman Heatley, whose ability to improvise apparatus for the large-scale production of penicillin was to prove invaluable.

With support for the project secured, work on penicillin now began to proceed at a faster pace under Florey's skilled administration. The entire staff of the Sir William Dunn School of Pathology now became involved. This meant considerable effort could be put into investigating the best means of producing mould juice and recovering the active material from it. Heatley dealt with this aspect, and he developed the cylinder-plate method of obtaining an accurate measure of the strength of numerous extracts that had to be prepared. It was found that the best yields were obtained when the culture fluid was not more than 1.5 cm deep. This meant that the cultures had to be grown in bottles laid on their sides; milk bottles proved useful at first, but were later replaced by specially manufactured glazed porcelain containers that bore an unmistakable resemblance to hospital bed-pans!

Early in 1940, Chain ran up against the problem that had defeated Raistrick, namely the recovery of penicillin from its solution in ether. Heatley solved this by a simple method that, unknown to anyone at Oxford, had been used at St Mary's by Lewis Holt, namely back-extraction of the penicillin into very slightly alkaline solution. It was, however, not until 1941 that Chain established the superiority of amyl acetate as an extraction solvent; this, too, had also been previously discovered by Holt. This solvent then replaced ether in Heatley's mechanized counter-current extraction system. This involved transference of penicillin in downward flowing streams of filtered, acidified mould juice into upward flowing streams of the immiscible solvent in the same glass tubes. A reversal of the procedure was utilized for the back-transfer of the penicillin to slightly alkaline, aqueous solution. Recovery of penicillin from this had thwarted Holt in 1934, but the following year the process of freeze-drying had been developed in Sweden, and it was used at Oxford to recover a brown, dry powder, which although far from pure, was certainly highly potent and suitable for biological studies. Whereas one millilitre of the mould juice contained 1 or 2 units of penicillin (i.e. arbitrary Oxford units), the brown powder typically contained around 5 units per milligram. By mid-March 1940, Chain had in his possession a supply of about 100 mg of this powder. In terms of antibacterial potency, he already knew it was at least twenty times as effective as the most potent of the sulphonamides against cultures of various pathogenic bacteria. A substantial portion of Chain's 100 mg of penicillin powder was then injected intraperitoneally into two mice, with no ill effects being observed. From here on, Florey took over the complex biological, toxicological and clinical investigations.

It seems remarkable that Florey was prepared to take the gamble of committing the entire resources of his department to the production and purification of penicillin whilst having no idea whether it would be effective when injected. Until this point in time it was only proven to be a local antiseptic. After all, until the discovery of 'Prontosil', which had been made public in 1935, practically all antiseptics had proved ineffective when injected. Clearly, its introduction by Domagk had revolutionized medical thinking, else Florey would never have seriously considered proceeding as he had now done.

The crucial experiment with penicillin that settled the issue of its therapeutic potential once and for all took place on Saturday 25th May 1940. At 11 a.m., Florey injected each of eight mice with a lethal amount of virulent streptococci. At noon, two of these were each injected subcutaneously with 10 mg of penicillin, and two others received half this amount. These latter two mice received four more similar injections during the next ten hours. Heatley stayed in the laboratory that night and watched the untreated mice die. One of the mice which had received only a single dose of penicillin died two days later, while all the others survived. This experiment greatly heartened all those engaged on the project, and Florey and his colleagues immediately proceeded to conduct extensive biological studies with the crude powder. In July, it was confirmed that penicillin was also effective in mice infected with either staphylococci or *Clostridium septicum*, the causative organism of gas gangrene.

The results of the investigations were published in the *Lancet* on 24th August 1940. Only brief details were given in the paper, but the fact that it appeared at all is evidence of the lack of importance attached to penicillin by outsiders in comparison with other projects considered to be of military importance, such as that concerning antimalarials. Ten days after the paper appeared Fleming visited Oxford to make his first contact with Florey's department. Florey must have been disappointed that this was the sole response to his paper, for he had hoped that publication of his preliminary findings might impress somebody in the pharmaceutical industry. Earlier that summer, he had approached Burroughs Wellcome to see whether the company would produce penicillin on a large enough scale for him to begin clinical trials. This offer was declined because their facilities were stretched to the limit in trying to prepare blood plasma and also meet the requirements of the armed forces for vaccines. It was felt that penicillin would not be so important for the war effort.

These were momentous days not only for the researchers at the Sir William Dunn School of Pathology, but for the entire country. Following the disastrous withdrawal of the British Expeditionary Force from France, invasion by the Germans seemed imminent. Florey knew that if the Wehrmacht approached Oxford, he would have to destroy all the equipment and all records of the penicillin project. However, to guard against his work being entirely in vain, he and some of his colleagues smeared spores of the *P. notatum* inside the linings of their coats, hoping that at least one of them would escape and begin the research anew elsewhere.

In January 1941, the scale of penicillin production was stepped up so that supplies could be provided for clinical investigations and for determination of the chemical structure. In effect, the School of Pathology now became a factory for the production of penicillin. Then came a breakthrough when Chain and Edward Abraham tried the new technique of adsorption chromatography. This enabled penicillin to be adsorbed on to a column of powdered alumina, down which it was poured in solution. The impurities passed down through the column with the solvent while the penicillin was retained on the alumina. It was subsequently recovered by solvent extraction of the alumina, the material obtained in this manner having an activity of 50 units per milligram.

On the 17th January 1941, a single intravenous injection of 100 mg of penicillin was administered to a woman dying of cancer. The sole adverse reaction was a bout of shivering following by a fever, and this was traced to pyrogenic impurities in the penicillin preparation. Tests on rabbits showed that further chromatography could remove these pyrogens. Studies on volunteers then provided valuable information concerning the best way of administering penicillin. It was learned that penicillin was destroyed by the acidity of gastric juice, and hence could not be given by mouth. Since the kidneys were found to excrete penicillin rapidly from the body, it was seen to be preferable for it to be given by slow intravenous drip in order to maintain adequate bactericidal levels in the blood and tissues.

The first attempt to use penicillin to treat a patient with a life-threatening infection took place at the Radcliffe Infirmary, Oxford, on the 12th February 1941. The patient, Albert Alexander, was a forty-three-year-old policeman dying from a mixed staphylococcal and streptococcal infection that had spread throughout his body and had already necessitated removal of his eye. He had not responded to sulphapyridine. Charles Fletcher initially administered 200 mg of penicillin intravenously, followed by 100 mg every three hours for the next five days. The patient began to respond within twenty-four hours, and his condition continued to improve rapidly. On the third day of treatment, Fletcher's original supply of penicillin ran out, but more was obtained from the School of Pathology, where all the urine voided by the patient had been extracted to recover as much penicillin as possible. By this expedient, it proved possible to continue the injections for a further three days. The policeman's health remained good for the next ten days, but a residual lung infection then flared up and he died on the 15th March. Before his death occurred, two other patients had already responded to penicillin treatment. The fourth patient was a child with a severe infection behind the eye that was inevitably fatal. This infection disappeared after large doses of penicillin were administered, but the child later died when an artery in the brain ruptured as a result of damage caused prior to treatment with penicillin. Two other very ill patients in this preliminary trial of penicillin recovered uneventfully.

The second paper on penicillin by the Oxford workers appeared in the *Lancet* on 16th August 1941. It gave details of penicillin production, animal results, and clinical reports. By the time it was published, Florey and Heatley had already been in the United States for six weeks seeking to arrange the large-scale production of penicillin for extensive clinical trials. Before departing, Florey had approached the Boots Company and ICI, but nothing had come of this. When the Rockefeller Foundation offered to pay the expenses for Florey and a colleague to visit the United States to discuss penicillin production, Florey referred the matter to the secretary of the Medical Research Council. He was then advised to proceed with the visit as it was obvious no British manufacturer was in a position to produce penicillin.

Soon after their arrival in the United States, Florey and Heatley met Charles Thom of the US Department of Agriculture. He was the mycologist who had redesignated Fleming's mould *P. notatum*. Thom pointed out that the only realistic way of producing large quantities of mould juice was in deep fermentation tanks similar to those used by brewers. He at once put Florey in touch with Robert Coghill, head of the fermentation division of the Department of Agriculture's Northern Regional Research Laboratory at Peoria, Illinois. Coghill generously agreed to attempt to grow the penicillin in deep culture tanks, so long as Heatley would remain for several months to work with Andrew Moyer. Production of penicillin began the next day, using the mould culture Florey had brought from Oxford. Within six weeks, Moyer and Heatley had increased the yield of penicillin twelve-fold simply by including corn steep liquor in the culture medium. This syrupy material was a cheap waste product obtained during corn-starch manufacture. Further improvements were achieved by strict control of the acidity of the culture medium as the mould was growing. This was very important since penicillin was only stable within a narrow range of pH (acidity).

A few weeks after arriving in the United States, Florey met officials of the Committee on Medical Research, a division of the Office of Scientific Research and Development. The director, Newton Richards, was professor of pharmacology at the University of Pennsylvania, and Florey had worked with him fifteen years earlier. Richards agreed to arrange a meeting between government officials and the research directors of Merck and Company, E. R. Squibb and Sons, Charles Pfizer and Company, and Lederle Laboratories. Merck agreed to proceed with penicillin production at once and also to exchange information with other interested parties, but the representatives of the other companies reserved their position. After further meetings, when Coghill reported on the progress made at Peoria, Squibb and Pfizer joined in the collaborative effort. A consortium calling itself the Midwest Group was formed when Abbott Laboratories, Eli Lilly and Company, Parke, Davis, and Company, and the Upjohn Company agreed to exchange information on penicillin. Wyeth Laboratories also took up penicillin production by growing the mould in cellars near Philadelphia, where mushrooms had been cultivated for the gourmet market. They became the largest producer of penicillin until deep fermentation processes were introduced.

Until February 1943, all the penicillin produced by American companies was submitted to the Committee on Medical Research, which arranged clinical trials. The first large-scale trial in America was, however, unrehearsed; it took place in the most dramatic conditions conceivable when, on the night of 28th November 1942, over five hundred people perished in a disastrous fire at the Coconut Grove night club in Boston. As soon as it was known that there were some 220 badly burned casualties, the Committee on Medical Research authorized the release of supplies of penicillin in an attempt to reduce the anticipated mortality amongst the survivors. In the event, the drug exceeded expectations, but the public were never told since penicillin was by now classified as a military secret.

While the American collaborative programme was getting under way, Florey's team at Oxford intensified its efforts to produce enough penicillin for a large clinical trial. Information was informally exchanged between the British and American scientists. Following approaches from Alexander Fleming, the British Ministry of Supply eventually established a Penicillin Chemical Committee late in 1942, to coordinate industrial manufacture in much the same way as had been done in the United States. Boots, Burroughs Wellcome, Glaxo,

ICI, Kemball Bishop, and May and Baker eventually began to make arrangements for production of penicillin, using the surface culture method. Nevertheless, when Florey and his wife published the results of using penicillin in 187 cases, in March 1943, most of the penicillin for this trial had been made at Oxford. Only enough had been available to treat seventeen of these patients by intravenous injection; the others received the drug by direct local application.

At Peoria, intensive effort was put into finding a strain of penicillin producing mould that would grow in deep tanks and also deliver a higher yield of penicillin. The US Army Transport Command delivered thousands of samples from all over the world, either in the form of soil or as cultures of soil organisms. Notwithstanding this, one of the best improvements in penicillin yield was obtained with a strain of *P. chrysogenum* growing on a cantaloup melon in the fruit market in Peoria! As part of a collaborative effort involving Minnesota, Stanford, and Wisconsin Universities, as well as the Carnegie Institution at Cold Spring Harbor, this strain was irradiated at Cold Spring with X-rays, producing mutants that provided even higher yields of penicillin. One of these, singled out from tens of thousands that had to be examined, was found at the University of Wisconsin to produce 500 units of penicillin per millilitre. It became the standard strain for wartime production of penicillin in America. The Wisconson group even improved on it, achieving a further mutation which yielded 900 units per millilitre! The Oxford process had yielded only 1–2 units per millilitre. Such were the difficulties that had to be overcome by the American microbiologists involved in this work, that it took two years of intense effort before large amounts of penicillin could be produced in deep fermentation tanks containing up to twelve thousand gallons of mould juice. When this was eventually achieved, the impact was staggering. During the first five months of 1943, American manufacturers delivered 400 million units of penicillin, but 20 000 million units were produced by deep culture during the next seven months. In January 1944, Charles Pfizer and Company prepared 4000 million units, but by the end of that year they had become the world's largest producer of penicillin, turning out 100 000 million units a month! When the invasion of occupied France began in June 1944, enough penicillin was available to satisfy all military requirements.

Several pharmaceutical companies in Britain and the United States were at first reluctant to commit themselves to the production of penicillin by a fermentation process. As recently as 1938, the Lederle division of the American Cyanamid Company had lost millions of dollars when a new plant for the production of pneumonia vaccines was rendered obsolete in only eight months by the introduction of sulphapyridine. It was also generally recognized that vitamins were more cheaply prepared by synthesis than by recovery from natural sources. Furthermore, it was widely believed that penicillin would soon be synthesized and manufactured on as large a scale as the sulphonamides. In the event, it turned out that the company which made the largest commitment to penicillin synthesis, Merck, lost out to rivals who stuck to fermentation processes.

Before progress towards penicillin synthesis could begin, it was necessary to determine its precise chemical structure. Preliminary work towards this end was initiated when Sir Robert Robinson and his colleagues at the Dyson Perrins chemistry laboratory at Oxford joined forces with Florey's group early in 1942. The first investigations had to be carried out with penicillin that was only about 50% pure. This permitted degradation studies that produced fragments of the molecule which could be identified. Unfortunately, the presence of sulphur was not detected, an oversight that delayed progress for several months.

Purification by means of crystallization was essential if the structure of penicillin were to be determined. Oskar Wintersteiner and his group at the Squibb Laboratories achieved this in the summer of 1943. The news was passed to Oxford. The British workers then managed to crystallize their penicillin, only to find it was different to that isolated in the United States. This led to the realization that there were variant forms of the antibiotic with different side-chains. The Oxford material was first named penicillin I, and the American product penicillin II. These names were subsequently changed to penicillin F (2-pentenylpenicillin) and penicillin G. The later received the approved name of benzylpenicillin. Five more penicillins were identified the following year. However, the one that came into routine clinical use was benzylpenicillin, which was the variant obtained from deep fermentation.

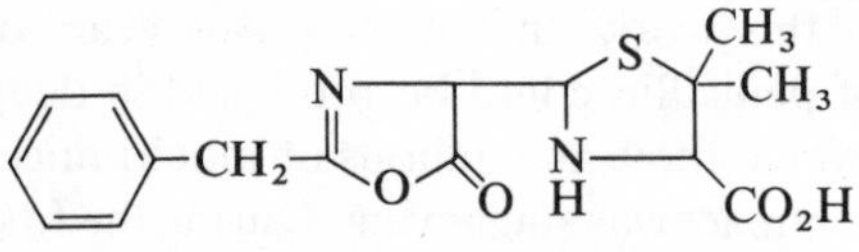

Oxazolone-Thiazolidine Structure for Penicillin

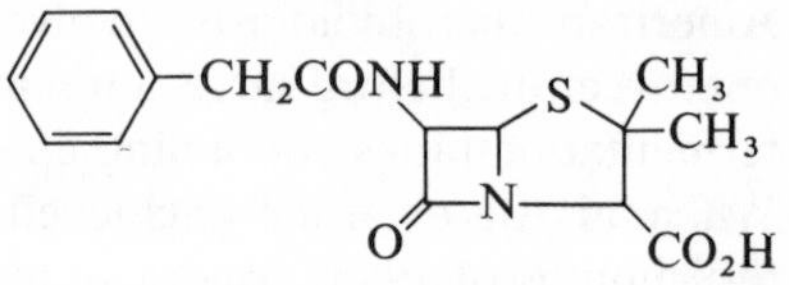

Benzylpenicillin

The Committee on Medical Research set up a project to deal with penicillin synthesis, sponsoring research in American universities, independent foundations, and industrial laboratories. Roger Adams of the University of Illinois was in overall charge of the project; he was generally considered to be the leading American organic chemist. Similar arrangements were made in the United Kingdom, where the Committee for Penicillin Synthesis was established. This arrangement allowed a formal exchange of information between British and American scientists, whilst maintaining strict secrecy about the chemical nature of penicillin. In this way it became possible, by late 1943, for chemists at Oxford and the Merck laboratories to propose that penicillin had one of two chemical structures. The first of these consisted of a five-membered oxazolone ring joined at one point to a thiazolidine ring, whilst the other had a rare four-membered beta-lactam ring fused at two points to the thiazolidine ring. Attempts were made to synthesize each of these compounds, more than a thousand chemists in thirty-nine university and industrial laboratories being involved. Most of their efforts were in vain, but the Merck and Oxford groups did manage to obtain trace amounts of synthetic penicillin. This still did not settle the issue of which chemical structure truly represented penicillin, since the possibility of interconversion existed. The issue was finally resolved through X-ray crystallographic studies conducted at Oxford by Dorothy Hodgkin in 1945. She confirmed that penicillin contained the beta-lactam ring system. However, with the ending of the war, the collaborative programme was wound up and all the laboratories involved in the synthetic work decided to abandon it.

SEMISYNTHETIC PENICILLINS

One of the leading Merck scientists, John Sheehan, managed to continue his work on the synthesis of penicillin after he left the company to become professor of chemistry at Massachusetts Institute of Technology. His research was financially supported by Bristol Laboratories (a division of the Bristol-Meyers Company) of Syracuse, New York. Sheehan's strategy was to devise new techniques that would make it possible to synthesize the unstable beta-lactam ring. This had been the stumbling block that had thwarted all previous attempts. In 1957, his persistence paid off when he synthesized phenoxymethylpenicillin (penicillin V). The overall yield was around one per cent, but within two years Sheehan had increased this to over 60 per cent. His synthesis also enabled him to prepare 6-aminopenicillanic acid, the key to making improved antibiotics with novel side-chains. Sheehan showed this could be done by reacting 6-aminopenicillanic acid with readily prepared acid chlorides. This was a major improvement on the previous method of obtaining new penicillins through addition of a chemical precursor to the liquor in which the *Penicillium* mould grew.

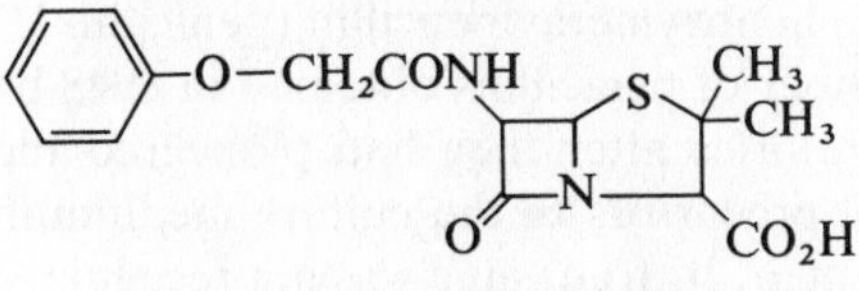

Phenoxymethylpenicillin

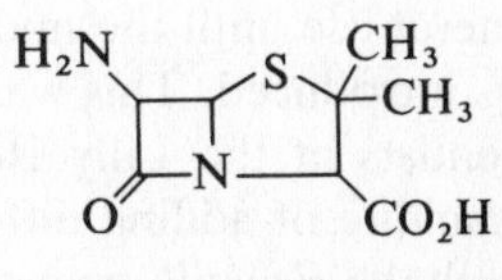

6-Aminopenicillanic Acid

Sheehan's synthesis would have been very difficult to scale up and develop for commercial application, but the problem did not arise. In 1958, Batchelor, Doyle, Nayler, and Rolinson of the recently established Beecham Research Laboratories at Betchworth, Surrey, made a remarkable discovery. They were total newcomers to the field of penicillin research and were advised by Ernst Chain to prepare 4-aminobenzylpenicillin (4-aminopenicillin G). Chain believed that it might be possible to make novel derivatives from this. The Beecham team agreed to do this, In the course of extracting the new penicillin, its acetyl derivative crystallized out of solution as a pure compound. Since the crystals could easily be converted to the desired 4-aminobenzylpenicillin, it was obvious that addition of an acetylating agent to the mould juice would facilitate the isolation process by converting all the 4-amino compound to the less soluble acetylamino derivative. When this was done, a discrepancy appeared in the microbiological assay of the mould juice, which now indicated enhanced antibacterial activity. In an inspired interpretation of this occurrence, the Beecham researchers realized that 6-aminopenicillanic acid must have been present in the mould juice. Only after it was converted to its 6-acetyl derivative did it exhibit sufficient activity to affect the assay result! Subsequent tests confirmed 6-aminopenicillanic acid was always present in mould juice, and was a stable substance, contrary to expectation. The Beecham team quickly exploited their discovery by developing methods of obtaining large quantities of this key intermediate by fermentation. Their first patent was applied for in August 1958.

Since Beecham Research Laboratories employed Sheehan's method of converting 6-aminopenicillanic acid into therapeutically useful penicillins, meetings were held with him and representatives of Bristol Laboratories. Early agreement was reached whereby both companies would collaborate in the development of semisynthetic penicillins. Harmony between the three interested parties rapidly dissipated. Bristol and Beecham went their own ways, and for the next twenty years there were interminable legal wrangles between Sheehan and Beecham over patent rights to the new semisynthetic penicillins. The arguments revolved around whether Sheehan deserved the credit for making 6-aminopenicillanic acid, which he had not actually isolated, by a commercially unrealistic method. The matter is too complex to be dealt with here, but suffice it to say that the US Board of Patent Interferences ruled in favour of Sheehan in 1979. He has written his own account of what transpired.

The ability to synthesize new penicillins at will permitted the shortcomings of benzylpenicillin (penicillin G) to be tackled. It had been the only penicillin in general use until the mid-fifties, when phenoxymethylpencillin (penicillin V) was introduced. This was one of a number of penicillins obtained in 1948 by scientists at the Lilly Research Laboratories after they had pioneered the technique of adding different chemical precursors to the culture medium in which the *Penicillium* mould was fermented. Its true value was not recognized until five years later, when a chemist in an Austrian penicillin production plant, E. Brandl, noticed that it had the unique property of being resistant to degradation by dilute acid. This meant that it could withstand exposure to gastric acid, thereby avoiding the extensive degradation of benzylpenicillin when taken by mouth. Eli Lilly and Company exploited this by introducing the potassium salt as the first reliable orally active penicillin ('V-Cil-K'). Not unexpectedly, the Bristol-Meyers Company based their first semisynthetic penicillin on phenoxymethylpenicillin. This was an analogue in which the methyl group in the side-chain was replaced by an ethyl group to give the acid-resistant phenethicillin ('Syncillin', 'Broxil'). It was introduced in 1959, its only advantage over phenoxymethylpenicillin being that equivalent doses gave higher blood levels. This advantage was offset by its higher price.

Methicillin ('Celbenin', 'Staphcillin') was introduced a few months later by Beecham Research Laboratories. It was the first penicillin to be insensitive towards penicillinase, an enzyme produced by resistant strains of staphylococci. The enzyme caused the beta-lactam ring to split. The Beecham chemists, led by Naylor and Doyle, found that stability towards attack by it was achieved merely by placing bulky substituents on the penicillin side-chain, in close proximity to the labile part of the beta-lactam ring; this is described as steric hindrance. As methicillin was still acid-sensitive and poorly absorbed from the gut, it could only be given by injection. In 1961, Beecham prepared more sterically hindered analogues and found that those containing an isoxazole ring in the side-chain were stable in acid. They introduced an orally active variant, oxacillin ('Prostaphlin'), that was marketed by Bristol Laboratories in the United States. For the British market, Beecham produced its chlorine-substituted derivative, cloxacillin ('Orbenin'). It and the similar flucloxacillin ('Floxapen', floxacillin) and dicloxacillin ('Diclocil') are now reserved for treating penicillin-resistant *Staphylococcus aureus* infections since their activity against other bacteria is inferior to alternative penicillins.

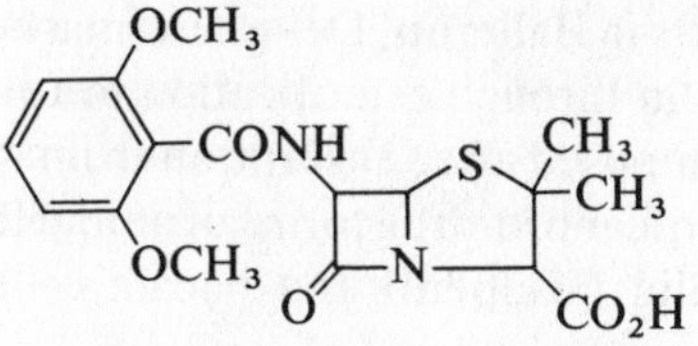

OCH_3 CONH S CH_3 CH_3 OCH_3 N O CO_2H

Methicillin

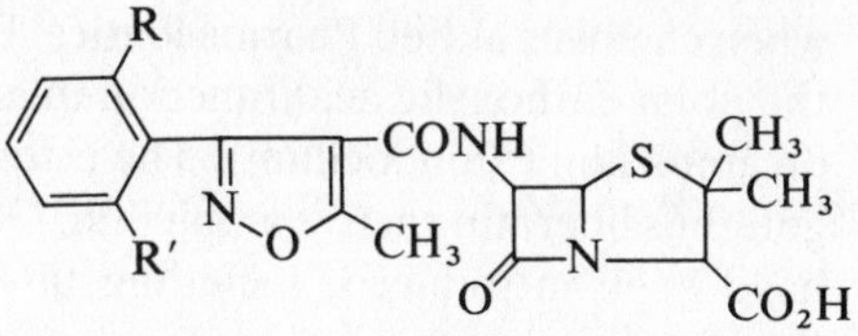

R CONH S CH_3 CH_3 N O CH_3 R' N O CO_2H

Oxacillin, R = R′ = —H
Cloxacillin, R = —Cl, R′ = —H
Dicloxacillin, R = R′ = —Cl
Flucloxacillin, R = —Cl, R′ = —F

Unlike benzylpenicillin, the early semisynthetic penicillins were not active against Gram-negative bacteria. In an attempt to prepare an orally active analogue of benzylpenicillin, the Beecham chemists introduced substituents into its side-chain in a manner akin to that appertaining to the acid-stable phenoxymethylpenicillin. Analysis of the acid stability of the resulting compounds indicated that this was enhanced by substituents which attracted electrons. The most suitable substituent proved to be the amino group, the compound featuring it being called ampicillin ('Penbritin'). It had a wider spectrum of activity against Gram-negative bacteria than had its parent compound, benzylpenicillin. The activity against Gram-positive bacteria was somewhat reduced. These differences were due to variation in the ability of the penicillins to penetrate into different types of bacteria. All penicillins act in the same manner by blocking bacterial cell wall synthesis.

Ampicillin was an outstanding success, but it had two drawbacks. The more serious was its inability to avoid destruction by penicillinase-producing bacteria. Many microorganisms that were once sensitive to it have become resistant

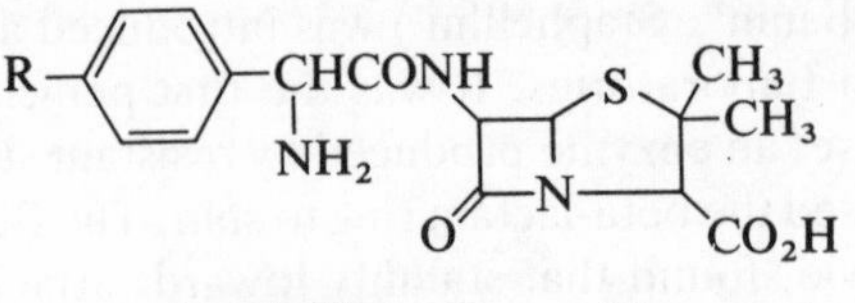

Ampicillin, R = —H
Amoxycillin, R = —OH

as a result of being able to produce penicillinase. There is no longer good justification for using ampicillin as a broad-spectrum antibiotic in the hope that it will eradicate infections caused by unidentified organisms. The other drawback was due to the dipolar nature of ampicillin, which reduced its ability to be absorbed from the gut. Less than half the dose was absorbed, and if food was present in the gut, the absorption was further reduced. Some patients experienced diarrhoea through upset of the normal bacterial flora in the gut by unabsorbed ampicillin. The closely related phenolic analogue, amoxycillin ('Amoxil'), was introduced by Beecham Research Laboratories in 1964. Its absorption was superior, as well as being unaffected by the presence of food in the gut. The problem of poor absorption was completely overcome in 1969 when chemists at Leo Pharmaceutical Products in Ballerup, Denmark, masked the polar carboxylic acid function in ampicillin through esterification to form pivampicillin ('Pondocillin'). The ester decomposed after absorption from the gut, thus liberating active ampicillin. Other similar pro-drug forms of ampicillin have been introduced, including talampicillin ('Talpen') and bacampicillin ('Ambaxin').

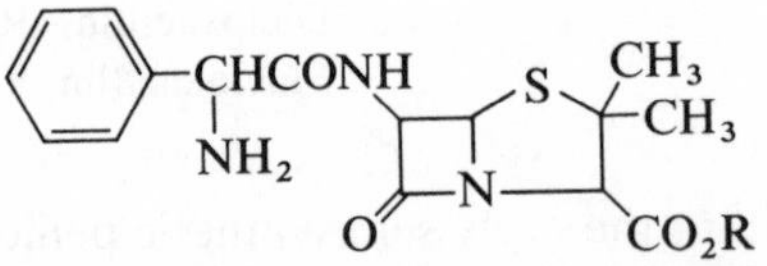

Pivampicillin, R = $-CH_2OC(=O)C(CH_3)_3$

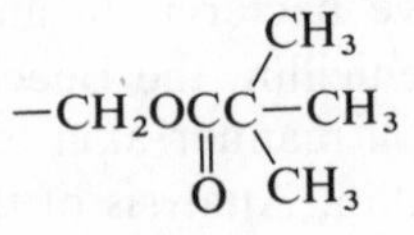

Talampicillin, R =

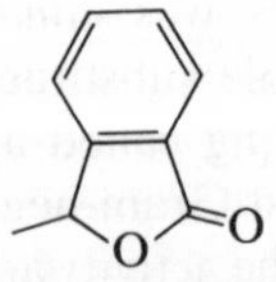

Bacampicillin, R = $-CH(CH_3)OC(=O)OCH_2CH_3$

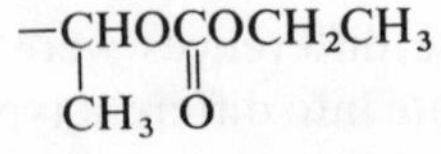

Beecham chemists reasoned that as the introduction of a basic amino group into the side-chain of benzylpenicillin afforded a derivative with a wider spectrum of activity, namely ampicillin, it would be worthwhile to examine the effect of introducing an acidic carboxyl function instead. The resulting compound, carbenicillin ('Pyopen'), was obtained in 1964 and it, as well as the closely related analogue prepared at the same time, ticarcillin ('Ticar'), turned out to be one of the first penicillins to have high activity against *Pseudomonas aeruginosa*. This was a particularly troublesome pathogen that could be

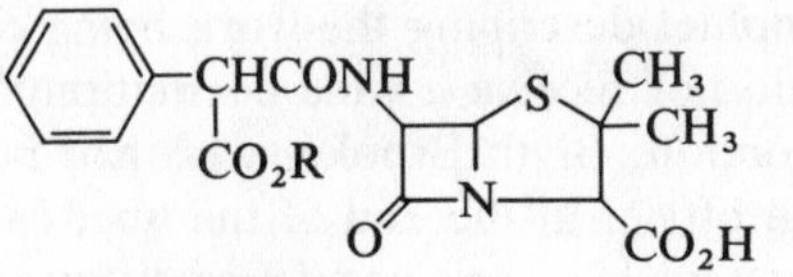

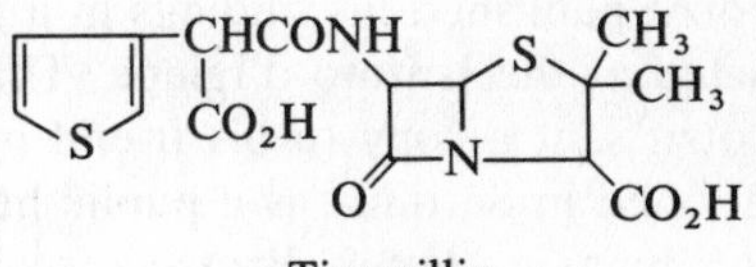

Carbenicillin, R = —H

Carfecillin, R = —C_6H_5

Ticarcillin

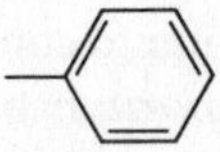

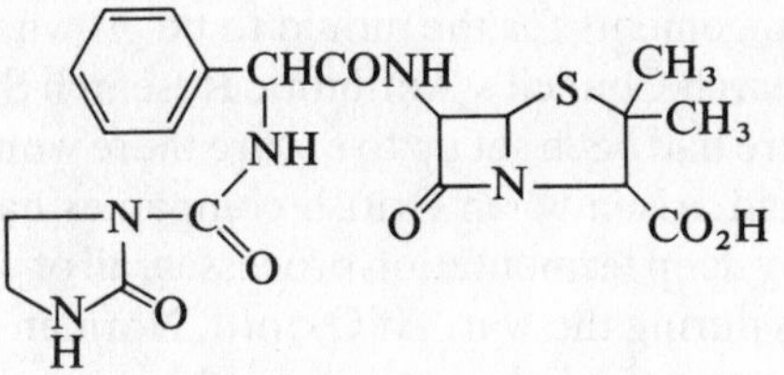

Azlocillin

life-threatening both in severely burned patients and in those whose immune system was in any way compromised. Because it featured a polar substituent on the benzyl side-chain, carbenicillin was poorly absorbed from the gut. Beecham chemists prepared carfecillin ('Uticillin') as a pro-drug in which one of the carboxyl groups was masked in the form of a phenyl ester so as to permit oral medication. Azlocillin ('Securopen'), prepared at Elberfield in 1971 by Bayer chemists, is one of the most effective analogues of ampicillin for use in *Pseudomonas* infections.

CEPHALOSPORINS

Following the clinical introduction of penicillin, Guiseppe Brotzu, former rector of Cagliari University and now director of the Istituto d'Igiene in Cagliari, Sardinia, began to search for antibiotic-producing organisms. Believing that the self-purification of sea-water near a local sewer might be due in some measure to microbial antagonism, Brotzu sampled the water and isolated a mould, *Cephalosporium acremonium*. It inhibited the growth of typhoid bacilli and also other pathogens growing on agar plates. Brotzu then prepared hundreds of subcultures of the mould until he had isolated a strain which conferred high antibacterial activity on filtrates of the mould juice. A crude extract was obtained from the juice by adding alcohol to precipitate inactive material. This had a wider spectrum of antibacterial action than had penicillin, being active against Gram-negative as well as Gram-positive bacteria. Preliminary clinical studies were carried out by direct application of filtered mould juice to boils and abscesses caused by staphylococci and streptococci. The results were encouraging. The concentrated extract was then injected into patients with typhoid, paratyphoid, and brucellosis, again with good results. In 1948, Brotzu published his findings in a pamphlet describing the work being conducted at the Istituto d'Igiene. This did not receive a wide circulation, but Brotzu sent a copy to his friend in London, Blyth Brooke, who had been stationed in Sardinia as a public health officer at the end of the war. In an accompanying letter, Brotzu explained that his work had failed to interest anyone in Italy. When Brook drew the attention of the Medical Research Council to Brotzu's discovery, they advised him to contact Sir Howard Florey. This resulted in a culture of the *Cephalosporium* being sent to Oxford in September 1948.

Florey made arrangements for the mould to be grown in deep culture tanks at the Medical Research Council's Antibiotic Research Station at Clevedon in Somerset. This centre had been set up to ensure there would be no repetition of the situation that had arisen when British companies had to pay royalties to produce penicillin by deep fermentation processes, all of which were developed in the United States during the war. At Oxford, Norman Heatley extracted an acidic antibiotic from the *Cephalosporium* mould juice, using organic solvents. By July 1949, Edward Abraham and H. S. Burton had isolated an antibiotic from Heatley's solvent extract, but were disappointed to find that it was active only against Gram-positive bacteria. Accordingly, they designated it cephalosporin P. Its chemical structure was not established until 1966. It was a steroidal antibiotic that clearly did not have the broad spectrum of antibacterial activity described by Brotzu, and was of little value. A second antibiotic was detected in mould juice that had already been extracted with organic solvent. Since this one had the activity of Brotzu's original material, it was named cephalosporin N, indicating activity against Gram-negative as well as Gram-positive organisms. It proved difficult to isolate, but three years later, Abraham and Guy Newton showed it to be an unstable penicillin. They renamed it penicillin N. Shortly after, Newton assisted the staff of the Antibiotic Research Station to increase the yield of the antibiotic so as finally to permit its isolation in fairly pure form by Abraham in 1954. It turned out to have only one-hundredth of the activity of benzylpenicillin against Gram-positive bacteria, but was much more active against Gram-negative organisms. When Newton and Abraham determined its exact chemical structure it became evident, for the first time, that the nature of the side-chain in a penicillin could have a marked effect on the spectrum of antibacterial activity. In the late fifties, Abbott Laboratories prepared sufficient supplies of penicillin N for its clinical value in typhoid to be established. The results were satsifactory, but the introduction of the semisynthetic penicillins had rendered it obsolete.

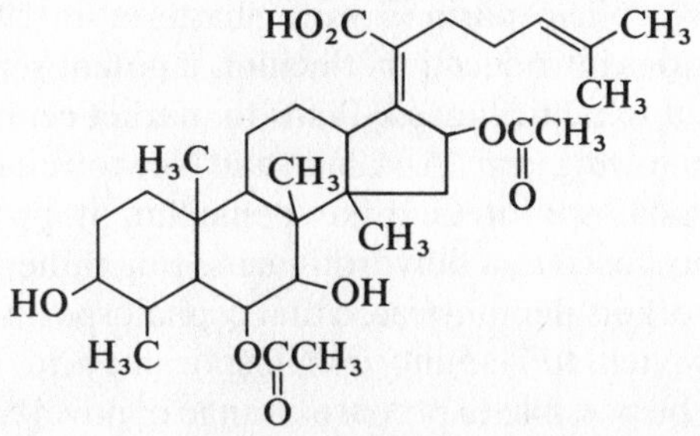

Cephalosporin P

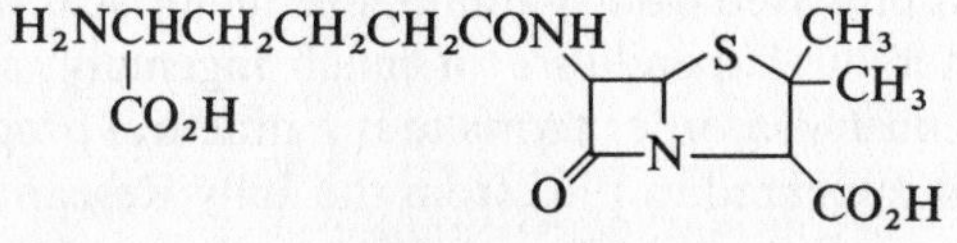

Penicillin N

Whilst completing degradation studies to confirm their proposed chemical structure for penicillin N, Newton and Abraham separated three contaminants from a crude sample of the antibiotic. The third of these contaminants was isolated as its crystalline sodium salt. Surprisingly, this exhibited weak antibiotic activity, and so was given the name cephalosporin C. Matters did not rest there. Abraham and Newton discovered that it was chemically related to penicillin, but had a much greater ability to withstand the destructive action of penicillinase. It was also less toxic than benzylpenicillin. This held out the promise of being able to inject large doses to destroy penicillinase-producing staphylococci that had become resistant to penicillin. Florey confirmed that cephalosporin C could indeed be used in this manner. However, major problems still had to be overcome before the new antibiotic could be produced on a scale large enough to meet the anticipated demand. These were overcome by the Antibiotic Research Station, and patents on cephalosporin C production were taken out by the National Research Development Corporation. This body had been set up in 1949 to exploit discoveries made in British universities and government laboratories. Licenses were issued to Glaxo Laboratories, the only British manufacturer to express an interest in cephalosporin C. Later, several foreign manufacturers were also given licenses. These provided a rich dividend for the British taxpayer when cephalosporin C became the main starting material for the production of semisynthetic cephalosporins.

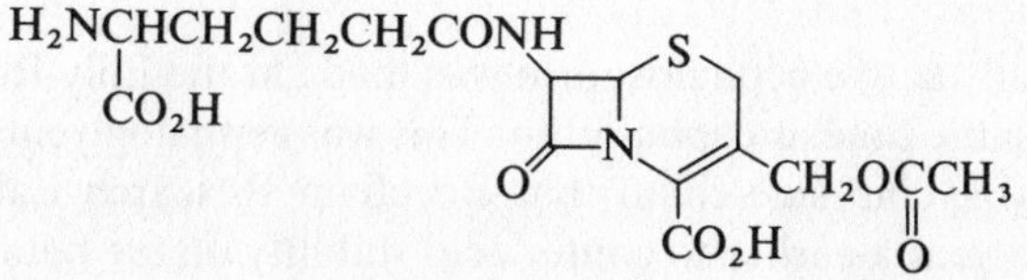

Cephalosporin C

Expectations for cephalosporin C were shattered in 1960 when Beecham Research Laboratories introduced methicillin, a potent semisynthetic penicillin that was resistant to penicillinase. Plans to market cephalosporin C had to be scrapped. The following year, Abraham and Newton elucidated the chemical structure of cephalosporin C. Like penicillin, it possessed a sensitive beta-lactam ring, but fused to a dihydrothiazine ring rather than a thiazolidine one. The Oxford workers demonstrated that cephalosporin C could, with care, be converted in low yield to 7-aminocephalosporanic acid. This meant that, in principle, it should be possible to prepare a range of novel cephalosporins from 7-aminocephalosporanic acid in much the same way as was then being done with the penicillins prepared from 6-aminopenicillanic acid. This objective was met, but only after the expenditure of much ingenuity and money by the pharmaceutical industry in order to make it a practical proposition. The first fruits of this effort appeared in 1962 from the Lilly Research Laboratories in Indianapolis. There, Robert Morin and his colleagues devised a sensitive chemical technique for obtaining the key intermediate and then converting it to cephalothin ('Keflin'), a potent antibiotic that was marketed two years later. Shortly after, Glaxo Laboratories introduced the closely related cephaloridine ('Ceporin'), which differed only insofar as it contained a pyridine ring in place of the acetyl ester originally present in cephalosporin C. This facile chemical modification had been introduced by the Oxford researchers, who found it altered the antibacterial spectrum. It also had the advantage of conferring a longer duration of action than the metabolically labile acetyl ester which was rapidly decomposed by esterase enzymes. Although cephalothin and cephaloridine were both broad-spectrum antibiotics, they have been overshadowed by newer cephalosporins, the former because of its relatively low activity and instability, the latter on account of its occasional toxicity to the kidneys.

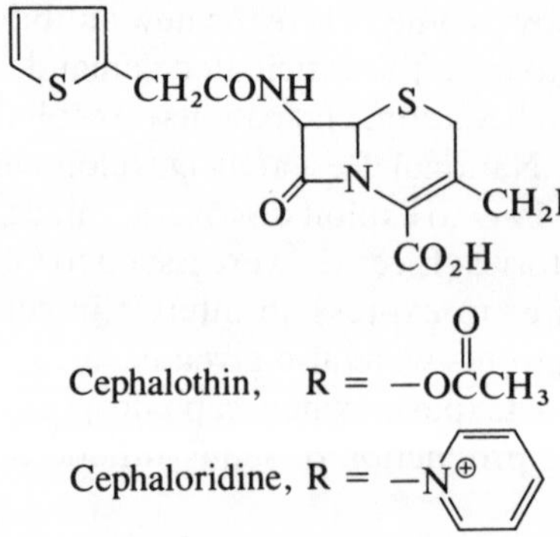

Cephalothin, $R = -O\overset{O}{\overset{\|}{C}}CH_3$

Cephaloridine, $R = -N^{\oplus}$ (pyridinium)

The first orally active cephalosporin was made at the Lilly Research Laboratories at the same time as cephalothin. This was cephaloglycin, which possessed the phenylglycine side-chain that Beecham Research Laboratories had selected only a year a earlier to confer acid stability on the beta-lactam ring of ampicillin. Cephaloglycin retained the labile acetyl ester present in the cepha-

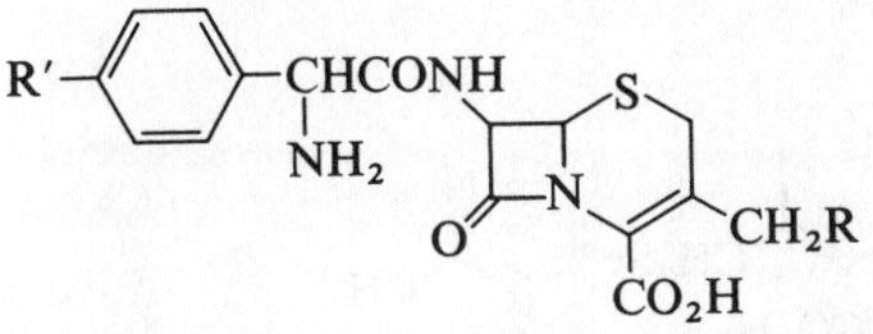

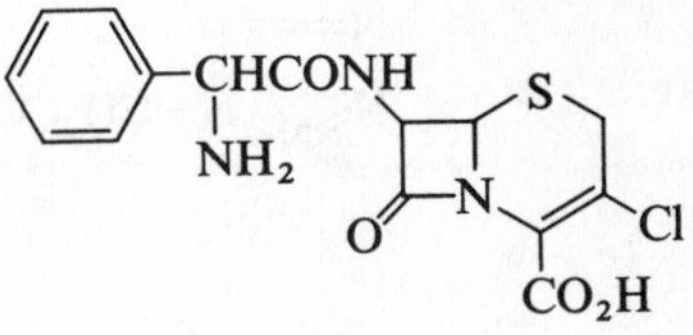

Cephaloglycin, R = $-OCOCH_3$, R′ = $-H$
Cephalexin, R = R′ = $-H$
Cefadroxil, R = $-H$, R′ = $-OH$

Cefaclor

losporin C from which it was synthesized via 7-aminocephalosporanic acid. This meant that it was rapidly metabolized, hence its clinical value was somewhat limited. The problem was overcome by replacing the acetyl ester function with a hydrogen atom. This was duly achieved when Lilly and Glaxo announced their synthesis of cephalexin ('Keflex', 'Ceporex'), introduced as an orally active, broad-spectrum antibiotic in 1967. By analogy with amoxycillin, the phenolic derivative, cefadroxil ('Baxan'), was developed in 1972. Another oral cephalosporin was introduced in 1974 when Lilly Research Laboratories synthesized cefaclor ('Distaclor'), another stable analogue of cephalexin. In general, the oral cephalosporins have not turned out to have any clearly demonstrable superiority over the cheaper penicillins or other broad-spectrum antibiotics.

A variety of so-called 'second generation' injectable cephalosporins have been developed, e.g. cephradine ('Velosef'), cephazolin ('Kefzol'), cefuroxime ('Zinacef'), and cefamandole ('Kefadol'). These do not feature the acid-stabilizing phenylglycine side-chain, hence they are unsuitable for oral use. However, they have a wider spectrum of antibiotic activity than the cephalosporins previously mentioned, allowing them to be employed to treat infections caused by Gram-negative bacteria. They are also of considerable value against certain bacteria that have become resistant to penicillins. There are also 'third generation' cephalosporins, e.g. cefotamine ('Claforan'), cefsulodin ('Monaspor'), ceftazidime ('Fortum'), and ceftizoxime ('Cefizox'), as well as many others currently on trial. These are limited in their range of application, but often exhibit outstanding activity against specific pathogens, such as *Pseudomonas aeruginosa*. There remains plentiful scope for further developments such as the discovery of an 'ultra-broad-spectrum' cephalosporin.

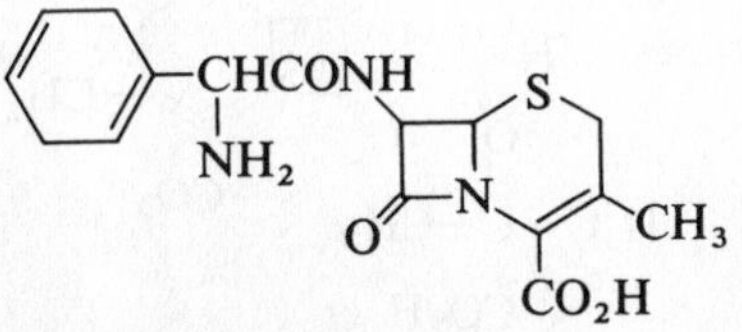

Cephradine

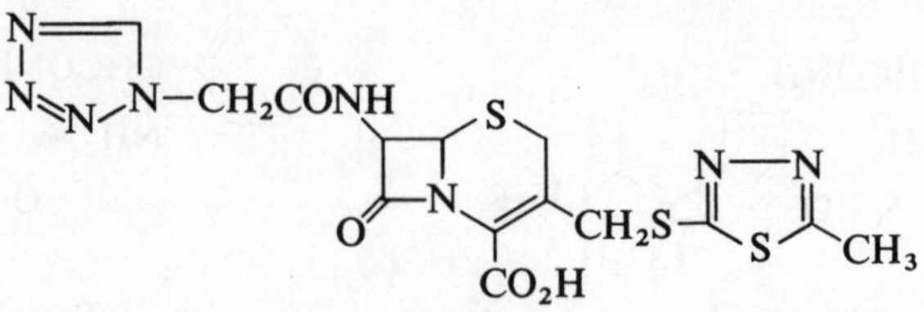

N
N N N–CH_2CONH S N–N
N CH$_2$S S CH$_3$
O CO$_2$H

Cephazolin

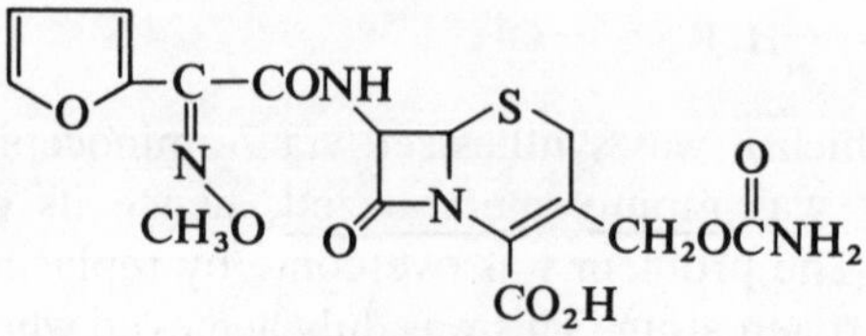

C–CONH S O
O N
CH$_3$O O N CH$_2$OCNH$_2$
CO$_2$H

Cefuroxime

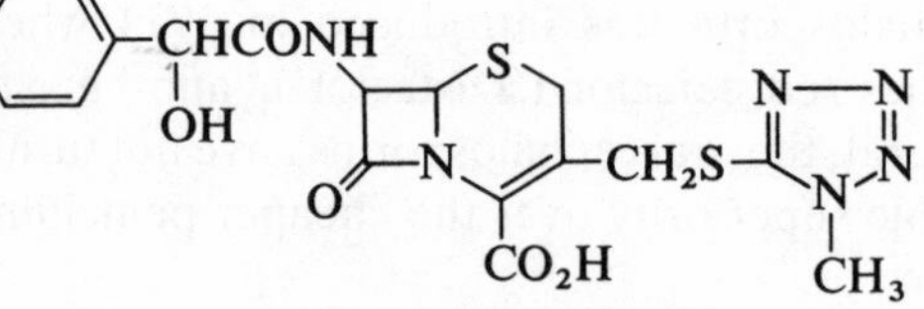

CHCONH S N–N
OH N
O N CH$_2$S N N
CO$_2$H CH$_3$

Cephamandole

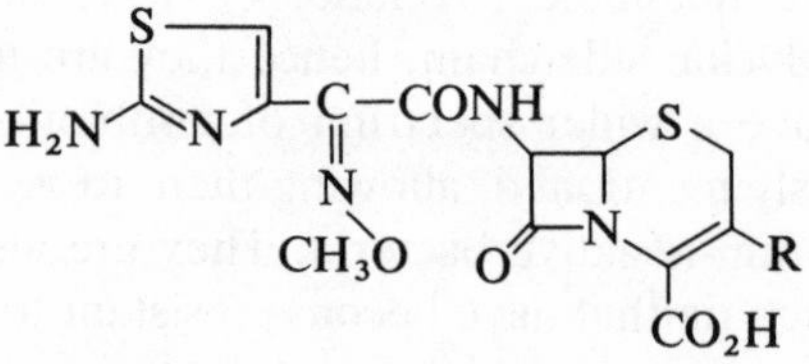

S
H$_2$N N C–CONH S
N
CH$_3$O O N R
CO$_2$H

Cefotaxime, R = $-CH_2OCCH_3$ (O)
Ceftizoxime, R = $-H$

S
H$_2$N N C–CONH S
N
O O N CH$_2$N$^{\oplus}$
H$_3$C–C–CH$_3$ CO$_2^{\ominus}$
CO$_2$H

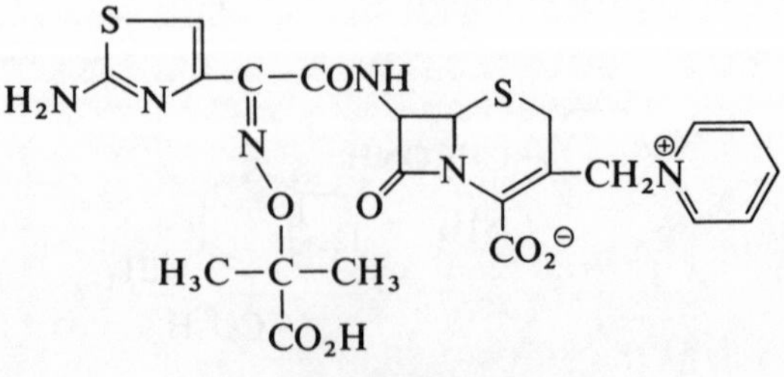

Ceftazidime

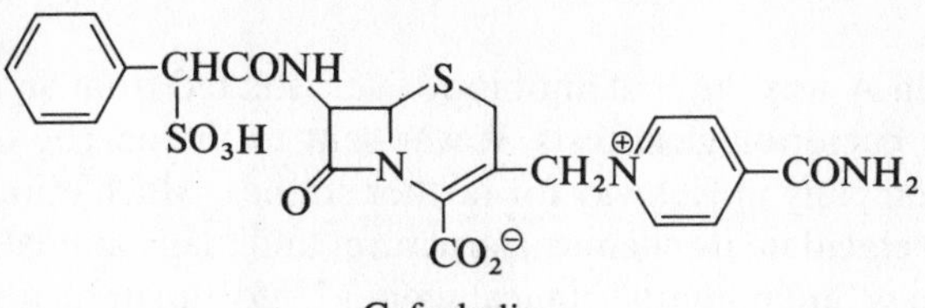

Cefsulodin

STREPTOMYCIN

During his work leading to the isolation of tyrothricin from a bacterium growing in the soil (see page 297), Rene Dubos was regularly in contact with Selman Waksman, his former teacher in the department of microbiology at the New Jersey Agricultural Experiment Station, Rutgers University. As a result of his many discussions with Dubos, Waksman became convinced that his own wide experience in dealing with soil microbes could be turned to good effect by seeking further antibiotic-producing organisms. He began in 1939 by investigating pyocyanase production by *Pseudomonas aeruginosa*, but found this did not meet his requirements when tested against a range of pathogenic bacteria. Nevertheless, the experience he and his students gained in the course of isolating this old antibiotic (see page 297) stood them in good stead when they moved on to examine fungi and actinomycetes. Waksman used the pyocyanase isolation procedure to discover a novel crystalline, ether-soluble antibiotic in 1940. This was actinomycin A, obtained from *Actinomyces antibioticus*. The actinomycetes are ubiquitous soil organisms which have features in common with both bacteria and fungi. The ability of *Streptothrix foersteri* to lyse (dissolve) certain microbes had been noted by G. Gasperini as long ago as 1890. Since 1923, Andre Gratia and his group at the Pasteur Institute in Brussels had published a long series of papers on the bacteriolytic properties of various species of *Streptothrix* and *Actinomyces*. Gratia and his colleague Maurice Welsch established that the lytic agent from these organisms, actinomycetin, was a soluble substance released into the culture in which the actinomycete was growing. The Belgian workers used actinomycetin preparations to liberate antigens from lysed bacterial cultures. The resulting 'mucolysates' were administered to patients in attempts to induce formation of antibodies that would produce immunity against the specific bacterium from which the mucolysate had been prepared. These clinical experiments were not particularly successful. They were still being conducted when Waksman had the idea that an actinomycete might produce a bacteriolytic substance safe enough to be used as an antibiotic.

Before starting his attempt to isolate an antibiotic produced by actinomycetes, Waksman and his students carried out a preliminary survey of these organisms. This confirmed findings of earlier investigators by establishing that out of 244 freshly isolated cultures from soil, over one hundred had some antimicrobial activity, and 49 were highly active. The results confirmed Waksman's long-held view that the disappearance of pathogenic bacteria that had found their way into the soil (e.g. in excreta) might be due to their destruction by chemical substances released from soil organisms.

Actinomycin A was the first antibiotic ever isolated from an actinomycete. After routine bacteriological tests, it was sent to the nearby laboratories of Merck and Company in Rahway for further studies. Much effort was spent in attempting to elucidate its chemical structure, and many animals were used to assess its toxicity and potential clinical scope. Unfortunately, it proved far too toxic for human application as an antibacterial agent. Nor was there convincing evidence to suggest that it had any worthwhile effect on experimental infections in animals.

Disappointing as the results with actinomycin A were, Waksman felt they confirmed the potential value of actinomycetes as a source of antibiotics. His position was not unlike that of Florey at the same period in time. Florey was turning his department at Oxford into a penicillin-producing factory before he had clinical evidence to justify such a bold step. Taking advantage of the academic freedom offered in a university environment, both men forged ahead towards their goals. Seeking funds to expand his investigations, Waksman wrote to the Committee on Medical Research in Washington shortly after it was established in 1941. His request for funding was rejected, but the chairman, Professor Newton Richards, asked Waksman's permission to approach the Commonwealth Fund on his behalf. Two weeks later, Waksman was awarded an annual grant of $9600 to support the search for antibiotics, this being renewable for six years. Further funds were forthcoming from Mrs Albert Lasker. Waksman, like Florey, and Ehrlich before them, thus received the main financial support for his chemotherapeutic research from charitable foundations. The similarity goes further, for all three subsequently witnessed their discoveries being rapidly brought to the clinic as a consequence of the readiness of major pharmaceutical firms to commit vast sums of money in the exploitation of these discoveries.

In 1941, Waksman isolated two more antibiotics, clavacin and fumigacin. They were less toxic than actinomycin A, but still unsuitable for clinical application. When it became evident that penicillin was active only against Gram-positive bacteria. Waksman's search centered on the need for an antibiotic to treat Gram-negative infections. He appeared to have achieved a breakthrough when he isolated streptothricin. It proved lethal to bacteria unaffected by penicillin, and at first appeared to be safe enough for human trials. Several industrial concerns were given cultures of the actinomycete that produced it, and pilot-plant production was begun. Unfortunately, chronic toxicity testing revealed that several days after the drug was injected into animals, a variety of toxic effects were caused as a result of degenerative changes in the kidney tubules. Work on streptothricin had to be abandoned.

Early in 1943, Waksman decided to concentrate his activities on finding an antibiotic that could be used to treat tuberculosis, one of the major scourges of mankind. Millions of deaths each year were caused by 'the white plague'. The problem here was that the causative organism, *Mycobacterium tuberculosis*, grew very slowly. This made screening of potential antitubercular drugs difficult at the best of times, and impracticable on the scale Waksman was operating on. Acting on a suggestion from his son, Byron Waksman, he instead screened his cultures of actinomycetes against the much faster growing *Mycobacterium phlei*, a non-pathogenic organism. Those few cultures that were active against this were then to be examined further in animals infected with tuberculosis. Waksman also began to enrich his numerous soil samples with *Mycobacterium tuberculosis* in order to favour elaboration of actinomycetes that produced an antitubercular antibiotic. In September 1943, having cultured thousands of strains of actinomycetes since the start of the project, Waksman found what he wanted. By a remarkable twist of fate, this was an actinomycete that he and his former professor had been the first to isolate twenty-eight years earlier at Rutgers whilst he was working for his doctorate, namely *Streptomyces griseus*. Within four months, the new antibiotic had been isolated in a concentrated form, and its activity against *Mycobacterium tuberculosis* had been confirmed. It also proved effective against microbes causing a variety of other diseases that had been unaffected by penicillin, including plague, brucellosis, and various forms of bacterial dysentery. Although its chemical properties turned out to be similar to those of streptothricin, the new antibiotic was free of the renal toxicity associated with that drug. It was given the name streptomycin.

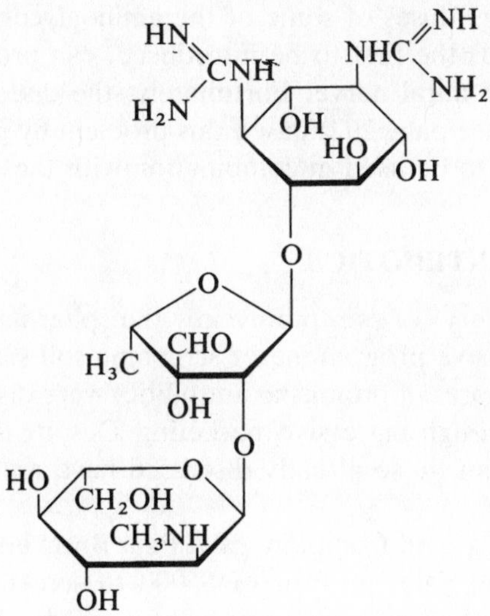

Streptomycin

Tests of the effect of streptomycin on guinea pigs infected with tuberculosis were carried out at the Mayo Clinic by Feldman and Hinshaw. A few months later the first clinical trials on patients with tuberculosis began. When the early results proved encouraging, it was agreed that the National Research Council should coordinate large-scale trials in order to hasten progress, such was the pressing need for a cure for tuberculosis. The coordinated programme cost nearly a million dollars to finance, but this was met entirely by the pharmaceutical industry. By the time the trials were completed and large-scale production of streptomycin had begun, the US industry alone had spent twenty times as much again on production plant. One firm, in particular, must be mentioned in this connection. In 1939, Waksman had obtained a grant to support his antibiotic project from Merck and Company, who had already provided him with a research fellowship in his department for the study of industrial fermentation. In return for Merck's support, Waksman had agreed to assign to the company any patentable processes that might arise from his research on antibiotics. The university would then receive two and a half per cent royalties on any sales. This arrangement might have presented no difficulty were it not for the fact that in the interim Rutgers University had changed its status from that of a private institution to become a State university. When it became clear that streptomycin was the first drug effective against tuberculosis and was, therefore, likely to be highly profitable, Merck agreed to withdraw their exclusive rights so as to avoid acute embarrassment for the university. Instead, the patents were vested in the Rutgers Research and Endowment Foundation. Licenses were then issued to many reputable manufacturers throughout the world, earning millions of dollars for Rutgers.

Clinical experience with streptomycin revealed that it was by no means as safe as penicillin. Large doses of some of the aminoglycoside antibiotics, of which streptomycin was the first to be introduced, can produce deafness by causing damage to the aural nerve. Fortunately, the discovery of synthetic antitubercular drugs (see page 292) eased this problem by permitting smaller doses of streptomycin to be used in combination with these.

MISCELLANEOUS ANTIBIOTICS

Following the discovery of streptomycin, the pharmaceutical industry embarked upon a massive programme of screening soil samples from every corner of the globe. Scores of promising antibiotics were discovered, and vast fortunes were made through aggressive marketing. Despite this, only a handful of antibiotics other than those already discussed have an important role in modern chemotherapy.

In 1943, Parke, Davis, and Company gave Paul Burckholder, an associate professor of botany at Yale, a grant of $5000 to screen soil samples for antibiotic activity against six selected types of bacteria. Burkholder examined more than 7000 samples that were sent to New Haven from all over the world, including one collected by his friend, Professor Gerald Langham, who was carrying out research near Caracas in Venezuela. This sample was one of four that contained organisms with interesting activity. In this instance, a hitherto unknown antibiotic-producing actinomycete was isolated, then given the name *Streptomyces venezuelae*. A culture of this was sent to Parke, Davis in Detroit, where John Ehrlich and Quentin Bartz isolated chloramphenicol ('Chloromycetin') in 1947, so named because of the presence of chlorine in its chemical structure. Three months later, David Gottlieb, Warren Anderson, and Herbert Carter independently isolated the same antibiotic at the University of Illinois. It turned out to be the first broad-spectrum antibiotic ever discovered, even having activity against rickettsial diseases such as typhus. The first patients to receive chloramphenicol were victims of an epidemic of this disease sweeping through Bolivia. In December 1947, twenty-two patients in the General Hospital at La Paz, of whom at least five were close to death, were given injections of the new antibiotic by Eugene Payne, a clinical researcher from Parke, Davis. All were cured. Similar results were obtained in Kuala Lumpur, in Indonesia, but this time there was some confusion that resulted in patients with typhoid also receiving injections of chloramphenicol. Their rapid recovery was unprecedented. Subsequent trials confirmed the drug to be an orally active, broad-spectrum antibiotic.

$$O_2N-C_6H_4-CH(OH)CH(NHCOCHCl_2)CH_2OH$$

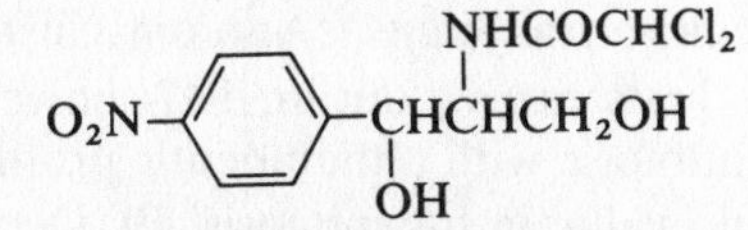

Chloramphenicol

In Detroit, Harold Crooks, Loren Long, John Controulis, and Mildred Rebstock rapidly elucidated the chemical structure of chloramphenicol, then synthesized it. The synthesis was scaled up without difficulty, with the result that the company hurriedly built manufacturing plants in Detroit and London. By 1949, large amounts of chloramphenicol were being synthesized, with sales that year exceeding nine million dollars. They increased five-fold over the next two years, turning Parke, Davis into the largest pharmaceutical company in the world at that time. Some eight million patients had been treated with this apparently safe antibiotic, when disaster struck. Reports that patients had died from delayed aplastic anaemia caused by chloramphenicol appeared in leading medical journals. The incidence of this has been estimated as between 1 in 20 000 and 1 in 100 000, with 80% of the victims dying. Had the drug not been rushed on to the market, the number of cases would hardly have caught the public eye. However, the *Journal of the American Medical Association* published a warning against the promiscuous use of chloramphenicol, then issued a press release that was widely published. The available evidence was studied by the Food and Drugs Administration, with the result that Parke, Davis were permitted to continue selling the drug, but with a warning label enclosed in the package. The label stated that prolonged or intermittent use could cause blood disorders. The issue was a highly emotional one, with lawsuits being brought against the company. Although sales dropped markedly for a year or two, the drug regained some of its popularity when the problem of resistant strains of staphylococci appeared. Eventually the arrival of the broad-spectrum penicillins largely replaced it for many purposes. Today, chloramphenicol is reserved for the treatment of certain severe conditions, including typhoid, salmonella, meningitis, and rickettsial infections. It is widely and safely used topically in treating eye and ear infections.

In 1943, a 71-year-old retired botany professor from the University of Wisconsin, Benjamin Duggar, joined Legerle Laboratories at Pearl River as a consultant, to investigate a plant alleged to have antimalarial activity. A year later he was asked by the research director, Yellapragada SubbaRow, to supervize the screening of hundreds of soil samples in order to find a safer antibiotic than streptomycin for treatment of tuberculosis. Like Paul Burkholder at Yale, Duggar then invited his academic associates to send him samples of soil from far and wide. In the summer of 1945, a sample was received from Professor William Albrecht of the University of Missouri, where Duggar had been a member of the faculty forty years earlier. This sample yielded an unknown, golden actinomycete with some antibiotic activity. Duggar named the organism *Streptomyces aureofaciens*. The antibiotic was isolated and given the approved name of chlortetracycline ('Aureomycin') after its chemical structure was elucidated. Tests carried out in 1947 showed it to be an orally active, broad-spectrum antibiotic with a therapeutic profile similar to that of chloramphenicol. It had no value in tuberculosis. By December 1948, large-scale production in fermentation tanks had begun and the antibiotic was put on the market shortly before chloramphenicol.

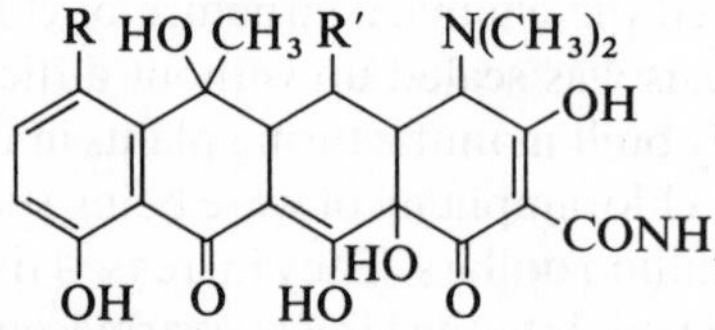

R HO CH_3 R' $N(CH_3)_2$
OH
$CONH_2$
HO
OH O HO O

Chlortetracycline, R = —Cl, R' = —H
Oxytetracycline, R = —H, R' = —OH
Tetracycline, R = —H, R' = —H

The introduction of chlortetracycline and chloramphenicol was viewed with concern by Charles Pfizer and Company, which considered its dominant market position was being severely undermined by these developments, together with the plummeting prices of penicillin and streptomycin. In retaliation, Pfizer organized a team of eleven researchers and forty-five assistants who examined almost one hundred thousand soil samples from around the world, all within eighteen months. Paradoxically, the antibiotic that emerged was obtained from an actinomycete, *Streptomyces rimosus*, cultured from a soil sample collected near the company's Terre Haute factory! This resulted in the isolation of oxytetracycline ('Terramycin') in record time. It had similar properties to chlortetracycline. A patent was applied for at the end of November 1949. The following spring, the company decided to market the antibiotic themselves, rather than supply it in bulk to others as it had done with all its previous products. Lacking a sales force of representatives who could call on physicians, Pfizer indulged in a much criticized campaign of direct advertising. The cost of developing oxytetracycline had been about four million dollars, but in two years the company spent almost double this amount on advertising the drug. This resulted in their obtaining a quarter of the American market for broad-spectrum antibiotics in 1951, roughly the same share as chloramphenicol. Chlortetracycline retained half of it until the introduction of tetracycline ('Achromycin', 'Tetracyn'). This new drug was the first semisynthetic antibiotic ever prepared. In 1952, Lloyd Conover of Pfizer discovered that an active substance was obtained when the chlorine atom of Lederle's chlortetracycline was removed by catalytic hydrogenation. It was believed that this new antibiotic caused fewer gastrointestinal upsets. Lederle then isolated tetracycline from the same actinomycete from which they were already producing chlortetracycline, whilst Bristol Laboratories obtained it from *Streptomyces viridifaciens*. These three companies all applied for patents on tetracycline in 1952–1953, Lederle and Pfizer agreeing to cross-license each other. All the applications were rejected, but Pfizer and Bristol fought the decision and succeeded in obtaining patents in 1955. An agreement then made between the tetracycline producers and their licensees was criticized by the US Federal Trade Commission in 1958, on the grounds that it was an attempt to eliminate competition and fix prices. This was denied by the companies concerned.

The early tetracyclines owed much of their commercial success to their oral activity. The first penicillin suitable for oral administration, phenoxymethylpenicillin, was not introduced until the mid-fifties. However, the proportion of the tetracyclines actually absorbed from the intestine was only a fraction of the total dose, leading to disturbance of the normal bacterial flora of the gut. In hospitalized patients, this gradually led to replacement of the flora by strains of bacteria resistant to the tetracyclines. Inevitably, this prompted a decline in the use of these antibiotics in hospital practice. Nevertheless, they remain the drugs of choice in the treatment of trachoma, psitacosis, and similar chlamydial infections, as well as in those caused by rickettsia, mycoplasma, and brucella.

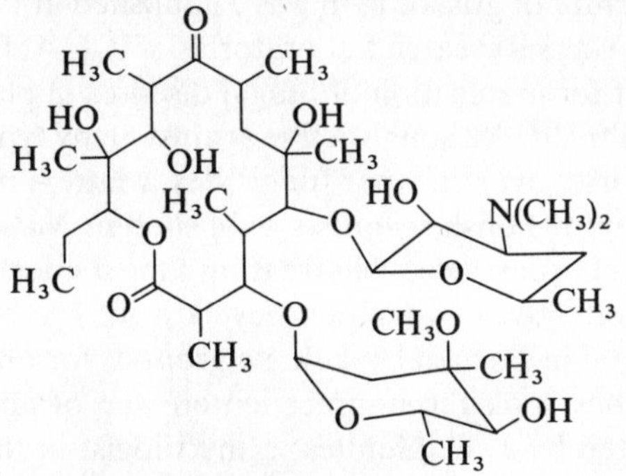

Erythromycin

Robert Bunch and James McGuire of Eli Lilly and Company isolated erythromycin ('Ilotycin'), in 1952, from a strain of *Streptomyces erythreus* cultured from a soil sample collected in Iloilo in the Philippines. Although its chemical structure bore no resemblance whatsoever to penicillin, its spectrum of action was similar. This meant it was of value in patients allergic to penicillin, and also in the treatment of penicillin-resistant staphylococcal infections, an increasing problem until the introduction of the semisynthetic penicillins in the early sixties.

During the thirties, the failure of newly planted conifers to grow on Wareham Heath in Dorset was attributed to the presence in the soil of a substance that was toxic to the fungi whose presence was essential for normal tree development. During the war, the matter was investigated further at ICI's Jealott Hill Research Station in Bracknell by P. W. Brian, H. G. Hemming, and J. C. McGowan, who showed that the common soil microbes were largely absent from Wareham Heath soil. In their place was an abundance of *Penicillia*, cultures of which were screened against a fungus known as *Botrytis allii*. It was found that a strain of *Penicillium janczewskii* caused a peculiar distortion of the germ tubes of the fungus, even at high dilutions. An active substance was extracted with chloroform and crystallized from alcohol, this being named 'curling factor' until identified as griseofulvin by John Grove and McGowan at the Butterwick Research Laboratory of ICI. This was a mould metabolite previously isolated from *Penicillium griseofulvum* in 1939 by Harold Raistrick and his colleagues at the London School of Hygiene and Tropical Medicine. They had not screened it for antibiotic activity.

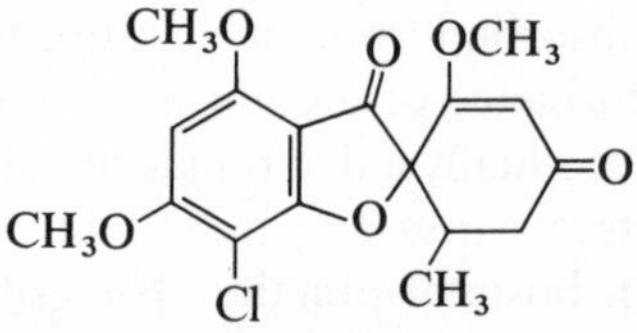

CH_3O O OCH_3 =O CH_3O O Cl CH_3

Griseofulvin

The chemical structure of griseofulvin was established in 1952 by Grove and his colleagues at the Akers Research Laboratories of ICI. At first, the company hoped to introduce it for eradication of fungal diseases in plants, but its early promise was never fulfilled. Although active against many types of fungi, it had no long-term superiority over cheaper fungicides. Matters might have rested there had not a group of Polish scientists published in *Nature*, in November 1957, a report of spectacular success in treating fungal infections with salicylhydroxamic acid, a derivative of 4-aminosalicyclic acid. Their clinical studies at the Municipal Hospital in Poznan broke new ground, for instead of applying the drug topically, they took the unprecedented step of administering it by mouth. This was noted by J. C. Gentles, a mycologist in the department of bacteriology at Anderson's College Medical School, in Glasgow. He administered griseofulvin by mouth to guinea pigs that had developed severe lesions after having been infected with *Microsporum canis*. The beneficial effects of the antibiotic were evident within only four days. Further studies were carried out at the University of Glasgow Veterinary College, where it was shown that griseofulvin could prevent and cure ringworm in cattle. Clinical trials were then organized in Vienna. These confirmed the animal studies and an announcement of this was made at a meeting of the Austrian Dermatological Society in November 1958. In April 1959, it was marketed by both ICI (as 'Fulcin') and Glaxo (as 'Grisovin'), the latter company having developed a fermentation plant for its economical production.

Although griseofulvin remains the drug of choice for treatment of intractable fungal infections of the skin, amphotericin ('Fungizone') is the most important antibiotic for use in systemic infections, especially those which threaten the lives of patients whose immune system is compromised as a consequence of intensive cancer chemotherapy. It was isolated in 1953 from a strain of *Streptomyces nodosus* growing in a sample of Orinoco river soil sent from Venezuela to the Squibb Institute for Medical Research. It proved to be very similar to nystatin ('Nystan'), an antifungal antibiotic which Squibb had introduced three years earlier after its discovered by Hazen and Brown during an extensive soil survey conducted by the New York State Department of Health. Both substances turned out to be polyene antibiotics, many further examples of which have since been isolated. Neither was absorbed from the gut, and nystatin was too toxic for systemic use. After the problems presented by its lack of water solubility were overcome, it was found that amphotericin

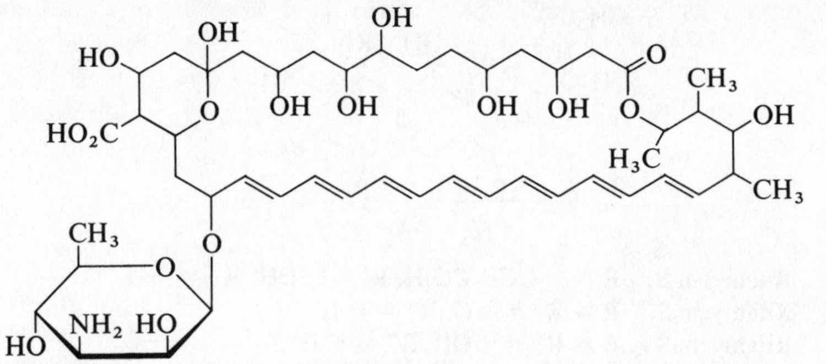

Amphotericin

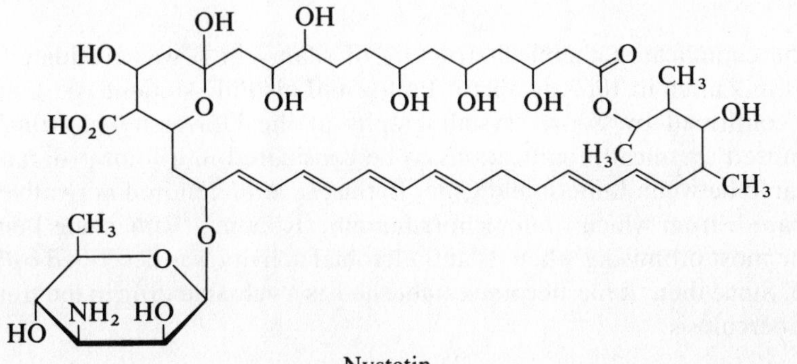

Nystatin

could be administered by intravenous infusion. It was still a toxic drug, but its use could be justified in serious fungal infections. Indeed, it is one of the most toxic drugs currently in clinical use; the toxicity is due to its ability to damage cell membranes.

In 1959, P. Sensi of the Lepetit Research Laboratories in Milan isolated a new class of antibiotics from the fermentation broth of *Streptomyces mediterranei*. These were given the generic name of rifamycins, this term being chosen from the title of a Jules Dassin gangster film, namely 'Rififi' (French argot: *rififi* = a struggle)! The only crystalline antibiotic isolated from the original mixture was rifamycin B, which constituted less than 10% of the total antibiotic titre. However, a strain of *Nocardia mediterranea* was found to produce high yields of this. Rifamycin B had low antibacterial activity, but on standing in aqueous solution it was oxidized to a much more potent antibacterial agent, rifamycin S. A reduction product of this, rifamycin SV, had high activity against cultures of Gram-positive bacteria and *Mycobacterium tuberculosis*. Although tests on infected animals were encouraging, clinical trials of rifamycin SV in tubercular patients were disappointing, principally because the drug was rapidly metabolized.

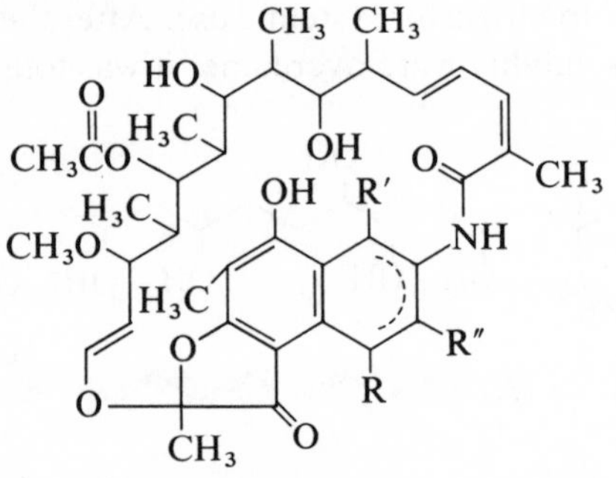

Rifamycin B, R = $-OCH_2CO_2H$, R′ = $-OH$, R″ = $-H$
Rifamycin S, R = R′ = $=O$, R″ = $-H$
Rifamycin SV, R = R′ = $-OH$, R″ = $-H$
Rifampicin, R = R′ = $-OH$, R″ = $-CH{=}N{-}N \quad N{-}CH_3$

The complicated chemical structure of rifamycin B was elucidated at the ETH in Zurich in 1963 by Victor Prelog and his PhD student, W. Oppolzer, and confirmed by X-ray crystallography at the University of Rome. This permitted chemical modifications to be considered in a joint programme of research between Lepetit and Ciba. In this, several hundred derivatives were prepared, from which rifampicin (rifampin, 'Rifadin', 'Rimactane') emerged as the most promising when its antimicrobial activity was described by Seni in 1966. Since then, it has become established as a valuable drug in the treatment of tuberculosis.

XVI

Cancer Chemotherapy

In October 1865, a Dr Lissauer of Bendorf reported in the *Berliner Klinische Wochenschrift* that he had ameliorated the condition of a young woman with acute leukaemia. She had presented with enlarged spleen and liver, together with a large number of white cells in her blood. Her disease was clearly advanced. Acting on the recommendation of a colleague, Lissauer had administered Fowler's Solution (potassium arsenite, see page 258) as a tonic, hoping it would improve his patient's appearance. Her general condition improved dramatically, with the result that she was discharged from hospital five months later, only to die soon after. Lissauer disappeared from the scene, but the use of Fowler's Solution in leukaemia persisted until after the introduction of the first modern cytotoxic drugs in the 1940s. It is pertinent to reflect that of all the acids, alkalis, and corrosive metal salts applied topically to tumours throughout the ages, that to which most frequent reference was made was arsenic. Egyptian, Greek, and Roman medical texts mentioned its value; in the eleventh century, Avicenna recommended it as a specific for cancer. From then onwards, salves containing arsenic formed the basis of most of the popular 'cures' for cancer.

The only other form of drug therapy of cancer prior to 1940 which is worth mention was the use of 'Coley's Mixed Toxins'. In 1891, a newly qualified New York surgeon at the Memorial Hospital, William Coley, observed the disappearance of a thrice-recurring inoperable neck tumour in a patient who had contracted erysipelas (a haemolytic streptococcal infection). On consulting the literature, he found descriptions of 38 similar cases dating back to 1868. His analysis of these revealed that three out of seventeen carcinomas and seven out of seventeen sarcomas disappeared after attacks of erysipelas, either from natural or deliberately induced infection. The remaining ten sarcomas showed signs of regression, some disappearing temporarily. Coley decided to expose inoperable cancer patients to others with erysipelas. As this turned out to be difficult, and he was apprehensive about the possible consequences, Coley instead inoculated them with suspensions of organisms killed by heating. These were ineffective. Coley then learned, in December 1892, of work being done in France indicating that the virulence of bacteria was sometimes increased if they were proliferating in the presence of *Serratia marcescens (Bacillus prodigiosus)* or its toxins. He therefore tried a mixture containing filtrates of cultures of these and streptococci, prepared for him at the New York College of Physicians and Surgeons. This met with only limited success. A change in the method of preparation in 1894 led to more encouraging results. During the next fifty years, at least fifteen different preparations of 'Coley's Mixed Toxins' were used; three of these proved much more successful than the rest. Unfortunately, the version marketed by Parke, Davis and Company from 1899 until 1907 was weak and variable, resulting in disillusionment in many quarters. It was replaced by an improved variant. The Lister Institute for Preventive Medicine then also manufactured 'Coley's Mixed Toxins' in London for the next forty years. It was, however, mainly around the turn of the century that there was most interest in Coley's work in leading medical circles. During this period, more than twenty others besides Coley himself claimed to have, on several occasions, eradicated tumours by the injection of these rather toxic preparations. Many others claimed total failure to achieve any measurable response. Although at least six hundred case histories were published, few discussed the use of the treatment in early cases of cancer, in which success would have been more likely. Nor were there many attempts to evaluate 'Coley's Mixed Toxins' in the absence of surgery, radiotherapy, or other treatment. Indeed, Coley's failure to arrange a properly controlled trial of his preparation accounted for its ultimate neglect, whatever its merits might have been.

HORMONE THERAPY

Following his isolation of testosterone, the principal male hormone (see page 195), Ernst Laquer and others administered it in unsuccessful attempts to treat enlarged prostate glands in elderly men. The tacit acceptance that overgrowth of this gland must be due to diminished production of male hormone was challenged by I. Wugmeister, a Polish endocrinologist working in Milan. He had read of the detection of female hormones in male urine and also of the ability of such hormones to inhibit release of gonad-stimulating hormone from the pituitary gland. He suggested that a deficiency of female hormone might permit excessive release of gonadotrophic hormone from the pituitary, causing prostatic enlargement. In 1937, he put his hypothesis to the test by treating 23 patients with large doses of oestrone. His patients showed marked improvement, with clear reduction in prostate size. Wugmeister published his findings in *Paris Médical*, where they were subsequently seen by Pierre Kahle and Emile Maltry of the School of Medicine at Louisiana State University, New Orleans. They carried out a similar study using the newly introduced stilboestrol, a synthetic hormone with greater oestrogenic potency than oestrone (see page 196). Their results, published in 1940, confirmed Wugmeister's claims and introduced this new form of therapy for benign prostatic overgrowth into the United States. On the basis of their findings, Kahle and Maltry decided to initiate a trial of stilboestrol in prostatic cancer. However, this was not completed until 1942, by which time others had already published evidence to show this type of treatment was highly effective.

While Kahle and Maltry were carrying out their investigations in New Orleans, the professor of surgery at the University of Chicago, Charles Huggins, made a far-reaching chance observation on dogs which had developed malignant prostatic tumours. He had been collecting prostatic fluid from dogs for several years as part of an enquiry into the biochemistry of seminal fluid. Some of his experiments involved castrating the dogs prior to measuring the effect of testosterone injections on prostatic fluid production. It was in the course of one such study that Huggins noticed that castration caused regression of tumours which had spontaneously arisen in elderly dogs. After confirming that castration consistently caused shrinkage of prostatic tumours in dogs, Huggins realized that this indicated these tumours were hormone dependent. He then administered stilboestrol in order to produce, as it were, chemical castration. Again, regression of the tumours was observed. The results were reported in the *Journal of Experimental Medicine* in 1940. A year later, Huggins published the results of a trial of stilboestrol in patients with prostatic cancer which had spread into their bones. These results were most encouraging, the patients exhibiting marked regression of tumours, and alleviation of pain. This was the first ever demonstration of a synthetic drug having induced an unquestionable improvement in a malignant disease. For this and his subsequent studies, Charles Huggins was awarded the Nobel Prize for physiology or medicine in 1966.

Huggins' pioneering investigations were confirmed by others. Studies of the effects of hormones on various types of tumours soon followed, notably that conducted by Alexander Haddow and his colleagues in the Chester Beatty Research Institute at the Royal Cancer Hospital, London. From there, in 1944, it was reported that stilboestrol and some of its analogues had beneficial effects in patients with breast cancer. This type of therapy was ultimately shown to be effective in up to forty per cent of post-menopausal women with mammary carcinoma. At present the drug of choice is the anti-oestrogen tamoxifen (see page 199), which the ICI pharmacologist Arthur Walpole suggested should be tried against mammary tumours. It was first used successfully in 1971 at the Christie Hospital and Holt Radium Institute in Manchester. Clinical experience with it has confirmed its lower incidence of side-effects than other hormones administered for the same purpose.

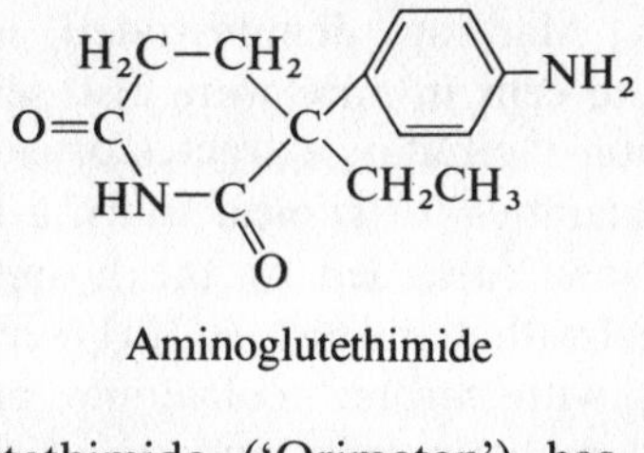

$H_2C{-}CH_2$ — NH_2 — $O{=}C$ — C — $HN{-}C$ — CH_2CH_3 — O

Aminoglutethimide

Recently, aminoglutethimide ('Orimeten') has emerged as a possible alternative to tamoxifen for the treatment of breast cancer. This was originally marketed in 1960 after Ciba researchers found it had stronger anticonvulsant, but weaker sedative-hypnotic, properties than glutethimide (see page 31). In 1963, Ralph Cash, a paediatrician at the Sinai Hospital in Detroit, reported that it had induced the typical signs of Addison's disease (adrenal insufficiency) in a young girl who had been receiving it for five months to control her epilepsy.

When similar reports from other doctors followed, laboratory studies were instituted. These revealed that aminoglutethimide blocked steroid biosynthesis, and the drug was withdrawn from the market in 1966. Cash demonstrated that aminoglutethimide inhibited the desmolase enzyme that removed the side-chain from cholesterol to form pregnenolone, a prerequisite for steroid hormone synthesis. Subsequently, it was administered to patients with Cushing's disease in the hope they might benefit from its ability to inhibit overproduction of corticosteroids, but results were disappointing. In 1969, Thomas Griffiths, Thomas Hall, Zeina Saba, and Joseph Barlow of Harvard Medical School administered aminoglutethimide to women with metastatic (spreading) breast cancer, supplementing the drug with dexamethasone to avoid pituitary release of ACTH as a result of diminished cortisone levels in the body. This gave remissions in three out of nine patients. Further joint investigations by researchers at Pennsylvania State University, Duke University, and the University of Oregon led to more successful results. The value of aminoglutethimide, especially in those women who have relapsed after initially responding to tamoxifen, is now accepted.

In 1943, the availability of a highly purified pituitary adrenotropic extract enabled Thomas Dougherty and Abraham White of the department of anatomy and physiological chemistry at Yale University to carry out fundamental studies on the consequences of adrenal stimulation in mice. Two conclusions of significance for the future of cancer treatment emerged early in their investigations. Firstly, adrenal stimulation caused shrinkage of the lymph nodes. This phenomenon was examined further at the Mayo Clinic by Edward Kendall and F. R. Heilman, who injected large doses of cortisone into mice bearing transplanted tumours of lymphatic origin. As with prostatic and some mammary tumours, these malignant cells proved sensitive to hormonal regulation. The second important observation by Dougherty and White was that adrenal stimulation induced a rapid, but short-lived, diminution in the number of white blood cells. L. W. Low of the Roscoe B. Jackson Memorial Laboratory at Bar Harbor in Maine, and Robert Spiers of the department of zoology at the University of Wisconsin, Madison, demonstrated, in 1947, that naturally occurring leukaemic white cells in mice were also sensitive to adrenal hormones released by injection of pituitary extract. Cortisone having now become available in sufficient quantities for clinical trials, a team of doctors at the Memorial Hospital in New York, led by the hospital director, Cornelius Rhoads, carried out a trial with it and also ACTH (see page 220). The results were dramatic; patients with chronic leukaemias or advanced Hodgkin's disease (malignancy of lymphatic cells) underwent remission and regained good health as the number of malignant cells declined. Professor Sidney Farber, a pathologist at the Harvard Medical School, reported similar results with acute leukaemia patients at the Children's Medical Center in Boston. Tragically, the remissions proved to be only temporary. Nonetheless, this was a major step towards the present situation where most children with acute lymphocytic leukaemia, and a majority of all patients with advanced Hodgkin's disease, can be cured by combining chemotherapy with radiotherapy. Cortisone itself has been superseded by prednisolone, which has fewer side-effects (see page 224).

BIOLOGICAL ALKYLATING AGENTS

Early in 1942, Yale University entered into a contract with the US Office of Scientific Research and Development, whereby Louis Goodman and Alfred Gilman agreed to investigate the pharmacological action of the recently developed nitrogen mustard chemical warfare agents. It did not take them long to establish that the damage done to an animal after a nitrogen mustard was absorbed through its skin into the circulation was of greater consequence than the blistering action on initial skin contact. The toxicity was most extensive in rapidly dividing cells, notably the blood-forming elements in the bone marrow, lymphoid tissue, and the epithelial linings of the gastrointestinal tract. So characteristic was this, that the Yale researchers actually assayed potential antidotes by measuring their influence on the rate of disappearance of lymphocytes and granulocytes from rabbit blood. The consistency of this phenomenon persuaded Goodman and Gilman to invite their colleague Thomas Dougherty to examine the influence of nitrogen mustards on transplanted lymphoid tumours in mice. After the necessary preliminary checks to establish a non-lethal dose range in normal mice, Dougherty administered a nitrogen mustard to a single mouse bearing a transplanted lymphoma that was expected to kill the animal within three weeks of transplantation. After only two injections had been administered, the tumour began to soften and regress, subsequently becoming unpalpable. On cessation of treatment, there was no sign of its return until a month had passed, when it gradually reappeared. A second course of injections afforded a shorter respite than before; the lymphoma ultimately killed the mouse eighty-four days after transplantation. This unprecedented prolongation of life was never matched in subsequent studies on a large group of mice bearing a variety of transplanted tumours, although good remissions were frequently obtained. This was correctly seen by Goodman and Gilman to indicate a varying susceptibility of different tumours to specific chemotherapeutic agents, a view that did not accord with the perceived wisdom of the time. There was, however, no doubting the therapeutic implication of their results on animals, and in August 1942, treatment of a patient in the New Haven Hospital began, under the supervision of an assistant professor of surgery at Yale, Gustav Lindskog. The patient, a forty-eight-year-old silversmith, was dying from a radiation-resistant lymphoma that had spread over his chest and face, preventing chewing or swallowing, and causing considerable pain and distress. The nitrogen mustard, code-named HN3 (it was actually 2,2′,2″-trichlorotriethylamine), was administered at a dose level corresponding to that previously used for mice. Belatedly, this was found to be somewhat high, resulting in a severe bone marrow damage, but nevertheless the patient survived a full ten-day course of injections. Despite his apparently hopeless condition at the onset of therapy, he responded as dramatically as the first mouse some months before. An improvement was detected within two days, when the tumour masses began to shrink. On the fourth day, the patient could once again swallow, whilst after two weeks there were no signs of any tumour masses. The bone marrow cells regenerated over a period of weeks, but so too did the tumour masses. A brief second course of injections was of some value, but the third course failed to prevent the lethal progress of the disease. A further six terminally ill patients with a variety of neoplastic diseases were treated at New Haven before the nitrogen mustard group at Yale was disbanded in July 1943. Earlier that year, Charles Spurr, Leon Jacobson, Taylor Smith, and E. S. Guzman Barron of the department of medicine at the University of Chicago began a full-scale clinical trial of another nitrogen mustard, then known as HN2 but later given the approved name of mustine (mechlorethamine). They examined its effects on fifty-nine patients with various blood dyscrasias, and obtained spectacular remissions in patients with Hodgkin's disease, amongst whom were several who had ceased to respond to X-ray therapy. On being informed of the results of this trial in August 1943, Cornelius Rhoads, director of the Memorial Hospital in New York, but on leave of absence to act as the chief of the Army Chemical Warfare Service based at Wedgewood Arsenal in Maryland, then arranged for Major Alfred Gilman to oversee further studies on a total of sixty patients at the University of Utah School of Medicine in Salt Lake City, where Louis Goodman joined Maxwell Winetrobe and Margaret McLennan, at Tufts College Medical School in Boston (William Dameshek), and at the University of Oregon Medical School in Portland (Morton Goodman). When their results had been collated, a contract was awarded for David Karnovsky and his colleagues to organize a clinical trial at the Memorial Hospital in order to establish the relative merits of HN2 and HN3 in patients with leukaemia, Hodgkin's disease, and brain tumours.

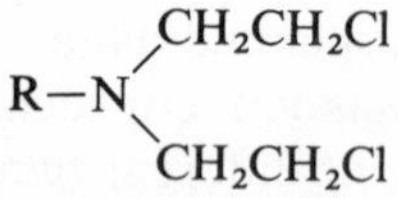

$$R{-}N\begin{cases} CH_2CH_2Cl \\ CH_2CH_2Cl \end{cases}$$

$$\text{HN3, } R = -CH_2CH_2Cl$$
$$\text{Mustine, } R = -CH_3$$

Because of wartime secrecy surrounding all work on nitrogen mustards, no information about any of the trials was released until 1946. It was then revealed that HN2 and HN3 had produced useful results in patients with Hodgkin's disease, lymphomas, or chronic leukaemias, although there was doubt concerning their superiority over X-rays properly applied to solid tumours. Poor results had been obtained with acute leukaemia, although partial remissions occurred in some cases. Patients with other neoplastic diseases failed to respond. On balance, HN2 (mustine) seemed a better drug than HN3.

During the war, information on nitrogen mustards had been freely ex-

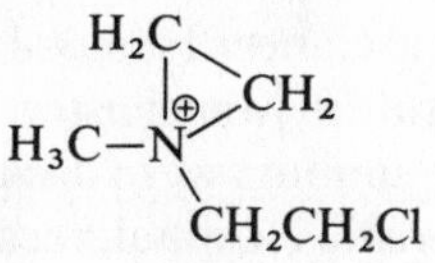

$$\begin{array}{l} \quad\ H_2C\diagdown \\ \quad\quad\ \ |\oplus\ \ CH_2 \\ H_3C-N\diagup \\ \quad\quad\ \ \diagdown \\ \quad\quad\quad CH_2CH_2Cl \end{array}$$

Ethyleneimonium Ion from Mustine

changed between British and American investigators holding government contracts. When George Hartley, Herbert Powell, and Henry Rydon, three leading chemists at Oxford University, obtained experimental proof that the chloroethyl side-chain in these compounds cyclized to form highly reactive ethyleneimonium ions (nowadays known as aziridine ions) which could rapidly alkylate vital tissue components, it was thereafter generally assumed that only drugs that could form such ions would be effective antitumour agents. That this was not necessarily so did not become apparent until 1948 when Alexander Haddow, George Kon, and Walter Ross, of the Chester Beatty Institute, discovered that aromatic nitrogen mustards were effective cytotoxic agents despite being unable to form ethyleneimonium ions. The following year, Reginald Goldacre, Anthony Loveless, and Walter Ross suggested that the cytotoxic action of both aliphatic and aromatic nitrogen mustards might be due to their ability to cross-link cellular components such as the chromosomal strands of nucleic acid, and that it was not essential for ethyleneimonium ions to be formed for this to occur. All that was necessary was the presence of two chemically reactive functional groups. Ross decided to investigate the action of a series of di-epoxides that could conceivably act as cross-linking agents in a manner akin to that of the nitrogen mustards. Before he had an opportunity to put this to the test, Professor John Speakman of the department of textile industries at the University of Leeds suggested to Professor Haddow that the biological properties of cross-linking agents used in textile technology should be examined, especially those of the di-epoxides! Speakman had become interested in the work at the Chester Beatty after examining the ability of aromatic nitrogen mustards to cross-link keratin fibres. A series of di-epoxides were then prepared by James Everett and Professor Kon; they did indeed turn out to act almost identically to the nitrogen mustards. However, Haddow's group were not alone in discovering this.

$$H_2NCO_2C_2H_5$$

Urethane

Francis Rose and James Hendry of the ICI Research Laboratories in Manchester had synthesized various urethanes (carbamate esters) after one of their colleagues, Wilfred Sexton, noted that certain phenylurethanes inhibited the growth of cereal seedlings. During the war, Haddow had tested Sexton's phenylurethanes on animal tumours and then went on to make the surprising discovery that urethane itself was more active. Clinical trials of the urethanes were initiated in 1943 at the Royal Cancer Hospital in London and the Christie Hospital and Holt Radium Institute in Manchester. The results in patients with breast cancer and various other tumours were disappointing. However, Edith Paterson, a radiologist at the Christie Hospital, recalled that, in 1925, Hawkins and Murphy had criticized the use of urethane as an anaesthetic in animal experiments designed to assess the effect of X-rays on leucocytes. They found it caused a reduction in the number of these. This took on an added significance in the light of the clinical trial of urethanes in cancer. Paterson suggested extending this to include an investigation into the effect of urethane. It turned out that although it had no influence on the course of malignant disease, the white blood cell count fell to between 40% and 70% of the initial value in five out of eleven patients. This promoted a trial of urethane in 38 patients with myeloid leukaemia and 26 with lymphocytic leukaemia. Two-thirds of the former, and half of the latter registered an immediate improvement in their overall condition, but no lasting benefit was observed despite initial marked effects on the blood picture.

After the results of the clinical trial of urethane in leukaemia were published in 1946, Rose and Hendry prepared a further thirty of its analogues and had Arthur Walpole screen them against a transplanted Walker carcinoma in rats at the company's Biological Laboratories in Wimslow. Only the di- or tri-substituted derivatives approached urethane in their activity, and it was assumed they were probably reacting with amino groups in some substance involved in the process of cell replication. However, on learning of the suggestion by the Chester Beatty scientists that nitrogen mustards probably acted as bifunctional cross-linking agents, Rose and Hendry concluded that their own bifunctional urethanes probably did likewise. They then examined polymethylolamides used as cross-linking agents in paper and textile technology, fields in which ICI had considerable expertise. The only compound with worthwhile cytotoxic activity turned out to be the product of condensing formaldehyde with melamine. The ICI researchers then tested epoxides and ethylene imines used to cross-link textile fibres. Of the former class, one of the most active compounds was diepoxybutane, consisting of an unresolved mixture of isomers. This had also been examined at the Chester Beatty, but it was not released for general clinical use. It was not until 1960 that Arthur Walpole found a di-epoxide suitable for clinical application, namely ethoglucid ('Epodyl'). It has been used mainly in the treatment of bladder cancer. In the late seventies, the Swedish manufacturer Leo Laboratories introduced treosulfan for use in ovarian cancer. It is a pro-drug that undergoes metabolic conversion to diepoxybutane.

The most potent of the cytotoxic agents investigated by the ICI researchers belonged to the third class of compounds which they examined. Tretamine (triethylenemelamine; TEM), which had been used in the textile industry to cross-link cellulose fibres, was of particular interest as it had three ethyleneimine groups attached to a central triazine nucleus. It was turned out to be suitable for treating lymphatic and myeloid leukaemias, as well as

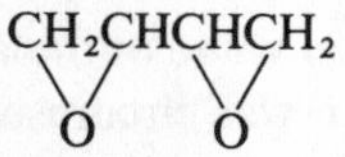

$CH_2CHCHCH_2$ (with two epoxide O bridges)

Diepoxybutane

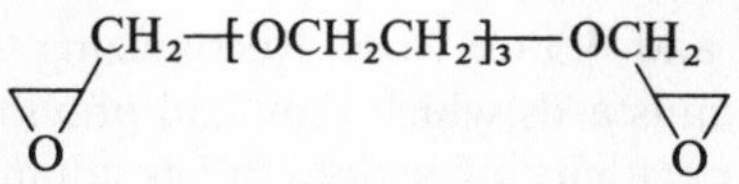

$CH_2 - [OCH_2CH_2]_3 - OCH_2$ (with terminal epoxide O rings)

Ethoglucid

SO_2CH_3 OH
$OCH_2 - CH - CH - CH_2O$
OH SO_2CH_3

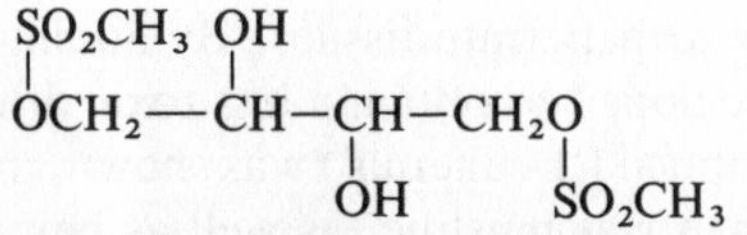

Treosulfan

Hodgkin's disease. It lacked the extremely high chemical reactivity of the nitrogen mustards and turned out to be the first biological alkylating agent that could be given by mouth. In 1950, after the ICI group had prepared, but not yet published, their first paper announcing this development, Joseph Burchenal and Chester Stock of the Sloan-Kettering Institute for Cancer Research, an adjunct of the Memorial Hospital in New York, in conjunction with Moses Crossley, the chief chemist at the Bound Brook laboratories of the Calco Chemical Division of the American Cyanamid Company, published a prior report describing the action of tretamine against experimental tumours. They, too, had searched for appropriate cross-linking agents after the Chester Beatty workers had published their proposal as to how the nitrogen mustards acted. The following year, the Sloan-Kettering researchers also introduced another biological cross-linking agent, thiotepa (triethylene thiophosphoramide), which had been specially synthesized by Erwin Kuh and Doris Seeger at the Bound Brook laboratories of the American Cyanamid Company. It is still used in the treatment of malignant effusions, as well as ovarian and bladder cancer.

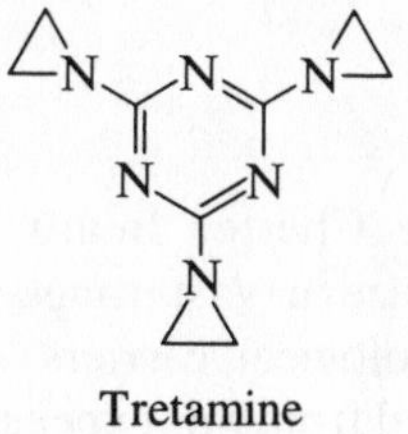

Tretamine

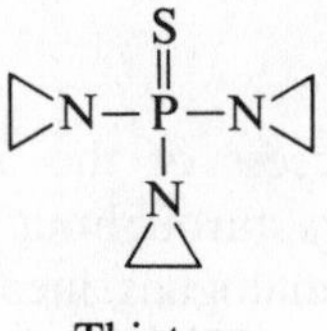

Thiotepa

Three clinically important cross-linking agents were synthesized and evaluated in the Chester Beatty laboratories, then marketed by the Burroughs Wellcome Company. The first was busulphan ('Myleran'), prepared in 1950 by Geoffrey Timmis, a former employee of Burroughs Wellcome. It was the most potent of a series of sulphonic acid esters that had been designed as alkylating agents, and it was orally active. Clinical studies completed in 1953 confirmed that it had a selective depressant action on the blood-forming cells of the bone marrow and was effective in the treatment of chronic myeloid leukaemia. The next important cross-linking agent was developed by James Everett, James

Roberts, and Walter Ross. Recognizing that the clinical value of the aromatic nitrogen mustards which they had previously prepared was circumscribed by lack of specificity of action and resultant toxicity, they exploited a concept enunciated half a century earlier by Paul Ehrlich, who had pointed out that introduction of acidic or basic groups into dyes and other molecules greatly influenced their ability to penetrate tissues. By adding acidic side-chains to aromatic mustards they hoped to obtain a less toxic drug. This objective was achieved when chlorambucil ('Leukeran') was shown, in 1952, to be less toxic to the bone marrow than was mustine, as well as being relatively free from other side-effects. Since then it has been widely used in the treatment of chronic lymphocytic leukaemia and ovarian cancer. The third valuable cross-linking agent from the Chester Beatty was melphalan ('Alkeran'), the synthesis of which was reported by Franz Bergel and John Stock in 1954. This represented a further refinement of the design approach that led to the synthesis of chlorambucil. This time, the extra moiety incorporated in an attempt to improve tissue selectively was an amino acid. Again, the approach succeeded, and melphalan proved to be a useful drug in the treatment of multiple myelomas.

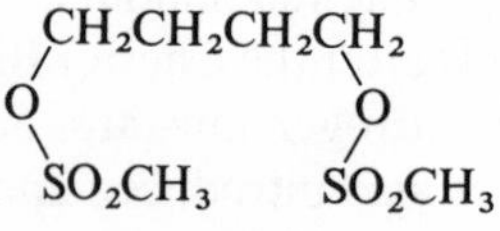

$CH_2CH_2CH_2CH_2$ (each end –O–SO_2CH_3)

Busulphan

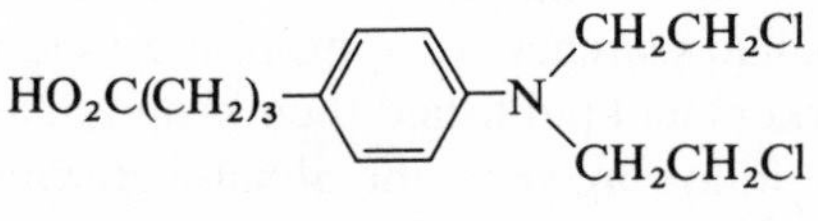

$HO_2C(CH_2)_3$–C_6H_4–$N(CH_2CH_2Cl)_2$

Chlorambucil

$H_2NCH(CO_2H)CH_2$–C_6H_4–$N(CH_2CH_2Cl)_2$

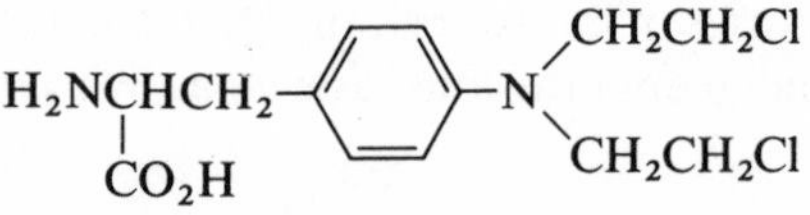

Melphalan

The success of the approach taken at the Chester Beatty encouraged researchers throughout the world to synthesize a vast range of nitrogen mustard analogues incorporating different biological carriers. Despite the expenditure of much effort, little has been gained from this expensive exercise. The only drug of note to have been developed is estramustine ('Estracyte'), which contains a nitrogen mustard function attached in the form of a urethane to oestradiol. It was developed by Niculescu-Duvaz, Cambanis, and Tarnauceanu at the Oncological Institute in Bucharest in 1966, and is used in the treatment of prostatic cancer.

The safest, and consequently the most frequently prescribed, cross-linking agent is cyclophosphamide ('Endoxana', 'Cytoxan'). It was developed in 1956 by Herbert Arnold, Friedrich Bourseaux, and Norbert Brock of Asta-Werke AG in Brackwede, Germany. The idea behind its synthesis was the same as that which had led Arnold to obtain a patent on the use of fosfestrol ('Honvan'), the diphosphate ester of stilboestrol. He believed that this was devoid of hormonal activity until it decomposed in the presence of acid phosphatase, an enzyme shown by Charles Huggins to be present in prostatic tumours. Since large amounts of this enzyme were released into the circulation in patients with prostatic cancer, it is hardly surprising that there is little evidence to support the contention that fosfestrol is superior to stilboestrol. In cyclophosphamide, a nitrogen mustard function was combined with a phosphoramide residue in the hope that there would be no significant alkylating activity until enzymic action decomposed the drug. Paradoxically, it turned out to have good activity against a wide variety of malignancies, but not against prostatic tumours. Ten years after its introduction, Norbert Brock found that this was because rather than being decomposed by acid phosphatase, as originally hypothesized, the drug was metabolized by liver enzymes and thereby converted to an active species. The isomer known as ifosphamide ('Mitoxana'), which was introduced by Asta-Werke in 1967, has similar therapeutic activity.

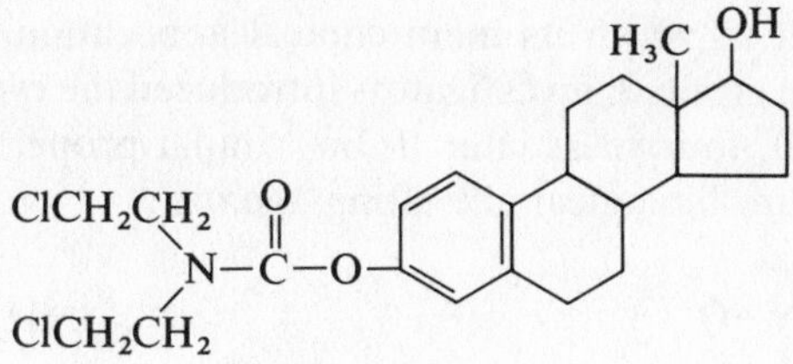

Estramustine

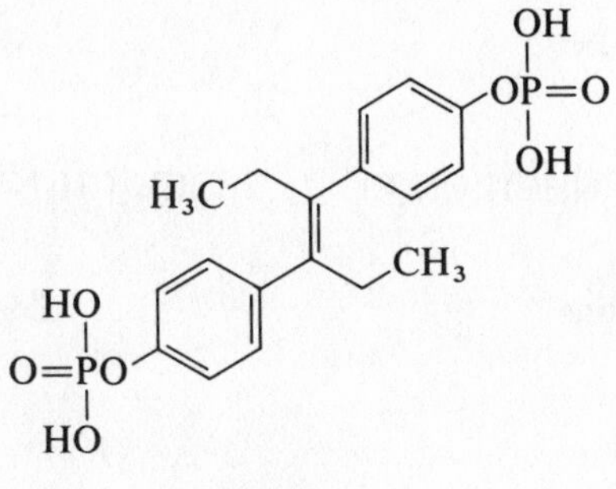

Fosfestrol

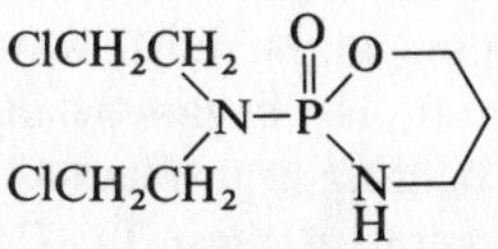

Cyclophosphamide

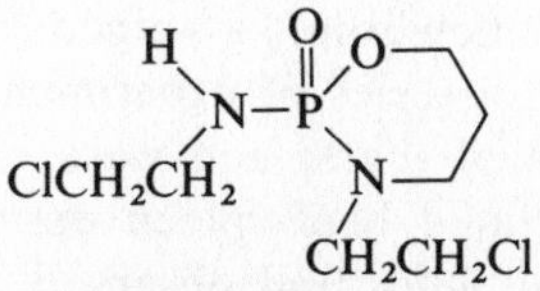

Ifosphamide

In 1955, following a decade of remarkable progress in the sphere of cancer chemotherapy, the United States Congress allocated large sums of money to set up a screening programme run by the Cancer Chemotherapy National

Service Center, which itself was part of the National Cancer Institute at Bethesda. The first contracts awarded by the CCNSC were to four screening centres that would confidentially test against animal tumours large numbers of compounds submitted by academic and industrial researchers. By the end of the decade, these centres were testing around a thousand chemicals each month. In 1959, at one of the centres, the Southern Research Institute in Birmingham, Alabama, it was discovered that an intermediate used in organic synthesis, 1-methyl-1-nitroso-3-nitroguanidine (MNNG), had antileukaemic activity in mice. Although human trials were disappointing, an immediate evaluation of structurally related compounds was launched at the Institute by Thomas Johnston, George McCaleb, and John Montgomery. When it transpired that 1-methyl-1-nitrosourea (MNU) was much more active, analogues of this were synthesized. This led to the discovery of high antileukaemic activity in mice with the twenty-third nitrosourea that was tested. This compound is now known as carmustine (BCNU; 1,3-bis (2-chloroethyl)-1-nitrosourea). Because of its high lipid solubility it was able to penetrate into the central nervous system, as a result of which its main clinical application has been in brain tumours. The same group of investigators introduced the cyclohexyl analogue, lomustine (CCNU), four years later. It has similar properties to carmustine. Both compounds are biological alkylating agents.

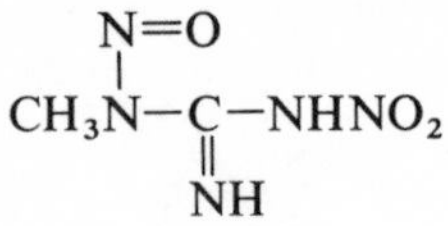

$$CH_3N(N{=}O)-C({=}NH)-NHNO_2$$

MNNG

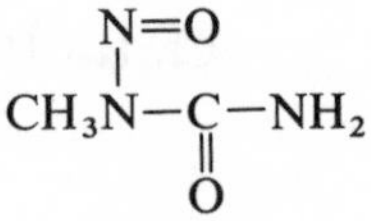

$$CH_3N(N{=}O)-C({=}O)-NH_2$$

MNU

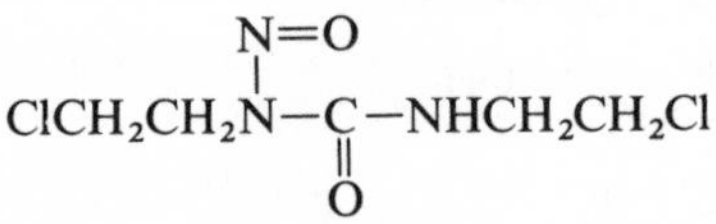

$$ClCH_2CH_2N(N{=}O)-C({=}O)-NHCH_2CH_2Cl$$

Carmustine

$$ClCH_2CH_2N(N{=}O)-C({=}O)-NH-C_6H_{11}$$

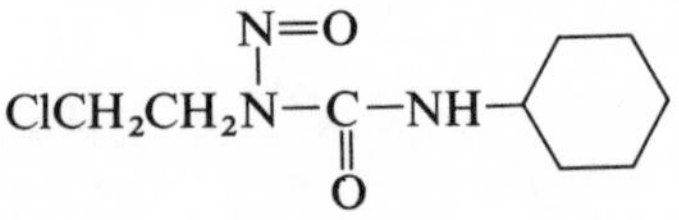

Lomustine

ANTIMETABOLITES

Richard Lewisohn, a surgeon at the Mount Sinai Hospital in New York, initiated a series of experiments in 1937, hoping to establish why primary tumours of the spleen were rarely encountered. He began by injecting a concentrated beef spleen extract into mice bearing transplanted tumours (sarcoma 180), and observed complete regression of these in 60% of the animals. The extract had to be given subcutaneously as it was highly irritant and caused thrombosis if injected into a vein. It had no effect on spontaneous (i.e. naturally occurring) tumours in mice. Noting that the spleens of the healed mice were often enlarged, Lewisohn prepared extracts of these too. Since the mouse spleen extract was dilute and non-irritant, it could be safely injected intravenously. Unlike the original beef extract, this induced complete regression of spontaneous breast tumours in 30% of mice. Although tumours reappeared in a quarter of the mice, never before had a non-toxic substance produced such a result.

Although encouraged by his findings, Lewisohn recognized that quantity production of the 'healed mouse extract' was impracticable. He began an extensive screening programme in 1939 to find other agents that could cause regression of tumours, being assisted by Rudolph and Cecilie Leuchtenberger, and Daniel Laszlo. An obvious step was to examine the recently discovered vitamins of the B group that were present in liver. Using yeast extract as a cheap, convenient source of these, the Mount Sinai researchers obtained positive results. Intensive investigations began in January 1941, requiring the use over the next three or four years of some twelve thousand mice bearing sarcoma 180, and half that number bearing spontaneous tumours. When the yeast extract was injected intravenously once daily for up to ten weeks, approximately one-third of spontaneous breast adenocarcinomas regressed completely. Measurable changes were also detected when autopsies were conducted on mice that received four intravenous injections over a forty-eight-hour period beginning a week or so after transplantation with sarcoma 180. With the entry of the United States into the war, it became impossible for Lewisohn to continue to obtain from Germany the brewer's yeast used in the preparation of his extracts. Seeking alternative sources of active material, he found that barley extract was as efficacious as yeast extract. By this time such was the nature of his findings, that the International Cancer Research Foundation, which had sponsored Lewisohn's investigations, asked Cornelius Rhoads to conduct an independent enquiry. This was duly entrusted to one of the senior scientists at the Memorial Hospital, Kanematsu Sugiura, in 1943. His attempts to confirm Lewisohn's findings were wholly unsuccessful, no significant difference being detected either in the growth of sarcoma 180 in treated and untreated mice, or in the regression of spontaneous breast cancer.

Early in 1944, it occurred to Lewisohn and his colleagues that the active substance in yeast and barley extracts might be the newly isolated folic acid (see page 241). The finding of antitumour activity in a folic acid concentrate did not settle the matter, since this contained many impurities. However, Lewisohn was able to obtain the support of Lederle Laboratories, as a result of which he was supplied with small amounts of the scarce crystalline fermentation *Lactobacillus casei* factor that had just been isolated. In an initial test on seven mice, each receiving a quarter of a microgram by intravenous injection, as high an inhibition of tumour growth as had ever before been seen was recorded. More extensive studies on mice with spontaneous breast cancer confirmed the initial results, these being published in January 1945. Subsequently, Lederle researchers discovered that the material supplied to Lewisohn was not identical to the liver *Lactobacillus casei* factor. The former was shown by them to be pteroyltriglutamic acid, whereas the liver factor was pteroylglutamic acid (now always described as folic acid). When the Mount Sinai Hospital researchers tested the liver *Lactobacillus casei* factor (i.e. folic acid), they found it to be ineffective against spontaneous breast cancer in mice. This remarkable result persuaded Lederle to synthesize both the di- and triglutamates for clinical trials. These were duly named 'Diopterin' and 'Teropterin', respectively (perhaps in anticipation of clinical success?). Both compounds were sent to Sidney Farber for clinical evaluation. He and his colleagues in three hospitals associated with the Harvard Medical School began by administering the drugs to ninety patients with a variety of malignancies that offered no hope of cure with established therapeutic procedures. The purpose of this Phase I trial was to determine the toxicology, appropriate dosage, and suitable routes of administration of the new folic acid conjugates in humans. After cautious initial studies, 'Teropterin' was given daily in doses up to 150 mg intramuscularly, or 500 mg intravenously, the average length of treatment being 35 days. Neither local nor systemic adverse effects were encountered. Although reluctant to draw conclusions about the efficacy of the treatment after so short a period of observation, Farber and his colleagues, in a brief preliminary report published in December 1947, suggested that the temporary improvements observed in some patients warranted further investigations.

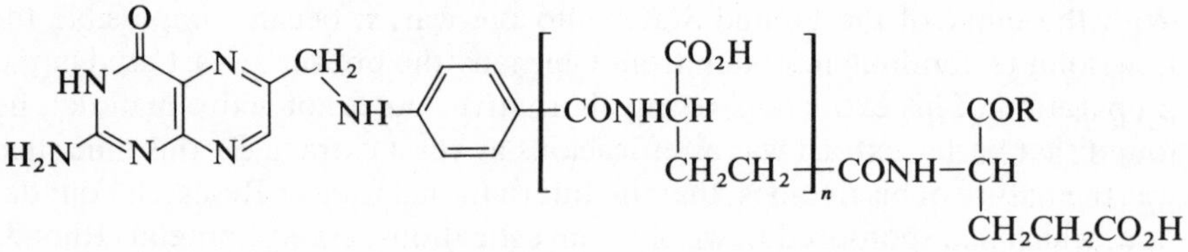

O; HN; H_2N; N; N; N; CH_2; NH; CO_2H; CONHCH; CH_2CH_2; n; CONH—CH; COR; $CH_2CH_2CO_2H$

Folic Acid, $n = 0$, R = —OH

Diopterin, $n = 0$, R = $-NHCH(CO_2H)CH_2CH_2CO_2H$

Teropterin, $n = 2$, R = —OH

During the Phase I trial, biopsies of bone marrow were routinely obtained from several patients with acute leukaemia. Unexpectedly, these indicated that administration of the folates had accelerated the progress of the disease towards its fatal termination. On reviewing the situation, Farber wondered whether this 'acceleration phenomenon' might not be put to advantage in either of two distinct ways. The first possibility was by using the folic acid conjugates to stimulate leukaemic cells to grow and divide, thereby rendering them more susceptible to nitrogen mustards. The second possibility was to administer one of the antagonists of folic acid, newly developed by SubbaRow and his colleagues at the Lederle Laboratories. A variety of antagonists of vitamins, hormones, and cell metabolites had been developed after the discovery in 1940 by Donald Woods of Oxford University that sulphonamides exerted their antibacterial action by antagonizing the role of 4-aminobenzoic acid, a growth factor for bacteria. Such antagonists were described as antimetabolites (see page 290). In 1947, Lederle researchers reported on the synthesis and biological properties of a crude methylated derivative of folic acid ('x-methyl pteroylglutamic acid', a mixture of methylated products) that was a weak antimetabolite. It had a depressant action on the blood-forming elements of the bone marrow in rats and mice, as was anticipated from prior knowledge of the effects of folic acid deficiency in animals. The extent of this, however, was greater than previously seen, the animals becoming not only anaemic, but also exhibiting a reduction in the number of white cells. When tested on a patient with chronic leukaemia, the antimetabolite was not potent enough to produce any worthwhile result. Further antimetabolites were subsequently prepared by Lederle Laboratories, the first of which, pteroylaspartic acid, was sent to Professor Farber.

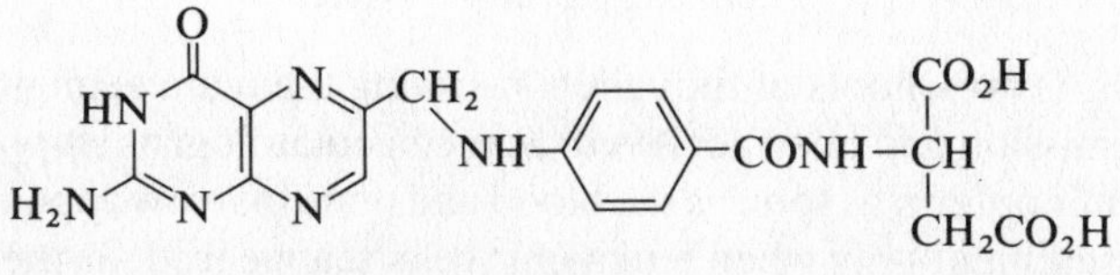

Pteroylaspartic Acid

At the Children's Medical Center in Boston on the 28th March 1947, Farber administered pteroylaspartic acid to a four-year-old girl dying of acute myeloid leukaemia. She received 40 mg daily intramuscular injections until her death a week later. Post-mortem examination revealed that although the treatment had not been started in time to save her life, the number of leukaemic cells in her bone marrow had been drastically reduced. Commenting on this, in their report in the *New England Journal of Medicine* a year later, Farber and his colleagues stated, 'A change of this magnitude in such a short time has not been encountered in the marrow of leukaemic children in our experience.' The deaths throughout the world, despite treatment with the best available therapy, of that four-year-old girl and thousands after her, were inevitable until the lessons painfully learned from them finally enabled Farber's successors gradually to conquer the most dreaded disease of childhood—in time to save the life of this grateful writer's own four-year-old son, who contracted acute lymphocytic leukaemia in 1972.

In 1947, Farber treated a further fourteen children with pteroylaspartic acid before receiving, in November, the first really potent folic acid antagonist, aminopterin (i.e. 4-aminofolic acid), synthesized by Doris Seeger, James Smith, and Martin Hultquist at the Bound Brook laboratories of the Calco Chemical Division of the American Cyanamid Company in New Jersey (Lederle Laboratories was the other main division). During the next six months, sixteen children with acute leukaemia were treated with aminopterin. Many of them had been moribund at the onset of therapy, yet complete remissions were obtained in ten cases. These were the first sustained remissions ever obtained in leukaemic patients. The white cell counts returned to normal levels, with either a marked reduction or complete disappearance of malignant blast cells (lymphoblasts). The red cell count also approached normal values, a surprising observation in view of the anaemia that had characterized folic acid dietary deficiency. Toxic effects were certainly present, the most frequent being a stomatitis affecting the rapidly dividing lining cells of the mouth, leading to painful ulceration.

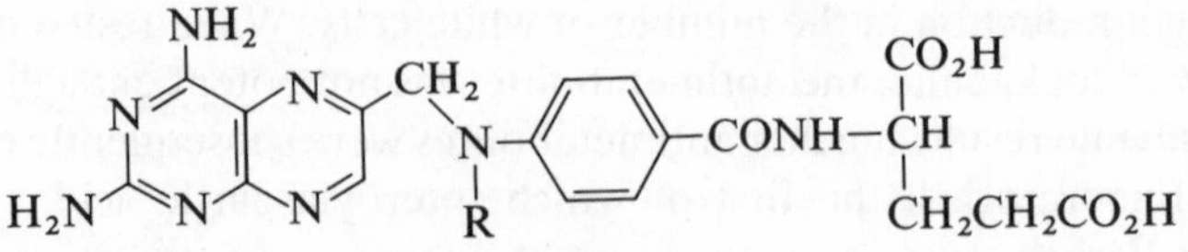

Aminopterin, R = $-H$
Methotrexate, R = $-CH_3$

There were two schools of thought concerning the management of children in whom remission had been achieved. Some haematologists stopped administration of aminopterin as soon as the blood and bone marrow appeared normal, restarting treatment only when leukaemic cells reappeared, as they inevitably did. This process was repeated until no further improvement could be achieved. Farber spurned this intermittent therapy, and opted for maintenance therapy which involved repetitive administration of aminopterin at regular intervals until resistance to the drug occurred. In this way, he kept his patients alive for an average of eight or nine months, and one in every hundred or so was cured. In the succeeding years, it became evident that Farber's approach was justified since remission did not represent elimination of the disease, but merely a reduction in the number of leukaemic cells in the body to a level at which they were no longer detectable. It has been estimated that something in the order of one thousand million leukaemic cells were present at this stage; without further treatment these repeatedly divide until symptoms were experienced when the number had increased a thousand-fold. Only by further exposure to regular cycles of therapy could the number of these remaining cells be kept under control. The problem facing Farber was to establish the maximum dosage of aminopterin that could be administered for several days without destruction of too many normal blood-forming elements in the bone marrow. When the white cell count fell below the critical level, the patient was exposed to the risk of death from overwhelming infection. To strike the right balance required constant monitoring of the bone marrow and blood. The task was made a little easier when the Bound Brook researchers developed a safer folic acid antagonist in the late summer of 1947. This drug, methotrexate ('A-Methopterin') superseded aminopterin.

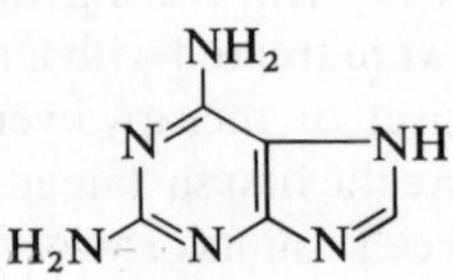

Diaminopurine

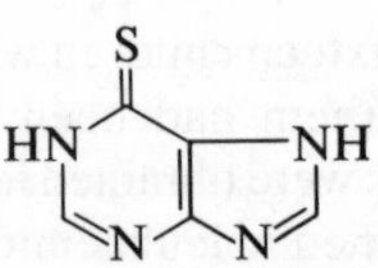

Mercaptopurine

The success of aminopterin led to the examination of other inhibitors of the growth of *Lactobacillus casei* as potential cytotoxic agents. Amongst these was a purine antimetabolite, 2,6-diaminopurine, which had been synthesized in 1948 by George Hitchings and his colleagues at the Wellcome Research Laboratories in Tuckahoe, New York (see page 290). It was evaluated at the Sloan-Kettering Institute, which was then the leading centre in the world for routine screening of potential anticancer agents against transplanted tumours. Subsequently, it was put on clinical trial at the Memorial Hospital by Joseph Burchenal and his colleagues. Although the occasional remission was seen, it was clearly inferior to aminopterin and methotrexate. However, in 1952, Gertrude Elion, Elizabeth Burgi, and George Hitchings introduced 6-mercaptopurine ('Puri-Nethol'), a much more potent purine antimetabolite that had originally been made purely as a chemical intermediate for the synthesis of more 6-aminopurines. When screening revealed its outstanding activity as an inhibitor of the growth of *Lactobacillus casei*, it was sent to the Sloan-Kettering Institute. Its promising activity led to a clinical trial at the Memorial Hospital, which established mercaptopurine as the safest and most effective antileukaemic agent to have been discovered, the average remission lasting about one year. Of particular significance was the fact that a remission was even induced in one patient who had relapsed after failing to respond to further treatment with a folic acid antagonist. This meant that there was no cross-resistance between the two types of compounds. This was to provide the basis for the introduction of combination chemotherapy. By 1956, it was evident that combination of cortisone with either methotrexate or mercaptopurine produced more remissions than any single drug, whilst also extending the duration of these remissions. By the end of the decade, the mean survival rate for children with acute leukaemia who had received intensive chemotherapy in specialized centres had risen beyond one and a half years. For those who received no therapy, the figure was less than three months.

An aggressive approach aimed at outright cure of leukaemia in children was pioneered by one of Faber's former associates, Donald Pinkel. Appointed medical director of the new St Jude's Children's Research Hospital in Memphis, Tennessee, in 1962, he administered different combinations of several cytotoxic drugs to randomly allocated groups of patients who had received no prior treatment. The progress of these groups revealed the optimum treatment schedules. One of the most important advances made by Pinkel was his recognition that the frequent deaths from meningeal leukaemia amongst long-term survivors was due to the persistence of small numbers of leukaemic cells present in the brain at the onset of the disease. These had not been destroyed because of the poor penetration of the central nervous system by most cytotoxic drugs. Simply by irradiating the craniums of children after they first entered remission, he virtually eliminated this complication. By the early seventies, more than half his patients were still alive five years after diagnosis of the disease. Today, the majority of children with acute lymphocytic leukaemia reach this stage, and most of these are cured.

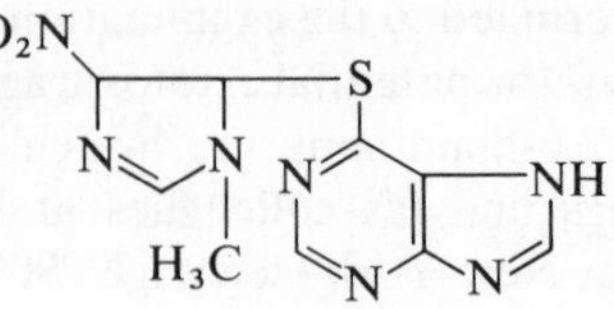

O_2N S N N N NH H_3C N N

Azathioprine

Much of the mercaptopurine administered to patients was rapidly metabolized before it could exert any therapeutic effect. This occurred through the action of the enzyme xanthine oxidase. Analogues containing substituents on the sulphur atom were synthesized in the hope of obtaining a pro-drug that would provide a sustained release of mercaptopurine. The most active of these was azathioprine ('Imuran'), synthesized in 1957, but it gave disappointing results in a clinical trial. Nevertheless, it later became an important drug in its own right as a result of the discovery of a new application for mercaptopurine. Professor Peter Medawar of University College in London had stimulated much interest in immunological tolerance with his Nobel Prize-winning work in the fifties. At Tufts University School of Medicine in Boston, William Dameshek had the idea that if a drug could be found that would be more effective than cortisone in depressing the immune response, it might be possible to carry out bone marrow transplantation in patients with aplastic anaemia, leukaemia, or radiation damage. With Robert Schwartz, he examined a variety of drugs to assess their effect on the ability of rabbits to produce antibodies against injected human serum albumin. Mercaptopurine turned out to be highly effective in this test, but when Roy Calne at Harvard Medical School tried it in dogs receiving kidney transplants, it was unable to induce the same degree of immunological tolerance. Notwithstanding, Calne found that the transplanted kidneys functioned for much longer than usual, no other drug having a comparable effect. In the light of this development, Hitchings set up a screening programme in which a variety of drugs were tested for their capacity to inhibit the haemagglutinin reaction in mice challenged with foreign red blood cells. From this screen, azathioprine emerged, in 1961, as the most effective drug. Calne pioneered its use in human transplant surgery, and it became the most commonly used cytotoxic immunosuppressant. The alternative has, until recently, been to employ a corticosteroid. However, in 1972, during a screening programme to detect immunosuppressant compounds, Jean Borel of the pre-clinical research department of Sandoz, in Basle, discovered an antifungal antibiotic known as cyclosporin A ('Sandimmun') had interesting properties. This was one of several closely related peptide antibiotics secreted by certain types of fungi imperfecti, and had been isolated three years earlier at the Sandoz Laboratories. After thorough examination, it was given to Professor Calne, now at the department of surgery in Cambridge University, for clinical trial in patients receiving kidney transplants at Addenbrooks Hospital, and also to clinicians at the Leukaemia Unit in the Royal Marsden Hospital, London, where its role in bone marrow transplantation was to be assessed. The first reports appeared in 1978, and extended trials have since been conducted on many patients. These have confirmed its value in dealing with the commonest cause of death after bone-marrow transplantation, namely attack on the tissues of the host by the transplanted cells. It has to be given for at least six months after transplantation.

Azathioprine was only one of several hundred analogues of mercaptopurine prepared at Tuckahoe in an attempt to find a superior antileukaemic agent. Although two dozen of these purine antimetabolites were sufficiently active against transplanted tumours in mice to warrant clinical investigation, none compared with mercaptopurine itself. In the light of this, an alternative strategy was adopted. Instead of trying to find an antimetabolite resistant to metabolism by xanthine oxidase, an inhibitor of this enzyme for use in conjunction with mercaptopurine was sought amongst the many purine analogues already synthesized by Burroughs Wellcome. Attention centered around pyrazolopyrimidines, isomers of purine in which carbon and nitrogen atoms in the ring system had been transposed in order to confer antimetabolite activity. Several of these had been shown, in 1957, to be xanthine oxidase inhibitors, so Hitchings and Gertrude Elion selected one, 4-hydroxypyrazolopyrimidine, which was known to be devoid of cytotoxic properties, and tested it for xanthine oxidase inhibitory activity. Having found it active against the isolated enzyme, they then demonstrated that it diminished the conversion of mercaptopurine and its analogues to thiouric acid. This resulted in a significant increase in the antitumour potency of mercaptopurine and in the immunosuppressant efficacy of azathioprine. Clinical trials on thirty-five patients were organized on behalf of the Southeastern Cancer Chemotherapy Cooperative Group by Wayne Rundles and his colleagues at Duke University School of Medicine in Durham, North Carolina, in 1962. The new compound proved to be remarkably safe and, as expected, it permitted the amount of mercaptopurine given to leukaemic patients to be reduced. Disappointingly, this reduction in dose was not accompanied by any therapeutic advantage. However, the development of the new drug had not been in vain, for one of the consequences of xanthine oxidase inhibition was that the amount of uric acid in the blood was diminished because the inhibited enzyme could no

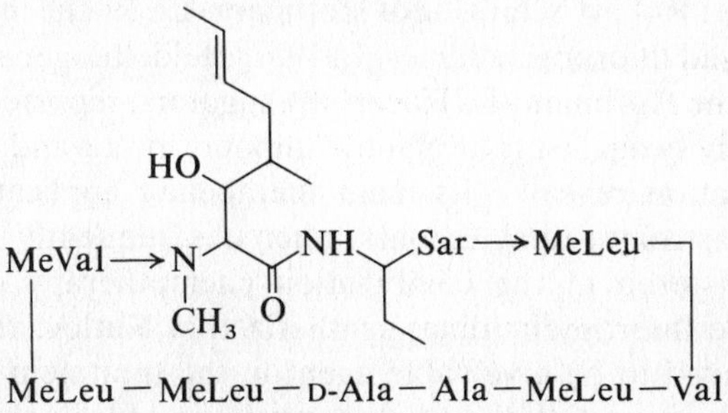

HO
MeVal → N — NH — Sar → MeLeu
CH_3 O
MeLeu — MeLeu — D-Ala — Ala — MeLeu — Val

Cyclosporin A

longer catalyse its formation from hypoxanthine; indeed, the ability of 4-hydroxypyrazolopyrimidine to inhibit the enzyme is due to its structural similarity to hypoxanthine. Taking advantage of this, Professor Rundles was able to use it successfully to lower uric acid levels in patients suffering from gout (see page 95). It was introduced for this purpose by Burroughs Wellcome in 1966, with the approved name of allopurinol ('Zyloric', 'Zyloprim').

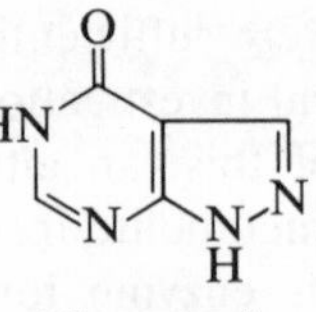

Allopurinol 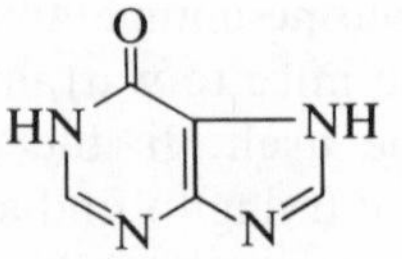 Hypoxanthine 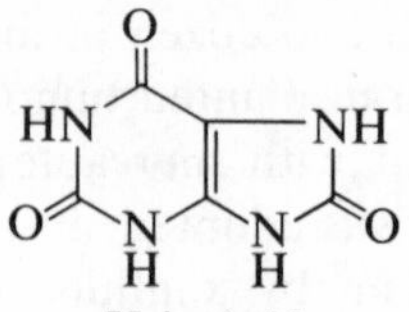 Uric Acid

The success of mercaptopurine and methotrexate stimulated the synthesis of many potential antimetabolites. Some of these had antitumour activity, but very few have proved to be clinically effective in the treatment of cancer. Two notable exceptions to this are fluorouracil and cytarabine. The first of these was introduced by Professor Charles Heidelberger of the McArdle Laboratory for Cancer Research at the University of Wisconsin, Madison. He was one of several scientists who sought a uracil antimetabolite after the publication, in 1954, of a report by Abraham Cantarow and Karl Paschkis, of the Jefferson Medical College in Philadelphia, revealing that radioactive uracil was more rapidly incorporated into liver tumours (induced by a carcinogenic chemical) than into normal rat liver cells. Heidelberger had previously carried out important work on the bichemical mode of action of the deadly poison fluoroacetic acid. Aware that the fluorine in this was ultimately responsible for it acting as an antimetabolite that blocked a vital enzyme involved in cell metabolism, he had the novel idea of incorporating a fluorine atom into uracil and other pyrimidines. This seemed particularly worth doing since Friedrich Weygand of Heidelberg University had just demonstrated that 5-bromo- and 5-iodouracil were incorporated into bacterial nucleic acids. These antimetabolites had been prepared in 1945 by Hitchings and his colleagues, who found them to have antibacterial activity. Heidelberger therefore asked Robert Duschinsky and Robert Schnitzer of Hoffmann-La Roche in Nutley to synthesize the proposed fluoropyrimidines. In 1957, Heidelberger and his colleagues, together with the Hoffmann-La Roche investigators, reported that fluorouracil had potent activity against transplanted tumours in rats and mice. Subsequent clinical application revealed its main therapeutic application to lie in the treatment of gastrointestinal tumours, when it is frequently used on its own. It is also administered in the combination chemotherapy of breast cancer. Another of the fluoropyrimidines synthesized at Nutley, flucytosine ('Alcobon'), has proved to be a valuable agent in the treatment of systemic yeast infections, such as candidiasis, cryptococcosis, and torulopsosis, which can be particularly hazardous in cancer patients whose immune system has been compromised as a consequence of exposure to intensive chemotherapy. Another halogenated pyrimidine metabolite is idoxuridine. This derivative of 5-iodouracil was introduced by William Prusoff of the department of pharmacology at Yale University in 1959. Although it is highly toxic, it can be topically applied to treat skin infections caused by the herpes virus.

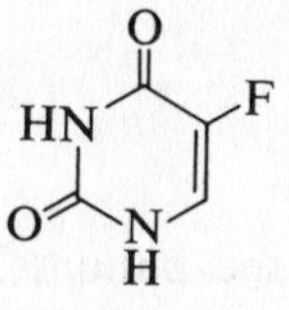

Fluorouracil

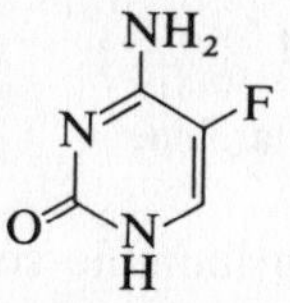

Flucytosine

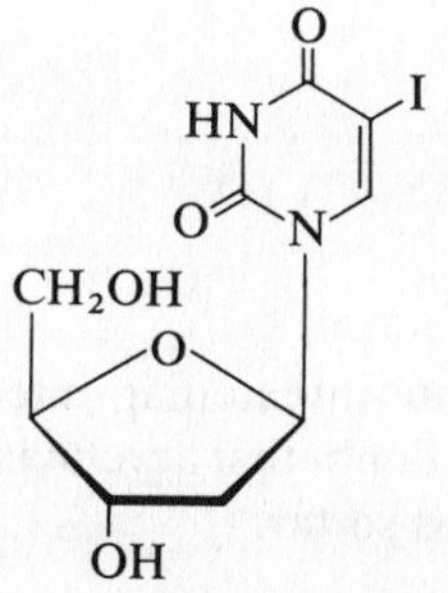

Idoxuridine

In 1951, Professor Werner Bergman and Robert Feeney isolated a new type of thymine derivative from a sponge collected in the shallow waters off Elliot Key in Florida. Four years later, Bergman and David Burke isolated a similar uracil derivative and established the structure of both these compounds, describing them as 'spongonucleosides'. Because of their close resemblance to the nucleosides which polymerize to form DNA, with a hydrogen in the deoxyribose component being replaced by a hydroxyl group, there was interest in the possible anticancer activity of these and related compounds. The only one that turned out to have useful clinical activity was cytarabine (cytosine arabinoside, spongocytidine, 'Cytosar'), which was synthesized in 1959 by Richard Walwick, Walden Roberts, and Charles Dekker in the biochemistry department of the University of California at Berkeley. It was shown to possess activity against a transplanted sarcoma 180 by John Evans and his colleagues of the Upjohn Company. Its main clinical use is in the treatment of acute leukaemia. However, in 1975 Gertrude Elion and her colleagues at the Wellcome Research Laboratories in North Carolina reported that the structurally related arabinosides of 2,6-diaminopurine and of guanine were active against DNA viruses. This led to the decision to examine other purine bases. As Howard Schaeffer, one of the Wellcome chemists, had previously established that the intact sugar ring of such compounds was not essential for binding to enzymes, acyclic analogues were prepared and tested as antiviral agents. In 1977, the Wellcome team reported that acyclovir (acyloguanosine, 'Zovirax') had outstanding activity against the herpes virus. Subsequently, it was shown that the virus selectively converted it into a triphosphate that, in turn, was incorporated into viral DNA thereby blocking further DNA synthesis. Acyclovir is now employed clinically, either topically to treat superficial herpes infections such as cold sores, or orally in palliation of genital herpes or for

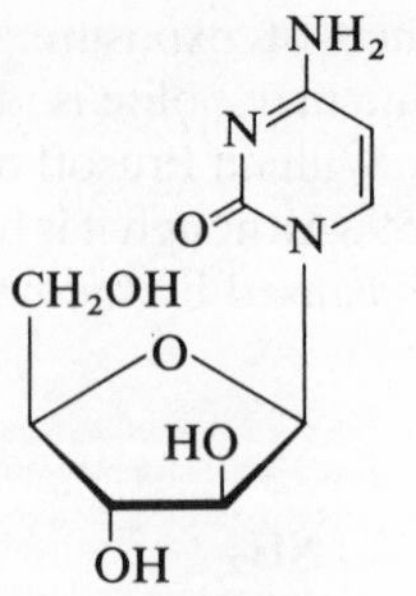

Cytarabine

treating life-threatening infection in immuno-compromised patients. There can be no doubt that acyclovir is the most important antiviral drug to have been introduced so far.

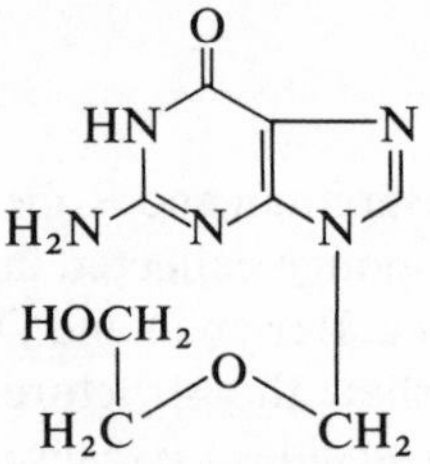

Acyclovir

ANTIBIOTICS

By the end of the Second World War, Cornelius Rhoads had virtually reorientated the entire nitrogen mustard programme of the Chemical Warfare Service to form the basis of an extensive research investigation into cancer chemotherapy at the Sloan-Kettering Institute. There, he established the transplanted Crocker sarcoma 180 as the basis for the largest anticancer drug screening programme in the world. With his renowned organizing genius, he won the confidence of academic and industrial researchers alike by agreeing to screen compounds on a confidential basis, with the assurance that any promising compound would be evaluated fully. Chester Stock, chief of the experimental chemotherapy division at the Sloan-Kettering Institute, published a preliminary account of the results obtained in this screening programme up to 1950. This revealed that of thirty-three antibiotics examined, only five had any activity against the sarcoma 180, namely actidione, actinomycin, illudin M, illudin S, and oxytetracycline. None of these was effective enough at a dose level that was considered safe for clinical application. By 1953, Stock and his colleagues had tested more than 1250 mould culture filtrates or partially purified substances, from which only five caused sufficient retardation of tumour growth to warrant further investigation. One of these was a culture filtrate from a *Streptomyces* submitted by Parke, Davis and Company. Further investigation resulted in the isolation of a new antitumour substance that was given the name of azaserine as it was an analogue of the amino acid serine, with two nitrogen atoms replacing methyl hydrogens. It has seen little clinical application. By 1955, when the Cancer Chemotherapy National Service Center took over the responsibility for the screening programme, more than twenty thousand chemicals and an undisclosed number of fermentation products had been screened. With the backing of the American government, the CCNSC stepped up the scale of the screening programme. During the next twenty years more than half a million chemicals and natural product extracts were examined. The cynic may see in this a latter-day version of the alchemists' endless search for the elixir of life!

In 1949, Hans Brockmann of the Organic Chemistry Institute at Göttingen University isolated an actinomycin from *Streptomyces chrysomallus* which was chemically related to, but different from the actinomycin A isolated by Selman Waksman in 1941 (see page 320). He named it actinomycin C, only to discover subsequently that it was a complex consisting of three components. These he later named actinomycin C_1, C_2, and C_3. Following the disclosure by Stock that Waksman's actinomycin had detectable activity against sarcoma 180, Brockmann sent the crude actinomycin C (also known as cactinomycin) to Christian Hackmann at the Bayer Company's Institute for Experimental Pathology at Elberfield. In 1952, Hackmann reported that the antibiotic inhibited tumour growth and Dr G. Schulte of the Knappschaftskrankenhaus in Recklinghausen found it effective in some patients with lymphatic tumours at dose levels which did not produce toxic effects. Although several reports disputing these findings were published by other clinicians, the drug was marketed for some years in the United Kingdom, principally for use in Hodgkin's disease, under the proprietary name of 'Sanamycin'. It was superseded by more effective agents.

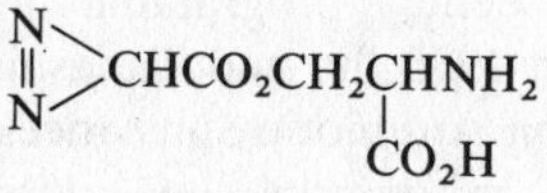

$$N_2CHCO_2CH_2CH(NH_2)CO_2H$$

Azaserine

In 1953, Waksman isolated actinomycin D from *Streptomyces parvullus*, the first actinomycin-producing organism that yielded a single antibiotic rather than a complex mixture. It turned out that this was identical to actinomycin C_1, both of these terms now being replaced by the approved name of dactinomycin. This pure antibiotic, which constituted about 10% of the actinomycin C complex, was superior to actinomycins C_2 or C_3 as an antitumour agent. In the summer of 1954, Selman Waksman and Sidney Farber discussed the prospects for actinomycin D. The outcome was that a preliminary trial was run on it in the mouse bioassay unit at The Children's Cancer Research Foundation in Boston,

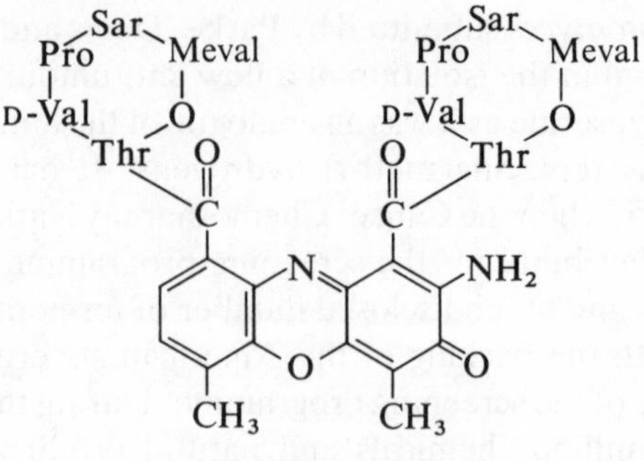

Dactinomycin

of which Farber was director. The drug showed marked antitumour activity against many transplanted mouse tumours, occasionally causing total regression of some of them. It was, on a weight basis, the most powerful agent Farber and his colleagues had yet tested against transplanted tumours. As soon as toxicological studies were completed, a trial was initiated in children with acute leukaemia. Disappointingly, the actinomycin D was clearly ineffective. However, encouraging results were achieved in children with advanced Wilm's tumour, rhabdomyosarcomas (muscle tumours), Ewing's tumour of the bone, and Hodgkin's disease. The prospects of survival for children with Wilm's tumour of the kidney have been completely transformed since then by a combination of radiotherapy and sustained chemotherapy with dactinomycin on its own or with vincristine. This is maintained for up to two years, with 90% of children being cured, which is three times as many as was the case prior to the introduction of dactinomycin. The main value of the drug in adults appears to be for the treatment of soft tissue sarcomas and testicular teratomas.

At the Institute of Microbiological Chemistry in Tokyo, Hamao Umezawa organized an antitumour screening programme utilizing the transplanted Yoshida sarcoma in mice. In 1956, he and Tadashi Yamamoto took out a patent, assigned to the Nippon Antibiotic Substances Association, after they detected potent activity in mitomycin, an antibiotic mixture obtained from *Streptomyces caespitosus* by the Kyowa Hakko Kogyo Company of Tokyo. Of several substances isolated from the mixture, mitomycin C proved to be the most effective, but its toxicity to the bone marrow severely limited its clinical application. It is sometimes employed in the treatment of gastrointestinal tumours. Another antibiotic that is rarely used because of bone marrow toxicity is mithramycin (aureolic acid), isolated from *Streptomyces argillaceus* by Walton Grundy and his colleagues of Abbott Laboratories. Its antitumour activity was disclosed in 1962. The principal value of it is in the emergency treatment of dangerously high levels of calcium in the blood, associated with malignant disease.

Daunorubicin (daunomycin, rubidomycin, 'Cerubidin') was isolated from *Streptomyces peucetius* in 1962 by Aurelio Di Marco and his colleagues of the

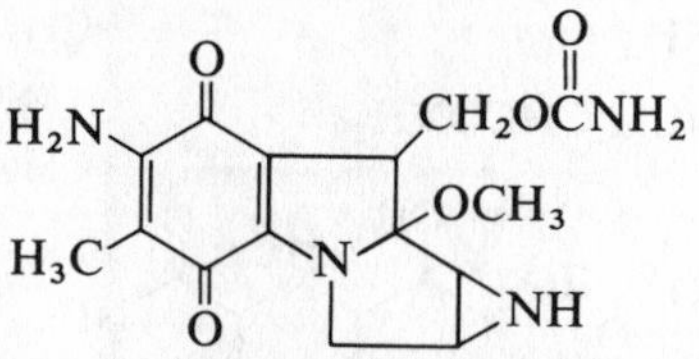

Mitomycin C

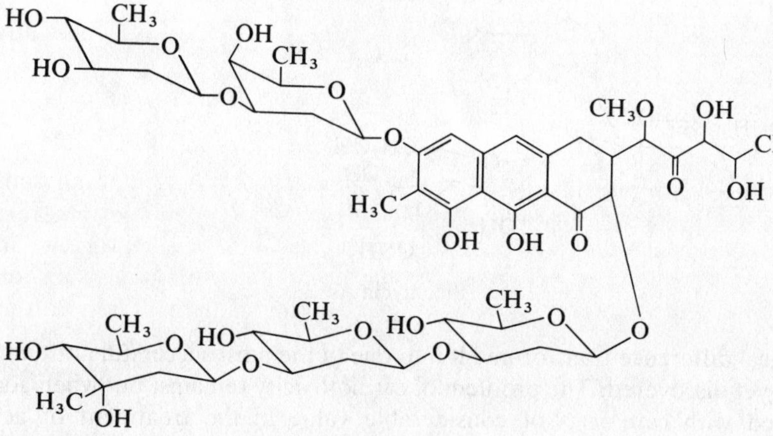

Mithramycin

Farmitalia Company in Milan. At the Istituto Nazionale dei Tumori it was shown to exhibit actinomycin-like activity against tumours. Shortly after, it was also isolated from *Streptomyces caeruleorubidus* at the Rhône-Poulenc laboratories. Its clinical value was limited by its severe toxicity to the heart, but it found some use in the combination chemotherapy of leukaemia, particularly in helping to induce remissions in cases where the patient failed to respond to safer drugs. Another antibiotic isolated by Farmitalia scientists from *Streptomyces peucetius* in 1967 was named doxorubicin ('Adriamycin'), this turning out to be the 14-hydroxy derivative of daunorubicin. This apparently minor

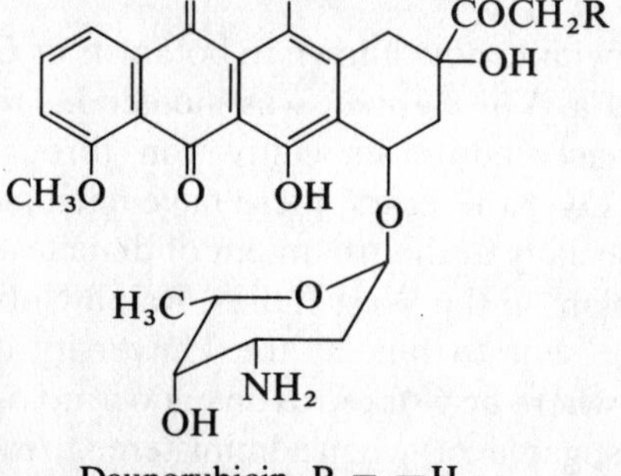

Daunorubicin, R = —H
Doxorubicin, R = —OH

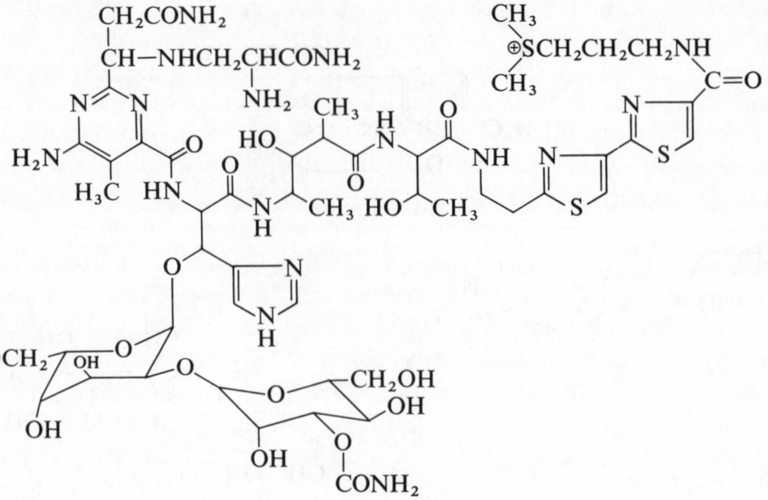

Bleomycin A_2

chemical difference transformed it into one of the most successful antitumour drugs yet discovered. The problem of cardiotoxicity remains, but when administered with care it is of considerable value in the treatment of acute leukaemias, lymphomas (non-Hodgkin's), and many solid tumours.

In 1962, Hamao Umezawa detected antitumour activity whilst screening culture filtrates of *Streptomyces verticillus*. He managed to isolate from this the bleomycins, a complex group of glycopeptides, of which bleomycin A2 is the main component of the product used clinically. The most striking feature of this is the fact that it is one of the few anticancer drugs which does not cause bone marrow depression, although it may damage the lung especially in elderly patients. Its main clinical value is in squamous cell carcinoma (cancer in flattened lining cells), and it is administered to treat head and neck tumours, non-Hodgkin's lymphoma, and testicular teratomas.

PLANT PRODUCTS

The species of periwinkle now known to botanists as *Catharanthus roseus*, but formerly described as *Vinca rosea*, was indigenous to tropical zones, but its attractive appearance led to its cultivation throughout the world as an ornamental plant. Over the years, there have been many reports of its use in folk medicine, especially in the treatment of diabetes. In 1949, on learning of such a use of the plant in the West Indies, Ralph Noble arranged for samples of the plant to be sent to him at the University of Western Ontario, in London, Ontario, where he worked as an endocrinologist. Finding no effect of the plant on blood sugar levels when administered to rats by mouth, Noble then injected an extract. Not only were the results once again negative, but the rats succumbed to infections. His suspicions aroused by this, Noble examined the blood of one of the rats receiving injections of the extract. This revealed that after an initial rise, the number of white cells was dramatically reduced. He confirmed this was due to damage to the bone marrow, a phenomenon that was becoming associated with the antileukaemic drugs then being introduced into medical practice. With the assistance of Charles Beer, a chemist sent to Ontario on an exchange fellowship by the British Empire Cancer Campaign, Noble separated an alkaloidal fraction with activity against transplanted tumours in mice. Beer announced the isolation of a pure alkaloid, vinblastine (originally known as vincaleukoblastine; 'Velbe') at a meeting of the New York Academy of Sciences in 1958. At the meeting, Noble and Beer were informed by Gordon Svoboda that he and his colleagues at the Eli Lilly Company had followed an almost identical approach after they too had discovered that *Cantharanthus roseus* lacked antidiabetic activity. On submitting extracts of the plant to the company's general pharmacological screening programme, their activity against a transplanted acute lymphocytic leukaemia in the mouse was revealed. In 1961, Svoboda reported the isolation of vincristine ('Oncovin'), another alkaloid with promising activity. Although vincristine differed from vinblastine solely in having a formyl group instead of a methyl group attached to the nitrogen of one of the indole rings, its toxicity and range of clinical applications was quite different. It is used mainly to treat leukaemias and lymphomas, particularly for the induction of remissions. It is also used against some solid tumours. Vinblastine is mainly of value against advanced teratomas and lymphomas. After the introduction of these two alkaloids, the Lilly chemists isolated seventy more alkaloids, but none of these attained clinical acceptability. However, a semisynthetic derivative of vinblastine developed in 1974 has been marketed, namely vindesine ('Eldisine').

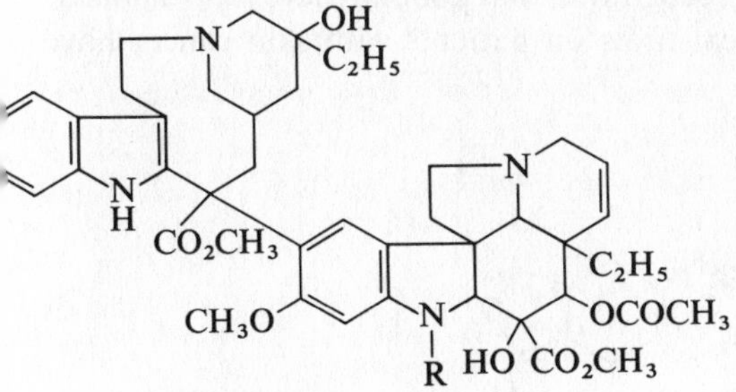

Vinblastine, R = $-CH_3$
Vincristine, R = $-CHO$

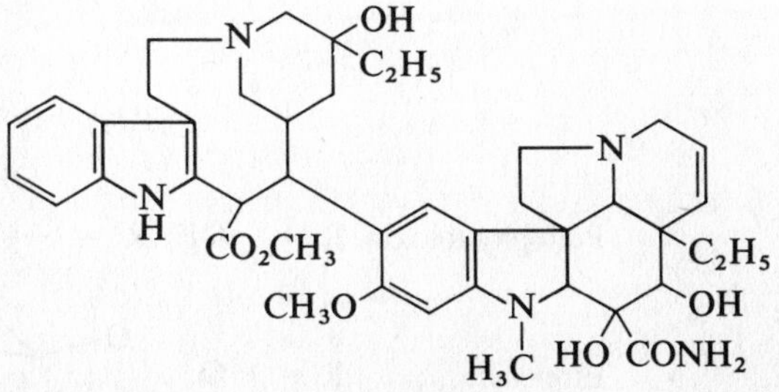

Vindesine

The vinca alkaloids act by binding to the microtubules that form the spindles to which the chromosomes are temporarily attached during cell division. This results in mitosis (division) being blocked during metaphase. Another mitotic poison is podophyllotoxin. This is the active principle in podophyllum, the dried rhizome and roots of the May apple (American mandrake), *Podophyllum peltatum*, a perennial found throughout the eastern side of North America. The juice expressed from the root was employed as a purgative and anthelmintic by the Wyandotte Indians. American doctors were sufficiently impressed by this for it to be included in the first edition of the *United States Pharmacopeia* in 1820. Fifteen years later, a young Ohio physicians, John King, introduced the highly potent popdophyllum resin. In so doing, he almost killed the first person to whom he administered it, a seventeen-year-old girl. By 1847, he had acquired enough confidence in it to persuade William Merrell, a Cincinnati pharmacist, to include it in the range of products his company supplied to the medical and pharmaceutical professions. It was sold as 'Podophyllin', under which name it quickly became established as one of the most popular purgatives, its fame spreading throughout the world. It was included in the *British Pharmacopoeia* in 1864. Today, its use as a purgative is frowned upon since this effect is induced by severe irritation of the intestinal mucosa. The irritant properties have also led to the other principal use of podophyllum. In *The Indian Doctor's Dispensatory*, published in Cincinnati in 1818, Peter Smith recommended the topical application of powdered podophyllum root as an escharotic. The use of such agents to deal with warts is still familiar, but throughout recorded medical history escharotics such as caustic alkalis, acids, or mercurial, arsenical, and antimonial salts were employed in attempts to remove superficial tumours. Podophyllum was used for this purpose because of its obvious irritant nature, but the recognition that podophyllotoxin is a mitotic poison explains why recent clinical trials on patients with skin cancer have given impressive results.

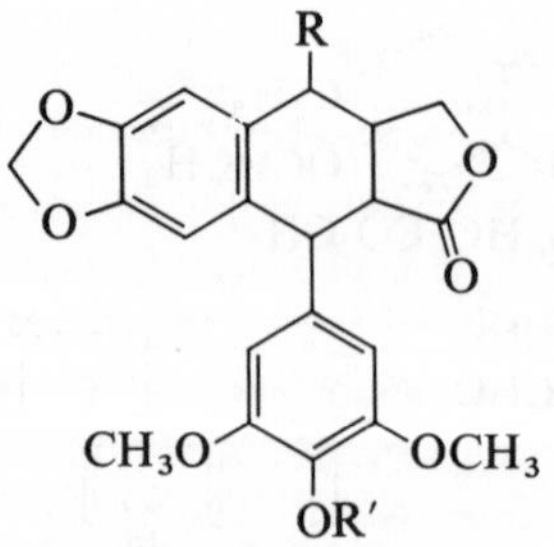

R

O O O O

CH_3O OCH_3

OR′

Podophyllotoxin, R = —OH, R′ = $—CH_3$

Etoposide, R = —O (glucoside with O, O, CH_3, OH, OH), R′ = —H

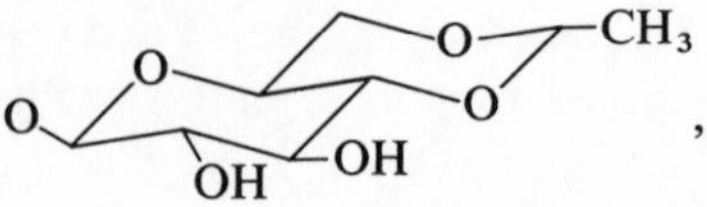

In 1970, H. Stahelin of the Sandoz Laboratories in Basle reported the preparation of semisynthetic analogues of podophyllotoxin which retained antitumour activity despite a reduction in their toxicity. Derivatives of the most active compound in this series were synthesized, leading to the introduction of etoposide (VP 16, 'Vespesid') three years later. Unexpectedly, the mode of its antitumour action was quite unlike that of podophyllotoxin since it appears to act by causing scission of a single strand of DNA. Etoposide seems to be of most value in non-small-cell lung cancer and certain types of testicular cancer resistant to other therapy.

MISCELLANEOUS AGENTS

Reports of liver toxicity following prolonged therapy with iproniazid, newly introduced as an antidepressant in 1957 (see page 186), persuaded Hoffmann-La Roche investigators in Basle to screen its possible successors very thoroughly. This led to the discovery that 1-methyl-2-benzylhydrazine had a pronounced tumour inhibitory effect. Screening of several hundred of its analogues that had been synthesized as potential antidepressants revealed that forty methylhydrazines were active antitumour agents. Two of these were selected for extended biological and clinical trials. In 1963, these revealed the value of procarbazine ('Natulan'), which had been synthesized by Werner Ballag and Emmanuel Grunberg. It was subsequently used in combination with mustine, vincristine ('Oncovin'), and prednisolone in the 'MOPP' regimen which has transformed the prospects for survival of patients with advanced Hodgkin's disease.

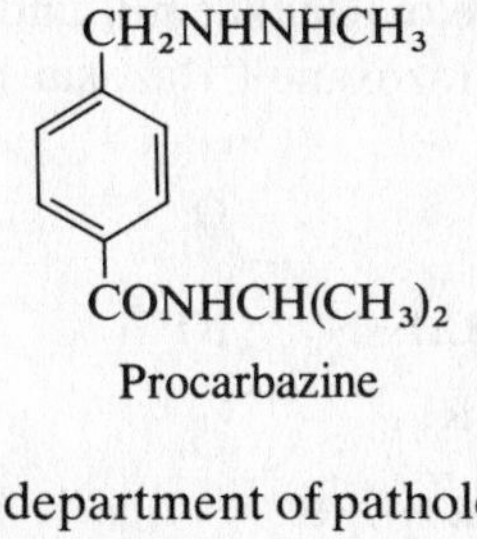

$CH_2NHNHCH_3$

$CONHCH(CH_3)_2$

Procarbazine

In 1953, John Kidd of the department of pathology at the Cornell University Medical School discovered that growth of transplanted lymphomas in mice and rats could be inhibited by guinea pig serum, but not by that from horses or rabbits. Later, this was followed up in the same laboratory by John Broome, who discovered that tumour cells grown in Eagle's tissue culture medium were unaffected by guinea pig serum when transplanted into mice, whereas those cells grown in Eagle's medium to which asparagine had been added were highly susceptible to the serum. This experiment had been carried out to see whether the presence of high levels of the enzyme asparaginase, known since 1922 to occur uniquely in guinea pig serum, might be in some way involved. After further tests with asparaginase, Broome concluded, in 1961, that if tumour cells were grown in a medium rich in asparagine, they lost the ability to synthesize this. Thus, on transplantation they became susceptible to the enzyme which destroyed this amino acid. The possibility of clinical exploitation of the sensitivity of some malignant cells to asparaginase was then pursued. The enzyme was obtainable from yeast and bacteria, but the activity against tumour cells varied according to the source of the asparaginase. Eventually an active preparation was obtained from *Escherichia coli*, this being known as colaspase. Unfortunately, clinical studies have revealed few tumours that are affected by colaspase. It is only used in combination with other drugs for the induction of remissions in acute lymphocytic leukaemia in children. In adults, side-effects are particularly troublesome.

During the late fifties and early sixties, several scientists had the idea that the growth of tumours might be affected by chelating agents which could inactivate the trace metals often associated with vital enzymes. Since the powerful chelating agent ethylenediaminetetraacetic acid (EDTA) was already known to be ineffective against tumours because it was too polar to penetrate into the cells, Andrew Creighton, Kurt Hellmann, and Susan Whitecross of the Imperial Cancer Research Fund's department of chemistry and cancer chemotherapy, in London, set out to prepare and evaluate less polar derivatives of EDTA. The methyl and ethyl esters proved to be inactive. An attempt to synthesize the tetra-amide led instead to cyclization, with the formation of a dipiperazine which had no chelating activity. Nevertheless, routine screening of this novel compound (ICRF 154) revealed it to have high antitumour activity. Systematic synthesis of analogues revealed that activity was only present in compounds very closely related to it. These findings were reported in 1969, and then ICRF 159 was put on clinical trial. This led to its introduction for the treatment of acute myeloid leukaemia and non-Hodgkin lymphomas, with the approved name of razoxane ('Razoxin').

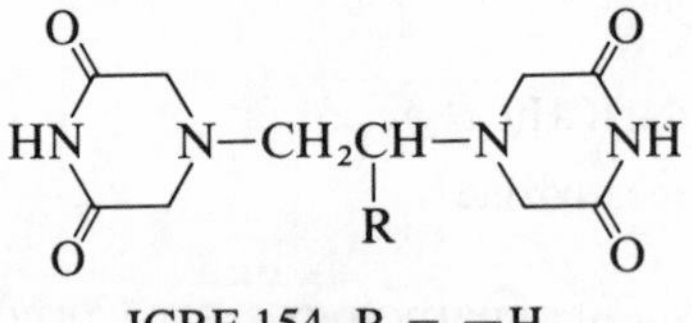

HN N—CH_2CH—N NH (R on CH; piperazinedione rings with O)

ICRF 154, R = —H
Razoxane, R = —CH_3

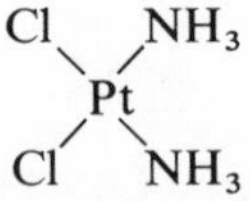

Cl NH_3 Pt Cl NH_3

Cisplatin

By an amusing twist of fate, the issue of *Nature* dated 26th April 1969 not only carried a report of the discovery of razoxane, a drug without chelating activity which was synthesized as part of a programme to develop a metal-chelating antitumour agent, but it also featured a report of the first metallic compound to have potent antitumour action, this also being a serendipitous discovery! In 1964, Barnet Rosenberg, Loretta Van Camp, and Thomas Krigas of the biophysics department at Michigan State University tried to find out what were the effects of an electric current on a suspension of bacteria. They detected no change in the growth rate, but observed abnormal elongation of the *Escherichia coli* cells, suggestive of interference with the cell division process. After extensive investigations, they came to the surprising conclusion that the effect was due to an electrolysis product being formed in the culture medium. Ultimately, it was established that the action of chloride ions on the platinum electrode was responsible for the formation of cytotoxic platinum ions. When the corresponding platinum salts with the *cis* configuration were added to bacterial cultures, they produced the same effects as had been seen before. When various platinum compounds were tested against transplanted tumours they inhibited their growth. Clinical trials began in the early seventies, with the outcome that cisplatin ('Neoplatin') has become established as a valuable agent in the treatment of malignant teratoma of the testis.

Bibliography

GENERAL

Balentin, H. *Geschichte der Pharmazie und Chemie*, Wissenschartlich Berlagstesellschalt, Stuttgart, 1950.

Boussel, P., Bonnemain, H. and Bove, F. J. *History of Pharmacy and Pharmaceutical Industry*, Asklepios Press, Lausanne, 1982.

Burger, A. *A Guide to the Chemical Basis of Drug Design*, John Wiley, Chichester, 1983.

Clendenning, L. *Source Book of Medical History*, Hoeber, London, 1942.

Doyle, P. A. *Readings in Pharmacy*, John Wiley, London, 1962.

Gaddum, J. H. Therapeutic discoveries, *J. Pharm. Pharmacol*, 1954, **6**, 497–512.

Gillispie, C. C. *Dictionary of Scientific Biography*, 16 vols. Scribner's, New York, 1970 onwards.

Grier, J. *A History of Pharmacy*, The Pharmaceutical Press, London, 1937.

Holmstedt, B. and Liljestrand, G. *Readings in Pharmacology*, Pergamon, London, 1963.

Hubbard, W. N. The origins of medicinals, *Advances in American Medicine*, vol 2 (ed. J. Z. Bowers, E. F. Purcell) Josiah Macy Jr. Foundation, New York, 1976, p. 685–721.

Issekutz, B. *Die Geschichte der Arzneimittelforschung*, Akademiai Kiado, Budapest, 1971.

Krantz, J. C., Jr. *Historical Medical Classics Involving New Drugs*, Williams and Wilkins, Baltimore, 1974.

Lloyd, J. U. *Origin and History of All Vegetable Drugs, Chemicals and Preparations. Volume 1. Vegetable Drugs*, Caxton Press, Cincinnati, 1921.

Mahoney, T. *The Merchants of Life: An Account of the American Pharmaceutical Industry*, Harper, New York, 1969.

Nobel Foundation. *Nobel Lectures in Physiology or Medicine 1922—1941*, Elsevier, London, 1965.

Nobel Foundation. *Nobel Lectures in Physiology or Medicine 1942–1962*, Elsevier, London, 1964.

Schneider, W. *Geschichte der pharmazeutischen Chemie*, Verlag Chemie, Weinheim, 1972.

Silverman, M. *War Against Disease*, Michael Joseph, London, 1943.

Silverman, M. *Magic in a Bottle*, Macmillan, New York, 1948 (revised edition of *War Against Disease*).

Slack, R. and Nineham, A. W. *Medical and Veterinary Chemicals*, 2 vols., Pergamon Press, Oxford, 1968.

CHAPTER ONE

Ackernecht, E. H. *Therapeutics from the Primitives to the 20th Century*, Macmillan, London, 1973.

Anft, B. Friedlieb Ferdinand Runge: a forgotten chemist of the nineteenth century, *J. Chem. Educt*, 1955, **32**, 566–574.

Beddoes, T. and Watt, J. *Considerations on the Medicinal Use and on The Production of Factitious Airs*, Bristol, 1795.
Berman, A. Pierre-Joseph Pelletier, *Dictionary of Scientific Biography*, vol. 10, p. 497–499.
Cartwright, F. F. *The English Pioneers of Anaesthesia*, Wright, Bristol, 1952.
Coenen, H. Über das Jahr der Morphiumentdeckung Sertürners im Paderborn, *Arch. Pharm*, 1954, **287**, 165–180.
Cullen, W. *A Treatise of the Materia Medica*, 2 vols., C. Elliot, Edinburgh, 1789.
Delepine, M. (translated by R. E. Oesper) Joseph Pelletier and Joseph Caventou, *J. Chem. Educt*, 1951, **28**, 454–461.
Duran-Reynals, M. L. *The Fever Bark Tree. The Pageant of Quinine*, Doubleday, New York, 1946.
Ebbell, B. *The Ebers Papyrus, the Greatest Egyptian Medical Document*, Levin and Munksgaard, Copenhagen, 1937.
Ernst, E. *Das Industrielle Geheimittel und Seine Werbung*, Würzburg, 1975.
Fluckiger, F. A. and Hanbury, D. *Pharmacographia. A History of the Principal Drugs of Vegetable Origin Met with in Great Britain and British India*, Macmillan, London, 1874.
Grmek, M. D. Francois Magendie, *Dictionary of Scientific Biography*, vol. 9, p. 6–11.
Gunther, R. T. *The Greek Herbal of Dioscorides* (translated by J. Goodyer), Hafner, New York, 1959.
Haggis, A. W. Fundamental errors in the early history of Cinchona, *Bull. Hist. Med*, 1941, **10**, 417–459; 568–592.
Hamarneh, S. A history of Arabic pharmacy, *Physis*, 1972, **14**, 5–54.
Hanzlik, P. J. 125th Anniversary of the discovery of morphine by Sertürner, *J. Am. Pharm. Ass*, 1929, **18**, 375–384.
Harrison, R. H. *Healing Herbs of the Bible*, E. J. Brill, Leiden, 1966.
Jones, H. N. Critical study of origins and early development of hypodermic medication, *J. Hist. Biol*, 1947, **2**, 201–249.
Jones, D. P. Christian Friedrich Samuel Hahneman, *Dictionary of Scientific Biography*, vol. 6, p. 17–18.
Kramer, S. N. First pharmacopeia in man's recorded history, *Am. J. Pharm*, 1954, **126**, 76–84.
Kreig, Margaret, *Green Medicine*, Harrap, London, 1965.
Krikorian, A. D. Were the opium poppy and opium known in the Ancient Near East? *J. Hist. Biol*, 1975, **8**, 95–114.
Kudlien, F. and Wilson, L. G. Galen, *Dictionary of Scientific Biography*, vol. 5, p. 27–237.
Leake, C. D. *The Old Egyptian Medical Papyri*, University of Kansas Press, Lawrence, 1952.
Lesch, J. Conceptual change in an empirical science: the discovery of the first alkaloid, *Historical Studies in the Physical Sciences*, 1981, **11**, 305–328.
Levy, M. *Early Arabic Pharmacology*, E. J. Brill, Leiden, 1973.
Lockemann, G. Friedrich Wilhelm Sertürner, the discoverer of morphine, *J. Chem. Educt*, 1951, **28**, 277–279.
Macht, D. I. The history of intravenous and subcutaneous administration of drugs, *J. Am. Med. Ass*, 1916, **66**, 856–860.
Magendie, F. *Formulaire pour la préparation et l'emploi de plusieres nouveaux médicaments, tels que la noix vomique, la morphine, etc*, Mequignon-Marvis, Paris, 1822.
Magendie, F. *Formulaire pour la préparation et l'emploi de plusieres nouveaux médicaments, tels que la noix vomique, la morphine, etc*, translation by J. Houlton, Philadelphia, 1843.
Moldenke, H. N. and Moldenke, A. *Plants of the Bible*, The Ronald Press, New York, 1952.

Olmsted, J. M. D. *Francois Magendie*, Schuman's, New York, 1944.
Pagel, W. Paracelsus, *Dictionary of Scientific Biography*, vol 10, p. 304–313.
Pereira, J. Cephaelis ipecacuanha, *The Elements of Materia Medica and Therapeutics*, Vol. 2, pt.2 Longman, Brown, Green, and Longman, London, 4th edn., 1857, p. 55–66.
Pereira, J. Cinchona, *The Elements of Materia Medica and Therapeutics*, Vol. 2, pt. 2 Longman, Brown, Green, and Longman, London, 4th edn., 1857, p. 70–152.
Pierson, S. Armand Séguin, *Dictionary of Scientic Biography*, vol. 12, p. 286–287.
Schmauderer, E. Friedrich Wilhelm Sertürner, *Dictionary of Scientific Biography*, vol. 12, p. 320–321.
Schofield, R. E. *The Lunar Society of Birmingham*, Oxford University Press, 1963.
Séguin, A. Premiere mémoire sur l'opium; lu a là Ire classe de l'Institut de France, le 24 decembre 1804, *Ann. de Chim*, 1814, **92**, 225.
Sertürner, F. Auszuge aus Briefen an den Herausgeber. (a) Säure im Opium. (b) Ein anderes Schreiben von Ebendenselben. Nachtrag zur Charakteristik der Säure im Opium, *J. Pharm. f. Arzte, Apotheker, u. Chemisten von D. J. B. Trommsdorff*, 1805, **13**, 29–30.
Sertürner, F. W. A. Ueber das Morphium, ein neue salzfahige Grundlage, und die Mekonsäure, als Hauptbestandtheile des Opiums, *Gilbert's Ann. d. Physik. (Leipzig)*, 1817, **25**, 56–89.
Siegel, R. E. and Poynter, F. N. L. Robert Talbor, Charles II, and Cinchona. A contemporary document, *Med. Hist*, 1962, **6**, 82–85.
Smeaton, W. A. Antoine François de Fourcroy, *Dictionary of Scientific Biography*, vol 5, p. 89–93.
Smeaton, W. A. Nicolas Louis Vauquelin, *Dictionary of Scientific Biography*, vol. 13, p. 596–598.
Smith, D. C. Quinine and fever: the development of the effective dosage, *J. Hist. Med*, 1976, **32**, 343–367.
Sigerist, H. Materia medica in the middle ages, *Bull. Hist. Med*, 1939, **7**, 417–423.
Sonnedecker, G. *Kremer and Urdang's History of Pharmacy*, 4th edn. J. B. Lippincott, Philadelphia, 1976.
Stannard, J. Hippocratic pharmacology, *Bull. Hist. Med*, 1961, **35**, 497–518.
Svoboda, G. H. Alkaloids: history, preparation, and use, *Kirk–Othmer Encyclopaedia of Chemical Technology*, 2nd edn, volume 1, p. 778–809.
Swain, T. (ed.) *Plants in the Development of Modern Medicine*, Harvard University Press, Cambridge, 1972.
Taylor, F. O. Forty-five years of manufacturing pharmacy, *J. Am. Pharm. Ass*, 1915, **4**, 468–484.
Taylor, N. *Plant Drugs that Changed the World*, Dodd, Mead, New York, 1965.
Temkin, C. L., Rosen G., Zilboorg, G. and Sigerist, H. E. *Four Treatises of Theophrastus von Hohenheim Called Paracelsus*, Johns Hopkins Press, Baltimore, 1941.
Temkin, O. Historical aspects of drug therapy, *Drugs in Our Society*, ed. P. Talalay Johns Hopkins Press, Baltimore, 1964, p. 3–16.
Tschirch, A. *Handbuch der Pharmakognosie*, 3 vols. C. H. Tauchnitz, Leipzig, 1917.
Wightman, W. P. D. William Cullen, *Dictionary of Scientific Biography*, vol. 3, p. 494–495.
Wright, A. D. The history of opium, *Med. Hist*, 1968, **18**, 62–70.
Wuest, H. M. A hundred years of alkaloid industry, *Chem. Ind.* (*London*), **1937**, 1084–1092.

CHAPTER TWO

Béhal, A. and Choay, E. Combinaisons du chloral avec la phényldiméthylpyrazolone (antipyrine), *J. Pharm. Chim*, **1890**, 539–542 [Dichloralphenazonel].

Bogue, J. Y. and Carrington, H. C. The evaluation of Mysoline—a new anticonvulsant drug, *Br. J. Pharmacol*, 1953, **8**, 230–236 [Primidonel].

Brunton, T. L. *An Introduction to Modern Therapeutics*, Macmillan and Co., London, 1892.

Butler, T. C. The introduction of chloral hydrate into medical practice, *Bull. Hist. Med*, 1970, **44**, 168–172.

Carter, M. K. The history of barbituric acid, *J. Chem. Educt*, 1951, **28**, 524–526.

Cervello, V. Physiological action of paraldehyde and contribution to the study of chloral hydrate, *Gazz. Chim. Ital*, 1883, **13**, 172.

Charonnat, R. Lechat, P. and Chareton, J. Sur les propriétés pharmacodynamiques d'un dérivé thiazolique, *Therapie*, 1957, **12**, 68–71 [Chlormethiazole].

Clouston, T. S. Experiments to determine the precise effect of bromide of potassium in epilepsy, *J. Ment. Sci*, 1868, **14**, 305–321.

Collins, G. W. and Leech, P. N. Origin of the name Veronal, *J. Am. Med. Ass*, 1931, **96**, 1869–1871.

Davis, B. and Pearce, D. R. An introduction to Althesin (CT 1341), *Postgrad. Med. J*, 1972, **June Suppl**, 13–17.

Duncum, B. M. *The Development of Inhalational Anaesthesia*, Oxford University Press, London, 1947.

Dundee, J. W. and Wyant, G. M. *Intravenous Anaesthesia*, Churchill Livingstone, Edinburgh, 1974, p. 3–5.

Farber, E. Emil Hermann Fischer, *Dictionary of Scientific Biography*, vol 5, p. 1–5.

Fischer, E. and Mering, J. von. Ueber eine neue Klasse von Schlafmitteln, *Ther. Gegenw*, 1903, **44**, 97–101 [Barbitone].

Gibson, W. R., Doran, W. J., Wood, W. C. and Swanson, E. E. Pharmacology of stereoisomers of 1-methyl-5-(1-methyl-2-pentenyl)-5-allyl barbituric acid, *J. Pharm. Exp. Ther*, 1959, **125**, 23–27 [Methohexitone].

Godefroi, E. F., Janssen, P. A., Van der Eycken, C. A. M., Van Heertum, A. H. M. T. and Niemegeers, C. J. E. DL-1-(1-Arylalkyl)imidazole-5-carboxylate esters. A novel type of hypnotic agents, *J. Med. Chem*, 1965, **8**, 220–223 [Etomidate].

Gujral, M. L., Saxena, P. N. and Tiwari, R. S. Comparative evaluation of quinazolones: a new class of hypnotics, *Ind. J. Med. Res*, 1955, 43, 637–641 [Methaqualone].

Haber, L. F. *The Chemical Industry 1900–1930*, Clarendon Press, Oxford, 1971.

Haber, L. F. *Chemical Industry During the Nineteenth Century*, Clarendon Press, Oxford, 1958.

Habermann, E. R. Rudolf Buchheim and the beginning of pharmacology as a science, *Ann. Rev. Pharmacol*, 1974, **14**, 1–8.

Hauptmann, A. Luminal bei Epilepsis, *Munch. Med. Wochenschr*, 1912, **54**, 1907–1909 [Phenobarbitone].

Heffter, A. Ueber die Einwirkung von Chloral auf Glucose, *Chem. Ber*, 1889, **22**, 1050–1051 [Chloralose].

Hems, B. A., Atkinson, R. M., Early, M. and Tomich, E. G. Trichloroethylphosphate, *Br. Med. J*, 1962, **1**, 1834–1835.

Hoesch, K. *Emil Fischer. Sein Leben und sein Werk*, Verlag Chemie, Berlin, 1921.

Hope, C. E. Clinical trial—the Minaxolone story, *Can. Anaesth. Soc. J*, 1981, **28**, 1–5.

Horatz, K., Frey, R. and Zindler, M. (eds.) *Die intravenose Kurznarkose mit dem neuen Phenoxyessigsäurederivat Propanidid (Epontol)*, Springer-Verlag, Berlin, 1965.

Janssen, P. A. J., Niemegeers, C. J. E., Schellekens, K. H. L. and Lenaerts, F. M. Etomidate, R-(+)-ethyl-1-(alpha-methyl-benzyl)imidazole-5-carboxylate (R 16659): a potent, short-acting and relatively atoxic intravenous hypnotic agent in rats, *Arzneimittel-Forschr*, 1971, **21**, 1234–1243.

Jarman, R. History of intravenous anaesthesia with six years' experience in the use of Pentothal, *Postgrad. Med. J*, 1941, **17**, 70–80.

Kast, A. Sulfonal, ein Neues Schlafmittel, *Berl. Klin. Wochenschr*, 1888, **25**, 309–314.

Kauffman, G. B. Adolf von Baeyer and the naming of barbituric acid, *J. Chem. Educt*, 1980, **57**, 222–223.

Kauffman, G. B. Solvents, serendipity, and seizures, *Educ. Chem*, 1982, **19**, 168–169.

Keys, T. E. *A History of Surgical Anaesthesia*, Dover, 1963.

Koch-Weser, J. and Schechter, P. Schmiedeberg in Strassburg 1872–1918: the making of modern pharmacology, *Life Sci*, 1978, **22**, 1361–1372.

Kohler, R. E. Roger Adams, *Dictionary of Scientific Biography*, vol 15, p. 1–3.

Krantz, J. C. Jr, Carr, C. J., Lu, G. and Bell, F. K. Anesthetic action of trifluoroethyl vinyl ether, *J. Pharm. Exp. Ther*, 1953, **108**, 488–495 [Fluroxene].

Kuchinsky, G. The influence of Dorpat on the emergence of pharmacology as a distinct discipline, *J. Hist. Med*, 1968, **23**, 258–271.

Kunz, W., Keller, H. and Muckter, H. *N*-Phthalyl-glutaminsaure-imid, *Arzneimittel. Forsch*, 1956, **6**, 426–430 [Thalidomide].

Kushner, S., Cassell, R. I., Morton, J. and Williams, J. H. Anticonvulsants. *N*-Benzylamides, *J. Org. Chem*, 1951, **16**, 1283–1288 [Beclamide].

Laubach, G. D., P'An, S. Y. and Rudel, H. Steroid anesthetic agent, *Science*, 1955, **122**, 78 [Hydroxydione].

Liebreich, O. *Das Chloralhydrat. Ein neues Hypnoticum und Anaestheticum und desen Anwendung in der Medicin*, Müller, Berlin, 1869.

Locock, C. cited in *Lancet*, 1857, **1**, 527–528 [Bromide].

Lundy, J. S. Intravenous anesthesia: preliminary report of the use of two new thiobarbiturates, *Proc. Staff Meet. Mayo Clinic*, 1935, **1**, 534–543 [Thiopentonel].

Markwood, L. N. European chemical industry in the nineteenth century, *J. Chem. Educt*, 1951, **28**, 348–352.

Mering, J. F. von. *Pharm. Central*, **1889**, 484, 494 [Chloralformamide].

Merritt, H. H., Putnam, T. J. and Schwab, D. M. A new series of anticonvulsant drugs tested by experiments in animals, *Trans. Am. Neurol. Ass*, 1937, **63**, 123–128.

Merritt, H. H. and Putnam, T. J. Sodium diphenylhydantoinate in the treatment of convulsive disorders, *J. Am. Med. Ass*, 1938, **111**, 1068–1072 [Phenytoin].

Meunier, H., Carraz, G., Meunier, Y., Eymard, P. and Aimard, M. Pharmacodynamic properties of dipropylacetic acid. I. Anticonvulsive action, *Therapie*, 1963, **18**, 435–438 [Valproic acid].

Meyer, H. H. Schmiedebergs Werk, *Arch. Exp. Path. Pharm*, 1922, **92**, 1–17.

Miller, C. A. and Long, L. Anticonvulsants. 1. An investigation of *N*-R-alpha-R1-phenylsuccinimides, *J. Am. Chem. Soc*, 1951, **73**, 4895–4898 [Phensuximide].

Page, I. H. and Coryllos, P. Isoamylethylbarbituric acid, Amytal, as an intravenous anesthetic, *J. Pharm, Exp. Ther*, 1926, **27**, 189.

Parascandola, J. Reflections on the history of pharmacology, *J. Hist. Med*, 1980, **35**, 131–140.

Pratt, T. W., Tatum, A. L., Hathaway, H. R. and Waters, R. M. Sodium ethyl(1-methyl butyl) thiobarbiturate: preliminary experimental and clinical study, *Am. J. Surg*, 1936, **31**, 464 [Thiopentone].

Putnam, T. J. The demonstration of the specific anticonvulsant action of diphenylhydantoin and related compounds, *Discoveries in Biological Psychiatry*, (ed. F. Ayd, B. Blackwell) Lippincott, Philadelphia, 1970.

Raventos, J. The action of fluothane, a new volatile anaesthetic, *J. Pharm. Exp. Ther*, 1956, **11**, 394–409 [Halothane].

Rice, J. W. The synthetic organic hypnotics, *J. Am. Pharm. Ass*, 1944, **33**, 289–297.

Richardson, B. W. *On Chloroform and Other Anaesthetics,* (ed. J. Snow) Churchill, London, 1858, p. xxvii–xxx.

Robbins, B. H. Preliminary studies of the anesthetic activity of fluorinated hydrocarbons, *J. Pharm. Exp. Ther*, 1946, **86**, 197–204.

Robinson, V. *Victory Over Pain*, Schuman, New York, 1946.

Schmiedeberg, O. On the pharmacological action and therapeutic application of some

ethereal salts of carbamic acid, *Practitioner*, 1885, **35**, 275–280, 328–332.
Selye, H. Anesthetic effect of steroid hormones, *Proc. Soc. Exp. Med. Biol*, 1941, **46**, 116–121.
Selye, H. Studies concerning correlation between anesthetic potency, hormonal activity and chemical structure among steroid compounds, *Curr. Res. Anesth. Analg.*, 1942, **21**, 41–47.
Smith, W. D. A. *Under the Influence: History of Nitrous Oxide and Oxygen Anaesthesia*, Macmillan, 1982.
Spielman, M. Some analgesic agents derived from oxazolidine-2,4-diones, *J. Am. Chem. Soc*, 1944, **66**, 1244–1245 [Troxidone].
Spinks, A. Chemistry and anaesthesia, *Chem. Ind. (London)*, **1977**, 475–485.
Suckling, C. W. Some chemical and physical factors in the development of fluothane, *Br. J. Anaesth*, 1957, **29**, 466–472 [Halothane].
Sunday Times Insight Team. *Suffer the Children. The Story of Thalidomide*, Andre Deutsch, London, 1979.
Sutton, J. A. The way to a steroid anaesthetic, *Glaxo Volume*, 1972, **36**, 5–16.
Sutton, J. A. A brief history of steroid anaesthesia before Althesin (CT 1341), *Postgrad. Med. J*, 1972, **June Suppl**, 9–13.
Tabern, D. L. and Volwiler, E. H. Sulfur-containing barbiturate hypnotics, *J. Am. Chem. Soc*, 1935, **57**, 1961–1963 [Thiopentone].
Tainter, M. L. and Marcelli, G. M. A. The rise of synthetic drugs in the American pharmaceutical industry, *Bull. N.Y. Acad. Med*, 1959, **35**, 387–405.
Taylor, F. L. Crawford William Long, *Ann. Med. Hist*, 1925, **7**, 267–296.
Taylor, F. L. *Crawford W. Long and the Discovery of Ether Anesthesia*, Hoeber, New York, 1928.
Thierfelder, H. and Mering, J. von. Physiological action of the tertiary alcohols, *Z. Physiol. Chem*, 1885, **9**, 511–517.
Thullier, M. J. and Domenjoz, R. Zur Pharmakologie der intravenosen Kurznarkose mit 2-Methyl-4-allylphenoxyessigsäure-*N*,N-diäthylamid (G29,505), *Anaesthesist*, 1957, **6**, 163–167 [Propanidid].
Volwiler, E. H. and Tabern, D. L. 5,5-Substituted barbituric acids, *J. Am. Chem. Soc*, 1930, **52**, 1676–1679 [Pentobarbitone].
Weese, H. and Scharpff, W. Evipan, ein neuartiges Einschlaffmittel, *Deut. Med. Wochenschr*, 1932, **58**, 1205–1207 [Hexobarbitone].

CHAPTER THREE

Andrews, G. and Solomon, D. *The Coca Leaf and Cocaine Papers*, Harcourt, Brace, Janovich, London, 1975.
Arnott, J. On cold as a means of producing insensibility, *Lancet*, 1848, **2**, 98–99, 287–288.
Braun, H. Bedeutung des Adrenaline für die Lokalanästhesie, *Langenbecks Arch. Klin. Chir.*, 1903, **69**, 150.
Braun, H. Ueber eine neue örtliche Anästhetica (Stovain, Alypin, Novocain), *Deut. Med. Wochenschr*, 1905, **31**, 1667–1671.
Corning, J. L. On the prolongation of the anaesthetic effect of the hydrochloride of cocaine when subcutaneously injected, *N. Y. Med. J*, 1885, **42**, 317–319.
Ehrlich, P. and Einhorn, A. Ueber die physiologische Wirkung der Verbindung der Cocainreihe, *Chem. Ber*, 1894, **27**, 1870–1873.
Einhorn, A. Ueber die Chemie der local Anästhetica, *Munch. Med. Wochenschr*, 1898, **46**, 1218–1220; 1899, **46**, 1254–1256.
Einhorn, A. Ueber neue Arzneimittel, *Justus Liebigs Ann Chem*, 1900, **311**, 26–34.
Einhorn, A. *Deut. Med. Wochenschr*, 1905, **31**, 1668.

Einhorn, A. Ueber neue Arzneimittel, 5, *Justus Liebigs Ann. Chem*, 1910, **371** 125–131 [Procaine].
Einhorn, A. and Heinz, R. Orthoform. Ein Lokalanästheticum für Wundschmerz, Brandwunden, Geschwure, etc., *Munch. Med. Wochenschr*, 1897, **44**, 931–934.
Einhorn, A. and Oppenheimer, M. Ueber die Glycocollverbindundungen der Ester aromatischer Amido- und Amidooxysäuren, *Justus Liebigs Ann. Chem*, 1900, **311**, 154–178 [Nirvanin].
Ekenstam, B., Egner, B. and Pettersson, G. Local anaesthetics I. *N*-Alkyl pyrrolidine and *N*-alkyl piperidine carboxylic acid amides, *Acta Chem. Scand*, 1957, **11**, 1183–1190 [Mepivacaine, bupivacaine].
Erdtman, H. and Löfgren, N. Ueber eine neue Gruppe von lokalanästhetische wirksamen Verbindungen. alpha-*N*-Dialkylaminosäureanilide, *Svensk. Kem. Tidskr*, 1937, **49**, 163–174.
Filehne, W. Die local-anästhesirende Wirkung von Benzoylderivaten, *Berl. Klin, Wochenschr*, 1887, **24**, 107–108.
Fourneau, E. *Organic Medicaments and their Preparation*, Churchill, London, 1925, p. 61 [Amylocainel].
Huisgen, R. Richard Willstätter, *J. Chem. Educt*, 1961, **38**, 10–15.
Koller, C. Historical notes on the beginning of local anaesthesia, *J. Am. Med. Ass*, 1928, **90**, 1742–1743.
Koller, C. Ueber die Verwendung des Cocain zur Anästhesirung am Auge, *Wien. Med. Wochenschr*, 1884, **34**, 1276–1278, 1309–1311.
Koller, C. History of cocaine as a local anaesthetic, *J. Am. Med. Ass*, 1941, **117**, 1284–1287.
Koller-Becker, H. Carl Koller and cocaine, *Psychoanal. Quart*, 1963, **32**, 309–373.
Liljestrand, G. Carl Koller and the development of local anaesthesia, *Acta. Physiol. Scand*, 1967, **Suppl. 299**, 1–30.
Liljestrand, G. The historical development of local anaesthesia, *Local Anaesthesia*, (ed. P. Lechet) Pergamon Press, Oxford, 1971, p. 1–38.
Löfgren, N. and Lundqvist, B. Studies on local anaesthetics II, *Svensk. Kem. Tidskr*, 1946, **58**, 206–217 [Lignocaine].
Merling, G. Eucaine, *Ber. Deut. Pharm. Ges*, 1896, **6**, 173–176.
Miescher, K. Studien über Lokalanästhetica, *Helv. Chim. Acta*, 1932, **15**, 163–190 [Cinchocaine].
Moréna Y Maiz, T. *Recherches Chimiques et Physiologiques sur L'Erythroxylum coca du P*érou et la Cocaine, L. Leclerc, Paris, 1868.
Niemann, A. *Ueber eine neue organische Base in den Cocabl*ättern, E. A. Huth, Göttingen, 1860.
Richardson, B. W. On a new and ready mode of producing local anaesthesia, *Med. Times Gaz*, 1886, **1**, 115–117.
Ritsert, E. *Berl. Klin. Wochenschr*, 1902, No. 17 [Benzocaine].
Simpson, J. Y. Local anaesthesia, *Lancet*, 1848, **2**, 39–42.
Thomson, A. J. Ether, *The London Dispensatory*, Longmans, London, 1837, p. 769.
Wrotnowska, D. Ernest Fourneau, *Dictionary of Scientific Biography*, vol. 5, p. 99–100.

CHAPTER FOUR

Bechold, H. and Ehrlich, P. Beziehungen zwischen chemischer Konstitution und Desinfektionswirkung, *Hoppe-Seyl. Z. physiol. Chem*, 1906, **47**, 173–199.
Binz, C. *Lectures on Pharmacology*, 2 vols., New Sydenham Society, London, 1895 (Translated by A. C. Latham from: *Vorlesungen* über Pharmakologie, published in Berlin by A. Hirschwald, 1884–6).
Block, S. S. Historical review, *Disinfection, Sterilization, and Preservation,* (ed. C. A.

Lawrence, S. S. Block) Lea and Febiger, Philadelpia, 1968, p. 3–8.

Calvert, F. C. On the therapeutic properties of carbolic acid, *Lancet*, 1863, **2**, 362–363.

Crellin, J. K. Internal antisepsis or the dawn of chemotherapy, *J. Hist. Med*, 1981, **36**, 9–18.

Crellin, J. K. The disinfectant studies by F. Crace Calvert and the introduction of phenol as a germicide, *Veroef. Int. Ges. Gesch. Pharm*, 1966, **28**, 61–67.

Dakin, H. D. On the use of certain antiseptic substances in the treatment of infected wounds, *Br. Med. J*, 1915, **2**, 318–320 [Chloramines].

Dodd, M. C. and Stillman, W. B. The in vitro bacteriostatic action of some simple furan derivatives, *J. Pharm. Exp. Ther*, 1944, **82**, 11–18 [Nitrofurans].

Dolman, C. Joseph Lister, *Dictionary of Scientific Biography*, vol. 8, p. 399–413.

Duaker, M. F. W. A history of early antiseptics, *J. Chem. Educt*, 1938, **15**, 58–61.

Fourneau, E. *Organic Medicaments and their Preparations*, Churchill, London, 1925.

Galdstone, I. *Behind the Sulfa Drugs*, Appleton-Century, New York, 1943.

Geison, G. L. Louis Pasteur, *Dictionary of Scientific Biography*, vol. 10, p. 350–416.

Godlee, R. *Lord Lister*, Macmillan, London, 1917.

Kelly, H. A. Jules Lemaire, *J. Am. Med. Ass*, 1901, **30**, 1083–1088.

Lechevalier, H. A. and Solotorovsky, M. *Three Centuries of Microbiology*, McGraw-Hill, London, 1965.

Lemaire, F. J. *Du Coaltar Saponine, Desinfectant Energique*, Germer-Bailliere, Paris, 1860.

Lesher, G. Y., Froelich, E. J. Gruett, M. D. Bailey, J. H. and Brundage, P. R. 1,8-Naphthyridine derivatives. A new class of chemotherapeutic agents, *J. Med. Pharm. Chem*, 1962, **5**, 1063–1065 [Nalidixic acid].

Lister, J. On the antiseptic principle in the practice of surgery, *Lancet*, 1867, **2**, 353–356.

Lister, J. On the antiseptic treatment in surgery, Br. Med. J. 1868, **2**, 53–56, 101–102, 461–463, 515–517; 1869, **1**, 301–304.

Lister, J. On a new method of treating compound fracture, abscesses, etc., with observations on the conditions of suppuration, *Lancet*, 1867, **1**, 320–329, 357–359, 387–389, 507–509; 1867, **2**, 95–96.

Pringle, J. Some experiments on substances resisting putrefaction, *Roy. Soc. Philos. Trans*, 1750, **46**, 480–488, 525–534, 550–558.

Risse, G. B. Ignaz Philipp Semmelweis, *Dictionary of Scientific Biography*, vol. 12, p. 294–297.

Sansom, A. E. On the uses of septicidal agents in disease, *Retrospect Med*, 1868, **57**, 6–11 [Sodium sulphite]

Schechter, D. C. and Swan, H. Jules Lemaire: a forgotten hero of surgery, *Surgery*, 1961, **49**, 817–826.

Upmalis, I. H. The introduction of Lister's treatment in Germany, *Bull. Hist. Med*, 1968, **42**, 221–240.

Wangensteen, O. H. and Wangensteen, S. D. Lister, his books, and evolvement of his antiseptic wound practice, *Bull. Hist. Med*, 1974, **48**, 100–128.

CHAPTER FIVE

Archer, S., Alberts, N. F., Harris, L. S., Pierson, A. K. and Bird, J. G. Pentazocine. Strong analgesics and analgesic antagonists in the benzomorphan series, *J. Med. Chem*, 1964, **7**, 123–127.

Armstrong, H. E. Chemical industry and Carl Duisberg, *Nature*, 1935, **135**, 1021–1025.

Bassett, J. R. Cairncross, K. D. Hackett, N. B. and Story, M. Studies on the peripheral pharmacology of fenzoxine, a potent antidepressant drug, *Br. J. Pharmacol*, 1969, **37**,, 69–78 [Nefopam]

Cahn, A. and Hepp, P. Das Antifebrin, ein neues Fiebermittel, *Centralbl. Klin. Med*, 1886, **7**, 561–564. [Acetanilide].
Crum Brown, A. and Fraser, T. R. On the physiological action of the ammonium bases derived from atropia and conia, *Trans. Roy. Soc. Edin*, 1868, **25**, 693–739.
Crum Brown, A. and Fraser, T. R. On the physiological action of the salts of the ammonium bases, derived from strychnia, brucia, thebaia, codeia, morphia, and nicotia, *Trans. Roy. Soc. Edin*, 1869, **25**, 151–203.
Dott, D. B. and Stockman, R. The chemistry and pharmacology of some of the morphine derivatives, *Year Book of Pharmacy*, **1887**, 538–548; **1888**, 349–355.
Dreser, H. Pharmakologisches über Aspirin (acetylsalicylsäure), *Pfluger's Archiv. Anat. Physiol* 1899, **76**, 306–318.
Dreser, H. Pharmakologisches über einige Morphinderivate, *Deut. Med. Wochenschr*, 1898, **24**, 185–186 [Diamorphine].
Duisberg, C. Zür Geschichte der Entdeckung des Phenacetins, *Angew. Chem*, 1913, **26**, 240.
Eddy, N. A new morphine-like analgesic, *J. Am. Pharm. Ass*, 1947, **8**, 536–540 [Methadone].
Eddy, N. B. and May, E. L. Origin and history of antagonists, *Narcotic Antagonists*, (ed. M. C. Braude, L. S. Harris, E. L. May, J. P. Smith, J. E. Villard) Raven Press, New York, 1974, p. 9–11.
Eisleb, O. and Schaumann, O. Dolantin, ein neuartiges Spasmolytikum und Analgetikum, *Deut. Med. Wochenschr*, 1939, **63**, 967–968 [Pethidine].
Ellmer, R. Acetylsalicylsäure, *Lernen Leisten*, 1978, **7**, 82–83.
Filehne, W. Weiteres über Kairin und analoge Körper, *Berl. Klin. Wochenschr*, 1883, **20**, 77–79.
Filehne, W. Ueber das Pyramidon, ein Antipyrinderivat, *Berl. Klin. Wochenschr*, 1896, **33**, 1061–1063 [Amidopyrinel].
Flinn, F. B. and Brodie, B. B. The effect on the pain threshold of *N*-acetyl p-aminophenol, a product derived in the body from acetanilide, *J. Pharm. Exp. Ther*, 1948, **94**, 76–77 [Paracetamol].
Freund, M. and Speyer, E. Transformation of thebaine into hydroxycodeinone and its derivatives, *J. Prakt. Chem*, 1916, **94**, 135–178 [Oxycodone].
Greenbaum, F. R. The gold treatment of tuberculosis, *Am. J. Pharm*, 1926, **98**, 471–475.
Grewe, R. Synthetic drugs with morphine action, *Angew. Chem*, 1947, **59**, 194–199 [Levorphanol].
Gulland, J. M. and Robinson, R. The morphine group. Part 1. A discussion of the constitution problem, *J. Chem. Soc*, 1923, **123**, 980–998.
Hartung, E. History of the use of colchicine and related medicaments in gout, *Ann. Rheum. Dis*, 1954, **13**, 190.
How, H. On some new basic products obtained by the decomposition of vegetable alkaloids, *Quart. J. Chem. Soc*, 1853, **6**, 125–139.
Janssen, P. A new series of potent analgesics, *J. Am. Chem. Soc*, 1956, **78**, 3862 [Dextromoramide].
Knorr, L. Ueber die Constitution der Chinizin-derivative, *Chem. Ber*, 1884, **17**, 2032–2038 [Phenazone].
Lester, D. and Greenberg, L. A. Metabolic rate of acetanilide and other aniline derivatives. II, Major metabolites of acetanilide appearing in the blood, *J. Pharm. Exp. Ther*, 1947, **90**, 68–75 [Paracetamol].
Maclagan, T. The treatment of rheumatism by salicin and salicylic acid, *Br. Med. J*, 1876, **1**, 627.
Maclagan, T. The treatment of acute rheumatism by salicin, *Lancet*, 1876, **1**, 342–343.
May, E. L. and Eddy, N. B. A new potent synthetic analgesic, *J. Org. Chem*, 1959, **24**, 294–295 [Phenazocine].

Mering, J. von. Physiological and therapeutic investigations on the action of some morphine derivatives, *Merck's Report*, **1898**, 5–24.

Monkovitch, I., Conway, T. T., Wong, H., Perron, Y. G., Pachter, I. J. and Belleau, B. Total synthesis and pharmacological activity of *N*-substituted 3,14-dihydroxymorphinans, *J. Am. Chem. Soc*, 1973, **95**, 7910–7912 [Butorphanol].

Nicholson, J. S. Ibuprofen, *Chronicles of Drug Discovery*, (ed. J. S. Bindra, D. Lednicer) John Wiley, Chichester, 1982.

Ofner, P. and Walton, E. Search for new analgesics. Part IV. Variations in the basic side-chain of Amidone, *J. Chem. Soc*, **1950**, 2158–2166 [Dipipanone].

Rechenberg, von H. K. (ed.) *Butazolidin (Phenylbutazone)*, Arnold, London, 1962.

Rodnan, G. P. and Benedek, T. G. The early history of antirheumatic drugs, *Arthritis and Rheumatism*, 1970, **13**, 145–165.

Schaumann, O. A new class of compounds with spasmolytic and central analgesic action derived from 1-methyl-4-phenylpiperidine-4-carboxylic acid, *Arch. Exp. Path. Pharm*, 1940, **196**, 109–136 [Pethidine].

Schaumann, O. 150 Jahre Morphin und 15 Jahre synthetische, Morphin-ahnlich wirkende Analgetica, *Angew. Chem*, 1954, **66**, 765–768.

Schneider, W. *Geschichte der pharmazeutischen Chemie*, Verlag Chemie, Weinheim, 1972.

Sharp, G. Colchicine studied historically, *Med. Mag*, (*London*), 1909, **18**, 506, 568.

Shen, T. Y., Windholz, T. B., Rosegay, A., Witzel, B. E., Wilson, A. N., Willett, J. D., Holtz, W. J., Ellis, R. L., Matzuk, A. R., Lucas, S., Stammer, C. H., Holly, F. W., Sarett, L. H., Risley, E. A., Nuss, G. W. and Winter, C. A. Non-steroidal anti-inflammatory agents, *J. Am. Chem. Soc*, 1963, **85**, 488–489 [Indomethacin].

Small, L., Fitch, H. M. and Smith, W. E. The addition of organomagnesium halides to pseudocodeine types. II. Preparation of nuclear alkylated morphine derivatives, *J. Am. Chem. Soc*, 1936, **58**, 1457–1463 [Metopon].

Stockman, R. and Dott, D. B. Physiological action of morphine and its derivatives, *Br. Med. J*, 1896, **2**, 189–192.

Weijlard, J. and Erikson, A. E. *N*-Allylnormorphine, *J. Am. Chem. Soc*, 1942, **64**, 869–870 [Nalorphine].

Winder, C. V., Wax, J., Scott, L., Scherrer, R. A., Jones, E. M. and Short, F. W. Antiinflammatory, antipyretic, antinocioceptive properties of *N*-(2,3-xylyl)anthranilic acid (mefenamic acid; CI-473), *J. Pharm. Exp. Ther*, 1962, **138**, 405–413.

Wiseman, E. H. and Lombardino, J. G. Piroxicam, *Chronicles of Drug Discovery*, (ed. J. S. Bindra, D. Lednicer) John Wiley, Chichester, 1982.

Witthauer, K. Aspirin, ein neues Salicylpreparat, *Die Heilkunde*, 1899, **3**, 396.

Wright, C. R. A. On the action of organic acids and their anhydrides on the natural alkaloids. Part 1 *J. Chem. Soc*, 1874, **12**, 1031–1043 [Diamorphine].

CHAPTER SIX

Abel, J. J. The active constituent of suprarenal capsules, *Proc. Am. Physiol. Soc*, **1898**, 3–5.

Adrian, E. D., Feldberg, W. and Kilby, B. A. Inhibitory action of fluorophosphonates on cholinesterase, *Nature*, 1946, **158**, 625.

Aeschlimann, J. A. The synthesis of new medicinal alkaloids, *J. Soc. Chem. Ind*, **1935**, 135–141 [Neostigmine].

Barcroft, H. and Talbot, J. F. Oliver and Schäfer's discovery of the cardiovascular action of suprarenal extract, *Postgrad. Med. J.* 1968, **44**, 6–8.

Barger, G. and Dale, H. H. Chemical structure and sympathomimetic action of amines, *J. Physiol*, 1910, **41**, 19–59.

Barger, G. and Walpole, G. S. Pressor substances in putrid meat, *J. Physiol*, 1909, **38**, 343–352 [Tyramine].
Black, J. W., Crowther, A. F., Shanks, R. G. and Dornhorst, A. C. A new adrenergic beta-receptor antagonist, *Lancet*, 1964, **1**, 1080–1081 [Propranolol].
Black, J. W. and Stphenson, J. S. Pharmacology of a new adrenergic beta-receptor blocking compound (Nethalide), *Lancet*, 1962, **2**, 311–314.
Blicke, F. F. and Monroe, E. Antispasmodics, *J. Am. Chem. Soc*, 1939, **61**, 91–93, 93–95 [Cyverine].
Bove, F. J. *The Story of Ergot* Karger, New York, 1970.
Braun, J. von, Braunsdorff, O., Räth, K. Beziehungen zwischen Konstitution und pharmakologische Wirkung bei Benzosäure- und Tropansäure-estern von Alkaminen, *Chem. Ber*, 1922, **55**, 1666.
Burtner, R. R. Antispasmodics, *Medicinal Chemistry*, vol 1, (ed. C. M. Suter) John Wiley, New York, 1951, p. 151–279.
Buth, W., Kulz, F. and Rosenmund, K. W. Über synthesen spasmolytische wirkenden-der Stoffe, *Chem. Ber*, 1939 **72**, 19–28 [Alverine].
Bynum, W. F. Henry Hallett Dale, *Dictionary of Scientific Biography*, vol. 15, p. 104–107.
Chen, K. K. and Schmidt, C. F. *Ephedrine and Related Substances*, Williams and Wilkins, Baltimore, 1930.
Chen, K. K. and Schmidt, C. F. The action of ephedrine, an alkaloid from Ma Huang, *Proc. Soc. Exp. Biol. Med*, 1923, **21**, 351–354.
Christison, R. On the properties of the ordeal-bean of Old Calabar, Western Africa, *Monthly J. Med. Sci., London and Edinburgh*, 1855, **20**, 193–204.
Craig, L. C., Shedlovsky, T., Gould, R. G. and Jacobs, W. A. The ergot alkaloids. XIV. The position of the double bond and the carboxyl group in lysergic acid and its isomers. The structure of the alkaloid, *J. Biol. Chem*, 1938, **125**, 289–298.
Dakin, H. D. Synthesis of a substance allied to adrenaline. Physiological activity of substances indirectly related to adrenaline, *Proc. Roy. Soc. London, Ser. B*, 1905, **76**, 491–497; 498–503.
Dale, H. H. The action of certain esters and ethers of choline, and their relation to muscarine, *J. Pharm. Exp. Ther*, 1914, **6**, 147–190 [Acetylcholine].
Davis, M. E., Adair, F. L., Rogers, G., Kharasch, M. S. and Legault, R. R. A new active principle in ergot and its effects on uterine motility, *Am. J. Obstet. Gynecol*, 1935, **29**, 155–167 [Ergometrine].
Dixon, W. E. and Taylor, F. Physiological action of the placenta, *Br. Med. J*, 1907, **2**, 1150.
Dudley, H. W. and Moir, J. C. The substance responsible for the traditional clinical effects of ergot, *Br. Med. J*, 1935, 520–523 [Ergometrine].
Euler, U. S. von. A specific sympathomimetic ergone in adrenergic nerve fibres (sympathin) and its relations to adrenaline and nor-adrenaline, *Acta Physiol. Scand*, 1946, **12**, 73–97.
Fest, C. and Schmidt, K. J. *The Chemistry of Organophosphorus Pesticides*, Springer-Verlag, New York, 1973.
Fraser, T. R. On the characters, actions and therapeutic uses of the ordeal bean of Calabar (Physostigma venenosum, Balfour), *Edinburgh Med. J*, 1863, **9**, 36–56, 123–132, 235–248.
Fromherz, K. Die parasympatisch hemmenden Nervendenwirkungen atropinartig gebauter Verbindungen, *Arch. Exp. Path Pharm*, 1933, **173**, 86–128.
Gaddum, J. H. The history of work on anticholinesterases, *Chem. Ind.* (*London*), **1954**, 266–268.
Gerrard, A. W. The alkaloid and active principle of Jaborandi, *Pharm J*, 1875, **5**, 865 [Pilocarpine].
Hardy, E. Sur le jaborandi (Polycarpus pinnatus), *Bull. Soc. Chim. Fr*, 1875, **24**, 497–501 [Pilocarpine].

Hartley, D., Jack, D., Luntz, L. and Ritchie, A. C. H. New class of selective stimulants of beta-adrenergic receptors, *Nature*, 1968, **219**, 861–862 [Salbutamol].

Hartung, W. H. Epinephrine and related compounds: influence of structure on physiological activity, *Chem. Rev*, 1931, **9**, 389–465.

Hofmann, A. Notes and documents concerning the discovery of LSD, *Agents Actions*, 1970, **1**, 148–150.

Hofmann, A. The discovery of LSD and subsequent investigations on naturally occurring hallucinogens, *Discoveries in Biological Psychiatry*, (ed. F. J. Ayd, B. Blackwell) Lippincott, Philadelphia, 1970, p. 91–106.

Holmstedt, B. Synthesis and pharmacology of dimethylamido-ethoxy-phosphoryl cyanide (Tabun), *Acta Physiol. Scand*, 1951, **Suppl. 90**, 1–20.

Hunt, R. Substances which lower blood pressure in suprarenal extracts, *Am. J. Physiol*, 1900, **3**, vi–vii [Acetylcholine].

Hunt, R. and Taveau, R. On the physiological action of certain choline derivatives and a new method for determining choline, *Br. Med. J*, 1906, **2**, 1788–1791.

Ing, H. R. Synthetic substitutes for atropine, *Br. Med. Bull*, 1947, **4**, 91–95 [Lachesine].

Jacobs, W. and Craig, L. The ergot alkaloids. II. The degradation of ergotinine with alkali. Lysergic acid, *J. Biol. Chem*, 1934, **104**, 547–551.

Kilby, B. A. and Kilby, M. The toxicity of alkylfluorophosphonates in man and animals, *Br. J. Pharmacol*, 1947, **2**, 234–240.

Konzett, H. Neue broncholytesche hochwirksame Korper der Adrenalinreibe, *Arch. Exp. Path. Pharm*, 1940, **197**, 27–40 [Isoprenaline].

Kreitmair, H. A new class of choline esters, *Arch. Exp. Path. Pharm*, 1932, **164**, 346–356.

Ladenburg, A. and Rugheimer, L. Künstliche Bildung der Tropasaure, *Chem. Ber*, 1880, **13**, 373–379 [Homatropine].

Lange, W. and Kreuger, G. von. Ueber ester der monofluorphosphorsäure, *Chem. Ber*, 1932, **65**, 1598–1601.

Larsen, A. A. Gould, W. A., Roth, H. R., Comer, W. T., Uloth, R. H., Dungan, K. W. and Lish, P. M. Sulfonamides II. Analogs of catecholamines, *J. Med. Chem*, 1967, **10**, 462–472 [Soterenol].

Le Count, D. J. Atenolol, *Chronicles of Drug Discovery*, (ed. J. S. Bindra, D. Lednicer) John Wiley, Chichester, 1982.

Leake, C. D. The long road for a drug from idea to use. The amphetamines, *Discoveries in Biological Psychiatry*, (ed. F. J. Ayd, B. Blackwell), Lippincott, Philadelphia, 1970, p. 68–83.

Levy, B. Alterations of adrenergic responses by *N*-isopropylmethoxamine, *J. Pharm. Exp. Ther*, 1964, **146**, 129–138.

Loewi, O. and Meyer, H. Ueber die Wirkung synthetischer, dem Adrenalin verwandter Stoffe, *Arch. Exper. Path. Pharm*, 1905, **53**, 213–226.

Loewi, O. Ueber humorale Uebertragbarkeit der Hertznervenwirkung, *Pflug. Arch. ges. Physiol*, 1921, **189**, 239–242.

Loewi, O. and Navratil, E. Ueber humorale Uebertragbarkeit der Hertznervenwirkung. XI. Mitteilung. Ueber den Mechanismus der Vaguswirkung von Physostigmin und Ergotamin, *Pflug. Arch. ges. Physiol*, 1926, **214**, 689–696.

Macht, D. I. Pharmacological investigation of papaverine, *Arch. Int. Med.* 1916, **17**, 786–806.

Major, R. T. and Cline, J. K. Preparation and properties of alpha- and beta-methacholine and gamma-homocholine, *J. Am. Chem. Soc*, 1932, **54**, 242–249.

Mannich, C. and Jacobsohn, W. Über Oxyphenyl-alkylamine und Dioxyphenyl-alkylamine, *Chem. Ber*, 1910, **43**, 189–197.

Mannich, C. and Walther, O. Synthesis of papaverine and related compounds, *Arch. Pharm*, 1927, **265**, 1–11.

Moir, J. C. Ergot: from St. Anthony's fire to the isolation of its active principle, ergometrine (ergonovine), *Am. J. Obstet. Gynecol*, 1974, **120**, 291–296.

Moran, N. C. and Perkins, M. E. Adrenergic blockade of the mammalian heart by a dichloro analogue of isoproterenol, *J. Pharm. Exp. Ther,* 1958, **124**, 223–237.
Nickerson, M. The pharmacology of adrenergic blockade, *Pharmacol. Rev*, 1949, **1**, 27–101.
Oliver, G. and Schäfer, E. A. On the physiological action of extracts of the suprarenal capsules, *J. Physiol*, 1894, **16**, 1–4.
Parascandola, J. John J. Abel and the early development of pharmacology at the Johns Hopkins University, *Bull. Hist.* Med, 1982, **56**, 512–527.
Pines, G., Miller, H. and Alles, G. Clinical observations on phenylaminoethanol sulphate, *J. Am. Med. Ass*, 1930, **94**, 790–791.
Powell, C. E. and Slater, I. H. Blocking of inhibitory adrenergic receptors by a dichloro analog of isoproterenol, *J. Pharm. Exp. Ther*, 1958, **122**, 480–488.
Rocke, A. J. George Barger, *Dictionary of Scientific Biography*, vol. 15, p. 10–11.
Rodin, F. H. Eserine: its history in the practice of ophthalmology, *Am. J. Ophthal*, 1947, **30**, 19–28.
Rosenberg, C. E. John Jacob Abel, *Dictionary of Scientific Biography*, vol. 1, p. 9–12.
Rosenheim, O. Pressor substances in placental extracts, *J. Physiol*, 1909, **38**, 337–342 [Tyramine].
Schleiffer, H. *Sacred Narcotic Plants of the New World Indians*, Collier Macmillan, London, 1973.
Scholz, C. R. Imidazole derivatives with sympathomimetic activity, *Ind. Eng. Chem*, 1945, **37**, 120–125.
Shanks, R. G. The discovery of beta-adrenoceptor blocking drugs, *Trends Pharm. Sci*, 1984, **5**, 405–409.
Simonart, A. On the action of certain derivatives of choline, *J. Pharm. Exp. Ther*, 1932, **46**, 157–193.
Smith, L. H. Cardio-selective beta-adrenergic blocking agents, *J. Appl. Chem. Biotechnol*, 1978, **28**, 201–212.
Starr, I., Elsom, K. A. and Reisinger, J. A. Acetyl-beta-methylcholine. The action on normal persons, *Am. J. Med. Sci*, 1933, **186**, 313–323.
Tullar, B. F. The resolution of DL-arterenol, *J. Am. Chem. Soc*, 1948, **70**, 2067–2068 [Noradrenaline]
Wasson, R. G., Hofmann, A. and Ruck, C. A. P. *The Road to Eleusis*, Harcourt Brace Jovanovich, London, 1978.
Weisser, U. Künstliches Adrenaline. Der mühevolle Weg zur ersten Hormonsynthese, *Med. Welt*, 1980, **31**, 40–44.

CHAPTER SEVEN

Barlow, R. B. and Ing, H. R. Curare-like action of polymethylene bis-quaternary ammonium salts, *Br. J. Pharmacol*, 1948, **3**, 298. [Decamethonium].
Bennett, A. E. The history of the introduction of curare into medicine, *Anesth. Analg. Curr. Res*, 1968, **47**, 484–492.
Bennett, A. E. Preventing traumatic complications in convulsive shock therapy by curare, *J. Am. Med. Ass*, 1940, **114**, 322–324.
Bernard, C. Action du curare et de la nicotine sur le système nerveux et sur le système musculaire, *C. R. Soc. Biol*, (*Paris*) 1850, **2**, 195.
Bernard, C. Analyse physiologique des proprietes des actions de curare et de la nicotine sur systèmes musculaire et nerveux au moyen du curare, *C. R. Hebd. Séances Acad. Sci*, 1856, **43**, 824–829.
Betcher, A. M. The civilizing of curare: a history of its development and introduction into anesthesia, *Anesth. Analg. Curr. Res*, 1977, **56**, 305–319.
Bolger, L., Brittain, R. T., Jack, D., Johnson, M. R., Martin, L. E., Mills, J., Poynter, D. and Tyers, M. B. Short-lasting, competitive neuromuscular blocking activity in a

series of azo-bisarylimidazo-[1,2-*a*]-pyridinium dihalides, *Nature*, 1972, **238**, 354–355 [Fazadinium].

Bovet, D., Courvoisier, S., Ducrot, R. and Horclois, J. R. Propriétés curarisantes du di-iodoéthylate de bis-(quinoéthoxy-8′) 1:5-pentane, *C. R. Hebd. Séances Acad. Sci*, 1946, **223**, 597–598.

Bovet, D., Depierre, F. and de Lestrange, Y. Curarizing properties of phenolic ethers with quaternary ammonium groups, *C. R. Hebd. Séances Acad. Sci*, 1947, **225**, 74–76.

Bovet, D., Depierre, F., Courvoisier, S. and de Lestrange, Y. Synthetic curarizing agents. II. Phenolic ethers with quaternary ammonium groups. The action of tris(diethylamino-ethoxy)benzene triiodoethylate (2559 F), *Arch. Int. Pharmacodyn*, 1949, **80**, 172–188 [Gallamine].

Bovet, D., Bovet-Nitti, F., Guarini, S., Longo, V. and Fusco, R. Synthetic curarizing agents. III. Succinylcholine and its aliphatic derivatives, *Arch. Int. Pharmacodyn*, 1951, **88**, 1–50.

Bovet, D., Bovet-Nitti, F. and Marini-Bettòlo, G. B. (eds.) *Curare and Curare-like Agents*, Elsevier, London, 1959.

Buckett, W. R., Hewett, C. L. and Savage, D. S. Pancuronium bromide and other steroidal neuromuscular blocking agents containing acetylcholine fragments, *J. Med. Chem*, 1973, **16**, 1116–1124.

De La Condamine, C. M. Relation abregée d'un voyage fait dans l'interieur de l'Amerique meridionale, *Mémoires de l'Academie des Sciences*, 1745, **62**, 391.

Everett, A. J., Lowe, L. A. and Wilkinson, S. Revision of the structure of (+)−tubocurarine chloride and (+)−chondrocurine, *J. Chem. Soc. Chem. Commun*, **1970**, 1020–1021.

Gill, R. C. Curare: misconceptions regarding the discovery and development of the present form of the drug, *Anesthesiology*, 1946, **7**, 14–24.

Glover, E. E. and Yorke, M. 1,1-Azoimidazo[1,2-*a*]pyridinium salts, *J. Chem. Soc. (C)*, **1971**, 3280–3281.

Griffith, H. R. and Johnson, G. E. The use of curare in general anesthesia, *Anesthesiology*, 1942, **3**, 418–420.

Grmek, M. D. Claude Bernard, *Dictionary of Scientific Biography*, vol. 2, p. 24–34.

King, H. Curare alkaloids. Part I. Tubocurarine, *J. Chem. Soc*, **1935**, 1381–1389.

McIntyre, A. R. *Curare. Its History, Nature, and Clinical Use*, University of Chicago Press, Chicago, 1947.

Olmsted, J. M. D. *Claude Bernard, Physiologist*, Harper, New York, 1939.

Paton, W. D. M. and Zaimis, E. J. The pharmacological actions of polymethylene bistrimethylammonium salts, *Br. J. Pharmacol*, 1949, **4**, 381–400. [Decamethonium].

Smith, P. *Arrows of Mercy*, Doubleday, New York, 1969.

Stenlake, J. B., Waigh, R. D. and Dewar, G. H. Atracurium besylate and related polyalkylene di-esters, *Eur. J. Med. Chem*, 1981, **16**, 515–524.

Stenlake, J. B., Waigh, R. D., Urwin, J., Dewar, G. and Coker, G. C. Atracurium: conception and inception, *Br. J. Anaesth*, 1983, **55**, 3s–10s.

Thomas, B. *Curare—its History and Usage*, Pitman Medical, London, 1960.

Walker, M. B. Treatment of myasthenia gravis with physostigmine, *Lancet*, 1934, **1**, 1200–1201.

Wintersteiner, O. and Dutcher, J. D. Curare alkaloids from Chondodendron tomentosum, *Science*, 1943, **97**, 467–470.

CHAPTER EIGHT

Ackernecht, E. Aspects of the history of therapeutics, II. Digitalis and some other panaceas, *Bull. Hist. Med*, 1962, **36**, 389–419.

Bergel, F. and Parkes, M. W. Drugs inhibiting symptomatic stimulators, *Prog. Org. Chem*, 1952, **1**, 173–218.

Best, C. H. Preparation of heparin and its use in the first clinical cases, *Circulation*, 1959, **19**, 79–86.

Beyer, K. H. Discovery of the thiazides: where biology and chemistry meet, *Persp. Biol. Med*, 1977, **20**, 410–420.

Beyer, K. H. Hypertension: from theory to practice, *Trends Pharmacol. Sci*, 1980, **1**, 114–121.

Bockmuhl, M. and Erhart, G. Über eine neue Klasse von spasmolytische und analgetische wirkenden Verbingunden. I, *Justus Liebigs Ann. Chem*, 1948, **561**, 52–85 [Fenpiprane].

Bossert, F. and Vater, W. Dihydropyridine, eine neue Gruppe stark Wirksamer Coronartherapeutika, *Naturwissenschaften*, 1971, **58**, 578 [Nifedipine].

Boura, A. L., Green, A. F., McCoubry, A., Laurence, D. R., Moulton, R. and Rosenheim, M. L. Darenthin. Hypotensive agent of new type, *Lancet*, 1959, **2**, 17–21 [Bretylium].

Boura, A. L. A. and Green, A. F. Noradrenergic neurone blocking agents, *J. Auton Pharmac*, 1981, **1**, 255–267.

Brunton, T. L. *A Textbook of Pharmacology, Therapeutics, and Materia Medica*, Macmillan, London, 1885.

Brunton, T. L. On the use of nitrite of amyl in angina pectoris, *Lancet*, 1867, **2**, 97–98.

Cushny, A. R. *The Action and Uses in Medicine of Digitalis and its Allies*, Longmans, Green, London, 1925.

Elderfield, R. C. The chemistry of the cardiac glycosides, *Chem. Rev*, 1935, **17**, 187–249.

Erhart, G. *et al*, *Arzneimittel-Forsch*, 1960, **10**, 569–588 [Prenylamine].

Estes, J. W. and White, P. D. William Withering and the purple foxglove, *Scient. Amer*, **1968**, 110–119.

Feit, P. W. Bumetanide, *Chronicles of Drug Discovery* (ed. J. S. Bindra, D. Lednicer) John Wiley, Chichester, 1982.

Feit, P. W. Structure—activity relationships of sulphamoyl diuretics, *Postgrad. Med. J*, 1975, **51**, **(Suppl. 6)** 9—13.

Fielden, R. The discovery of the noradrenergic neurone blocking action of TM 10 (xylocholine), *J. Auton. Pharmac*, 1981, **1**, 251–254.

Friend, D. G. Digitalis after two centuries, *Arch. Surg*, 1976, **111**, 14–19.

Fulton, J. F. Charles Darwin (1758—1778) and the history of the early use of digitalis, *Bull. N.Y. Acad. Med*, 1934, **10**, 496–506.

Gross, F., Druffy, J. and Meier, R. Eine neue Gruppe blutdrucksenkokender Substanzen von besonderem Wirkungscharkter, *Experentia*, 1950, **6**, 19 [Hydrallazine].

Hess, H.-J., Cronin, T. H. and Scriabine, A. Antihypertensive 2-amino-4(3H)-quinazolinones, *J. Med. Chem*, 1968, **11**, 130–136 [Prazosin].

Hey, P. and Willey, G. Choline 2:6-xylyl ether bromide; an active quaternary local anaesthetic, *Br. J. Pharmacol*, 1954, **9**, 471–475 [Xylocholine].

Holmes, L. C. and DiCarlo, F. Nitroglycerin: the explosive drug, *J. Chem. Educt*, 1971, **48**, 573–576.

Howell, W. H. and Holt, E. Two new factors in blood coagulation: heparin and antithrombin, *Am. J. Physiol*, 1918, **47**, 328–341.

Jacoby, C. Ueber Hirudin, *Deut. Med. Wochenschr*, 1904, **30**, 1786.

Jaques, L. B. The discovery of heparin, *Sem. Thrombosis Hemostasis*, 1978, **4**, 350–353.

Jorpes, J. E. and Bergstrom, S. Heparin: a mucoitin polysulphonic acid, *J. Biol. Chem*, 1937, **118**, 447–457.

Lee, G. E., Wragg, W. R., Corne, S. J., Edge, N. D. and Reading, H. W. 1:2:2:6:6-Pentamethylpiperidine: a new hypotensive drug, *Nature*, 1958, **181**, 1717–1719 [Pempidine].

Leonard, N. J. and Hauck, F. P. Unsaturated amines. X. The mercuric acetate route to substituted piperidines, Δ^2-tetrahydropyridines and Δ^2-tetrahydroanabasines, *J. Am. Chem. Soc*, 1957, **79**, 5279–5292 [Pempidine].

Libman, D. D., Pain, D. L. and Slack, R. Some bisquaternary salts, *J. Chem. Soc*, **1952**, 2305–2307 [Pentolinium].

Link, K. P. The discovery of dicumarol and its sequels, *Circulation*, 1959, **19**, 97.

Link, K. P. The anticoagulant from spoiled clover hay, *Harvey Lectures*, 1944, **39**, 162–216.

Loev, B., Goodman, M. M., Snader, K. M., Tedeschi, R. and Macko, E. Hantzch-type dihydropyridine hypotensive agents, *J. Med. Chem*, 1974, **17**, 956–965 [Nifedipine].

Mann, T. and Keilin, D. Sulphanilamide as a specific inhibitor of carbonic anhydrase, *Nature*, 1940, **146**, 164–165.

Mautz, F. R. Reduction of cardiac irritability by the epicardial and systemic administration of drugs as a protection in cardiac surgery, *J. Thoracic Surg*, 1936, **5**, 612–628 [Procaine].

Maxwell, R. A., Müller, R. P. and Plummer, A. J. [2-(Octahydro-1-azovinyl)-ethyl]-guanidine sulfate (CIBA 5864–SU), a new synthetic antihypertensive, *Experientia*, 1959, **15**, 267 [Guanethidine].

May, L. A. *Withering on the Foxglove and Other Classics in Pharmacology*, Dabor Science Publications, New York, 1977.

McLean, J. The discovery of heparin, *Circulation*, 1959, **19**, 75–78.

Meunier, P., Mentzer, C. and Molho, D. Sur l'action antivitaminique K (hemorrhagique) d'une indanedione, *C. R. Hebd. Séances Acad. Sci*, 1947, **224**, 1666–1667 [Phenindione].

Monro, A. M. A new synthesis of guanidines, *Chem. Ind*, (*London*) **1964**, 1806–1807 [Guanoxan].

Moorman, L. J. William Withering, his work, his health, his friends, *Bull. Hist. Med*, 1942, **12**, 355–366.

Murrell, W. Nitro-glycerine as a remedy for angina pectoris, *Lancet*, 1879, **1**, 80–81, 113, 151, 223.

Novello, F. C. and Sprague, J. M. Benzothiadizine as novel diuretics, *J. Am. Chem. Soc*, 1957, **79**, 2028–2029 [Chlorothiazide].

Oates, J. A., Gillespie, L., Udenfriend, S. and Szoerdsma, A. Decarboxylase inhibition and blood pressure reduction by alpha-methyl-3,4-dihydroxyphenylalanine, *Science*, 1960, **131**, 1890–1891.

Ondetti, M. A. Design of specific inhibitors of angiotensin-converting enzyme: new class of orally active antihypertensive agents, *Science*, 1977, **196**, 441–444 [Captopril].

Paterson, Locock. The history of cardiac glycosides, *Appl. Ther*, 1967, **9**, 60–65.

Peck, T. W. and Wilkinson, K. D. *William Withering of Birmingham*, Wright, Bristol, 1950.

Pereira, J. Arsenious acid, *The Elements of Materia Medica and Therapeutics*, Vol. 1, Longman, Brown, Green, and Longmans, London, 4th edn, 1854, p. 682–721.

Pereira, J. Digitalis purpurea, *The Elements of Materia Medica and Therapeutics*, Vol. 2, pt. 1, Longman, Brown, Green, and Longmans, London, 4th edn, 1855, p. 529–538.

Quick, A. J. The development and use of the prothrombin tests, *Circulation*, 1959, **19**, 92–96.

Randall, L. O., Peterson, W. G. and Lehman, G. The ganglionic blocking action of thiophanium derivatives, *J. Pharm. Exp. Ther*, 1949, **97**, 48–57 [Trimetaphan].

Roblin, R. O. and Clapp, J. W. The preparation of heterocyclic sulphonamides, *J. Am. Chem. Soc*, 1950, **72**, 4890 [Acetazolamide].

Rubin, A. A., Roth, F. E., Weinberg, M. M., Toplis, J. G., Sherlock, M. H., Sperber, N. and Black, J. A new class of antihypertensive agents, *Science*, 1961, **133**, 2067 [Diazoxide].

Ruddy, A. W. and Buckley, J. S. Antispasmodics. *N*-(3-phenylpropyl)-amines and 3-amino-1-phenyl-1-propanols, *J. Am. Chem. Soc*, 1950, **72**, 718–721 [Fenpiprane].

Saxl, P. and Heilig, R. Ueber die diuretische Wirkung von Novasurol und anderen Quecksilberpraparaten, *Wein. Klin. Wochenschr*, 1920, **33**, 943 [Merbaphen].
Schaeffer, A., Blumenfield, S., Pitman, E. and Dix, H. Procaine amide: its effect on auricular arrhythmias, *Am. Heart J*, 1951, **42**, 115–123.
Schultes, E. M., Cragoe, E. J., Jr, Bicking, J. B., Bolhofer, W. and Sprague, J. M. α,β-Unsaturated ketone derivatives of aryloxyacetic acids, a new class of diuretics, *J. Med. Pharm. Chem*, 1962, **5**, 660–662 [Ethacrynic acid].
Smith, S. Digoxin. A new digitalis glycoside, *J. Chem. Soc*, **1930**, 508–510.
Spinks, A. and Young, E. W. P. Polyalkylpiperidines: a new class of ganglion-blocking agents, *Nature*, 1958, **181**, 1397–1398 [Pempidine].
Stahle, H. Clonidine, *Chronicles of Drug Discovery* (ed. J. S. Bindra, D. Lednicer) John Wiley, Chichester, 1982.
Stein, G. A., Bronner, H. A. and Pfister, K. Alpha-methyl alpha-amino acids. II. Derivatives of DL-phenylethylamine, *J. Am. Chem. Soc*, 1955, **77**, 700–703 [Methyldopa].
Stein, G. A., Sletzinger, M., Arnold, H., Reinhold, D., Gaines, W. and Pfister, K. The reaction of camphene with hydrogen cyanide, *J. Am. Chem. Soc*, 1956, **78**, 1514–1515 [Mecamylamine].
Stoll, A. *The Cardiac Glycosides*, The Pharmaceutical Press, London, 1937.
Thornton, J. L. Sir Thomas Lauder Brunton 1844–1916, *St. Barts. Hosp. J*, 1967, **71**, 289–293.
Vogl, A. The discovery of the organic mercurial diuretics, *Am. Heart J*, 1950, **39**, 881–883.
Werner, W. 1,2,3,4-Tetrahydroquinoline derivatives with antihypertensive properties, *J. Med. Chem*, 1965, **8**, 125–126 [Debrisoquine].
Wien, R. Hypotensive agents, *Prog. Med. Chem*, 1961, **1**, 34–71.
Willius, F. A. and Keys, T. E. Abridged reprint of Withering's 'An Account of the Foxglove', *Cardiac Classics*, Henry Kimpton, London, 1941, p. 232–252.
Withering, W. *An Account of the Foxglove, and Some of its Medicinal Uses*, G. G. J. and J. Robinson, Birmingham, 1785.

CHAPTER NINE

Adamson, D. W. Aminoalkyl tertiary carbinols and derived products. Part 1, *J. Chem. Soc*, **1949**, S144–S146 [Triprolidine].
Black, J. W., Duncan, W. A. M., Durant, G. J., Ganellin, C. R. and Parsons, M. E. Definition and antagonism of histamine H_2-receptors, *Nature*, 1972, **236**, 385–390.
Bovet, D., Horclois, R. and Walther, F. Antihistaminic properties of 2-[(*p*-methoxybenzyl)(2-dimethylaminoethyl)amino]pyridine (RP 2786), *C. R. Soc. Biol*, (*Paris*), 1944, **138**, 99–100 [Mepyraminel].
Bovet, D. and Staub, A.–M. Action protectrice des ethers phenoliques au cours de l'intoxication histaminique, *C. R. Soc. Biol*, (*Paris*), 1937, **124**, 547–549.
Bradshaw, J., Brittain, R. T., Clitherow, J. W., Daly, M. J., Jack, D. and Price, B. J. Ranitidine (AH 19065): a new potent, selective histamine H_2-receptor antagonist, *Br. J. Pharmacol*, 1979, **66**, 464P.
Brimblecombe, R. W., Duncan, W. A., Durant, G. J., Ganellin, C. R., Parsons, M. E. and Black, J. W. The pharmacology of cimetidine, a new histamine H_2-receptor antagonist, *Br. J. Pharmacol*, 1975, **53**, 435P–436P.
Cox, J. S. G. Disodium cromoglycate (FPL 670) (Intal): a specific inhibitor of reaginic antibody-antigen mechanisms, *Nature*, 1967, **216**, 1328.
Dale, H. H. Local vasodilator reactions – histamine, *Lancet*, 1929, **1**, 1233–1237.
Dale, H. H. and Laidlaw, P. P. The physiological action of beta-iminazolylethylamine, *J. Physiol*, 1910, **41**, 318–344.

Fourneau, E. and Bovet, D. Recherches sur l'action sympathicolytique de nouveaux dérivés du dioxane, *C. R. Soc. Biol*, (*Paris*), 1933, **113**, 388–390 [Prosympal].
Gaddum, J. H. Histamine *Br. Med. J*, 1948, **1**, 867–873.
Ganellin, C. R. Cimetidine, *Chronicles of Drug Discovery*, (ed. J. S. Bindra, D. Lednicer) John Wiley, Chichester, 1982.
Ganellin, R. Medicinal chemistry and dynamic structure–activity analysis in the discovery of drugs acting at histamine H_2 receptors, *J. Med. Chem*, 1981, **24**, 913–920.
Gay, L. N. and Carling, P. E. The prevention and treatment of motion sickness. I. Seasickness, *Science*, 1949, **109**, 359 [Dimenhydrinate].
Halpern, B. N. and Walther, F. Influence of ionic balance of the bath fluid on histamine—antihistamine antagonists in the uterine horn of the guinea pig, *C. R. Soc. Biol*, (*Paris*), 1945, **139**, 402–440 [Phenbenzamine].
Howell, J. B. L. and Altounyan, R. E. C. A double-blind trial of disodium cromoglycate in the treatment of allergic bronchial asthma, *Lancet*, 1967, **2**, 539–542.
Huttrer, C. P., Djerassi, C., Beears, W. L., Mayer, R. L. and Scholz, C. R. Heterocyclic amines with antihistaminic activity, *J. Am. Chem. Soc*, 1946, **68**, 1999–2001 [Tripelennamine].
Kaufman, G. B. Asthma, anti-allergens and aerosols. The discovery and development of disodium cromoglycate (Cromolyn Sodium), *Educ. Chem*, 1984, **21**, 42–45.
Loew, E. R. Pharmacology of antihistamine compounds, *Physiol. Revs*, 1947, **27**, 542–573.
Loewe, E. R., Kaiser, M. E. and Moore, V. Synthetic benzhydryl alkamine ethers effective in preventing experimental asthma in guinea pigs exposed to atomised histamine, *J. Pharm. Exp. Ther*, 1945, **83**, 120–129 [Diphenhydramine].
Riley, J. F. Histamine and Sir Henry Dale, *Br. Med. J*, 1965, **1**, 1488–1490.
Staub, A. – M. Recherches sur quelque bases synthetiques antagonistes de l'histaminen, *Ann. Inst. Pasteur*, 1939, **63**, 400–436, 485–524.
Ungar, G., Parrot, J. – L. and Bovet, D. Inhibition des effets de l'histamine sur l'intestine isolé du cobaye par quelques substances sympathicomimétiques et sympathicolytiques, *C. R. Soc. Biol*, (*Paris*), 1937, **124**, 445–446.
Viaud, P. Les amines dérivées, de la phenothiazine, *J. Pharm. Pharmacol*, 1954, **6**, 361–389.

CHAPTER TEN

Barbeau, A. Biochemistry and treatment of Parkinson's disease, *Union Med. Canada* 1961, **96**, 1000–1001 [Levodopa].
Bein, H. J. Biological research in the pharmaceutical industry with reserpine, *Discoveries in Biological Psychiatry*, (ed. F. J. Ayd, B. Blackwell) Lippincott, Philadelphia, 1970, p. 142–154.
Berger, F. M. Anxiety and the discovery of the tranquilizers, *Discoveries in Biological Psychiatry* (ed. F. J. Ayd, B. Blackwell) Lippincott, Philadelphia, 1970, p. 115–129.
Berger, F. M. and Bradley, W. The pharmacological properties of α,β-dihydroxy-γ-(2-methylphenoxy)-propane (Myanesin), *Br. J. Pharmacol*, 1946, **1**, 265–272.
Birkmayer, W. and Hornykiewicz, O. Der L-Dioxyphenylalanine—Effect beim Parkinson-syndrom des menschen. Zur pathogenese und behandlung der Parkinson-Akinese, *Arch. Psychiat*, 1962, **203**, 560–574 [Levodopa].
Burkard, W., Gey, K. and Pletscher, A. A new inhibitor of decarboxylase of aromatic amino acids, *Experientia*, 1962, **18**, 411–412 [Benserazide].
Cade, J. F. J. The story of lithium, *Discoveries in Biological Psychiatry* (ed. F. J. Ayd, B. Blackwell) Lippincott, Philadelphia, 1970, p. 218–229.
Cade, J. F. J. Lithium salts in the treatment of psychotic excitement, *Med. J. Austral*, 1949, **36**, 349–352.

Caldwell, A. E. *Origins of Psychopharmacology. From CPZ to LSD*, Thomas, Springfield, 1970.

Carlsson, A., Cotzias, G. C., Van Hoert, M. H. and Schiffer, L. M. Dihydroxyphenylanaline and 5-hydroxtyryptamine as reserpine antagonists, *Nature*, 1957, **180**, 1200.

Charpentier, P. Sur la constitution d'un dimethylamino-*N*-phenothiazine, *C. R. Hebd. Séances Acad. Sci*, 1947, **225**, 306–308 [Promazine].

Chopra, R. N., Gupta, J. C. and Mukerjee, B. The pharmacological action of an alkaloid obtained from Rauwolfia serpentina Benth, *Indian J. Med Res*, 1933, **21**, 261–271.

Cohen, I. M. The benzodiazepines, *Discoveries in Biological Psychiatry*, (ed. F. J. Ayd, B. Blackwell) Lippincott, Philadelphia, 1970, p. 130–141.

Cotzias, G. C., Van Woert, M. H. and Schiffer, L. M. Aromatic amino acids and modification of parkinsonism, *New Eng. J. Med*, 1967, **276**, 374–379 [Levodopa].

Delay, J., Deniker, P. and Harl, J. M. Utilisation en therapeutique psychiatrique d'une phenothiazine d'action centrale elective (4560 RP), *Ann. Medicopsychol*, (*Paris*), 1952, **110 pt 2**, 112–117 [Chlorpromazine].

Deniker, P. Introduction of neuroleptic chemotherapy into psychiatry, *Discoveries in Biological Psychiatry*, (ed. F. J. Ayd, B. Blackwell) Lippincott, Philadelphia, 1970, p. 155–163.

Dikshit, R. K. The story of Rauwolfia, *Trends Pharm. Sci*, 1980, **1**(**12**), vii–x.

Janssen, P. A. The butyrophenone story, *Discoveries in Biological Psychiatry*, (ed. F. J. Ayd, B. Blackwell) Lippincott, Philadelphia, 1970, p. 165–179.

Janssen, P. A. Pimozide, a chemically novel, highly potent, and orally long-acting neuroleptic drug, *Arzneimittel.—Forschung*, 1968, **18**, 261–279.

Janssen, P. A. J., Van de Westeringh, C., Jagneau, A. W. M., Demoen, P. J. A., Hermans, B. K. F., Van Daele, G. H. P., Schellek, K. H. L., Van der Eycken, C. A. M. and Niemegerm, C. J. E. Chemistry and pharmacology of CNS depressants related to 4-(4-hydroxy-4-phenylpiperidino)butyrophenone. Part 1. Synthesis and screening data in mice, *J. Med. Pharm. Chem*, 1959, **1**, 281–297 [Haloperidol].

Kauffman, G. B. The discovery of iproniazid and its role in antidepressant therapy, *J. Chem. Educt*, 1979, **56**, 35–36.

Kline, N. S. Monoamine oxidase inhibitors: An unfinished picturesque tale, *Discoveries in Biological Psychiatry*, (ed. F. J. Ayd, B. Blackwell) Lippincott, Philadelphia, 1970, p. 194–204.

Kuhn, R. The imipramine story, *Discoveries in Biological Psychiatry*, (ed. F. J. Ayd, B. Blackwell) Lippincott, Philadelphia, 1970, p. 205–217.

Laborit, H., Therapeutique neuroplegique et hibernation artificielle: essai d'éclairecissement d'une équivoque, *Presse Med*, 1954, **62**, 359–362 [Chlorpromazine].

Laville, C. Chemistry and pharmacology of sulpiride, *Lille Med. 3rd. Series*, 1972, **17 Suppl**, 1–13.

Ludwig, B. and Piech, E. C. Some anticonvulsant agents derived from 1,3-propanediols, *J. Am. Chem. Soc*, 1951, **73**, 5779–5781 [Meprobamate].

Mallion, K. B., Todd, A. H., Turner, R. W., Bainbridge, J. G., Greenwood, D. T., Madinaveita, J., Somerville, A. R. and Whittle, B. A. 2-(2-Ethoxyphenoxymethyl)tetrahydro-1,4-oxazine hydrochloride, a potent psychotropic agent, *Nature*, 1972, **238**, 157 [Viloxazine].

Müller, J. M., Schlittler, E. and Bein, H. J. Reserpine, der sedative Wirkstoff aus Rauwolfia serpentina Benth, *Experientia*, 1952, **8**, 338.

Petersen, P. V., Lassen, N. L., Holm, T., Kopf, R. and Nielsen, I. M. Chemical constitution and pharmacological activity of some thiaxanthene analogs of chlorpromazine, promazine and meprazine, *Arzneimittel.—Forschung*, 1958, **8**, 395–397.

Ravn, J. The history of the thioxanthenes, *Discoveries in Biological Psychiatry*, (ed. F. J. Ayd, B. Blackwell) Lippincott, Philadelphia, 1970, p. 180–193.

Sletzinger, M., Chemerda, J. M. and Bollinger, F. W. Potent decarboxylase inhibitors. Analogs of methyldopa, *J. Med. Chem*, 1963, **6**, 101–103 [Carbidopa].

Snyder, H. R., Jr., Davis, C. S., Bickerton, R. K. and Halliday, R. P. 1-[(5-Arylfurfurylidene)amino]hydantoins. A new class of muscle relaxants, *J. Med. Chem*, 1967, **10**, 807–810 [Dantrolene].
Sternbach, L. H. The discovery of Librium, *Agents Actions*, 1972, **2**, 193–196.
Swazey, J. P. *Chlorpromazine in Psychiatry: a Study of Therapeutic Innovation*, MIT Press, Cambridge, 1974.
Viaud, P. Les amines dérivées de la phenothiazine, *J. Pharm. Pharmacol*, 1954, **6**, 361–389.

CHAPTER ELEVEN

Abel, J. J. Crystalline insulin, *Proc. Nat. Acad. Sci*, 1926, **12**, 132–136.
Allen, E. (ed.) *Sex and Internal Secretions*, Williams and Wilkins, Baltimore, 1939.
Allen, E. and Doisy, E. An ovarian hormone: Preliminary report on its localization, extraction and partial purification and action in test animals, *J. Am. Med. Ass*, 1923, **81**, 819.
Allen, W. M. and Corner, G. W. Maintenance of pregnancy in rabbits after very early castration, by corpus luteum extracts, *Proc. Soc. Exp. Biol. Med*, 1930, **27**, 403.
Allen, W. M. and Wintersteiner, O. Crystalline progestin, *Science*, 1934, **80**, 190.
Anderson, G. W., Halverstadt, I. F., Miller, W. H. and Roblin, R. O. Antithyroid compounds. Synthesis of 5- and 6-substituted 2-thiouracils from beta-oxoesters and thiourea, *J. Am. Chem. Soc*, 1945, **67**, 2197–2200.
Ascheim, S. and Zondek, B. Hypophysenvorderlappen Hormon und Ovarialhormon im Harn von Schwangeren, *Klin. Wochenschr*, 1927, **6**, 1322.
Astwood, E. B. Treatment of hyperthyroidism with thiourea and thiouracil, *J. Am. Med. Ass*, 1943, **122**, 78–81.
Astwood, E. B. The chemical nature of compounds which inhibit the function of the thyroid gland, *J. Pharm. Exp. Ther*, 1943, **78**, 79–89.
Banting, F. G. and Best, C. H. Internal secretion of pancreas, *J. Lab. Clin. Med*, 1922, **7**, 251–266.
Barger, G. and White, F. D. The constitution of galegine, *Biochem. J*, 1923, **17**, 827–835.
Baumann, E. Ueber das normale Vorkommen von Jod in Thierkörper, *Z. Physiol. Chem*, 1895–1896, **21**, 319–330; 481–493.
Bernal, J. D. A crystallographic examination of oestrin, *J. Soc. Chem. Ind*, 1932, **51**, 259.
Best, C. H. Nineteen hundred twenty-one in Toronto, *Diabetes*, 1972, **21, Suppl. 2**, 385–395.
Birch, A. J. Hydroaromatic steroid hormones. Part 1. 10-Nortestosterone, *J. Chem. Soc*, **1950**, 367–368.
Bliss, M. Banting's, Best's, and Collip's accounts of the discovery of insulin, *Bull. Hist. Med*, 1982, **56**, 554–568.
Borrell, M. Brown-Séquard's organotherapy and its appearance in America at the end of the nineteenth century, *Bull. Hist. Med*, 1976, **50**, 309–320.
Bouin, P. and Ancel, P. Recherches sur les fonctions du corps jaune gestatif. I. Sur le déterminisme de la préparation de l'uterus à la fixation de l'oeuf, *J. Physiol. Path. Gen*, 1910, **12**, 1.
Brown-Séquard, C. E. Expérience demonstrant la puissance dynamogenique chez l'homme d'un liquide extract de testicules d'animaux, *Arch. Physiol. Norm. Path*, 1889, **21**, 651, 740.
Butenandt, A. Über die chemische Untersuchung der Sexuallhormone, *Angew. Chem*, 1931, **44**, 905.
Butenandt, A. Über Progynon ein krystallisertes weibliches Sexualhormon, *Naturwissenschaft*, 1929, **17**, 878.
Butenandt, A. The discovery of oestrone, *Trends Biol. Sci*, 1979, **4**, 215–216.

Butenandt, A. and Westphal, U. Zur Isolierung und Charakterisierung des Corpus-luteum-Hormons, *Chem. Ber*, 1934, **67**, 1440.

Campbell, N. R., Dodds, E. C. and Lawson, W. Oestrogenic activity of Di-Anol, a derivative of *p*-propenyl-phenol, *Nature*, 1938, **141**, 78–79.

Chalmers, J. R., Dickson, G. T., Elks, J. and Hems, B. A. The synthesis of thyroxine and related substances. Part V. A synthesis of L-thyroxine from L-tyrosine, *J. Chem. Soc*, **1949**, 3424–3433.

Coindet, J. F. C. Iodine, on its application as a medicine, *Quart. J. Sci*, 1821, **11**, 408.

Collip, J. B., Anderson, E. M. and Thomson, D. L. The adrenotropic hormone of the anterior pituitary lobe, *Lancet*, 1933, **2**, 347–348.

Colton, F. B., Nysted, L. N., Riegel, B. and Raymond, A. L. 17-Alkyl-19-nortestosterone, *J. Am. Chem. Soc*, 1957, **79**, 1123–1127 [Norethandrolone].

Cook, J. W., Dodds, E. C. and Hewett, C. L. A synthetic oestrus-exciting compound, *Nature*, 1933, **131**, 56–57.

Corner, G. W. *The Hormones in Human Reproduction*, Oxford University Press, London, 1946.

Djerassi, C., Miramontes, L. E. and Rosenkranz, G. Steroids. LIV. Synthesis of 19-nor-17-α-ethynyltestosterone and 19-nor-17-α-methyltestosterone, *J. Am. Chem. Soc*, 1954, **76**, 4092–4094.

Dodds, C. Oral contraceptives: The past and future, *Clin. Pharmacol. Ther*, 1969, **10**, 147–162.

Dodds, E. C., Goldberg, L., Lawson, W. and Robinson, R. Oestrogenic activity of certain synthetic compounds, *Nature*, 1938, **141**, 247–248.

Dodds, E. C. and Lawson, W. Oestrogenic activity of *p*-hydroxy propenyl benzene (anol), *Nature*, 1937, **139**, 1068–1069.

Doisy, E. A., Veller, C. D. and Thayer, S. A. Folliculin from the urine of pregnant women, *Am. J. Physiol*, 1929, **90**, 329.

Doisy, E. A., Veler, C. D. and Thayer, S. The preparation of the crystalline ovarian hormone from the urine of pregnant women, *J. Biol. Chem*, 1930, **86**, 499–509.

Ehrenstein, M. Investigations on steroids. VIII. Lower homologs of hormones of the pregnane series: 10-nor-11-desoxy-corticosterone acetate and 10-norprogesterone, *J. Org. Chem*, 1944, **9**, 435–456.

Fellner, O. O. Experimentelle Untersuchungen über die Wirkung von Gewebsextrakten aus der Plazenta und den weiblichen Sexualorganen auf das Genitale, *Arch. Gynäk*, 1913, **100**, 641.

Fieser, L. F. and Fieser, M. *Steroids*, Chapman & Hall, London, 1959.

Fraenkel, L. Die Function des Corpus-luteum, *Arch. Gynäk*, 1903, **68**, 438.

Frank, E., Northmann, M. and Wagner, A. Synthetic compounds having an insulin-like activity in the normal and in the diabetic organism, *Klin. Wochenschr*, 1926, **5**, 2100–2107.

Frank, R. T. and Rosenbloom, J. Physiologically active substances contained in the placenta and in the corpus luteum, *Surg. Gynec. Obst*, 1915, **21**, 646.

Franke, H. and Fuchs, J. Ein neues antidiabetisches Prinzip. Ergebnisse klinischer Untersuchungen, *Deut. Med. Wochenschr*, 1955, **80**, 1449–1452.

Girard, A. and Sandulesco, G. Sur une nouvelle série de réactifs du groupe carbonyl, leur utilisation à l'extraction des substances cétoniques et la à charactérisation microchimique des aldéhydes et cétones, *Helv. Chim. Acta*, 1936, **19**, 1095.

Greenblatt, R. B., Barfield, W. E., Jungck, E. C. and Ray, A. W. Induction of ovulation with ML/41. Preliminary report, *J. Am. Med. Ass*, 1961, **178**, 101–104 [Clomiphene].

Gross, J. and Pitt-Rivers, R. 3:5:3-Triiodothyronine. 2. Physiological activity, *Biochem. J*, 1953, **53**, 652–657.

Gull, W. O. On a cretinoid state supervening in adult life in women, *Trans. Clin. Soc. Lond*, 1873–1874, **7**, 180–185.

Harper, M. J. K. and Walpole, A. L. Contrasting endocrine activities of cis and trans isomers in a series of substituted triphenylmethanes, *Nature*, 1966, **212**, 87 [Tamoxifen].

Harrington, C. R. Isolation of throxine from the thyroid gland, *Biochem. J*, 1926, **20**, 293–313.

Hartmann, M. and Wettstein, A. Ein krystallisiertes Hormon aus Corpus-luteum, *Helv. Chim. Acta*, 1934, **17**, 878.

Hughes, A. F. W. A history of endocrinology, *J. Hist. Med*, **1977**, 292–313.

Inhoffen, H. H., Logemann, W., Hohlweg, W. and Serini, A. Sex hormone series, *Chem. Ber*, 1938, **71**, 1024–1032.

Inhoffen, H. H. and Hohlweg, W. New female glandular derivatives active per os, *Naturwissen*, 1938, **26**, 96 [Ethinyloestradiol].

Iscovesco, H. Le lipoide utero-stimulant de l'ovaire, *C. R. Soc. Biol*, (*Paris*), 1912, **73**, 104.

Jones, R. G., Kornfeld, E. C. and McLaughlin, K. C. Studies on imidazolines. IV. The synthesis and antithyroid activity of some 1-substituted-2-mercaptoimidazoles, *J. Am. Chem. Soc*, 1949, **71**, 4000–4002.

Junkmann, K. Gestagens of prolonged action, *Arch. Exp. Path. Pharm*, 1954, **223**, 244–253.

Kathol, J., Logemann, W. and Serini, A. Transformation from the androstane to the pregnane series, *Naturwissenschaften*, 1937, **25**, 682.

Kendall, E. C. The isolation in crystalline form of the compound containing iodine which occurs in the thyroid; its chemical nature and physiological activity, *J. Am. Med. Ass*, 1915, **64**, 2042–2043.

Kendall, E. C. The crystalline compound containing iodine which occurs in the thyroid, *Endocrinology*, 1917, **1**, 153–169.

Kendall, E. C. *Thyroxine*, The Chemical Catalog Co., New York, 1929.

Kendall, E. C. *Cortisone*, Scribner, New York, 1971.

Knauer, E. Die ovarien Transplantation, *Arch. Gyn*äk, 1900, **60**, 322.

Koch, F. C., Moore, C. R. and Gallagher, T. F. The effect of extract of testes in correcting the castrate condition in the fowl and in mammals, *13th Int. Physiol. Congr., 1929*, abstract 148.

Kochakian, C. D. and Murlin, J. R. The effect of male hormone on the protein and energy metabolism of castrate dogs, *J. Nutrition*, 1935, **10**, 437.

Kohn, L. A. Goiter, iodine, and George. Goler: The Rochester experiment, *Bull. Hist. Med*, 1975, **49**, 389.

Laqueur, E., David, K., Dingemanse, E., Freud, J. and de Jongh, S. E. Über männliches Hormon. Unterschied von Androsteron aus Harn und Testosteron aus Testis, *Acta Brev. Neerland*, 1935, **4**, 5.

Lugol, J. G. A. *Memoire sur l'Emploi de l'Iode dans les Maladies Scrofuleuses*, Baillière, Paris, 1829.

Mackenzie, J. B., Mackenzie, C. G. and McCollum, E. V. The effect of sulphaguanidine on the thyroid of the rat, *Science*, 1941, **94**, 518–519.

Macleod, J. J., Banting, F. G. and Best, C. H. Internal secretion of the pancreas, *Amer. J. Physiol*, 1922, **59**, 479.

Macleod, J. J. R. History of the researches leading to the discovery of insulin, *Bull. Hist. Med*, 1978, **52**, 295–312.

Maisel, A. Q. *The Hormone Quest*, Random House, New York, 1965.

Marine, D. Etiology and prevention of simple goitre, *Medicine*, 1924, **3**, 453–479.

Marrian, G. F. Early work on the chemistry of pregnanediol and the oestrogenic hormones, *J. Endocrinol*, 1966, **35**, vi–xvi.

Mason, H. L., Hoehn, W. H. and Kendall, E. C. Chemical studies of the suprarenal cortex, IV. Structures of compounds C, D, E, F, and G, *J. Biol. Chem*, 1938, **124**, 459.

McGee, L. C. The effect of the injection of a lipoid fraction of bull testicle in capons, *Proc. Inst. Med. Chicago*, 1927, **6**, 242.

Mering, J. von and Minkowsky, O. Diabetes mellitus nach Pankreasextirpation, *Arch, Exp. Path. Pharm*, 1889, **26**, 371.

Murnaghan, J. H. John Jacob Abel and the crystallization of insulin, *Persp. Biol. Med*, 1967, **11**, 334–380.

Murray, G. R. Note on the treatment of myxoedema by hypodermic injections of an extract of the thyroid gland of a sheep, *Br. Med. J*, 1891, **2**, 796–797.

Murray, I. The search for insulin, *Scot. Med. J*, 1969, **14**, 286–293.

Murray, I. Paulesco and the isolation of insulin, *J. Hist. Med*, 1971, **26**, 150–157.

Noller, K. L. and Fish, C. R. Diethylstilbestrol usage: its interesting past, important present, and questionable future, *Med. Clin. N. Amer*, 1974, **58**, 793–810.

Oesper, R. Adolf Butenandt, *J. Chem. Educt*, 1949, **26**, 91.

Oesper, R. Tadeus Reichstein, *J. Chem. Educt*, 1949, **26**, 529–530.

Palopoli, F. P., Feil, V. J., Allen, R. E., Holtkamp, D. E. and Richardson, A. Jr. Substituted aminoalkoxytriarylhaloethylenes, *J. Med. Chem*, 1967, **10**, 84–86.

Paulesco, N. C. Recherche sur le rôle du pancreas dans l'assiucitation nutritive, *Arch. Int. Physiol*, 1921, **17**, 85.

Petrow, V. The contraceptive progestagens, *Chem. Revs* 1970, **70**, 713–726.

Pincus, G. Progestational agents and the control of fertility, *Vitam. Horm. (N.Y.)* 1959, **17**, 307.

Pitt-Rivers, R. The thyroid hormones: historical aspects, *Hormonal Proteins and Peptides. Vol. 6* (ed. C. H. Li) Academic Press, New York, 1978, p. 391–422.

Querido, A. History of iodine prophylaxis with regard to cretinism and deaf-mutism, *Adv. Exp. Med*, 1972, **30**, 191–199.

Richards, D. W. The Effects of Pancreas Extract on Depancreatized Dogs. Ernest L. Scott's thesis of 1911, *Persp. Biol. Med*, 1966, **10**, 84–95.

Robson, J. M., Schonberg, A. and Fahim, H. A. Duration of action of natural and synthetic oestrogens, *Nature*, 1938, **142**, 292–293.

Robson, J. M. and Schonberg, A. A new synthetic oestrogen with prolonged action when given orally, *Nature*, 1942, **150**, 22–23.

Ruzicka, L., Goldberg, M. W., Meyer, J., Brungger, H. and Eichenbergr, F. Über die Synthesis des Testikelhormons (Androsteron) und Stereoisomerer desselben durch Abbau hydrierter Sterine, *Helv. Chim. Acta*, 1934, **17**, 1395.

Ruzicka, L. and Wettstein, A. Über die künstliche Herstellung des Testikelhormons Testosteron (Androsten-3-on-17-ol), *Helv. Chim. Acta*, 1935, **18**, 1264.

Ruzicka, L. and Hofmann, K. Über die Anlagerung von Acetylen an die 17-standige Ketogrupe bei trans-Androsteron und Δ^5-trans-Dehydro-androsteron, *Helv. Chim. Acta,* 1937, **20**, 1280–1282.

Schwenk, E. and Hildebrandt, F. Ein neues isomeres Follikelhormon aus Stutenharn, *Naturwissenschaft*, 1932, **20**, 658.

Scott, E. L. On the influence of intravenous injections of an extract of the pancreas on experimental pancreatic diabetes, *Am. J. Physiol*, 1912, **29**, 306.

Shapiro, S. L., Parino, V. and Freedman, L. Hypoglycemic agents. I. Chemical properties of beta-phenethylbiguanide. A new hypoglycemic agent, *J. Am. Chem. Soc*, 1959, **81**, 2220–2225 [Phenformin].

Shapiro, S. L., Parino, V. and Freedman, L. Hypoglycemic agents. III. *N*-Alkyl and aralkylbiguanides, *J. Am. Chem. Soc*, 1959, **81**, 3728–3736 [Metformin].

Slotta, K., Ruschig, H. and Fels, E. Reindarstellung der Hormone aus dem Corpus-luteum, *Chem. Ber*, 1934, **67**, 1270.

Slotta, R. and Tschesche, R. Die blutzucker-senkende Wirkung der Biguanides, *Chem. Ber*, 1929, **62**, 1398–1405.

Smith, H., Hughes, G. A., Douglas, G. H., Hartley, D., McLoughlin, B. J., Siddall, J. B., Wendt, G. R., Buzby, G. C., Jr, Herst, D. R., Ledig, K. W., McMenamin, J. R.,

Pattison, T. W., Suida, J., Tokolics, J., Edgren, R. A., *et al.* Totally synthetic (±)-13-alkyl-3-hydroxy and methoxy-gona-1,3,5(10)-trien-17-ones and related compounds, *Experientia*, 1963, **19**, 394–396.
Solmssen, V. Synthetic estrogens and the relationship between their structure and their activity, *Chem. Rev*, 1945, **37**, 481–598.
Thorp, J. M. and Waring, W. S. Modification of metabolism and distribution of lipids by ethyl chlorophenoxyisobutyrate, *Nature*, 1962, **194**, 948–949.
Ungar, G., Freedman, L. and Shapiro, S. L. Pharmacological study of a new oral hypoglycaemic drug, *Proc. Soc. Exp. Biol. Med*, 1957, **95**, 190–192 [Metformin, phenformin].
Vigneaud, V. du. Trail of sulfur research: from insulin to oxytocin, *Science*, 1956, **123**, 967–974.
Watanabe, C. K. Studies in the metabolic changes induced by the administration of guanidine bases. I. The influence of injected guanidine hydrochloride upon blood sugar content, *J. Biol. Chem*, 1918, **33**, 253–265.
Wrenshall, G. A., Hetenyi, G. and Feasby, W. R. *The Story of Insulin*, Bodley Head, London, 1962.
Zuelzer, G. Ueber Versuch einer specifischen Fermenttherapie des Diabetes, *Z. Exp. Path. Ther*, 1908, **5**, 307.

CHAPTER TWELVE

Andersag, H. and Westphal, K. Über die Synthesis des antineuritische Vitamins, *Chem. Ber*, 1937, **70**, 2035–2054.
Baldwin, E. Frederick Gowland Hopkins, *Dictionary of Scientific Biography*, vol. 6, p. 498–502.
Becker, S. L. Elmer Verner McCollum, *Dictionary of Scientific Biography*, vol. 8, p. 590–591.
Chick, H. The discovery of vitamins, *Prog. Food Nutrition*, 1975, **1**, 1–20.
Chick, H. and Roscoe, M. The dual nature of water-soluble vitamin B. II. The effect upon young rats of vitamin B_2 deficiency and a method for the biological assay of vitamin B_2, *Biochem. J*, 1928, **22**, 790–799.
Cline, J. K. Williams, R. R. and Finkelstein, J. Studies of crystalline vitamin B_1. Synthesis of vitamin B_1, *J. Am. Chem. Soc*, 1937, **59**, 1052–1054.
Corner, G. W. *George Hoyt Whipple and his Friends: the Life-story of a Nobel Prize Pathologist*, Lippincott, Philadelphia, 1963.
Drummond, J. C. The nomenclature of the so-called accessory food factors (vitamins), *Biochem. J*, 1920, **14**, 660.
Eijkman, C. *Geneesk, Tijdschr. Nederland.—Indie*, 1890, **30**, 295.
Eijkman, C. Eine Beri Beri-ahnliche Krankheit der Huhner, *Virchow's Arch. Path. Anat*, 1897, **148**, 523–532.
Elvehjem, C. A., Madden, R. J., Strong, F. M. and Woolley, D. W. The isolation and identification of the anti-black tongue factor, *J. Biol. Chem*, 1938, **123**, 137–149.
Funk, C. The etiology of the deficiency diseases, *J. State Med*, 1912, **20**, 341–368.
Goldblith, S. A. and Joslyn, M. A. *Milestones in Nutrition*, Avi Publishing, Westport, 1964.
Grijns, G. *Geneesk. Tijdschr. Nederland.—Indie*, 1901, **41**, 3.
György, P. Reminiscences on the discovery and significance of some of the B vitamins, *J. Nutrition*, 1967, **91: Suppl. 1**, 5–9.
Harden, A. and Zilva, S. S. Antiscorbutic factor in lemon juice, *Biochem. J*, 1918, **12**, 259–269.
Hirst, E. L. and Zilva, S. S. Ascorbic acid as the anti-scorbutic factor, *Biochem. J*, 1933, **27**, 1271–1278.

Holst, A. and Frolich, T. Experimental studies relating to ship beriberi and scurvy, *J. Hygiene*, 1907, **7**, 634.
Hopkins, F. G. Feeding experiments illustrating the importance of accessory factors in normal dietaries, *J. Physiol*, 1912, **44**, 425–460.
Hopkins, F. G. The analyst and the medical man, *Analyst*, (*London*), 1906, **31**, 385.
Hutchings, B. L., Stockstad, E. L. R., Bohonos, N. and Slobodkin, N. H. Isolation of a new Lactobacillus casei factor, *Science*, 1944, **99**, 371.
Jansen, B. C. Early nutritional researches on beriberi leading to the discovery of vitamin B_1, *Nutrition Abstracts and Reviews*, 1956, **26**, 1–14.
Jansen, B. C. P. and Donath, W. F. The isolation of the anti-beri beri vitamin, *Geneesk. Tidskr. Nederland.—Indie*, 1927, **66**, 810–827.
Jukes, T. H. Vitamin K—a reminiscence, *Trends Biol. Sci*, 1980, **5**, 140–141.
Jukes, T. H. The discovery of folic acid, *Trends Biol. Sci*, 1980, **5**, 112–113.
Kuhn, R., György, P. and Wagner-Jauregg, T. Über eine neue Klasse von Naturfarbstoffen, *Chem. Ber*, 1933, **66**, 317–320.
Lepkovsky, S. The isolation of factor one in crystalline form, *J. Biol. Chem*, 1938, **124**, 125–128.
Lepkovsky, S., Jukes, T. H. and Krause, M. E. The multiple nature of the third factor of the vitamin B complex, *J. Biol. Chem*, 1936, **115**, 557–566.
Lind, J. *Treatise on Scurvy* [containing a reprint of the first edition of A Treatise of the Scurvy, with additional notes] (ed. C. P. Stewart, D. Guthrie), University Press, Edinburgh, 1953.
Lindeboom, G. Christiaan Eijkman, *Dictionary of Scientific Biography*, Vol. 4, p. 310–312.
Lorenz, A. Some pre-Lind writers on scurvy, *Proc. Nutrition Soc*, 1953, **12**, 306.
Lorenz, A. J. The conquest of scurvy, *J. Amer. Dietetic Ass*, 1954, **30**, 665–670.
Lunin, N. Ueber die Bedeutung der anorganischen Salze fur die Ernahrung des Thieres, *Z. Physiol. Chem*, 1881, **5**, 31–39.
Lusk, G. A history of metabolism, *Endocrinology and Metabolism*, Vol. 3, Appleton, New York, 1922, p. 3–78.
McCollum, E. V. The necessity of certain lipins in the diet during growth, *J. Biol. Chem*, 1913, **15**, 167–175.
McCollum, E. V., Simmonds, N. and Becker, J. E. Studies on experimental rickets. XXI. An experimental demonstration of the existence of a vitamin which promotes calcium deposition, *J. Biol. Chem*, 1922, **53**, 293–312.
McCollum, E. V. *A History of Nutrition*, Houghton Mifflin, Boston, 1957.
McCollum, E. V. The paths to the discovery of vitamins A and D, *J. Nutrition*, 1967, **91: Suppl.** 1, 11–16.
Mellanby, E. A further demonstration of the part played by accessory factor in the aetiology of rickets, Proc. Physiol. Soc, 14th December 1918, lii–liv.
Minot, G. R. and Murphy, W. P. A diet rich in liver in the treatment of pernicious anemia, *J. Am. Med. Ass*, 1927, **89**, 759–766.
Mitchell, H. H. On the identity of the water-soluble growth-promoting vitamin and the antineuritic vitamin, *J. Biol. Chem*, 1919, **40**, 399–413.
Morton, R. A. The history of vitamin research. Selected aspects, *Int. Z. Vit. Forschung*, 1968, **38**, 5–44.
Pekelharing, C. A. *Geneesk. Tijdschr. Nederland.—Indie*, 1905, **2**, 3.
Pfiffner, J. J., Binkley, S. B., Bloom, E. S., Brown, R. A., Bird, O. P., Emmett, A. D., Hogan, A. G. and O'Dell, B. L. Isolation of the antianemic factor (vitamin B_c) in crystalline form, *Science*, 1943, **97**, 404–405.
Roddin, L. H. *James Lind, Founder of Nautical Medicine*, Schuman, New York, 1950.
Sherman, H. C. and Smith, S. L. *The Vitamins*, The Chemical Catalog Co., New York, 1931.
Snell, E. E. and Peterson, W. H. Growth factor for bacteria. X. Additional factor required by certain lactic acid bacteria, *J. Bacteriol*, 1940, **39**, 273–285.

Svirbely, J. L. and Szent-Györgyi, A. The chemical nature of vitamin C, *Biochem. J*, 1932, **26**, 865–870; **27**, 279–285.
Todd, A. R. and Bergel, F. Aneurin. Part VII. A synthesis of aneurin, *J. Chem. Soc*, **1937**, 364–367.
Waugh, W. A. Unlocking another door to nature's secrets – vitamin C, *J. Chem. Educt*, 1934, **11**, 69–72.
Waugh, W. A. and King, C. G. Isolation and identification of vitamin C, *J. Biol. Chem*, 1932, **97**, 325–331.
Whipple, G. H. and Robscheit-Robbins, F. S. Blood regeneration in severe anemia II. Favourable influence of liver, heart, and skeletal muscle in diet, *Am. J. Physiol*, 1925, **72**, 395–407.
Woolley, D. W., Waisman, H. A. and Elvehjem, C. A. Nature and partial synthesis of the chick antidermatitis factor, *J. Am. Chem. Soc*, 1939, **61**, 977–978.
Zilva, S. S. The antiscorbutic factor of lemon juice. II, *Biochem. J*, 1924, **18**, 632–637.

CHAPTER THIRTEEN

Browning, C. H., The chemotherapy of trypanosomiasis, *Ann. N. Y. Acad. Sci*, 1954, **59**, 198–213.
Browning, C. H. Paul Ehrlich – memories of 1905–1907, *Br. Med. J*, 1954, **1**, 664–665.
Browning, C. H. Exerpts from the history of chemotherapy, *Scot. Med. J*, 1967, **12**, 310–313.
Collier, H. O. J. *Chemotherapy of Infections*, Chapman and Hall, London, 1954.
Cowen, D. L. Ehrlich the man, the scientist, *Am. J. Pharm. Ed*, 1962, **26**, 4–11.
Dale, H. H. Paul Ehrlich, *Br. Med. J*, 1954, **1**, 659–663.
Dale, H. H. *The Collected Papers of Paul Ehrlich, Vol. 3*, (ed. F. Himmelweit), Pergamon, London, 1960, p. 1–8.
Dolman, C. Paul Ehrlich, *Dictionary of Scientific Biography* vol. 4, p. 295–305.
Dowling, H. F. Comparison and contrasts between the early arsphenamine and early antibiotic periods, *Bull. Hist. Med*, 1973, **47**, 236–249.
Dressel, J. (translated by R. E. Oesper). The discovery of Germanin by Oskar Dressel and Richard Kothe, *J. Chem. Educt*, 1961, **38**, 620–621.
Duthie, E. S. *Molecules Against Bacteria*, Sigma, London, 1946.
Ehrlich, P. Closing Notes to The Experimental Chemotherapy of Spirilloses, *The Collected Papers of Paul Ehrlich, Vol. 3*, (ed. F. Himmelweit), Pergamon, London, 1960, p. 282–309.
Elderfield, R. C., Gensler, W. J., Head, J. D., Hageman, H. A., Kremer, C. B., Wright, J. B., Holley, A. D., Williamson, B., Galbreath, J., Wiederhold, L., Frohardt, R., Kupchan, S. M., Williamson, T. A. and Birstein, O. Alkylaminoalkyl derivatives of 8-aminoquinoline, *J. Am. Chem. Soc*, 1946, **68**, 1524 [Primaquine].
Fischl, V. and Schlossberger, H. *Handbook of Chemotherapy*, H. G. Roebuck, Philadelphia, 1936.
Galdstone, I. *Behind the Sulfa Drugs*, Appleton-Century, New York, 1943.
Himmelweit, F. (ed.) *The Collected Papers of Paul Ehrlich, Vol. 3*, Pergamon, London, 1960.
Hitchings, G. H. Daraprim as an antagonist of folic and folinic acid, *Trans. Roy. Soc. Trop. Med. Hyg*, 1952, **46**, 467–473.
Jancsó, von N. and Jancsó von H. Chemotherapeutische Schnellfestigung von Trypanosomen durch Ausschaltung der natürlichen Abwehrkräfte, *Z. Immun. Forschr*, 1935, **85**, 81.
Kikuth, W. Zur Weiterentwicklung synthetisch dargestellter Malariamittel. I. Über die chemotherapeutische Wirkung des Atebrin, *Deut. Med. Wochenschr*, 1932, **58**, 520, 530–531.
Kruif, P. de Paul Ehrlich: the magic bullet, *Microbe Hunters*, Jonathen Cape, London, 1930, p. 295–320.

Lewis, W. L. and Stiegler, H. W. The beta-chlorovinyl-arsines and their derivatives, *J. Am. Chem. Soc*, 1925, **47**, 2546–2556.
Livingstone, D. Arsenic as a remedy for the tse-tse bite, *Br. Med. J*, 1858, **1**, 360–361.
Lourie, E. M. and Yorke, W. The trypanocidal action of certain aromatic diamidines, *Ann. Trop. Med*, 1939, **33**, 289–304.
Marquardt, M. *Paul Ehrlich*, Heinemann, London, 1949.
Marquardt, M. Paul Ehrlich. Some reminiscences, *Br. Med. J*, 1954, **1**, 665–667.
McMahon, J. E. History of chemotherapy of bilharziasis. Development of schistosomicides, *East African Med. J*, 1976, **53**, 295–299.
Morrison, H. Carl Weigert, *Ann. Med. Hist*, 1924, **6**, 163–177.
Nicolle, P. La vie, la personnalité et l'oeuvre de Maurice Nicolle, *Bull. Soc. Pathol. Exotique*, 1962, **55**, 188–200.
Parascandola, J. The theoretical basis of Paul Ehrlich's chemotherapy, *J. Hist. Med*, **1981**, 19–43.
Pereira, J. Arsenious acid, *The Elements of Materia Medica and Therapeutics, Vol. 1*, Longman, Brown, Green, and Longmans, London, 4th edn, 1854, p. 682–721.
Peters, R. A., Stocken, L. A. and Thompson, R. H. S. British anti-Lewisite, *Nature*, 1945, **156**, 616–619.
Pinkus, H. Paul Ehrlich and his impact on dermatosyphilology, *A. M. A. Arch. Dermatol*, 1955, **72**, 113–119.
Poindexter, H. A. Observations on the defense mechanism in Trypanosoma equiperdum and Trypanosoma lewisi infections in guinea pigs and rats, *Am. J. Trop. Med*, 1933, **13**, 555–575.
Poindexter, H. A. Further observations on the relation of certain carbohydrates to Trypanosoma equiperdum metabolism, *Parasitol*, 1935, **21**, 292–301.
Procopio, J. Harold Wolferstan Thomas: cientista canadense a servico de medecina no Amazonas, *Revista Brasileira de Medicina* 1953, **10**, 371–374.
Richards, H. C. Oxamniquine, *Chronicles of Drug Discovery*, (ed. J. S. Bindra, D. Lednicer) John Wiley, Chichester, 1982.
Rose, F. L. R. A chemotherapeutic search in retrospect, *J. Chem. Soc*, **1951**, 2770–2788. [Proguanil].
Schulemann, W. Synthetic anti-malarial preparations, *Proc. Roy. Soc. Med*, 1932, **25**, 897–905.
Sonnedecker, G. The concept of chemotherapy, *Am. J. Pharm*, 1962, **26**, 1–3.
Surrey, A. and Hammer, H. F. Some 7-substituted 4-aminoquinoline derivatives, *J. Am. Chem. Soc*, 1946, **68**, 113–116 [Chloroquine].
Thomas, H. W. Some experiments in the treatment of trypanosomiasis, *Br. Med. J*, 1905, **1**, 1140–1141.
Thomas, H. W. and Breinl, A. The experimental treatment of trypanosomiasis in animals, *Proc. Roy. Soc., Ser. B*, 1905, 76, 513.
Thomson, A. T. Acidum Arseniosum, *The London Dispensatory*, Longmans, London, 1837, p. 217.
Tréfouël, J. Le rôle de Maurice Nicolle en chimotherapie anti-trypanosomes, *Bull. Soc. Pathol. Exot*, 1962, **55**, 200–207.
Ward, P. S. The American reception of Salvarsan, *J. Hist. Med*, 1981, 44–62.

CHAPTER FOURTEEN

Albert, A. *The Acridines*, Arnold, London, 1951.
Benda, L. Über das 3,6-Diamino-acridin, *Chem. Ber*, 1912, **45**, 1787–1799.
Browning, C. H. Exerpts from the history of chemotherapy, *Scot. Med. J*, 1967, **12**, 310–313.
Browning, C. H., Gulbransen, R., Kennaway, E. L. and Thornton, L. H. D. Flavine

and brilliant green. Powerful antiseptics with low toxicity to the tissues: their use in infected wounds, *Br. Med. J*, 1917, **1**, 73–76.

Browning, C. H. and Gilmour, W. Bactericidal action and chemical constitution of basic benzol derivatives, *J. Path. Bact*, 1913, **18**, 144.

Burchall, J. J. and Hitchings, G. H. Inhibitor binding analysis of dihydrofolate reductases from various species, *Molec. Pharmacol*, 1965, **1**, 126–136 [Trimethoprim].

Caldwell, W. T., Kornfeld, E. C. and Donnelly, C. K. Substituted 2-sulphamidopyrimidines, *J. Am. Chem. Soc*, 1941, **63**, 2188–2190.

Churchman, J. The selective bactericidal action of gentian violet, *J. Exp. Med*, 1912, **16**, 221–247.

Clark, J. H., English, J. P., Jansen, G. R., Mason, H. W., Rogers, M. M. and Taft, W. E. 3-Sulfanilamido-6-alkoxypyridazines and related compounds, *J. Am. Chem. Soc*, 1958, **80**, 980–983 [Sulphamethoxypyridazine].

Colebrook, L. Gerhard Domagk, *Biographical Memoirs of Fellows of the Royal Society*, 1964, **10**, 39–50.

Colebrook, L. and Kenny, M. Treatment of human puerperal infections, and of experimental infections in mice, with Prontosil, *Lancet*, 1936, **1**, 1279–1286.

Dohrn, M. and Diedrich, P. Albucid, a new sulfanilic acid derivative, *Munch. Med. Wochenschr*, 1938, **85**, 2017–2018.

Domagk, G. Ein Beitrag zur Chemotherapie der bakteriellen Infektionen, *Deut. Med. Wochenschr*, 1935, **61**, 250.

Dutcher, J. The discovery and development of amphotericin B, *Dis. Chest*, 1968, **54**, Suppl. 1, 296–298.

Eisenberg, P. The action of dyes on bacteria, and the effect of vital staining on growth of bacteria, *Zbl. Bakteriol., Parasiten. u. Infektkr*, 1913, **71**, 420.

Feldt, A. Zur Chemotherapy der Tuberkulose mit Gold, *Deut. Med. Wochenschr*, 1913, **39**, 549–551.

Foerster, *Zentrallbl. f. Haut u. Geschlechtskr*, 1933, **45**, 549 [Prontosil].

Fosbinder, R. and Walter, L. A. Sulphanilamido derivatives of heterocyclic amines, *J. Am. Chem. Soc*, 1939, **16**, 2032–2033 [Sulphathiazole].

Fuller, A. T. Is *p*-aminobenzenesulphonamide the active agent in Prontosil therapy? *Lancet*, 1937, **1**, 194–198.

Galdstone, I. *Behind the Sulfa Drugs*, Appleton-Century, New York, 1943.

Geiling, E. M. and Cannon, P. R. Pathologic effects of Elixir of Sulfanilamide (diethylene glycol) poisoning, *J. Am. Med. Ass*, 1938, **111**, 919–926.

Hitchings, G. H. Daraprim as an antagonist of folic and folinic acid, *Trans. Roy. Soc. Trop. Med. Hyg*, 1952, **46**, 467–473.

Hitchings, G. H. Chemotherapy and comparative biochemistry, *Cancer Res*, 1969, **29**, 1895–1903.

Hitchings, G. H., Falco, E. A. and Sherwood, M. B. The effect of pyrimidines on the growth of L. casei, *Science*, 1945, **102**, 251–252.

Hitchings, G. H., Elion, G. B., VanderWerff, H. and Falco, E. A. Pyrimidine derivatives as antagonists of P.G.A. *J. Biol. Chem*, 1948, **174**, 765–766.

Horlein, H. The development of chemotherapy for bacterial diseases, *Practioner*, 1937, **139**, 635–649.

Horlein, H. The chemotherapy of infectious disease caused by protozoa and bacteria, *Proc. Roy. Soc. Med*, 1936, **29**, 313–324.

Jacobs, W. A. and Heidelberger, M. Syntheses in the Cinchona series. II. Quaternary salts, *J. Am. Chem. Soc*, 1919, **41**, 2090–2120.

Levaditi, C. and Vaisman, A. Actions curative et preventive du chlorhydrate de 4′-sulfamido-2,4-diamino-azobenzene dans l'infection streptococcique experimental, *C. R. Hebd. Séances Acad. Sci*, 1935, **200**, 1694–1696.

Long, P. H. and Bliss, E. A. *The Clinical and Experimental Use of Sulfanilamide, Sulfapyridine and Allied Compounds*, Macmillan, New York, 1939, p. 1–13.

Mietzsch, F. Zur Chemotherapie der bakteriellien Infektionskrankheiten, *Chem. Ber*, 1938, **71**, 15–28.
Morgenroth, J. and Kaufmann, M. The disinfectant action of quinine alkaloids on streptococci, *Berl. Klin. Wochenschr*, 1916, **53**, 794–796.
Morgenroth, J. and Levy, R. Chemotherapie der Pneumokokkeninfektion, *Berl. Klin. Wochenschr*, 1911, **48**, 1560.
Oesper, R. Gerhard Domagk and chemotherapy, *J. Chem. Educt*, 1954, **31**, 188–191.
Ostromislensky, I. Note on bacteriostatic azo compounds, *J. Am. Chem. Soc*. 1934, **56**, 1713–1714 [Pyridium].
Posner, E. Gerhard Domagk, *Dictionary of Scientific Biography*, vol. 4, p. 153–156.
Rich, A. R. and Follis, R. H. Jr. The inhibitory effect of sulfanilamide on the development of experimental tuberculosis in the guinea pig, *Bull. Johns Hopkins Hosp*, 1938, **62**, 77–84.
Roblin, R. O., Williams, J. H., Winner, P. S. and English, J. P. Some sulphanilamido heterocycles, *J. Am. Chem. Soc*, 1940, **62**, 2002–2005 [Sulphadiazine].
Schnitzer, R. Chemotherapie bacterieller Infektionen, *Deut. Med. Wochenschr*, 1929, **55**, 1888–1889.
Schnitzer, R. J. Chemotherapy of bacterial infections, *Ann. N. Y. Acad. Sci*, 1954, **59**, 227–242.
Tréfouël, J., Tréfouël, J., Nitti, F. and Bovet, D. Activité de *p*-aminophenylsulfamide sur les infections streptococciques de la souris et du lapin, *C. R. Soc. Biol*, (*Paris*), 1935, **120**, 756–758.
Whitby, L. E. H. Chemotherapy of pneumococcal and other infections with 2-(*p*-aminobenzenesulphonamido)pyridine, *Lancet*, 1938, **1**, 1210–1212 [Sulphapyridine].
Woods, D. D. The relation of *p*-aminobenzoic acid to the mechanism of the action of sulphonamide, *Br. J. Exp. Path*, 1940, **21**, 74–90.
Woods, D. D. and Fildes, P. The anti-sulphanilamide activity (in vitro) of *p*-aminobenzoic acid and related compounds, *Chem. Ind*, (*London*), 1940, **59**, 133–134.

CHAPTER FIFTEEN

Abraham, E. P. Cephaloridine: historical remarks, *Postgrad. Med. J*, 1967, **43**, Suppl. 9–10.
Abraham, E. P., Chain, E., Fletcher, C. M., Gardner, A. D., Heatley, N. G., Jennings, M. A. and Florey, H. W. Further observations on penicillin, *Lancet*, 1941, **2**, 177–189.
Abraham, E. P. and Chain, E. An enzyme from bacteria able to destroy penicillin, *Nature*, 1940, **146**, 837.
Abraham, E. P. and Loder, P. B. *Cephalosporins and Penicillins*, (ed. E. H. Flynn) Academic Press, New York, 1972.
Acred, P., Brown, D. M., Knudson, E. T., Rolinson, G. N. and Sutherland, R. New semi-synthetic penicillins active against Pseudomonas pyocyanea, *Nature*, 1967, **215**, 25–30 [Carbenicillin].
Alkiewicz, J., Eckstein, Z., Halweg, H., Krakowka, P. and Urbanski, T. Fungicidal activity of some hydroxamic acids, *Nature*, 1957, **180**, 1204–1205.
Baldry, P. *The Battle Against Bacteria. A Fresh Look*, Cambridge University Press, London, 1976.
Bartz, Q. Isolation and characterization of Chloromycetin, *J, Biol. Chem*, 1948, **172**, 445–450.
Batchelor, F. R., Doyle, F. P., Nayler, J. H. and Rollinson, G. N. Synthesis of penicillin: 6-aminopenicillanic acid in penicillin fermentation, *Nature*, 1959, **183**, 257–258.
Brian, P. W., Hemming, H. G. and McGowan, J. C. Origin of a toxicity to Mycorrhiza in Wareham Heath, *Nature*, 1945, **155**, 637–638.

Brunel, J. Antibiosis from Pasteur to Fleming, *J. Hist. Med*, 1951, **6**, 287–301.

Calder, R. Science in industry: as prescribed, *Discovery*, 1976, **21**, 541–545 [Griseofulvin].

Cantani, A. Tentavi di bacterioterapia, *Centr. Med. Wiss*, 1885, **23**, 513.

Chain, E. B., Florey, H. W. Gardner, A. D., Heatley, N. G., Jennings, M. A. Orr-Ewing, J. and Sanders, A. G. Penicillin as a chemotherapeutic agent, *Lancet*, 1940, **2**, 226–228.

Chauvette, R. R., Pennington, P. A., Ryan, C. W., Cooper, R. D. G., Jose, F. L., Wright, I. G., Van Heyningen, E. M. and Huffman, G. W. Conversion of penicillins to cephalexin, *J. Org. Chem*, 1971, **36**, 1259–1263.

Controulis, J., Rebstock, M. C. and Crooks, H. M. Jr. Chloramphenicol (Chlormycetin). V. Synthesis, *J. Am. Chem. Soc*, 1949, **71**, 2463–2468.

Doyle, F. P., Long, A. A., Nayler, J. H. C. and Stove, E. P. New penicillins stable towards both acid and penicillinase, *Nature*, 1961, **192**, 1183–1184 [Oxacillin].

Doyle, F. P., Nayler, J. H., Smith, H. and Stove, E. R. Some novel acid-stable penicillins, *Nature*, 1961, **191**, 1091–1092 [Ampicillin].

Dubos, R. Bactericidal agent extracted from a soil bacillus. Preparation of the agent. Its activity in vitro, *J. Exp. Med*, 1939, **70**, 1–10 [Tyrothricin].

Dubos, R. and Cettaneo, C. Bactericidal agent extracted from a soil bacillus. III Preparation and activity of a protein-free fraction, *J. Exp. Med*, 1939, **70**, 249–256 [Tyrothricin].

Duchesne, E. *Contribution a l'*étudé de la concurrence vitale chez les micro-organismes; antagonisme entre les microbes, Lyon, 1897.

Duggar, B. M. Aureomycin, a product of the continuing search for new antibiotics, *Ann. N. Y. Acad. Sci*, 1948, **51**, 177–181.

Ehrlich, J., Bartz, Q. R., Smith, R. M., Joslyn, D. A. and Burkholder, P. R. Chloromycetin, a new antibiotic from a soil actinomycete, *Science*, 1947, **106**, 417.

Elder, A. E. *The History of Penicillin Production*, American Institute of Chemical Engineers, New York, 1970.

Finlay, A. C., Hobby, G. L., P'an, S. Y., Regna, P. P., Routien, J. P., Seely, D. B., Shull, G. M., Sobin, B. A., Solomons, I. A., Vinson, J. W. and Kane, J. H. Terramycin, a new antibiotic, *Science*, 1950, **111**, 85.

Fleming, A. On the antibacterial action of cultures of a penicillium, with special reference to their use in the isolation of B. Influenzae, *Br. J. Exp. Med*, 1929, **10**, 226–236.

Florey, H. W., Chain, E. B., Heatley, N. G., Jennings, M. A., Sanders, A. G., Abraham, E. P. and Florey, M. E. *Antibiotics,* Vol. 1, Oxford University Press, 1949, p. 1–73.

Florey, H. W., Chain, E. B., Heatley, N. G. Jennings, M. A., Sanders, A. G., Abraham, E. P. and Florey, M. E. *Antibiotics,* Vol. 2, Oxford University Press, 1949, p. 631–671.

Florey, H. W. and Abraham, E. P. The work on penicillin at Oxford, *J. Hist. Med*, 1951, **6**, 302–317.

Fraser-Moodie, W. Struggle against infection, *Proc. Roy. Soc. Med*, 1971, **64**, 87.

Freudenreich, E. du. De l'antagonisme des bacteries et de l'immunite qu'il confere aux milieux de culture, *Ann. Inst. Pasteur*, 1888, **2**, 200–206.

Garrod, L. P., Lambert, H. P. L. and O'Grady, F. *Antibiotics and Chemotherapy*, Livingstone, Edinburgh, 1973.

Gentles, J. C. Experimental ringworm in guinea pigs: oral treatment with griseofulvin, *Nature*, 1958, **182**, 476–477.

Gerzon, K., Flynn, E. H., Sigal, M. V., Wiley, P. F., Monahan, R. and Quarek, U. C. Erythromycin. VIII. Structure of dihydroerythrronolide, *J. Am. Chem. Soc*, 1956, **78**, 6396.

Goldsmith, M. *The Road to Penicillin*, Drummond, London, 1946.
Gosio, B. Ricerch batteriologische e chimiche sulle alterazioni del mais. Contributo all'e etiologia della pellagra. (Memoria 2a), *Riv. Igiene Sanit. Publ*, 1896, **7**, 825.
Gratia, A. and Dath, S. Moisissures et microbes Bacteriophages, *C. R. Soc. Biol*, (*Paris*), 1925, **92**, 461.
Grove, J. F. Griseofulvin, *Quart. Revs*, **1963**, 1–19.
Hare, R. *The Birth of Penicillin*, Allen and Unwin, London, 1970.
Hitchings, G. H. Chemotherapy and comparative biochemistry, *Cancer Res*, 1969, **29**, 1895–1903.
Joffe, J. S. Early observations on antibiotic substances in Penicillium glaucum and other organisms against a virus, *Science*, 1945, **102**, 623.
Lazell, H. G. *From Pills to Penicillin. The Beecham Story*, Heinemann, London, 1975.
Lister, J. A contribution to the germ theory of putrefaction and other fermentative changes, and to the natural history of torulae and bacteria, *Trans. Roy. Soc. Edin*, 1875, **27**, 313.
Macfarlane, G. *Alexander Fleming: The Man and the Myth*, Chatto and Windus, London, 1984.
Macfarlane, G. *Howard Florey. The Making of a Great Scientist*, Oxford University Press, 1979.
Masters, D. *Miracle Drug*, Eyre & Spotiswood, 1946.
McKeen, J. E. The role of industry in the mass production of antibiotics, *The Impact of the Antibiotics on Medicine and Surgery*, (ed. I. Goldstein) International University Press, New York, 1958, p. 88–97.
Minieri, P. P., Firman, M. C., Mistretta, A. G., Abbey, A. A., Bricker, C. F., Rigler, N. E. and Sokol, H. A new broad spectrum antibiotic product of the tetracycline group, *Antibiot. Ann. 1953–1954*, Proc. Symposium Antibiotics (Washington D.C.) 1958, 81–87.
Morin, R. B., Jackson, B. G., Flynn, E. H. and Roeske, R. W. Chemistry of cephalosporin antibiotics. I 7-Aminocephalosporanic acid from cephalosporin C, *J. Am. Chem. Soc*, 1962, **84**, 3400–3401 [Cephalothin].
Newton, G. G. F. and Abraham, E. P. Cephalosporin C, a new antibiotic containing sulphur and D-alpha-aminoadipic acid, *Nature*, 1955, **175**, 548.
Pasteur, L. and Joubert, J. Charbon et septicemie, *C. R. Hebd. Séances Acad. Sci*, 1877, **85**, 101.
Perron, Y. G., Minor, W. F., Holdrege, C. T., Gottstein, W. J., Godfrey, J. C., Crast, L. B., Babel, R. B. and Chene, L. Derivatives of 6-aminopenicillanic acid. I. Partially synthetic penicillins prepared from alpha-aryloxyalkanoic acids, *J. Am. Chem. Soc*, 1960, **82**, 3934–3935 [Phenethicillin].
Richards, A. N. Production of penicillin in the United States (1941–1946), *Nature*, 1964, **201**, 441–445.
Schatz, A., Bugie, E. and Waksman, S. A. Streptomycin, a substance exhibiting antibiotic activity against Gram positive and Gram negative bacteria, *Proc. Soc. Exp. Biol. Med*, 1944, **55**, 66–69.
Selwyn, S. The discovery and evolution of the penicillins and cephalosporins, *The beta-Lactam Antibiotics: Penicillins and Cephalosporins in Perspective*, Hodder and Stoughton, London, 1980, p. 1–55.
Sensi, P. Rifampicin, *Chronicles of Drug Discovery*, (ed. J. S. Bindra, D. Lednicer) John Wiley, Chichester, 1982.
Sheehan, J. C. *The Enchanted Ring. The Untold Story of Penicillin*, MIT Press, London, 1982.
Sheehan, J. C. and Henry-Logan, K. R. The total synthesis of penicillin, *J. Am. Chem. Soc*, 1957, **79**, 1262–1263.
Soper, Q. F., Whitehead, C. W., Behrens, O. K., Corse, J. J. and Jones, R. G. Biosynthesis of penicillins. VII Oxy- and mercaptoacetic acids, *J. Am. Chem. Soc*, 1948, **70**, 2849–2855 [Phenoxymethylpenicillin].

Spencer, J. L., Flynn, E. H., Roeske, R. W., Siu, F. Y. and Chauvette, R. R. Synthesis of cephaloglycin and some homologs, *J. Med. Chem*, 1966, **6**, 746–748.

Stevenson, L. G. Antibacterial and antibiotic concepts in early bacteriological studies and in Ehrlich's chemotherapy, *The Impact of the Antibiotics on Medicine and Surgery* (ed. I. Goldstein) International University Press, New York, 1958, p. 38–57.

Sturli, A. Ueber ein in schimmelpilzen (Penicillium glaucum) Vorkommen gift, *Wien. Klin. Wochenschr*, 1908, **21**, 711.

Tartakovskii, M. G. Ekssudatny tiff ili chuma kur, *Qark. Veter. Nauk*, 1904, **34**, 545–575, 617–666.

Taylor, F. S. *The Conquest of Bacteria*, Secker and Warburg, London, 1940.

Waksman, S. A. Streptomyin. Isolation, properties, and utilisation, *J. Hist. Med*, 1951, **6**, 318–347.

Waksman, S. A. *My Life with the Microbes*, Robert Hale, London, 1958.

Waksman, S. A. History of the word 'antibiotic', *J. Hist. Med*, **1973**, 284–286.

Waksman, S. A. *The Literature on Streptomycin 1944—1952*, Rutgers University Press, 1952.

Wilson, D. *Penicillin in Perspective*, Faber & Faber, London, 1976.

CHAPTER SIXTEEN

Armstrong, J. G., Dyke, R. and Fouts, P. J. Initial clinical experience with leurocristine, a new alkaloid from Vinca rosea Linn, *Proc. Am. Ass. Cancer Res*, 1962, **3**, 301.

Arnold, A. M. and Whitehouse, J. M. A. Etoposide: a new anti-cancer agent, *Lancet*, 1981, **2**, 912–915.

Arnold, H., Bourseaux, F. and Brock, N. Chemotherapeutic action of a cyclic nitrogen mustard phosphoramide ester (B 518-ASTA) in experimental tumours of the rat, *Nature*, 1958, **181**, 931 [Cyclophosphamide].

Barnett, C. J., Cullinan, G. J., Gerzon, K., Hoying, R. C., Jones, W. E., Newlon, W. M., Poore, G. A., Robison, R. E., Sweeney, M. J. and Todd, G. C. Structure–activity relationships of dimeric Catharanthus alkaloids. 1. Deacetylvinblastine amide (vindesine), *J. Med. Chem*, 1978, **21**, 88–96.

Beer, C. T. The leukopaenic action of extracts of Vinca rosea, *Br. Empire Cancer Camp. Ann. Rep*, 1955, **33**, 487–488.

Bergel, F. and Stock, J. A. Cyto-active amino-acid and peptide derivatives. Part 1. Substituted phenylalanines, *J. Chem. Soc*, **1954**, 2409–2417 [Melphalan].

Bergman, W. and Feeney, R. J. Contributions to the study of marine products. The nucleosides of sponges. I, *J. Org. Chem*, 1951, **16**, 981–987.

Bollag, W. and Grunberg, E. Tumour inhibitory effects of a new class of cytotoxic agents: methylhydrazine derivatives, *Experientia*, 1963, **19**, 130–131 [Procarbazine].

Borel, J. F. Essentials of cyclosporin A. A novel type of antilymphocytic agent, *Trends Pharmacol. Sci*, 1980, **1**, 146–149.

Borel, J. F. Biological effects of cyclosporin A: a new antilymphocytic agent, *Agents Actions*, 1976, **6**, 468–475.

Brain, K. R. The oncolytic potential of periwinkles, *MIMS Magazine*, **198**, 2 (1st June), 56–61.

Brockmann, H. and Grubhofer, N. Actinomycin C, *Naturwissenshcaft*, 1949, **36**, 376–377.

Broome, J. D. Evidence that the L-asparaginase activity of guinea pig serum is responsible for its antilymphoma effects, *Nature*, 1961, **191**, 1114–1115.

Burchenal, J. H. The historical development of cancer chemotherapy, *Sem. Oncology*, 1977, **4**, 135–146.

Burchenal, J. H., Crossley, M. L., Stock, C. C. *et al.* The action of certain ethyleneimine (aziridine) derivatives on mouse leukemia, *Arch. Biochem*, 1950, **26**, 321.

Burchenal, J. H., Johnstone, S. C., Parker, R. P., Crossley, M. L., Kuh, E. and Seeger, D. R. Effects of *N*-ethylene substituted phosphoramides on transplantable mouse leukemia, *Cancer Res*, 1952, **12**, 251–252.

Burchenal, J. H., Bendwich, A., Brown, G. B., Elion, G. B., Hitchings, G. H., Rhoads, C. P. and Stock, C. C. Preliminary studies on the effect of 2,6-diaminopurine on transplanted mouse leukemia, *Cancer*, 1949, **2**, 119–120.

Calne, R. Y., Alexander, G. P. J. and Murray, J. E. A study of drugs in prolonging survival of homologous renal transplants in dogs, *Ann. N. Y. Acad. Sci*, 1962, **99**, 743–761.

Cash, R., Brough, A. J., Cohen, M. N. P. and Satoh, P. S. Aminoglutethimide (Elipten-Ciba) as an inhibitor of adrenal steroidogenesis: mechanism of action and therapeutic trial, *J. Clin. Endocrin. Metab*, 1967, **27**, 1239.

Clarkson, B., Krakoff, I., Burchenall, J., *et al*, Clinical results of treatment with E. coli asparaginase in adults with leukemia, lymphoma, and solid tumours, *Cancer*, 1970, **25**, 279–305.

Cole, M. P., Jones, C. T. A. and Todd, I. D. A new anti-oestrogenic agent in late breast cancer. An early clinical appraisal of ICI 46474, *Br. J. Cancer*, 1971, **25**, 270–275 [Tamoxifen].

Coley Nauts, H. and Coley, B. L. A review of the treatment of malignant tumours by Coley bacteria toxins, *Approaches to Tumour Chemotherapy* (ed. F. R. Moulton) American Association for the Advancement of Science, Washington, 1947, p. 217–235.

Creighton, A. M. Hellmann, K. and Whitecross, S. Antitumour activity in a series of bisdiketopiperazines, *Nature*, 1969, **222**, 384–385 [Razoxane].

Di Marco, A., Gaetani, M., Orezzi, P., Scarpinato, B. M., Silvestrini, R., Soldati, M., Dasdia, T. and Valentini, L. Daunomycin, a new antibiotic from the rhodomycin group, *Nature*, 1964, **201**, 706–707.

Di Marco, A., Gaetani, M. and Scarpinato, B. Adriamycin, (NSC-123, 127): a new antibiotic with antitumour activity, *Cancer Chemother. Rep*, 1969, **53**, 33.

Dougherty, T. F. and White, A. Effect of pititary adrenotropic hormone on lymphoid tissue, *Proc. Soc. Exp. Biol. Med*, 1943, **53**, 132–133.

Duschinsky, R. and Pleven, E. The synthesis of 5-fluoropyrimidines, *J. Am. Chem. Soc*, 1957, **79**, 4559–4560 [Fluorouracil].

E.O.R.T.C., Epipodophyllotoxin VP 16213 in treatment of acute leukaemia, haematosarcoma, and in solid tumours, *Br. Med. J*, 1973, **3**, 199–202.

Elion, G. B., Burgi, E. and Hitchings, G. B. Studies on condensed pyrimidine systems. IX. The synthesis of some 6-substituted purines, *J. Am. Chem. Soc*, 1952, **74**, 411–414 [Mercaptopurine].

Elion, G. B., Furman, P. A., Fyfe, J. A., Miranda, P. de, Beauchamp, L., and Schaeffer, H. J. Selectivity of action of antiherpetic agent, 9-(2-hydroxyethoxymethyl) guanine, *Proc. Natl. Acad. Sci. USA*, 1977, **74**, 5716–5720 [Acyclovir]

Evans, J. S., Musser, E. A., Mengel, G. D., Forsblad, K. R. and Hunter, J. Antitumour activity of 1-beta-D-arabinofuranosylcytosine hydrochloride, *Proc. Soc. Exp. Biol. Med*, 1961, **106**, 350–353 [Cytarabine].

Everett, J., Robertson, J. J. and Ross, W. C. J. Aryl-2-halogenalkylamines. Part XII. Some carboxylic derivatives of *NN*-di-2-chloroethylamines, *J. Chem. Soc*, **1953**, 2386–2390 [Chlorambucil].

Everett, J. L. and Kon, G. A. R. The preparation of some cytotoxic epoxides, *J. Chem. Soc*, **1950**, 3131–3135.

Farber, S. Carcinolytic action of antibiotics: puromycin and actinomycin D, *Am. J. Pathol*, 1955, **31**, 582.

Farber, S., Toch, R., Sears, E. M. and Pinkel, D. Advances in chemotherapy of cancer in man, *Adv. Cancer Res*, 1956, **4**, 1–71.

Farber, S. Chemotherapy in the treatment of leukemia and Wilms' tumour, *J. Am. Med. Ass*, 1966, **198**, 826.

Farber, S., Cutler, E. C., Hawkins, J. W., Harrison, J. H., Peirce, E. C. and Lenz, G. G. The action of pteroylglutamic acid conjugates on man, *Science*, 1947, **106**, 619–621.

Farber, S., Diamond, L. K., Mercer, R. D., Sylvester, R. F. and Wolff, J. A. Temporary remissions in acute leukemia in children produced by folic acid antagonist, 4-aminopteroyl-glutamic acid, *New Eng. J. Med*, 1948, **238**, 787–793.

Gilman, A. The initial clinical trial of nitrogen mustard, *Am. J. Surg*, 1963, **105**, 574–578.

Goldacre, R. J., Loveless, A. and Ross, W. C. J. Mode of production of chromosome abnormalities by the nitrogen mustards, *Nature*, 1949, **163**, 667–669.

Goodman, L. S., Winetrobe, M. W., McLennan, M. T., Dameshek, W., Goodman, M. J. and Gilman, A. Use of methyl-bis(beta-chloroethyl)amine hydrochloride and tris(beta-chloroethyl) amine, *Approaches to Tumour Chemotherapy*, (ed. F. R. Moulton) American Association for the Advancement of Science, Washington, 1947, p. 338–346.

Grundy, W. E., Goldstein, A. W., Rickher, C. J., Hanes, M. E., Warren, H. B. and Sylvester, J. C. Aureolic acid, a new antibiotic. I. Microbiol studies, *Antibiot. Chemother*, 1953, **3**, 1215–1217 [Mithramycin].

Hackmann, C. Experimentelle Untersuchungen über die Wirkung von Actinomycin C (HBF386) bei bosartigen Geschwulsten, *Z. Krebsforsch*, 1952, **58**, 607.

Haddow, A. Note on the chemotherapy of cancer, *Br. Med. Bull.*, 1946–7, **4**, 417–426.

Haddow, A., Watkinson, J.M. and Paterson E. Influence of synthetic oestrogens upon advanced malignant disease, *Br. Med. J.*, 1944, **2**, 393–398.

Haddow, A., Kon, G.A.R. and Ross, W.C.J. Effects upon tumours of various haloalkylarylamines, *Nature*, 1948, **162**, 824–825.

Haddow, A. and Timmis, G.M. Myleran in chronic myeloid leukaemia, *Lancet*, 1953, **1**, 207–208.

Hall, T., Barlow, J., Griffiths, C. and Saba, Z. Treatment of metastatic breast cancer with aminoglutethimide, *Clinical Research*, 1969, **17**, 402.

Harper, M.J.K. and Walpole, A.L. Contrasting endocrine activities of *cis* and *trans* isomers in a series of substituted triphenylmethanes, *Nature*, 1966, **212**, 87 [Tamoxifen].

Heidelberger, C., Chaudhuri, N.K., Danneberg, P., Mooren, D., Griesbach, L., Duschinsky, R., Schnitzer, R.J., Pleven, E. and Scheiner, J. Fluorinated pyrimidines. A new class of tumour-inhibitory compounds, *Nature*, 1957, **179**, 663–666.

Heilman, F.R. and Kendell, E.C. The influence of 11-dehydro-17-hydroxycorticosterone on the growth of a malignant tumour in the mouse, *Endocrinology*, 1944, **34**, 416–420.

Hitchings, G.H., Elion, G.B., Vanderwerff, H. and Falco, E.A. Pyrimidine derivatives as antagonists of P.G.A., *J. Biol. Chem.*, 1948, **174**, 765–766.

Huggins, C. Anti-androgenic treatment of prostatic carcinoma in man, *Approaches to Tumour Chemotherapy*, (ed. F.R. Moulton), American Association for the Advancement of Science, Washington, 1947, p. 379–383.

Huggins, C. and Clark, P.J. Quantitative studies on prostatic secretion. II. The effect of castration and of estrogen injection on the normal and on the hyperplastic prostate glands of dogs, *J. Exp. Med.*, 1940, **72**, 747–762.

Huggins, C. and Hodges, C.V. Studies on prostatic cancer. I. The effect of castration, of estrogen and of androgen injection on serum phosphatase in metastatic carcinoma of the prostate, *Cancer Res.*, 1941, **1**, 293–297.

Hughes, S.W.M. and Burley, D.M. Aminoglutethimide: a side effect turned to therapeutic advantage, *Postgrad. Med. J.*, 1970, **46**, 409–416.

Hutchings, B.L., Mowat, J.H., Oleson, J.J., Stokstad, E.L.R., Boothe, J.J., Waller,

C.W., Angier, R.B., Semb, J. and Subbarow, Y. Pteroylaspartic acid, an antagonist for pteroylglutamic acid, *J. Biol. Chem*, 1947, **170**, 323–328.

Johnson, I.S., Wright, H.F. and Svoboda, G.H. Experimental basi
C.W., Angier, R.B., Semb, J. and Subbarow, Y. Pteroylaspartic acid, an antagonist for pteroylglutamic acid, *J. Biol. Chem*, 1947, **170**, 323–328.

Johnson, I.S., Wright, H.F. and Svoboda, G.H. Experimental basis for experimental evaluation of anti-tumour principles derived from Vinca rosea Linn, *J. Lab. Clin. Med.*, 1959, **54**, 830.

Johnston, T.P., McCaleb, G.S. and Montgomery, J.A. The synthesis of antineoplastic agents. XXXII. *N*-Nitrosoureas, *J. Med. Chem.*, 1963, **6**, 669–681.

Johnston, T.P., McCaleb, G.S., Opliger, P.S. and Montgomery, J.A. The synthesis of potential anticancer agents. XXXVI. *N*-Nitrosoureas. II. Haloalkyl derivatives, *J. Med. Chem.*, 1966, **9**, 892–911.

Kahle, P. and Maltry, E. Treatment of hyperplasia of the prostate with diethylstilboestrol and diethylstilboestrol dipropionate, *New Orleans Med. Surg. J.*, 1940, **93**, 121–131.

Kidd, J.G. Regression of transplanted lymphomas induced in vivo by means of normal guinea pig serum, *J. Exp. Med.*, 1953, **93**, 565–582.

Law, L.W. and Spiers, R. Responses of spontaneous lymphoid leukemias in mice to injection of adrenal cortical extracts, *Proc. Soc. Exp. Biol. Med.*, 1947, **66**, 226–230.

Leuchtenberger, R., Leuchtenberger, C., Laszlo, D. and Lewisohn, R. Folic acid, a tumour growth inhibitor, *Science*, 1945, **101**, 45.

Lewisohn, R. Review of the work of the laboratory, 1937–1945, *Approaches to Tumour Chemotherapy* (ed. F.R. Moulton), American Association for the Advancement of Science, Washington, 1947, p. 139–147.

Lewisohn, R., Leuchtenberger, C., Leuchtenberger, R. and Laszlo, D. Effect of intravenous injection of yeast extract on spontaneous breast adenocarcinoma in mice, *Proc. Soc. Exp. Biol. Med.*, 1940, **43**, 558–561.

Lewisohn, R., Leuchtenberger, C., Leuchtenberger, R. and Keresztesy, J.C. The influence of liver L. casei factor on spontaneous breast cancer in mice, *Science*, 1946, **104**, 436–437.

Lissauer, Zwei Falle von Leucaemia, *Berl. Klin. Wochenschr.*, 1865, **2**, 403–404.

Lloyd, J.U. Podophylllum, *Origin and History of all the Pharmacopeial Vegetable Drugs, Chemicals and Preparations. Vol. 1*, Caxton Press, Cincinnati, 1921, p. 248–256.

Loveless, A. and Revell, S. New evidence on the mode of action of the mitotic poisons, *Nature*, 1949, **164**, 938–944.

MacGregor, A.B. The search for a chemical cure for cancer, *Med. Hist.*, 1966, **10**, 374–385.

Neuss, N., Gorman, M., Hargrove, W., Cone, N.J., Biemann, K., Buchi, G. and Manning, R.E. Vinca alkaloids. XXI. The structure of the oncolytic alkaloids vinblastine (VBL) and vincristine (VCR), *J. Am. Chem. Soc.*, 1964, **86**, 1440–1442.

Niculescu-Duvaz, I., Cambani, A. and Tarnauceanu, E. Potential anticancer agents. II. Urethane-type nitrogen mustards of some natural sex hormones, *J. Med. Chem.*, 1967, **10**, 172–174 [Estramustine].

Noble, R.L. *et al.* Role of chance observations on the effects of vincaleukoblastine with special reference to Hodgkin's disease, *Proceedings Canadian Cancer Research Conference, 1966*, p. 373.

Paterson, E., Ap Thomas, I., Haddow, A. and Watkins, J.M. A further report on the action of urethane in leukaemia, *Approaches to Tumour Chemotherapy* (ed. F.R. Moulton), American Association for the Advancement of Science, Washington, 1947, p. 401–415.

Prusoff, W.H. Synthesis and biological activities of iododeoxyridine, an analogue of thymidine, *Biochem. Biophys. Acta*, 1959, **32**, 295–296 [Idoxuridine].

Rao, K.V., Cullen, W.P. and Sobin, B.A. A new antibiotic with antitumour properties, *Antibiot. Chemother.*, 1962, **12**, 182–186 [Mithramycin].

Rose, F.L., Hendry, J.A. and Walpole, A.L. New cytotoxic agents with tumour-inhibitory activity, *Nature*, 1950, **165**, 993–996 [Tretamine].

Rosenberg, B., Van Camp, L. and Krigas, T. Inhibition of cell division in E. coli by electrolysis products from a platinum electrode, *Nature*, 1965, **205**, 698–699.

Rosenberg, B., Van Camp, L., Trosko, J.E. and Mansour, V.H. Platinum compounds: a new class of potential antitumour agents, *Nature*, 1969, **222**, 385–386.

Rundles, R.W. Allopurinol in the treatment of gout, *Ann. Int. Med.*, 1966, **64**, 229–258.

Rundles, R.W., Wyngaarden, J.B., Hitchings, G.H., Elion, G.B. and Silberman, R. Effects of a xanthine oxidase inhibitor on thiopurine metabolism, hyperuricemia, and gout, *Trans. Ass. Am. Physicians*, 1963, **76**, 126–140.

Schwarz, R. and Dameshek, W. Drug induced immunological tolerance, *Nature*, 1959, **183**, 1682–1683.

Seeger, D.R., Smith, J.M. and Hultquist, M.E. Antagonist for pteroylglutamic acid, *J. Am. Chem. Soc.*, 1947, **69**, 2567 [Aminopterin].

Sigiura, K. Effect of intravenous injection of yeast and barley extracts and L. casei factor . . ., *Approaches to Tumour Chemotherapy*, (ed. F.R. Moulton) American Association for the Advancement of Science, Washington, 1947, p. 208–213.

Simone, J., Aur, R.J.A., Hustu, O. and Pinkel, D. Total therapy studies of acute lymphocytic leukemia in children, *Cancer*, 1972, **30**, 1488–1494.

Spurr, C.L., Jacobson, L.O., Smith, T.R. and Guzman Baron, E.S. The clinical application of methyl-*bis*(beta-chloroethyl)amine hydrochloride to the treatment of lymphomas and allied dyscrasias, *Approaches to Tumour Chemotherapy* (ed. F.R. Moulton) American Association for the Advancement of Science, Washington, 1947, p. 306–318.

Svoboda, G.H. Alkaloids of Vinca rosea. IX. Extraction and characterization of leurosidine and leurocristine, *Lloydia*, 1961, **24**, 173–178.

Svoboda, G.H., Johnson, I.S., Gorman, M. and Neuss, N. Current status of research on the alkaloids of Vinca rosea Linn. (Catharanthus roseus G. Don), *J. Pharm. Sci.*, 1962, **51**, 707–720.

Umezawa, H. Bleomycin and other antitumour antibiotics of high molecular weight, *Antimicrob. Ag. Chemother.*, 1965, **19**, 1079–1085.

Umezawa, H., Maeda, K., Takuchi, T. and Okami, Y. New antibiotics, bleomycin A and B, *J. Antibiot.* (*Tokyo*), 1966, **19**, 200–209.

Wakaki, S., Marumo, H., Tomioka, K., Shimizu, G., Kato, E., Kamada, H., Kudo, S. and Fujimoto, Y. Isolation of new fractions of antitumour mitomycins, *Antibiot. Chemother.*, 1958, **8**, 228–240.

Waksman, S.A. (ed.) *Actinomycin*, Interscience, New York, 1968.

Wugmeister, I. Le traitment de l'hypertrophie de la prostate par doses massives de Folliculine, *Paris Med.*, 1937, **1**, 535–536.

Zeller, P., Gutmann, H., Hegedus, B., Kaiser, A., Langeman, A. and Muller, M. Methylhydrazine derivatives, a new class of cytotoxic agents, *Experientia*, 1963, **19**, 129 [Procarbazine].

Index